Learning Guide

for Tortora and Grabowski:
Principles of Anatomy and Physiology

Learning Guide

for Tortora and Grabowski:
Principles of Anatomy and Physiology Eighth Edition

Kathleen Schmidt Prezbindowski

*College of Mount St. Joseph and
The University of Cincinnati*

Gerard J. Tortora

*Biology Coordinator
Bergen Community College*

HarperCollinsCollegePublishers

Executive Editor: Bonnie Roesch
Developmental Editor: Cyndy Taylor
Project Coordination and Text Design: Electronic Publishing Services Inc.
Electronic Production Manager: Mike Kemper
Manufacturing Manager: Helene G. Landers
Cover Designer: Nancy Sabato
Electronic Page Makeup: Electronic Publishing Services Inc.
Printer and Binder: Malloy Lithographing, Inc.
Cover Printer: Malloy Lithographing, Inc.

Learning Guide for Tortora and Grabowski: Principles of Anatomy and Physiology, Eighth Edition

Library of Congress Cataloging-in-Publication Data

Prezbindowski, Kathleen Schmidt.
    Learning guide for Tortora and Grabowski: principles of anatomy
and physiology / Kathleen Schmidt Prezbindowski, Gerard J. Tortora.
—8th ed.
        p.   cm.
    ISBN 0-673-99356-6
    1. Human physiology—Problems, exercises, etc. 2. Human anatomy—
Problems, exercises, etc.   I. Tortora, Gerard J.   II. Title.
III. Title: Principles of anatomy and physiology
QP34.5.T67   1996 Suppl.
612'.0076—dc20
                                                                95-40981
                                                                CIP

95 96 97 98 99   9 8 7 6 5 4 3 2 1

To Laurie, Amy, and Joost

Contents

Preface

This *Learning Guide* is intended to be used in conjunction with Tortora and Grabowski's *Principles of Anatomy and Physiology*, Eighth Edition. The emphasis is on active learning, not passive reading, and the student examines each concept through a variety of activities and exercises. By approaching each concept several times from different points of view, the student sees how the ideas of the text apply to real, clinical situations. In this way the student may be helped not simply to study but to *learn*.

The 29 chapters of the *Learning Guide* parallel those of the Tortora and Grabowski text. Each chapter begins with an *Overview* that introduces content succinctly and a *Framework* that visually organizes the interrelationships among key concepts and terms. *Objectives* from the text are included for convenient previewing and reviewing. Objectives are arranged according to major chapter divisions in the *Topic Outline*. *Wordbytes* present word roots, prefixes, and suffixes that facilitate understanding and recollection of key terms. *Checkpoints* offer a diversity of learning activities that follow the exact sequence of topics in the chapter, with cross-references to pages in the text. This format makes the *Learning Guide* "user friendly" or "student friendly" and enhances the effectiveness of the excellent text. *Answers to Selected Checkpoints* provide feedback to students. *Writing Across the Curriculum* provides students with opportunities for improving writing skills as students compose short essays on chapter concepts. A 25-question *Mastery Test* includes both objective and subjective questions as a final tool for student self-assessment of learning. *Answers to the Mastery Test* immediately follow the Mastery Test.

The Big Picture is a new feature in the eighth edition. *The Big Picture: Looking Ahead* guides students into brief ventures in upcoming chapters where students will see further examples of concepts being introduced. These brief glimpses into the future are designed to reduce anxiety: when students reach these chapters later in the course, they will recognize familiar content. *The Big Picture: Looking Back* helps integrate concepts as students recognize that they are building upon information from previous chapters.

This edition of the *Learning Guide* is especially designed to fit the needs of people with different learning styles. Visual learners will identify the *Frameworks* as assets in organizing concepts and key terms. Visual and tactile learners will welcome the increased number of coloring and labeling exercises with color-code ovals. Further variety in *Checkpoints* is offered by short definitions, fill-ins, tables for completion, and multiple-choice, matching, and arrange-in-correct-sequence activities. Learning on multiple levels is made possible by *For Extra Review* sections that provide enrichment or assistance with particularly difficult topics. *Clinical Challenges* introduce applications especially relevant for students in allied health. These are based on my own experiences as an R.N. and on discussions with colleagues. These challenging exercises enable students to apply abstract concepts to actual clinical settings.

The eighth edition of the *Guide* also provides readily available feedback for self-assessment. Answers to most *Checkpoints* are included at the end of the chapter. The placement of *Answers to the Mastery Test* immediately following the *Mastery Test* offers readily accessible feedback.

A number of persons have contributed to the revision of this book. I would particularly like to thank Richard D. Harrington, Rivier College and C. Aubrey Morris, Pensacola Junior College for their helpful comments and suggestions while the manuscript was being revised.

I offer particular thanks to the students and faculty at the College of Mount St. Joseph and the University of Cincinnati for their ongoing sharing and support of teaching and learning. I thank my daughters, Amy and Laurie Prezbindowski, my parents and sister, Norine, Harry, and Maureen Schmidt, and close friends for their constant encouragement and caring. I am thankful for the expertise, enthusiasm, and patience of the production staff at HarperCollins, especially to Cyndy Taylor, developmental editor.

Kathleen Schmidt Prezbindowski

To the Student

This *Learning Guide* is designed to help you do exactly what its title indicates—*learn*. It will serve as a step-by-step aid to help you bridge the gap between goals (objectives) and accomplishment (learning). The 29 chapters of the *Learning Guide* parallel the 29 chapters of Tortora and Grabowski's *Principles of Anatomy and Physiology*, Eighth Edition. Each chapter consists of the following parts: *Overview, Framework, Topic Outline with Objectives, Wordbytes, Checkpoints, Answers to Selected Checkpoints, Writing Across the Curriculum*, and *Mastery Test with Answers*. Take a moment to turn to Chapter 1 and look at the major headings in this *Learning Guide*. Now consider each one.

Overview. As you begin your study of each chapter, read this brief introduction, which presents the topics of the chapter. The overview is a look at the "forest" (or the "big picture") before examination of the individual "trees" (or details).

Framework. The Framework arranges key concepts and terms in an organizational chart that demonstrates interrelationships and lists key terms. Refer back to the Framework frequently to keep sight of the interrelationships.

Topic Outline and Objectives. Objectives that are identical to those in the text organize the chapter into "bite-size" pieces that will help you identify the principal subtopics that you must understand. Preview objectives as you start the chapter, and refer back to them to check off as you complete each one. Objectives are arranged according to major chapter divisions in the *Topic Outline*.

Wordbytes present word roots, prefixes, and suffixes that will help you master terminology of the chapter and facilitate understanding and remembering of key terms. Study these; then check your knowledge. (Cover meanings and write them in the lines provided.) Try to think of additional terms in which these important bits of words are used.

Checkpoints offer a variety of learning activities arranged according to the sequence of topics in the chapter. At the start of each new Checkpoint section, note the related pages in the Tortora and Grabowski text. Read these pages carefully and then complete this group of Checkpoints. Checkpoints are designed to help you to "handle" each new concept, almost as if you were working with clay. Through a variety of exercises and activities, you can achieve a depth of understanding that comes only through active participation, not passive reading alone. You can challenge and verify your knowledge of key concepts by completing the Checkpoints along your path of learning.

A diversity of learning activities is included in Checkpoints. You will focus on specific facts in the chapter by doing exercises that require you to provide definitions and comparisons, and you will fill in blanks in paragraphs with key terms. You will also color and label figures. Fine-point, color, felt-tip pens or pencils will be helpful so that you can easily fill in related color-coded ovals. Your understanding will also be tested with matching exercises, multiple-choice questions, and arrange-in-correct-sequence activities.

Three special types of Checkpoints are *For Extra Review, Clinical Challenges,* and *The Big Picture. For Extra Review* provides additional help for difficult topics or refers you by page number to specific activities earlier in the *Learning Guide* so that you may better integrate related information. Applications for enrichment and interest in areas particularly relevant to nursing and other health science students are found in selected *Clinical Challenge* exercises. *The Big Picture: Looking Ahead* (or *Looking Back*) refers you to relevant sections of the *Guide* or text to help you integrate concepts within the entire text. When you come upon a *Big Picture* Checkpoint, take a few moments to turn to those sections of the book that may reinforce content you studied earlier and also reduce your anxiety about upcoming content.

When you have difficulty with an exercise, refer to the related pages (listed with each Checkpoint of the *Learning Guide*) of *Principles of Anatomy and Physiology* before proceeding further. You will find specific references by page number for many text figures and exhibits.

Answers to Selected Checkpoints. As you glance through the activities, you will notice that a number of questions are marked with solid boxes (■). Answers to these Checkpoints are given at the end of each chapter, in order to provide you with some immediate feedback about your progress. Incorrect answers alert you to review related objectives in the *Learning Guide* and corresponding text pages. Each *Learning Guide* (LG) figure included in *Answers to Selected Checkpoints* is identified with an "A," such as Figure LG 13.1A. Activities without answers are also purposely included in this book. The intention is to encourage you to verify some answers independently by consulting your textbook and also to stimulate discussion with students or instructors.

Writing Across the Curriculum provides several suggestions for essays that may improve your writing skills as you enhance your understanding of concepts in this chapter.

Mastery Test. This 25-question self-test provides an opportunity for a final review of the chapter as a whole. Its format will assist you in preparing for standardized tests or course exams that are objective in nature, because the first 20 questions are multiple-choice, true-false, and arrangement questions. The final five questions of each test help you to evaluate your learning with fill-in and short-answer questions. Answers for each mastery test are placed at the end of each chapter.

I wish you success and enjoyment in *learning* concepts relevant to your own anatomy and physiology and in applying this information in clinical settings.

Kathleen Schmidt Prezbindowski

About the Authors

Kathleen Schmidt Prezbindowski, PH.D., R.N., is a professor of biology at the College of Mount St. Joseph, where she was recently named Distinguished Teacher of the Year. Also a visiting professor at the University of Cincinnati, she has taught anatomy and physiology, pathophysiology, and biology of aging for over 25 years. Within the past decade, Dr. Prezbindowski earned a B.S.N. and also M.S.N. in both Gerontological Nursing and Psychiatric/Mental Health Nursing. Through a grant from the U.S. Administration on Aging, she produced the first video series on the human body ever designed especially for older persons, as well as a training manual for health promotion and aging. She is also the author of two other learning guides.

Gerard J. Tortora is a professor of biology and teaches human anatomy and physiology and microbiology at Bergen Community College in Paramus, New Jersey. He has just completed 33 years as a teacher, the past 27 at Bergen CC. He received his B.S. in biology from Fairleigh Dickinson University in 1962 and his M.A. in biology from Montclair State College in 1965. He has also taken graduate courses in education and science at Columbia University and Rutgers University. He belongs to numerous biology organizations, such as the American Association for the Advancement of Science (AAAS), *The Human Anatomy and Physiology Society* (HAPS), the American Association of Microbiology (ASM), and the Metropolitan Association of College and University Biologists (MACUB). Jerry is the author of a number of best-selling anatomy and physiology, anatomy, and microbiology textbooks and several laboratory manuals.

UNIT 1

Organization of the Human Body

FRAMEWORK 1

An Introductory Tour of the Human Body

GETTING STARTED

- **ANATOMY AND PHYSIOLOGY DEFINED (A)**
  - Branches of anatomy and physiology

WHERE TO GO, HOW TO GET THERE

- **LEVELS OF STRUCTURAL ORGANIZATION (B)**
  - Chemicals
  - Cells
  - Tissues (4)
  - Organs
  - Systems (11)
  - Organisms

- **ANATOMICAL AND DIRECTIONAL TERMS, PLANES, AND SECTIONS (E)**
  - Anatomical position
  - Regional names
  - Directional terms
  - Planes, sections

HOW IT WORKS: FUNCTIONS

- **LIFE PROCESSES (C)**
  - Metabolism
  - Responsiveness
  - Movement
  - Growth
  - Differentiation
  - Reproduction

- **HOMEOSTASIS (D)**
  - ECF
  - ICF
  - Stressor
  - Feedback systems (loops)

- **BODY CAVITIES, REGIONS, AND QUADRANTS (F)**
  - Dorsal
  - Ventral
    - thoracic
    - abdominopelvic
      - nine regions
      - four quadrants

WHEN THE BODY DOES NOT WORK

- **MEDICAL IMAGING (G)**
  - Radiography
  - CT
  - DSA
  - PET
  - MRI
  - US

An Introduction to the Human Body

CHAPTER

1

You are about to embark on a tour, one that you will enjoy for at least several months, and hopefully for many years. Your destination: the far-reaching corners and fascinating treasures of the human body. Imagine the human body as an enormous, vibrant world, abundant in architectural wonders, and bustling with activity. You are the traveler, the visitor, the learner on this tour.

Just as the world is composed of individual buildings, cities, countries, and continents, so the human body is organized into cells, tissues, organs, and systems. In fact, a close-up look at a cell (comparable to a detailed study of a building) will reveal structural intricacies—the array of chemicals that comprise cells. Any tour requires a sense of direction and knowledge of the language: directional and anatomical terms that guide the traveler through the body. A tour is sure to enlighten the visitor about the culture in each region—in the human body, the functions of each body part or system. By the time you leave each area (body part or system), you will have gained a sense of how that area contributes to overall stability and well-being (homeostasis). You will also learn about how feedback systems (loops) contribute to homeostasis.

In planning a trip, an itinerary serves as a useful guide. A kind of itinerary is found below in the objectives (same as in your text) listed according to a topic outline that can help you to organize your study. It can also give you a well deserved feeling of accomplishment as you check off the box next to each objective after you complete it. Then, at the end of the trip/ chapter, look back at this itinerary/outline to put all aspects of the entire chapter in context. Also look at the Framework for Chapter 1, study it carefully, and refer back to it as often as you wish. It contains the key terms of the chapter and serves as your map and your itinerary. Bon voyage!

TOPIC OUTLINE AND OBJECTIVES

A. Anatomy and physiology defined

☐ 1. Define anatomy and physiology and their subdivisions.

B. Levels of structural organization

☐ 2. Define each of the following levels of structural organization that make up the human body: chemical, cellular, tissue, organ, system, and organismic.

☐ 3. Identify the principal systems of the human body, list representative organs of each system, and describe the functions of each system.

C. Life processes

☐ 4. List and define several important life processes of humans.

D. Homeostasis: maintaining physiological limits

☐ 5. Define homeostasis and explain the effects of stress on homeostasis.

☐ 6. Define the components of a feedback system.

☐ 7. Contrast the operation of negative and positive feedback systems.

☐ 8. Explain the relationship between homeostasis and disease.

E. Anatomical and directional terms, planes, and sections

☐ 9. Define the anatomical position and compare common and anatomical terms used to describe various regions of the human body.

☐ 10. Define several directional terms and anatomical planes used in association with the human body.

F. Body cavities, regions, and quadrants

☐ 11. List by name and location the principal body cavities and the organs contained within them.

☐ 12. Name and describe the nine abdominopelvic regions and four quadrants.

G. Medical imaging

☐ 13. Describe the principle and importance of selected medical imaging techniques in the diagnosis of disease.

WORDBYTES

Now become familiar with the language of this chapter by studying each wordbyte, its meaning, and an example of its use within a term. After you study the entire list, self-check your understanding by writing the meaning of each wordbyte on the line. As you continue through the *Learning Guide,* identify and write in other examples of terms that contain the same wordbyte.

Wordbyte	Self-check	Meaning	Example(s)
ana-	_____	up	*ana*tomy
ante-	_____	before	*ante*rior
homeo-	_____	same	*homeo*stasis
inter-	_____	between	*inter*cellular
intra-	_____	within	*intra*cellular
-logy	_____	study of	physio*logy*
-meter	_____	measure	milli*meter*
pariet-	_____	wall	*pariet*al
physio-	_____	function	*physio*logy
post-	_____	after	*post*erior
sagitta-	_____	arrow	*sagitta*l
stasis, stat-	_____	stand, stay	homeo*stasis*
-tomo	_____	to cut	ana*tomy*
viscero-	_____	body organs	*viscera*l

CHECKPOINTS

A. Anatomy and physiology defined   (pages 3–4)

■ **A1.** Anatomy is the study of _____ and physiology is the study of

_____.

A2. An intimate relationship exists between structure and function: one determines the other. Explain how the structure of each of the following determines the functions for which it can be used.

a. Spoon/fork

b. Hand/foot

c. Incisors (front teeth)/molars

■ **A3.** Identify the subdivisions of anatomy and physiology by selecting the term that best fits the definitions provided below.

C. Cytology	H. Histology
DA. Developmental anatomy	I. Immunology
Embr. Embryology	PA. Pathological anatomy
Endo. Endocrinology	PP. Pathophysiology
EP. Exercise physiology	RP. Renal physiology
GA. Gross anatomy	

_____ a. Study of cells

_____ b. Study of tissues

_____ c. Study of structures that can be examined without the aid of a microscope, such as study of an entire cadaver

_____ d. Study of development during the first two months of life in utero

_____ e. Study of function of kidneys

_____ f. Study of functions of hormones

_____ g. Study of functional changes associated with disease

_____ h. Study of the defense mechanisms of the body

B. Levels of structural organization (page 4)

■ **B1.** Arrange the terms in the box from highest to lowest in level of organization. Write terms on the lines provided below. One is done for you.

Cell	Organism
Chemical	System
Organ	Tissue

a. _____ (highest)

b. **System** _____

c. _____

d. _____

e. _____

f. _____ (lowest)

■ **B2.** List the four basic tissue types.

_____   _____

_____   _____

B3. Complete Table LG 1.1 on page 7 describing systems of the body. Name two or more organs in each system. Then list one or more functions of each system.

■ **B4.** *The Big Picture: Looking Ahead.* Look ahead to upcoming chapters as you identify organs that function as parts of more than one system. Text page numbers in parentheses may guide you to answers.

a. The pancreas is part of the _____ system because the pancreas secretes enzymes that digest foods (page 776). Also, because it produces hormones

such as insulin and glucagon, the pancreas is also part of the _____ system (page 533). List two other organs that are parts of these same systems:

_____ and _____ (pages 541 and 789).

b. Because the pituitary releases many hormones, such as human growth hormone (hGH)

and antidiuretic hormone (ADH), it is part of the _____ system (page 509). Yet the connections of the pituitary with the brain make it a component of

the _____ system (page 405).

c. Because ovaries produce female sex cells (ova) and also hormones (estrogens and

progesterone), ovaries are components of both the female _____

system (page 937) and the _____ system (page 539).

Table LG 1.1   Systems of the human body

System	Organs	Functions
a.	Skin, hair, nails	
b. Skeletal		
c. Muscular		
d.		Regulates body by nerve impulses
e.	Glands that produce hormones	
f.	Blood, heart, blood vessels	
g. Lymphatic		
h.		Supplies oxygen, removes carbon dioxide, regulates acid-base balance
i.		Breaks down food and eliminates solid wastes
j.	Kidney, ureters, urinary bladder, urethra	
k. Reproductive		

d. The large bone of your thigh (the femur) provides strong support for your body and it

   also forms blood cells, so the femur is part of two systems: _____

   (page 206) and _____ (page 561).

e. The urethra of the male serves as a passageway for both urine (page 881) and semen

   (page 809), so it is part of both the _____ system and the

   _____ system.

C. Life processes (pages 4–7)

C1. Six characteristics distinguish you from nonliving things. List these characteristics below and give a brief definition of each. One is done for you.

a. _____

b. _____

c. _____

d. <u>**Growth: increase in size and complexity**</u>_____

e. _____

f. _____

■ **C2.** _____

Refer to the list of terms in the box, all related to life processes. Demonstrate your understanding of these terms by selecting the term that best fits each description provided below.

A. Anabolism	M. Metabolism
C. Catabolism	Respo. Responsiveness
D. Differentiation	Repro. Reproduction

_____ a. Chemical processes that involve the breakdown of large, complex molecules into smaller ones with release of energy

_____ b. Energy-requiring chemical processes that build structural and functional components of the body

_____ c. Sum of all the chemical processes [both (a) and (b) above] in the body

_____ d. Formation of new cells for growth, repair, or replacement, or for production of a new individual

_____ e. Changes that cells undergo during development from unspecialized ancestor cells to specialized cells such as muscle or bone cells

_____ f. Ability to detect and respond to changes in the environment, for example, production of insulin in response to elevated blood glucose level

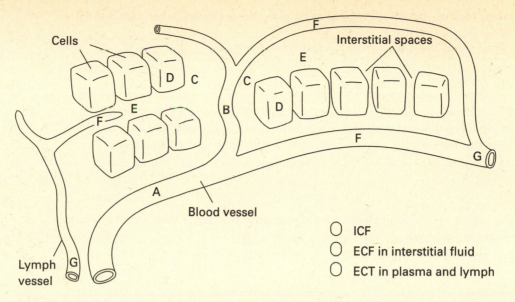

Cells

Interstitial spaces

Blood vessel

Lymph vessel

○ ICF
○ ECF in interstitial fluid
○ ECT in plasma and lymph

Figure LG 1.1 Internal environment of the body: types of fluid. Complete as directed. Areas labeled A to G refer to Checkpoint D2.

D. Homeostasis: maintaining physiological limits   (pages 7–10)

■ **D1.**   Describe the types of fluid in the body and their relationships to homeostasis by completing these statements and Figure LG 1.1.

a. Fluid inside of cells is known as _____ fluid (ICF). Color areas containing this type of fluid yellow.

b. Fluid in spaces between cells is called _____. It surrounds and bathes cells and is one form of *(intracellular? extracellular?)* fluid. Color the spaces containing this fluid light green.

c. Another form of extracellular fluid is that located in _____

vessels and _____ vessels. Color these areas dark green.

d. The body's "internal environment" (that is, surrounding cells) is

_____-cellular fluid (ECF) (all green areas in your figure). The condition of maintaining ECF in relative constancy is known as

_____.

■ **D2.**   Refer again to Figure LG 1.1. Show the pattern of circulation of body fluids by drawing arrows connecting letters in the figure in alphabetical order (*A* → *B* → *C,* etc.). Now fill in the blanks below describing this lettered pathway:

A (arteries and arterioles)   →   B (blood capillaries)   →

C (_____) → D (_____) →

E (_____) → F (_____) →

G (_____).

■ **D3.** List six qualities of your ECF that are maintained under optimal conditions when your body is in homeostasis.

D4. Consider how your internal environment is affected by your external environment. Which of the six qualities of your ECF listed above would be altered if you were subjected to the following conditions? Compare your answers with those of a study partner.

a. Taking a two-day hike at high altitude without appropriate food or clothing

b. Invasion of your body by an infectious microorganism that causes fever and diarrhea over a prolonged period of time

■ **D5.** Are you seated right now? For safety, hold on to the back of a chair or a desk, and then stand up quickly. As you do, conceptualize what happens to your blood as you change position: as a response to gravity, blood tends to rush to the lower part of your body. If nothing happened to offset that action, your blood pressure (BP) would drop and you would likely faint because of insufficient blood supply to the brain. This usually does not happen because homeostatic mechanisms act to quickly elevate blood pressure back to normal.

Match the answers in the box with factors and functions that helped maintain your blood pressure when you stood up. (*Hint:* Figure 1.4 [page 10 in your text] describes a situation that is just the opposite of this one.)

CCen.	Control center	O.	Output
CCond.	Controlled condition	Recep.	Receptors
E.	Effector	Respo.	Response
I.	Input	S.	Stimulus

_____ a. Your blood pressure (BP): a factor that must be maintained homeostatically

_____ b. Standing up; blood flowed by gravity to lower parts of your body

_____ c. Pressure-sensitive nerve cells in large arteries in your neck and chest

_____ d. Nerve impulses bearing the message "BP is too low" (since less blood remained in the upper part of your body as you were standing)

_____ e. Your brain

_____ f. Nerve impulses from your brain to your heart, bearing the message, "Beat faster!"

_____ g. Your heart (which beat faster)

_____ h. Elevated BP

■ **D6.** *A clinical challenge.* In some people (for example, the elderly, the malnourished, and those with certain neurological disorders), nerve messages necessary for maintaining blood pressure may occur more slowly. As a result, blood pressure may drop as the individual suddenly changes position (from lying to sitting or sitting to standing). This condition, known as postural hypotension, may lead to faintness and falls. What might a person subject to postural hypotension do to minimize the chance of falling?

■ **D7.** Consider this example of a feedback system loop. Ten workers on an assembly line are producing handmade shoes. As shoes come off the assembly line, they pile up around the last worker, Worker Ten. When this happens, Worker Ten calls out, "We have hundreds of shoes piled up here." This information (input) is heard by (fed back to) Worker One, who determines when more shoes should be begun. Worker One is therefore the ultimate controller of output. This controller could respond in either of two ways. In a *negative feed-back system,* Worker One says, "We have an excess of unsold shoes. Let us slow down or stop production until the excess (stress) is relieved—until these shoes are sold." What might Worker One's response be if this were a *positive feedback system?*

■ **D8.** Distinguish the two types of feedback mechanisms in this exercise.

a. Increase in nerve impulses to arterioles (causing them to constrict so that more blood flows into major arteries and increases blood pressure) as a response to messages to the brain that blood pressure is too low is an example of a *(negative? positive?)* feedback mechanism. This is true because the decreased blood pressure is *(reversed? decreased even more?).*

b. The increase of uterine contraction during labor as a response to increase of the hor-mone oxytocin is an example of a *(negative? positive?)* feedback mechanism because the uterine contractions are *(decreased? enhanced?).*

c. Most feedback mechanisms in the human body are *(negative? positive?).* Why do you think this might be advantageous and directed toward maintenance of homeostasis?

■ **D9.** Circle the terms below that are classified as signs rather than as symptoms.

chest pain　　fever　　headache　　diarrhea　　swollen ankles　　skin rash　　toothache

D10. *A clinical challenge.* Contrast the two terms in each set.

a. *Local disease/systemic disease*

b. *Medical history/diagnosis*

c. *Epidemiology/pharmacology*

E. Anatomical and directional terms, planes, and sections (pages 11–14)

■ **E1.** How would you describe anatomical position? Assume that position yourself.

■ **E2.** Complete the table relating common terms to anatomical terms. (See Figure 1.5, page 11 in your text, to check your answers.) For extra practice use common and anatomical terms to identify each region of your own body and that of a study partner.

Common Term	Anatomical Term
a.	Axillary
b. Fingers	
c. Arm	
d.	Popliteal
e.	Cephalic
f. Mouth	
g.	Inguinal
h. Chest	
i.	Cervical
j.	Antebrachial
k. Buttock	
l.	Calcaneal

■ **E3.** Using your own body, a skeleton, a torso, or Figure 1.6 (page 13 in your text), determine relationships among body parts. Write the correct directional term(s) to complete each of these statements.

a. The liver is ____________________ to the diaphragm.

b. Fingers (phalanges) are located ____________________ to wrist bones (carpals).

c. The skin on the dorsal surface of your body can also be said to be located on your

____________________ surface.

d. The great (big) toe is ____________________ to the little toe.

e. The little toe is _____ to the great toe.

f. The skin on your leg is _____ to muscle tissue in your leg.

g. Muscles of your arm are _____ to skin on your arm.

h. When you float face down in a pool, you are lying on your _____ surface.

i. The lungs and heart are located _____ to the abdominal organs.

j. Because the stomach and the spleen are both located on the left side of the abdomen,

   they could be described as _____-lateral.

k. The _____ pleura covers the external surface of the lungs.

■ **E4.** Match each of the following planes with the phrase telling how the body would be divided by such a plane.

> F. Frontal (coronal)   S. Sagittal
> M. Midsagittal (median)   T. Transverse

_____ a. Into superior and inferior portions   _____ c. Into anterior and posterior portions

_____ b. Into equal right and left portions   _____ d. Into right and left portions

F. Body cavities, regions, and quadrants   (pages 14–19)

■ **F1.** After you have studied Figures 1.9–1.11 (pages 16–19 in your text), complete this exercise about body cavities. Circle the correct answer in each statement.

a. The (dorsal? ventral?) cavity consists of the cranial cavity and the vertebral canal.

b. The viscera, including such structures as the heart, lungs, and intestines, are all located in the (dorsal? ventral?) cavity.

c. Of the two body cavities, the (dorsal? ventral?) appears to be better protected by bone.

d. Pleural, mediastinal, and pericardial are terms that refer to regions of the (thorax? abdominopelvis?).

e. The (heart? lungs? esophagus and trachea?) are located in the pleural cavities.

f. The division between the abdomen and the pelvis is marked by (the diaphragm? an imaginary line from the symphysis pubis to the superior border of the sacrum?).

g. The stomach, pancreas, small intestine, and most of the large intestine are located in the (abdomen? pelvis?).

h. The urinary bladder, rectum, and internal reproductive organs are located in the (abdominal? pelvic?) cavity.

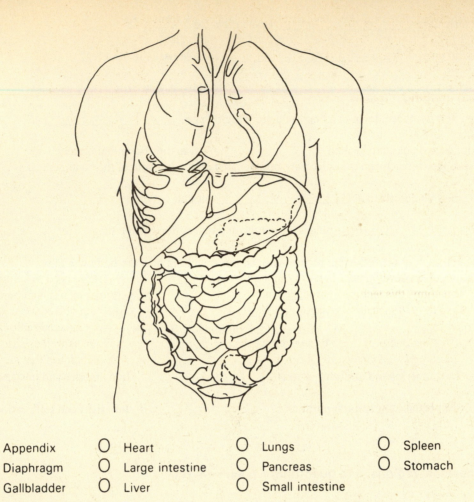

○ Appendix   ○ Heart   ○ Lungs   ○ Spleen

○ Diaphragm   ○ Large intestine   ○ Pancreas   ○ Stomach

○ Gallbladder   ○ Liver   ○ Small intestine

Figure LG 1.2   Regions of the ventral body cavity. Complete as directed in Checkpoint F2.

■ **F2.**   Complete Figure LG 1.2 according to the directions below.

   a. Color each of the organs listed on the figure. Select different colors for each organ, and be sure to use the same color for the related color code oval (○).

   b. Next, label each organ on the figure.

   c. *For extra review.* Draw lines dividing the abdomen into the nine regions, and note which organs are in each region.

■ **F3.**   Using the names of the nine abdominal regions and quadrants, complete these statements. Refer to the figure you just completed and to Figures 1.11 and 1.12 (pages 18 and 19 in your text).

   a. From superior to inferior, the three abdominal regions on the right side are

   _____, _____, and _____.

   b. The stomach is located primarily in the two regions named _____

   and _____.

   c. The navel is located in the _____ region.

   d. The region immediately superior to the urinary bladder is named the

   _____ region.

   e. The liver and gallbladder are located in the _____ quadrant.

F4. List at least three reasons why an autopsy might be performed.

G. Medical imaging   (pages 19–22)

■ **G1.** Match the radiographic anatomy techniques listed in the box with the correct descriptions.

CR.	Conventional radiograph	MRI.	Magnetic resonance imaging
CT.	Computed tomography	PET.	Positron emission tomography
DSA.	Digital subtraction angiography	US.	Ultrasound
DSR.	Dynamic spatial reconstruction		

_____ a. A recent development in radiographic anatomy, this technique produces moving three-dimensional images of body organs.

_____ b. This method provides a cross-sectional picture of an area of the body by means of an x-ray source moving in an arc. Results are processed by a computer and displayed on a video monitor.

_____ c. This technique produces a two-dimensional image (radiograph) via a single barrage of x-rays. Images of organs overlap, making diagnosis difficult.

_____ d. A method used to study blood vessels by comparing a region before and after injection of a contrast substance (dye).

_____ e. Utilizes injected radioisotopes that assess function as well as structure.

_____ f. Two techniques that are noninvasive and do not utilize ionizing radiation (two answers).

A1. Structure, function.

A3. (a) C. (b) H. (c) GA. (d) Embr (also a part of DA). (e) RP. (f) Endo. (g) PP. (h) I.

B1. From highest to lowest: organism, (system), organ, tissue, cell, chemical.

B2. Epithelial tissue, connective tissue, muscle tissue, nervous tissue.

B4. (a) Digestive; endocrine; stomach, small intestine. (b) Endocrine; nervous. (c) Reproductive, endocrine. (d) Skeletal, cardiovascular. (e) Urinary, reproductive.

C2. (a) C. (b) A. (c) M. (d) Repro. (e) D. (f) Respo.

D1. (a) Intracellular. (b) Interstitial (or intercellular or tissue) fluid, extracellular. (c) Blood, lymph. (d) Extra-; homeostasis.

D2. C, interstitial (intercellular) fluid; D, intracellular fluid; E, interstitial (intercellular) fluid again; F, blood or lymph capillaries; G, venules and veins or lymph vessels.

D3. Concentration of gases (such as oxygen or carbon dioxide), nutrients (such as proteins and fats), ions (such as sodium and potassium), and water; blood pressure and temperature.

D5. (a) CCond. (b) S. (c) Recep. (d) I. (e) CCen. (f) O. (g) E. (h) Respo.

D6. Change position gradually, for example, by sitting on the side of the bed before getting out of bed. Support body weight on a nightstand or chair for a moment after standing. Wear support hose. Drink adequate fluid.

D7. "Hurray! We have produced hundreds of shoes; let us go for thousands! Do not worry that the shoes are not selling. Step up production even more."

D8. (a) Negative; reversed. (b) Positive, enhanced. (c) Negative; they permit continual fine-tuning so changes do not become excessive in one direction.

D9. Fever, diarrhea, swollen ankles, skin rash.

E1. Stand with your arms at your sides and palms of your hands facing forward.

E2. (a) Armpit. (b) Phalanges. (c) Brachial. (d) Hollow area at back of knee. (e) Head. (f) Oral. (g) Groin. (h) Thoracic. (i) Neck. (j) Forearm. (k) Gluteal. (l) Heel.

E3. (a) Inferior (caudal). (b) Distal. (c) Posterior. (d) Medial. (e) Lateral. (f) Superficial. (g) Deep. (h) Anterior (ventral). (i) Superior (cephalic or cranial) (j) Ipsi-. (k) Visceral.

E4. (a) T. (b) M. (c) F. (d) S.

F1. (a) Dorsal. (b) Ventral. (c) Dorsal. (d) Thorax. (e) Lungs. (f) An imaginary line from the symphysis pubis to the superior border of the sacrum. (g) Abdomen. (h) Pelvic.

F2. (a–c) See Figure LG 1.2A.

F3. (a) Right hypochondriac, right lumbar, right iliac (inguinal). (b) Left hypochondriac, epigastric. (c) Umbilical. (d) Hypogastric (pubic). (e) Right upper.

G1. (a) DSR. (b) CT. (c) CR. (d) DSA. (e) PET. (f) MRI, US.

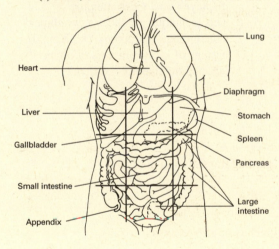

Figure LG 1.2A   Regions of the ventral body cavity.

WRITING ACROSS THE CURRICULUM: CHAPTER 1

1. Think of a person you know who has a chronic (ongoing) illness. Explain how changes in anatomy (abnormal body structure) in that person are associated with changes in physiology (inadequate or altered body function).
2. Cancer is a disease of abnormal cell production. Explain how alterations at the cellular level affect higher levels of organization, such as specific organs, systems, and the entire organism.
3. Suppose you decide to run a quarter mile at your top speed. Tell how your body would respond to this stress created by exercise. List according to system the changes your body would probably make in order to maintain homeostasis.

MASTERY TEST: CHAPTER 1

Questions 1–11: Circle the letter preceding the one best answer to each question.

1. In a negative feedback system, when blood pressure decreases slightly, the body will respond by causing a number of changes which tend to:
   A. Lower blood pressure
   B. Raise blood pressure

2. The following structures are all located in the ventral cavity *except:*
   A. Spinal cord
   B. Urinary bladder
   C. Heart
   D. Gallbladder
   E. Esophagus

3. Which pair of common/anatomical terms is mismatched?
   A. Eye/ocular
   B. Skull/cranial
   C. Armpit/brachial
   D. Neck/cervical
   E. Buttock/gluteal

4. Which science studies the "why, when, and where" regarding transmission of diseases?
   A. Endocrinology
   B. Histology
   C. Cytology
   D. Pharmacology
   E. Epidemiology

5. Which is most inferiorly located?
   A. Abdomen
   B. Pelvic cavity
   C. Mediastinum
   D. Diaphragm
   E. Pleural cavity

6. The spleen, tonsils, and thymus are all organs in which system?
   A. Nervous
   B. Lymphatic
   C. Cardiovascular
   D. Digestive
   E. Endocrine

7. Which of the following structures is located totally outside of the upper right quadrant of the abdomen?
   A. Liver
   B. Gallbladder
   C. Transverse colon
   D. Spleen
   E. Pancreas

8. Choose the FALSE statement. (It may help you to mark each statement T for true or F for false as you read it.)
   A. Stress disturbs homeostasis.
   B. Homeostasis is a condition in which the body's internal environment remains relatively constant.
   C. The body's internal environment is best described as extracellular fluid.
   D. Extracellular fluid consists of plasma and intracellular fluid.

9. The system responsible for providing support, protection, and leverage; for storage of minerals; and for production of most blood cells is:
   A. Urinary
   B. Integumentary
   C. Muscular
   D. Reproductive
   E. Skeletal

10. Which is most proximally located?
    A. Ankle
    B. Hip
    C. Knee
    D. Toe

11. Which of the following describes systemic anatomy?
    A. Study of a specific region of the body such as the chest or arm
    B. Study of markings of the surface of the body
    C. Study of the body according to specific systems such as skeletal or respiratory
    D. Study of structural changes associated with disease

T F 12. In order to assume the anatomical position, you should <u>lie down with arms at your
sides and palms turned backwards</u>.

T F 13. The appendix is usually located in the <u>left iliac</u> region of the abdomen.

T F 14. <u>Anatomy</u> is the study of how structures function.

T F 15. In a negative feedback system, the body will respond to an increased blood glucose
level by <u>increasing blood glucose to an even higher level</u>.

T F 16. A <u>tissue</u> is a group of cells and their intercellular substances that are similar in both
origin and function.

T F 17. The right kidney is located <u>mostly in the right iliac region of the abdomen</u>.

T F 18. The study of the microscopic structure of cells is <u>cytology</u>.

T F 19. A sagittal section is <u>parallel to and lateral to</u> a midsagittal section.

T F 20. The term homeostasis means that the body exists <u>in a static state</u>.

Questions 21–25: Fill-ins. Write the word that best fits the description.

_____ 21. Region of thorax located between the lungs. Contains heart, thymus, esophagus.

_____ 22. Name the fluid located within cells.

_____ 23. Life process in which unspecialized cells, as in the early embryo, change into
specialized cells such as nerve or bone.

_____ 24. A painless diagnostic technique in which sound waves are generated and detected
by a transducer, resulting in a sonogram.

_____ 25. Name one or more extracellular fluids.

Multiple Choice

1. B
2. A
3. C
4. E
5. B
6. B
7. D
8. D
9. E
10. B
11. C

True–False

12. F. Stand erect facing observer, arms at sides, palms forward
13. F. Right iliac
14. F. Physiology
15. F. Decreasing blood glucose to a lower level
16. T
17. F. At the intersection of four regions of the abdomen: right hypochondriac, right lumbar, epigastric, and umbilical
18. T
19. T
20. F. In a state of dynamic equilibrium

Fill-ins

21. Mediastinum
22. Intracellular
23. Differentiation
24. Ultrasound (US)
25. Plasma, interstitial fluid, lymph, serous fluid, mucus, fluid within eyes or ears, cerebrospinal fluid (CSF)

FRAMEWORK 2
The Chemical Level of Organization

TYPES OF CHEMICALS

INTRODUCTION (A)

ENERGY
- Potential
- Kinetic
- Radiant
- Electrical
- Heat
- Chemical
- Mechanical

BASIC CHEMISTRY
- Elements
- Atoms
  - protons
  - neutrons
  - electrons
- Atomic number
- Mass number
- Isotopes

CHEMICAL REACTIONS (B)

CONCEPTS
- Electron shell
  - valence
- Molecules
- Compounds

CHEMICAL BONDS
- Ionic
  - cations
  - anions
- Covalent
  - single
  - double
  - triple
- Hydrogen

CHEMICAL REACTIONS
- Activation energy
- Orientation
- Concentration
- Temperature

TYPES
- Synthesis
  - anabolism
- Decomposition
  - catabolism
- Exchange
- Reversible
- Oxidation
- Reduction

INORGANIC (C)

Inorganic Acids, Bases, Salts
- pH
- Buffer

Water

ORGANIC (D)

Carbohydrates
- Energy source
- $H:O = 2:1$
- Mono-, di, polysaccharide

Lipids
- Water insoluble
- Types
  - fats
  - phospholipids
  - steroids
  - vitamins A, D, E, K
  - eicosanoids

Proteins
- Amino acids
- Peptide bonds

Nucleic acids
- DNA
- RNA
- Nucleotides

ATP, ADP

The Chemical Level of Organization

CHAPTER 2

Chemicals comprise the ultrastructure of the human body. They are the submicroscopic particles of which cells, tissues, organs, and systems are constructed. Chemicals are the minute entities that participate in all of the reactions that underlie bodily functions. Every complex compound that molds a living organism is composed of relatively simple atoms held together by chemical bonds. Two classes of chemicals contribute to structure and function: inorganic compounds, primarily water, but also inorganic acids, bases, and salts, and organic compounds including carbohydrates, lipids, proteins, nucleic acids, and the energy-storing compound ATP.

As you begin your study of the chemistry of life, carefully examine the Chapter 2 Framework and note the key terms associated with each section. Also refer to the Chapter 2 Topic Outline and check off objectives as you meet each one.

TOPIC OUTLINE AND OBJECTIVES

A. Introduction: matter and energy, elements and atoms

☐ 1. Distinguish between matter and energy and describe five forms of energy.
☐ 2. Identify by name and symbol the principal chemical elements of the human body.
☐ 3. Describe the structure of an atom.

B. Chemical reactions

☐ 4. Explain how ionic, covalent, and hydrogen bonds form.
☐ 5. Define a chemical reaction and explain the basic differences between synthesis, decomposition, exchange, reversible, oxidation–reduction, exergonic, and endergonic chemical reactions.

C. Inorganic compounds: water, inorganic acids, bases, and salts; pH and buffers

☐ 6. List and compare the properties of inorganic acids, bases, salts, and water.
☐ 7. Define pH and explain the role of buffer systems in homeostasis.

D. Organic compounds

☐ 8. Compare the structure and functions of carbohydrates, lipids, proteins, deoxyribonucleic acid (DNA), ribonucleic acid (RNA), and adenosine triphosphate (ATP).
☐ 9. Describe the characteristics and importance of enzymes.

21

WORDBYTES

Now become familiar with the language of this chapter by studying each wordbyte, its meaning, and an example of its use within a term. After you study the entire list, self-check your understanding by writing the meaning of each wordbyte on the line provided. As you continue through the *Learning Guide,* identify and write in additional terms that contain the same wordbyte.

Wordbyte	Self-check	Meaning	Example(s)
-ase	_____	enzyme	lip*ase*
de-	_____	remove	*de*hydrogenase
di-	_____	two	*di*saccharide
hex-	_____	six	*hex*ose
mono-	_____	one	*mono*saccharide
pent-	_____	five	*pent*ose
poly-	_____	many	*poly*peptide
prot-	_____	first	*prot*ein
-saccharide	_____	sugar	mono*saccharide*
tri-	_____	three	*tri*phosphate

CHECKPOINTS

A. Introduction: matter and energy, elements and atoms (pages 26–28)

A1. Why is it important for you to understand the "language of chemistry" in order to communicate about life processes?

■ **A2.** Complete this exercise about matter and energy.

a. Matter exists in three states: solid, _____, or

_____.

b. When you are on a mountaintop, your *(mass? weight?)* is less than when you are standing on an ocean beach at sea level.

c. _____ is the capacity to do work. Inactive or stored energy is known as *(kinetic? potential?)* energy, whereas the energy of motion is

_____ energy.

d. *(Chemical? Electrical? Radiant?)* energy is the energy released or absorbed as chemicals are either broken apart or formed. An action potential (impulse) in a nerve or

muscle cell is an example of _____ energy. Energy that travels in waves, such as microwaves, ultraviolet (UV) rays, or x-rays, is classified as

_____ energy. Mechanical energy is the energy used to perform

_____.

■ **A3.** Complete this exercise about matter and chemical elements.

a. All matter is made up of building units called chemical _____.

b. Of the total 109 different chemical elements found in nature, four elements make up

96% of the mass of the human body. These elements are _____ (O),

_____ (C), hydrogen (_____), and

_____ (_____).

c. Circle the nine chemical symbols of elements listed here that together make up 3.9% of human body mass:
Ca Cl Co Cu Fe I K Mg Mn Na P S Si Sn Zn

d. Define this term: *trace elements*.

Now write on these lines the symbols of six trace elements from the list of elements

above (in d): _____ _____ _____ _____ _____ _____

■ **A4.** Refer to Exhibit 2.1 (page 27) in your text. Match the elements listed in the box with descriptions of their functions in the body. The first one has been done for you.

C. Carbon	K. Potassium
Ca. Calcium	Na. Sodium
Cl. Chlorine	N. Nitrogen
Fe. Iron	O. Oxygen
F. Fluorine	P. Phosphorus
I. Iodine	

____C____ a. Found in every organic molecule

_____ b. Component of all protein molecules, as well as DNA and RNA

_____ c. Vital to normal thyroid gland function

_____ d. Essential component of hemoglobin, which gives red color to blood

_____ e. Composes part of every water molecule; functions in cellular respiration

_____ f. Contributes hardness to bone and teeth; required for blood clotting and for muscle contraction

_____ g. Component of ATP, DNA, and RNA; also component of bone and teeth

_____ h. Anion (negatively charged ion) in table salt (NaCl)

_____ i. Cation (positively charged ion) in table salt (NaCl); helps control water balance in the body; important in nerve and muscle function

_____ j. A trace element that contributes to tooth structure

_____ k. Most abundant positively charged ion (cation) in intracellular fluid (inside of cells)

A5. Contrast *element* with *atom*.

■ **A6.** Complete this exercise about atomic structure.

a. An atom consists of two main parts: _____ and

_____ .

b. Within the nucleus are positively charged particles called _____ and

uncharged (neutral) particles called _____ .

c. Electrons possess *(positive? negative?)* charges. The number of electrons forming a charged cloud around the nucleus is *(greater than? equal to? smaller than?)* the number of protons in the nucleus of the atom.

d. Atoms are measured in *(grams? daltons?)*. A proton and a neutron each have the mass

of approximately _____ dalton. Electrons have a mass of about _____ dalton.

■ **A7.** Refer to Figure 2.2 in your text (page 29) and complete this exercise about atomic structure.

a. The atomic number of potassium (K) is _____. Locate this number below the

symbol K on the figure. This means that a potassium atom has _____ protons and

_____ electrons.

b. Identify the atomic mass number associated with potassium in the figure. This

number is _____, and it represents the average of the mass numbers of the potassium atom. This indicates the total number of protons and neutrons in the potassium nucleus.

■ **A8.** Complete this exercise about isotopes.

a. If you subtract the atomic number from the mass number, you will identify the number

of neutrons in an atom. Most potassium atoms have _____ neutrons. Different - isotopes of potassium all have 19 protons, but have varying numbers of *(electrons? neutrons?)*.

b. The isotope ^{14}C differs from ^{12}C in that ^{14}C has _____ neutrons.

c. Radioisotopes have a nuclear structure that is unstable and can decay, emitting

_____ . Each radioactive isotope has its own distinctive

_____ , which is the time required for the radioactive isotope to emit half of its original amount of radiation.

d. A radioactive form of iodine, ^{131}I, is most often used to indicate size, location, and activity of the *(adrenal? pituitary? thyroid?)* gland. Imaging of the heart and its coronary blood flow most often uses an isotope of *(lead [Pb]? thallium [Tl]? uranium [U]?)*.

B. Chemical reactions   (pages 28–34)

■ **B1.**   The type of chemical reacting (or bonding) that occurs between atoms depends on the number of electrons possessed by the bonding atoms. Refer to Figure 2.2 in your text (page 29) and do this exercise about electrons and bonding.

a. Electrons are arranged in electron shells. The electron shell closest to the nucleus has a

   maximum capacity of _____ electrons, whereas the second shell holds a

   maximum of _____ electrons.

b. The combining capacity (or _____) refers to the number of extra or

   deficient electrons in the outermost shell (or _____ shell). (Visualize t hat this is the valence shell that atom A presents to nearby atom B, which may be a possible "bonding partner.") In other words, the valence of an atom tells the

   number of _____ that an atom attempts to gain, lose, or share in an effort to have its valence shell complete.

c. Find helium (He) in Figure 2.2. How many electrons does a helium atom need to gain, lose, or share to make its valence shell complete? _____. In other words, it does not "need" to react with any other atoms; for this reason, helium is said to be

   _____.

d. Now find potassium (K) in the same figure. Potassium has one *(extra? missing?)* electron in its valence shell. It is therefore more likely to become an electron *(donor? acceptor?)*. When potassium gives up the electron (which is a negative entity), potassium becomes more *(positive? negative?)*. In other words, potassium forms K^+, which is a(n) *(anion? cation?)*.

e. Chlorine has a valence of *(+1? −1?)*, indicating that it has one *(extra? missing?)* electron in its valence shell. As potassium chloride is formed, chlorine *(accepts? gives up?)* an electron and so becomes the *(anion? cation?)* Cl^-.

f. In summary, ions are atoms that have *(gained or lost? shared?)* one or more electrons into or from their valence shells. A cation, such as K^+, Na^+, or H^+, is an atom that has *(gained? lost?)* an electron (a negative entity) by ionic bonding. A(n) *(anion? cation?)*, such as Cl^-, has gained an electron from another atom. Ions in solution, such

   as table salt (Na^+Cl^-) in water, are called _____.

■ **B2.**   Circle the molecules that are classified as compounds.

   H   Glucose

   H_2 N_2

   H_2O A molecule that contains two or more different kinds of atoms

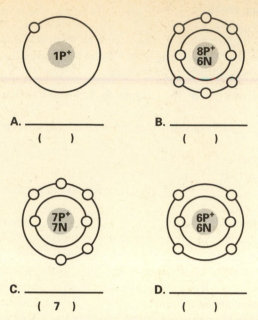

A. _____
( )

B. _____
( )

C. _____
(7)

D. _____
( )

Figure LG 2.1 Atomic structure of four common atoms in the human body. Complete as directed in Checkpoint B3.

■ **B3.** Refer to Figure LG 2.1 and complete this exercise.

a. Write atomic numbers in the parentheses ( ) under each atom. (*Hint:* Count the number of electrons or protons = atomic number.) One is done for you.

b. Write the name or symbol for each atom on the line under that atom. (*Hint:* You may identify these by atomic number in Figure 2.2, page 29 in your text.)

c. Based on the number of electrons that each of these atoms is missing from a complete outer electron shell, draw the number of covalent bonds that each of the atoms can be expected to form. Note that this is the valence, or combining capacity, of these atoms. One is done for you.

$$\text{H} \qquad \text{O} \qquad \text{N} \qquad \text{C}$$

d. *For extra review.* On separate paper draw the atomic structure of an ammonia molecule (NH_3). Note that this is a covalently bonded compound.

■ **B4.** Match the descriptions with the types of bonds listed in the box.

C. Covalent	H. Hydrogen	I. Ionic

_____ a. Atoms lose electrons if they have just one or two electrons in their valence shells; they gain electrons if they need just one or two electrons to complete the valence shell, as in K^+Cl^- (see Activity B1 above).

_____ b. This bond is a bridgelike, weak link between a hydrogen atom and another atom such as oxygen or nitrogen.

_____ c. A pair of electrons is *shared* in this type of bond, for example, in a C–H bond; the two electrons (one from C and one from H) orbit alternately around the two nuclei (of C and of H).

_____ d. These bonds are only about 5% as strong as covalent bonds and so are easily formed and broken. They are vital for holding large molecules like protein or DNA in proper configurations.

_____ e. Double bonds, such as $O{=}O$ (O_2), or triple bonds, such as $N{\equiv}N$ (N_2).

_____ f. This type of bond is most easily formed by carbon (C) because its outer orbit is half-filled.

_____ g. These bonds may be polar, such as H_2O (H — OH), or nonpolar, such as $O{=}O$ (O_2).

B5. Write a paragraph describing four factors that are likely to enhance collisions between atoms, leading to chemical reactions.

■ **B6.** Identify the kind of chemical reaction described in each statement below.

D. Decomposition reaction	R. Reversible reaction
E. Exchange reaction	S. Synthesis reaction

_____ a. The end product can revert to the original combining molecules.

_____ b. Two or more reactants combine to form an end product; for example, many glucose molecules bond to form glycogen.

_____ c. Such a reaction is partly synthesis and partly decomposition.

_____ d. All chemical reactions involve making or breaking of bonds. This type of reaction involves only breaking of bonds.

_____ e. This type of reaction is catabolic, as in digestion of foods, such as starch digestion to glucose.

■ **B7.** We recognize the importance of oxygen for our survival. Oxygen is ultimately required for processes called biological *(oxidation? reduction?)* reactions that provide energy for all types of body activities. Describe such reactions in this exercise.

a. Most biological oxidation reactions involve *(addition? removal?)* of hydrogen atoms from chemicals in foods we have eaten. Such reactions are known as *(hydrogenation? dehydrogenation?)* reactions. For example, high energy sugars we eat, with their many C–H (carbon-to-hydrogen) bonds, will lose H's and ultimately form C–O (carbon-to-oxygen) bonds, as in the low energy waste we exhale, carbon dioxide (CO_2).

b. Look at the figures of the 3-carbon sugars, lactic acid and pyruvic acid, on page 33 of your text. Count the atoms in each chemical and write them here. (The exercise is already started for you.)

	Carbons (C)	Hydrogens (H)	Oxygens (O)
Lactic acid	_____	6	_____
Pyruvic acid	3	_____	_____

c. Which atom has more hydrogens? *(Lactic? Pyruvic?)* acid. This acid loses _____ hydrogen atoms (or is dehydrogenated) as it is converted to pyruvic acid. In this way lactic acid is *(oxidized? reduced?)* to form pyruvic acid.

d. Because energy is released from this reaction, oxidation is called an _____-gonic reaction. Energy released from chemical bonds in foods we eat is eventually stored in the

high energy bonds of the molecule named _____: In fact, the chemical energy from

one molecule of glucose can be used to form _____ molecules of ATP.

■ **B8.** *The Big Picture: Looking Ahead.* You can expand on what you have just learned about oxidation if you turn to Figure 25.3, page 815 of your text. There you will see a number of examples of biological oxidation in metabolism (or utilization) of the foods.

a. Think of the structure of each hydrogen atom: it consists of *(one? two?)* proton(s) with *(one? two?)* electrons(s). Suppose a hydrogen atom suddenly loses its electron—which takes off to whirl around another hydrogen's proton. What happens to that first

hydrogen? It becomes: _____.
A. H^+ (a hydrogen ion which is simply a proton with no orbiting electrons)
B. H^- (a hydride ion which is a proton with two orbiting electrons)

b. It is possible then for a chemical, such as a sugar, to lose two hydrogens in the ionized forms, one as H^+ and one as H^-. In fact, this is exactly what happens in biological oxidation: removal not simply of two hydrogen atoms (2H), but removal of a pair of

hydrogens in the _____ forms, that is, (H^++H^-).

c. Whenever one chemical gives up ($H^+ + H^-$), another chemical must *gain* those two ionized hydrogens; this type of reaction (the opposite of oxidation) is called

_____. In fact, oxidation reactions must always be

_____ with reduction reactions.

d. In summary, oxidation reactions involve *(gain? loss?)* of hydrogens (or their equiv-

alents) with release of energy; oxidation reactions are _____-gonic, so they result

in a(n) _____-crease of ATP in the body. Reduction reactions involve *(gain? loss?)*

of hydrogens (or equivalents) and are _____-gonic.

C. Inorganic compounds: water, inorganic acids, bases, and salts; pH and buffers
(pages 34–38)

■ **C1.** Write I next to phrases that describe inorganic compounds and write O next to phrases that describe organic compounds.

_____ a. Always contain carbon and hydrogen; held together almost entirely by covalent bonds

_____ b. Tend to be very large molecules that serve as good building blocks for body structures

_____ c. The class that includes carbohydrates, proteins, fats, and nucleic acids

_____ d. The class that includes water, the most abundant compound in the body

■ **C2.** Fill in the blanks in this exercise about acids, bases, and salts.

a. NaOH (sodium hydroxide) is an example of a(n) _____, that is, a chemical that breaks apart into hydroxide ions (OH^-) and one or more cations.

b. H_2SO_4 (hydrogen sulfate) is an example of a(n) _____, that is, a chemical that dissociates into hydrogen ions (H^+) and one or more anions.

c. Salts, such as NaCl (sodium chloride), are chemicals that dissolve in water to form

cations and ions, neither of which is _____ or _____.

■ **C3.** Contrast solutions and suspensions in this exercise.

a. An oil-based salad dressing is a *(solution? suspension?)* that is formed as the dressing is shaken. Upon sitting for a few seconds, the lighter weight oil rises to the top as the herbs settle out. Blood is another example of a *(solution? suspension?)* because blood cells are likely to settle out of blood plasma if the blood is centrifuged.

b. Coffee, tea, and tears are examples of *(solutions? suspensions?)*; water serves as the *(solute? solvent?)* in which ions and other *(solutes? solvents?)* (such as crystals of coffee or tea) are dissolved.

■ **C4.** Concentrations of solutions may be expressed in several ways. Describe them in this Checkpoint. (*Hint:* Refer to atomic masses for Na and Cl in Figure 2.2, page 29 in your text.)

a. A mole per liter (1 mol/liter) means that a number of *(milligrams? grams?)* equal to the molecular weight of the chemical should be dissolved in enough solvent to total one

   _____ of solution. Because the atomic weights of Na and Cl are

   _____ and _____, then *(12.45? 58.44? 1000?)* g of NaCl would be needed to prepare a liter of a "one mole per liter" (1 mol/liter) solution of NaCl.

b. A one millimole per liter solution (1 mmol/liter) has *(1000 times the? 10 times the? the same? one-thousandth the?)* concentration of a one mole per liter (1 mol/liter) solution.

   To prepare a liter of a 1 mmol/liter solution of NaCl, how much NaCl is needed? _____.

■ **C5.** Do this exercise on roles of water in the body.

a. The atoms of water have *(polar covalent? nonpolar covalent? ionic?)* bonds. This means that the electrons of the hydrogens are shared with oxygen in such a way that they spend more time around *(hydrogen? oxygen?)* and less time around

   _____. As a result, *(each hydrogen? oxygen?)* has a slightly negative charge, whereas *(each hydrogen? oxygen?)* has a slightly positive charge. These charges account for the polarity of water.

b. When added to water, salt is readily dissolved as the Na^+ ions are attracted to the *(positively charged hydrogens? negatively charged oxygens?)* and Cl^- ions are attracted to the *(positively charged hydrogens? negatively charged oxygens?)* of water.

c. Besides NaCl, name several other chemicals in the human body that are dissolved in the excellent solvent, water.

d. Water is useful in digesting foods via the breakdown process called

   _____-lysis. Water also serves as a lubricant; give two or more examples of this function.

e. List two mechanisms by which water helps to cool the body.

■ **C6.** Draw a diagram showing the pH scale. Label the scale from 0 to 14. Indicate by arrows increasing acidity (H^+ concentration) and increasing alkalinity (OH^- concentration).

■ **C7.** Choose the correct answers regarding pH.

a. Which pH is most acid?

    A. 4   C. 10

    B. 7

b. Which pH has the highest concentration of OH^- ions?

    A. 4   C. 10

    B. 7

c. Which solution has pH closest to neutral?

    A. Gastric juice (digestive juices of the stomach)

    B. Blood

    C. Lemon juice

    D. Milk of magnesia

d. A solution with pH 8 has 10 times *(more? fewer?)* H^+ ions than a solution with pH 7.

e. A solution with pH 5 has *(10? 20? 100?)* times *(more? fewer?)* H^+ ions than a solution with pH 7.

C8. Write a sentence or two explaining each of the following:

a. How pH is related to homeostasis

b. How buffers help to maintain homeostasis

D. Organic compounds (pages 38–50)

D1. Organic compounds all contain carbon (C) and hydrogen (H). State two reasons why carbon is an ideal element to serve as the primary structural component for living systems.

■ **D2.** Carbohydrates carry out important functions in your body.

a. State the principal function of carbohydrates.

b. State two secondary roles of carbohydrates.

■ **D3.** Complete these statements about carbohydrates.

a. The ratio of hydrogen to oxygen in all carbohydrates is _____:_____. The

   general formula for carbohydrates is _____.

b. The chemical formula for a *hexose* is _____. Two common hexoses

   are _____ and _____.

c. One common pentose sugar found in DNA is _____. Its name and its
   formula ($C_5H_{10}O_4$) indicate that it lacks one oxygen atom from the typical pentose

   formula which is _____.
d. When two glucose molecules ($C_6H_{12}O_6$) combine, a disaccharide is formed. Its

   formula is _____, indicating that in a synthesis reaction such as this,
   a molecule of water is *(added? removed?)*. Refer to Figure LG 2.2. Identify which reaction *(1? 2?)* on that figure demonstrates synthesis.
e. Continued dehydration synthesis leads to enormous carbohydrates called

   _____. One such carbohydrate is _____.
f. When a disaccharide such as sucrose is broken, water is introduced to split the bond
   linking the two monosaccharides. Such a decomposition reaction is termed

   _____, which literally means "splitting using water." (See Figure
   2.8, page 39 in your text.) On Figure LG 2.2, hydrolysis is shown by reaction *(1? 2?)*.

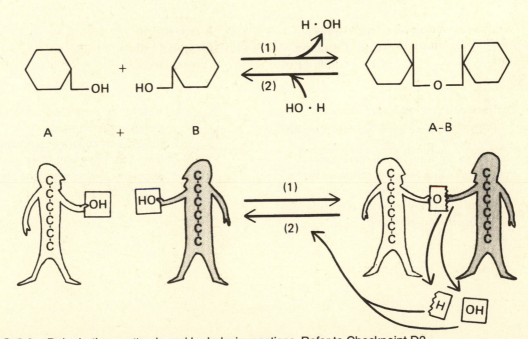

Figure LG 2.2   Dehydration synthesis and hydrolysis reactions. Refer to Checkpoint D3.

■ **D4.** Contrast carbohydrates with lipids in this exercise. Write C next to any statement true of carbohydrates and write L next to any statement true of lipids.

_____ a. These compounds are insoluble in water (hydrophobic). (*Hint:* Water alone is ineffective for washing such compounds out of clothes.)

_____ b. These compounds are organic.

_____ c. These substances have a hydrogen to oxygen ratio of about 2:1.

_____ d. Very few oxygen atoms (compared to the numbers of carbon and hydrogen atoms) are contained in these compounds. (Look carefully at Figure 2.9, page 41 in your text.)

■ **D5.** Circle the answer that correctly completes each statement.
  a. A triglyceride consists of:
    A. Three glucoses bonded together
    B. Three fatty acids bonded to a glycerol
  b. A typical fatty acid contains (see Figure 2.9, page 41 in your text):
    A. About 3 carbons
    B. About 16 carbons
  c. A polyunsaturated fat contains _____ than a saturated fat.
    A. More hydrogen atoms
    B. Fewer hydrogen atoms
  d. Saturated fats are more likely to be derived from:
    A. Plants, for example, corn oil or peanut oil
    B. Animals, as in beef or cheese
  e. Oils that are classified as saturated fats include:
    A. Tropical oils such as coconut oil and palm oil
    B. Vegetable oils such as safflower oil, sunflower oil, and corn oil
  f. A diet designed to reduce the risk for atherosclerosis is a diet low in:
    A. Saturated fats
    B. Polyunsaturated fats

■ **D6.** Match the types of lipids listed in the box with their functions given below.

B. Bile salts	E. Estrogens
Car. Carotene	Pho. Phospholipids
Cho. Cholesterol	Pro. Prostaglandins

_____ a. Present in egg yolk and carrots; leads to formation of vitamin A, a chemical necessary for good vision

_____ b. The major lipid in cell membranes

_____ c. Present in all cells, but infamous for its relationship to "hardening of the arteries"

_____ d. Substances that break up (emulsify) fats before their digestion

_____ e. Steroid sex hormones produced in large quantities by females

_____ f. Important regulatory compounds with many functions, including modifying responses to hormones and regulating body temperature

■ **D7.** Complete this exercise about fat-soluble vitamins.

a. Name four fat-soluble vitamins (consult Exhibit 2.5, page 40 in your text, for help):

_____ _____ _____ _____

b. Because these four vitamins are absorbed into the body along with fats, any problem with fat absorption is likely to lead to:

c. The fat-soluble vitamin that helps with calcium absorption and so is necessary for

normal bone growth is vitamin _____.

d. Vitamin _____ may be administered to a patient before surgery in order to prevent excessive bleeding because this vitamin helps in the formation of clotting factors.

e. Normal vision is associated with an adequate amount of vitamin _____.

f. Vitamin _____ has a variety of possible functions including the promotion of wound healing and prevention of scarring.

■ **D8.** Match the types of proteins listed in the box with their functions given below.

Cat. Catalytic	R. Regulatory
Con. Contractile	S. Structural
I. Immunological	T. Transport

_____ a. Hemoglobin in red blood cells

_____ b. Proteins in muscle tissue, such as actin and myosin

_____ c. Keratin in skin and hair; collagen in bone

_____ d. Hormones such as insulin

_____ e. Defensive chemicals such as antibodies and interleukins

_____ f. Enzymes such as lipase, which digests lipids

■ **D9.** Contrast proteins with the other organic compounds you have studied so far by completing this exercise.

a. Carbohydrates, lipids, and proteins all contain carbon, hydrogen, and oxygen. A fourth

element, _____, makes up a substantial portion of proteins.

b. Just as large carbohydrates are composed of repeating units (the

_____), proteins are composed of building blocks called

_____.

c. As in the synthesis of carbohydrates or fats, when two amino acids bond together, a water molecule must be (*added? removed?*). This is another example of (*hydrolysis? dehydration?*) synthesis.

d. The product of such a reaction is called a _____-peptide. (See Figure 2.13, page 44 in your text.) When many amino acids are linked together in this way, a

_____-peptide results. One or more polypeptide chains form a

_____.

e. Refer to Figure 2.12 on page 44 of your text. Which part of the molecular structure varies among the 20 different kinds of amino acids to result in their unique identities? *(Amino group? Side chain = R group? Carboxyl group?)*

D10. There are _____ different kinds of amino acids forming the human body, much like a 20-letter alphabet that forms protein "words." It is the specific sequence of amino acids that determines the nature of the protein formed. In order to see the significance of this fact, do the following activity. Arrange the five letters listed below in several ways so that you form different five-letter words. Use all five letters in each word.

e i l m s

___ ___ ___ ___ ___   ___ ___ ___ ___ ___

___ ___ ___ ___ ___   ___ ___ ___ ___ ___

Note that although all of your words contain the same letters, the sequence of letters differs in each case. The resulting words have different meanings, or functions. Such is the case with proteins, too, except that protein "words" may be thousands of "letters" (amino acids) long, involving all or most of the 20-amino acid "alphabet." Proteins must be synthesized with accuracy. Think of the drastic consequences of omission or misplacement of even a single amino acid. (How different are *limes, miles, smile, slime!*)

■ **D11.** Match the description below with the four levels of protein organization in the box.

P. Primary structure	S. Secondary structure
Q. Quaternary structure	T. Tertiary structure

_____ a. Specific sequence of amino acids

_____ b. Twisted and folded arrangement (as in spirals or pleated sheets) due to hydrogen bonding

_____ c. Irregular three-dimensional shape; disulfide bonds play a role in this level of structure

_____ d. Relative arrangement of two or more polypeptide chains

■ **D12.** Describe changes in protein structure by completing this activity.

a. Replacement of one amino acid (valine) with another (glutamate) in the protein

portion of hemoglobin results in a condition known as _____.

b. Cooking an egg white results in destruction (or _____) of the protein albumin. Extreme increases in body temperature, for example, increases caused

by _____, can denature body proteins such as those in skin or blood plasma.

■ **D13.** Do this exercise about roles of enzymes related to chemical reactions.

a. Explain why it is necessary for chemical reactions to take place rapidly within the human body.

b. Increase in body temperature will affect rates of chemical reactions. Discuss pros and cons of increasing body temperature.

c. Explain how enzymes increase chemical reaction rates without requiring a raised body temperature.

d. Because enzymes speed up chemical reactions (yet are not altered themselves), they are

known as _____.

e. Chemically, all enzymes consist mostly of *(carbohydrates? lipids? proteins?)*. Each

enzyme reacts with a specific molecule called a _____.

f. The substrate interacts with a specific region on the enzyme called the

_____, producing an intermediate known as an

_____–_____ complex. Very quickly the *(enzyme?
substrate?)* is transformed in some way (for example, is broken down or transferred to
another substrate), while the *(enzyme? substrate?)* is recycled for further

use. A _____ number of 1 to 600,000 molecules of product per second may be expected for each enzyme molecule.

g. Whole enzymes are sometimes known as *(apo? holo?)*-enzymes. These consist of a protein called the *(apo? holo?)*-enzyme, as well as a nonprotein cofactor.

h. Cofactors may be of two types. Some are metal ions, such as _____.
So, for example, a protein (apoenzyme) plus calcium ion (Ca^{2+}) may serve as a holo-

enzyme. The other type of cofactor is known as a(n) _____-enzyme. Most coenzymes are derived from *(vitamins? minerals?)*.

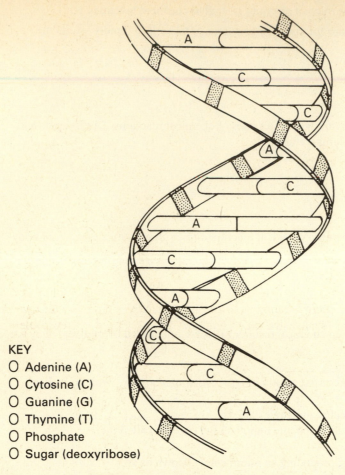

KEY
O Adenine (A)
O Cytosine (C)
O Guanine (G)
O Thymine (T)
O Phosphate
O Sugar (deoxyribose)

Figure LG 2.3   Structure of DNA. Complete as directed in Checkpoint D14.

■ **D14.** Refer to Figure LG 2.3 and fill in the blanks in this exercise about DNA.

a. DNA has a structure much like a ladder, but one that is arranged in a spiral; this

   arrangement is called a double _____. The sides of the ladder are
   formed of *(bases? alternating sugars and phosphates?)*. Color these according to the
   color code ovals on the figure.

b. "Rungs" of this DNA ladder are formed of _____. There are *(2? 4?
   8? 20?)* different bases in DNA; their sequence "writes" the genetic code. Note the dire
   consequences of omission or substitution in this "four-letter" alphabet.

c. Bases in DNA are paired _____ (A) to _____

   (_____) and _____ (C) to _____ (_____).
   Label complementary nucleotides on the right side of the DNA in the figure. Then
   color all nucleotides in the figure according to the color code ovals.

d. A DNA _____ is one monomer of DNA, and it consists of three

   parts: _____ (a sugar), a base, and a _____.
   Circle one nucleotide in the figure.

e. Describe the structure and function of a gene.

D15. How does a strand of RNA differ from DNA?

■ **D16.** Describe the structure and significance of ATP by completing these statements.

a. ATP stands for adenosine triphosphate. Adenosine consists of a base that is a

   component of DNA, that is, _____, along with the five-carbon sugar

   named _____.

b. The "TP" of ATP stands for _____. The final *(one? two? three?)*
   phosphates are bonded to the molecule by high energy bonds.

c. When the terminal phosphate is broken, a great deal of energy is released as ATP is

   split into _____ + _____.

d. ATP, the body's primary energy-storing molecule, is constantly reformed by the reverse
   of this reaction as energy is made available from foods you eat. Write this reversible
   reaction.

e. The anaerobic phase of cellular respiration yields _____ molecules of ATP, whereas

   the aerobic phase yields _____ molecules of ATP.

ANSWERS TO SELECTED CHECKPOINTS: CHAPTER 2

A2. (a) Liquid, gas. (b) Weight. (c) Energy; potential, kinetic. (d) Chemical; electrical; radiant; movements.

A3. (a) Elements. (b) Oxygen, carbon, H, nitrogen (N). (c) Ca, Cl, Fe, I, K, Mg, Na, P, S. (d) Thirteen elements found in very small amounts, together comprising only 0.1% of the total body mass; Co, Cu, Mn, Si, Sn, Zn.

A4. (b) N. (c) I. (d) Fe. (e) O. (f) Ca. (g) P. (h) Cl. (i) Na. (j) F. (k) K.

A6. (a) Nucleus, electrons. (b) Protons (p^+), neutrons (n^0). (c) Negative; equal to. (d) Daltons; 1; 1/2000.

A7. (a) 19; 19, 19. (b) 39.098 (rounded off to 39).

A8. (a) 20 (39–19); neutrons. *Note:* varied masses of isotopes account for the uneven mass number, such as 39.098 for potassium. (b) 8. (c) Radiation; half-life. (d) Thyroid; thallium [Tl].

B1. (a) 2, 8. (b) Valence, valence; electrons. (c) Zero (since the natural state of helium is to contain two electrons in its already-complete outermost energy level); inert. (d) Extra; donor; positive; cation. (e) −1, missing; accepts, anion. (f) Gained or lost; lost; anion; electrolytes.

B2. H_2O, glucose, a molecule that contains two or more different kinds of atoms.

B3. (a, b: see Figure LG 2.1A)

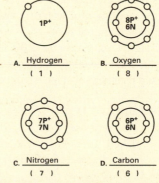

Figure LG 2.1A Atomic structure of four common atoms in the human body.

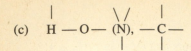

(c)

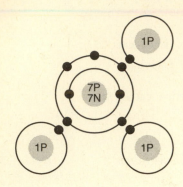

(d) Atomic structure of ammonia (NH_3):

B4. (a) I. (b) H. (c) C. (d) H. (e) C. (f) C. (g) C.

B6. (a) R. (b) S. (c) E. (d) D. (e) D.

B7. Oxidation. (a) Removal; dehydrogenation. (b) Lactic: 3 C, 6 H, 3 O; pyruvic: 3 C, 4 H, 3 O. (c) Lactic; 2; oxidized. (d) Exer-; ATP; 36–38.

B8. (a) One, one; A. (b) Ionized. (c) Reduction; coupled. (d) Loss, exer-, in; gain, ender.

C1. (a–c) O. (d) I.

C2. (a) Base. (b) Acid. (c) H^+ or OH^-.

C3. (a) Suspension; suspension. (b) Solutions, solvent, solutes.

C4. (a) Grams, liter; 22.99, 35.45, 58.44. (b) One thousandth the; 58.44 mg = 0.05844 g (or 0.058 g).

C5. (a) Polar covalent; oxygen, hydrogen; oxygen, each hydrogen. (b) Negatively charged oxygens, positively charged hydrogens. (c) Some of the O_2 and CO_2 gas molecules in blood, wastes such as urea, small sugars such as glucose. (d) Hydro; water in mucus permits foods and wastes to slide along the digestive tract and wastes to be flushed out in the urinary tract; within serous fluid, water permits organs to slide over one another; water in synovial fluid at joints facilitates smooth movements of bones. (e) Water absorbs heat readily, and evaporation of water (in sweating) releases large amounts of heat.

C7. (a) A. (b) C. (c) B. (d) Fewer. (e) 100, more.

D2. (a) Provide a readily available source of energy. (b) Converted to other substances; serve as food reserves.

D3. (a) 2:1; $(CH_2O)_n$. (b) $C_6H_{12}O_6$; glucose, fructose, galactose. (c) Deoxyribose; $C_5H_{10}O_5$. (d) $C_{12}H_{22}O_{11}$, removed; 1. (e) Polysaccharides; starch, glycogen, cellulose. (f) Hydrolysis; 2.

D4. (a) L. (b) C, L. (c) C. (d) L.

D5. (a) B. (b) B. (c) B. (d) B. (e) A. (f) A.

D6. (a) Car. (b) Pho. (c) Cho. (d) B. (e) E. (f) Pro.

D7. (a) A, D, E, K. (b) Symptoms of deficiencies of fat-soluble vitamins. (c) D. (d) K. (e) A. (f) E.

D8. (a) T. (b) Con. (c) S. (d) R. (e) I. (f) Cat.

D9. (a) Nitrogen. (b) Monosaccharides, amino acids. (c) Removed; dehydration. (d) Di; poly; protein. (e) Side chain = R group.

D11. (a) P. (b) S. (c) T. (d) Q.

D12. (a) Sickle cell anemia. (b) Denaturation; burns or fever (hyperthermia).

D13. (a) Thousands of different types of chemical reactions are necessary daily to maintain homeostasis. (b) Chemical reactions occur more rapidly at higher temperature (one reason for increased catabolism and weight loss when you have a fever). However, very high temperatures will destroy body tissues. (c) Enzymes orient molecules so that they are more likely to react, lower activation energy needed for reactions, and increase frequency of collisions. (d) Catalysts. (e) Proteins; substrate. (f) Active site, enzyme–substrate; substrate, enzyme; turnover. (g) Holo; apo. (h) Magnesium (Mg^{2+}), zinc (Zn^+),

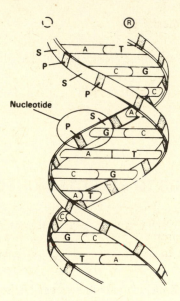

or calcium (Ca^{2+}); co; vitamins.

Figure LG 2.3A Structure of DNA. P, phosphate; S, sugar (deoxyribose); A, adenine; T, thymine; G, guanine; C, cytosine.

D14. (a) Helix; alternating sugars and phosphates; see Figure LG 2.2A. (b) Bases; 4. (c) Adenine (A), thymine (T), cytosine (C), guanine (G); see Figure LG 2.2A. (d) Nucleotide, deoxyribose, phosphate; see Figure LG 2.2A. (e) Segment of DNA that determines one trait or regulates one activity.

D16. (a) Adenine, ribose. (b) Triphosphate; two. (c) ADP + phosphate (P). (d) ATP $\rightleftharpoons$ ADP + P + energy. (e) 2, 36–38.

1. Write several functions of water that explain why this chemical is vital to living organisms.
2. Contrast the roles of carbohydrates and proteins in the human body.
3. Describe three factors that affect enzyme action: specificity, efficiency, and control.

MASTERY TEST: CHAPTER 2

Questions 1–12: Circle the letter preceding the one best answer to each question.

1. All of the following statements are true *except:*
   A. A reaction in which two amino acids join to form a dipeptide is called a dehydration synthesis reaction.
   B. A reaction in which a disaccharide is digested to form two monosaccharides is known as a hydrolysis reaction.
   C. About 65 to 75% of living matter consists of organic matter.
   D. Strong acids ionize more easily than weak acids do.

2. Choose the one *true* statement.
   A. A pH of 7.5 is more acidic than a pH of 6.5.
   B. Anabolism consists of a variety of decomposition reactions.
   C. An atom such as chlorine (Cl), with seven electrons in its outer orbit, is likely to be an electron donor (rather than electron receptor) in ionic bond formation.
   D. Polyunsaturated fats are more likely to reduce cholesterol level than saturated fats are.

3. In the formation of the ionically bonded salt NaCl, Na^+ has:
   A. Gained an electron from Cl^-
   B. Lost an electron to Cl^-
   C. Shared an electron with Cl^-
   D. Formed an isotope of Na^+

4. Over 99% of living cells consist of just six elements. Choose the element that is NOT one of these six.
   A. Calcium
   B. Hydrogen
   C. Carbon
   D. Iodine
   E. Nitrogen
   F. Oxygen
   G. Phosphorus

5. Which of the following describes the structure of a nucleotide?
   A. Base-base
   B. Phosphate-sugar-base
   C. Enzyme
   D. Dipeptide
   E. Adenine-ribose

6. Which of the following groups of chemicals includes only polysaccharides?
   A. Glycogen, starch
   B. Glycogen, glucose, galactose
   C. Glucose, fructose
   D. RNA, DNA
   E. Sucrose, polypeptide

7. $C_6H_{12}O_6$ is most likely the chemical formula for:
   A. Amino acid
   B. Fatty acid
   C. Hexose
   D. Polysaccharide
   E. Ribose

8. All of the following answers consist of correctly paired terms and descriptions related to enzymes *except:*
   A. Active site—a place on an enzyme which fits the substrate
   B. Substrate—molecule(s) upon which the enzyme acts
   C. -ase—ending of most enzyme names
   D. Holoenzyme—another name for a cofactor

9. Which of the following substances are used mainly for structure and regulatory functions and are not normally used as energy sources?
   A. Lipids
   B. Proteins
   C. Carbohydrates

10. Which is a component of DNA but not of RNA?
    A. Adenine
    B. Phosphate
    C. Guanine
    D. Ribose
    E. Thymine

11. All of the following answers consist of pairs of proteins and their correct functions *except:*
    A. Contractile—actin and myosin
    B. Immunological—collagen of connective tissue
    C. Catalytic—enzymes
    D. Transport—hemoglobin

12. All of the following answers consist of pairs of elements and their correct chemical symbols *except:*
    A. Nitrogen (N)
    B. Sodium (Na)
    C. Carbon (C)
    D. Calcium (Ca)
    E. Potassium (P)

Questions 13–20: Circle T (true) or F (false). If the statement is false, change the underlined word or phrase so that the statement is correct.

T F 13. Oxygen, water, NaCl, and glucose are <u>inorganic</u> compounds.

T F 14. There are about <u>four</u> different kinds of amino acids found in human proteins.

T F 15. The number of protons always equals the number of <u>neutrons</u> in an atom.

T F 16. <u>K$^+$ and Cl$^-$ are both cations.</u>

T F 17. Oxygen can form <u>three bonds since it requires three electrons</u> to fill its outer electron shell.

T F 18. Carbohydrates constitute about <u>18 to 25%</u> of the total body weight.

T F 19. ATP contains <u>more</u> energy than ADP.

T F 20. <u>Prostaglandins, ATP, steroids, and fats</u> are all classified as lipids.

Questions 21–25: Fill-ins. Identify each organic compound below. Write the names of the compounds on the lines provided.

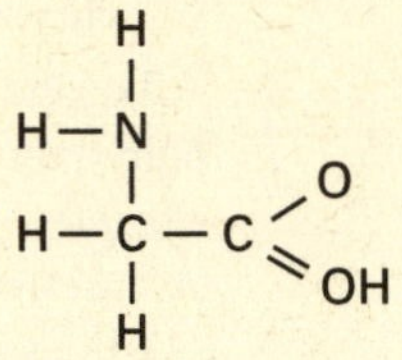

21. _____

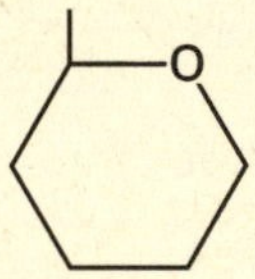

22. _____

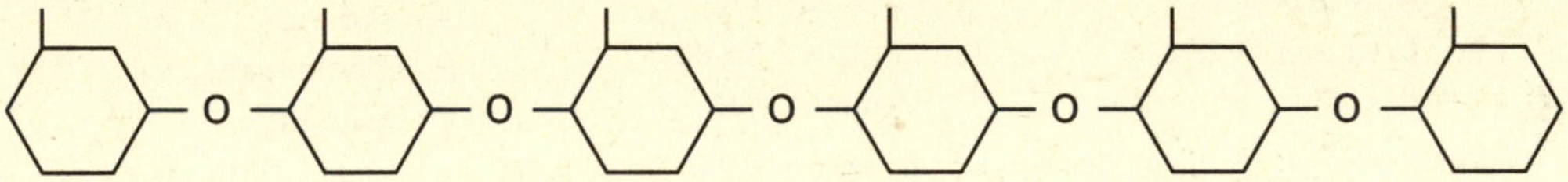

23. _____

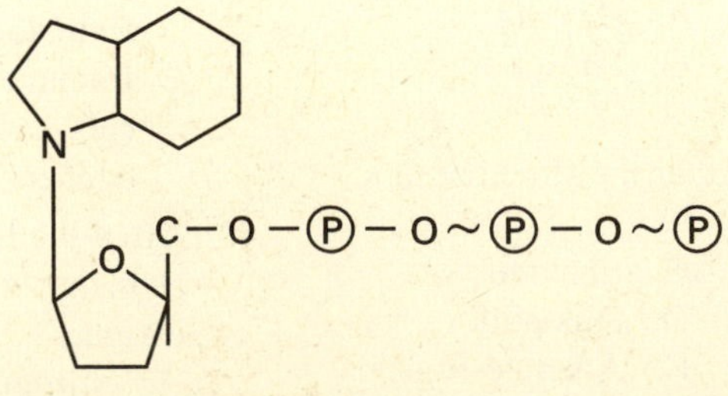

24. _____

```
      H          H   H   H   H   H   H   H   H   H   H   H   H   H   H   H
      |          |   |   |   |   |   |   |   |   |   |   |   |   |   |   |
  H—C—O—C—C—C—C—C—C—C—C—C—C—C—C—C—C—C—H
      |          ||  |   |   |   |   |   |   |   |   |   |   |   |   |   |   |
      O          O   H   H   H   H   H   H   H   H   H   H   H   H   H   H   H


                 H   H   H   H   H   H   H   H   H   H   H   H   H   H   H
                 |   |   |   |   |   |   |   |   |   |   |   |   |   |   |
  H—C—O—C—C—C—C—C—C—C—C—C—C—C—C—C—C—C—H
      |          ||  |   |   |   |   |   |   |   |   |   |   |   |   |   |   |
      O          O   H   H   H   H   H   H   H   H   H   H   H   H   H   H   H


                 H   H   H   H   H   H   H       H   H   H       H   H   H   H
                 |   |   |   |   |   |   |       |   |   |       |   |   |   |
  H—C—O—C—C—C—C—C—C—C=C—C—C—C=C—C—C—C—C—H
      |          ||  |   |   |   |   |           |   |           |   |   |   |
      H          O   H   H   H   H               H   H           H   H   H   H
```

25. _____

ANSWERS TO MASTERY TEST: ★ CHAPTER 2

Multiple Choice

1. C
2. D
3. B
4. D
5. B
6. A
7. C
8. D
9. B
10. E
11. B
12. E

True–False

13. F. Water and NaCl (Oxygen is not a compound; glucose is organic.)
14. F. 20
15. F. Electrons
16. F. K^+ is a cation.
17. F. Two bonds because it requires two electrons
18. F. 2–3%
19. T
20. F. Prostaglandins, steroids, and fats

Fill-ins

21. Amino acid
22. Monosaccharide or hexose or glucose
23. Polysaccharide or starch or glycogen
24. Adenosine triphosphate (ATP)
25. Fat or lipid or triglyceride (unsaturated)

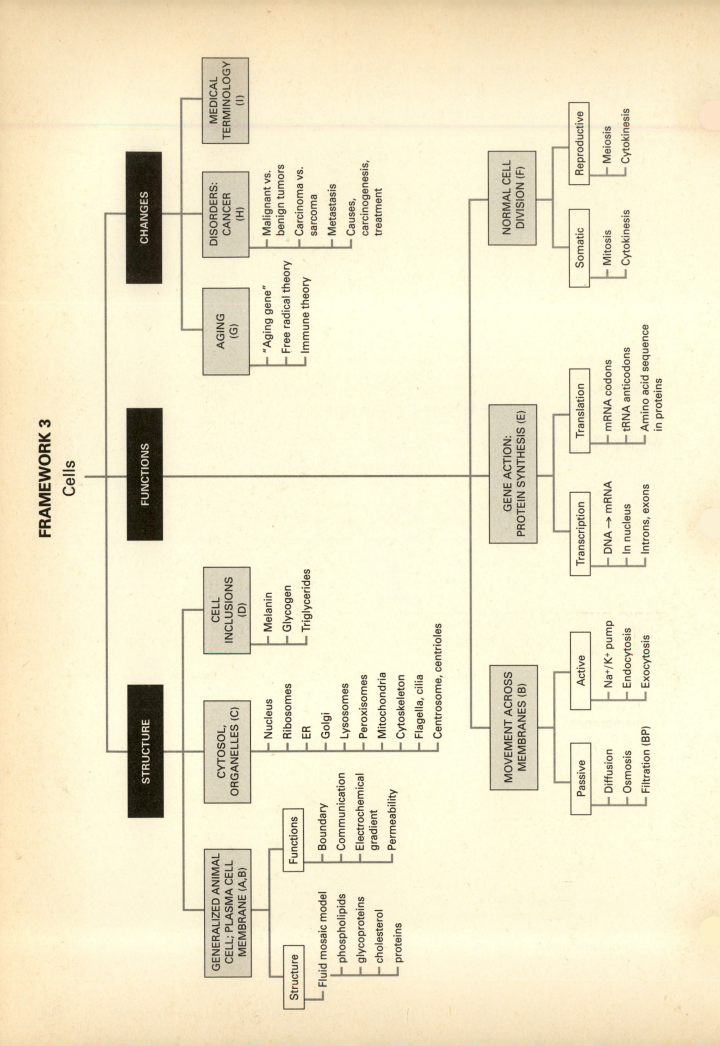

FRAMEWORK 3
Cells

The Cellular Level of Organization

CHAPTER

3

Cells are the basic structural units of the human body, much as buildings comprise the cities of the world. Every cell possesses certain characteristics similar to those in all other cells of the body. For example, membranes surround and compartmentalize cells, just as walls support virtually all buildings. Yet individual cell types exhibit variations that make them uniquely designed to meet specific body needs. In an architectural tour, the structural components of cathedral or mansion, lighthouse or fort, can be recognized as specific for activities of each type of building. So too the number and type of organelles differ in muscle or nerve, cartilage or bone cells. Cells carry out significant functions such as movement of substances into, out of, and throughout the cell; synthesis of the chemicals that cells need (like proteins, lipids, and carbohydrates); and cell division for growth, maintenance, repair, and formation of offspring. Just as buildings change with time and destructive forces, cells exhibit aging changes. In fact, sometimes their existence is haphazard and harmful, as in cancer.

As you begin your study of cells, carefully examine the Chapter 3 Topic Outline and Objectives and note relationships among concepts and key terms in the Framework.

TOPIC OUTLINE AND OBJECTIVES

A. Generalized animal cell

☐ 1. Define a cell and list its principal parts.

B. Movement of materials across plasma membranes

☐ 2. Explain the structure and functions of the plasma membrane.

☐ 3. Describe the various passive and active processes by which materials move across plasma membranes.

C. Cytosol, organelles

☐ 4. Describe the structure and functions of the following cellular structures: cytosol, nucleus, ribosomes, endoplasmic reticulum, Golgi complex, lysosomes, peroxisomes, mitochondria, cytoskeleton, flagella, cilia, and centrosome.

D. Cell inclusions

☐ 5. Define a cell inclusion and give several examples.

E. Gene action: protein synthesis

☐ 6. Define a gene and explain the sequence of events involved in protein synthesis.

F. Normal cell division: somatic and reproductive

☐ 7. Discuss the stages, events, and significance of somatic and reproductive cell division.

G. Cells and aging

☐ 8. Explain the relationship of aging to cells.

H. Disorders: cancer

☐ **9.** Describe cancer as a homeostatic imbalance of cell division and describe the growth and spread of malignant tumors, causes of cancer, carcinogenesis, and treatment of cancer.

I. Medical terminology

☐ **10.** Define medical terminology associated with cells.

WORDBYTES

Now become familiar with the language of this chapter by studying each wordbyte, its meaning, and an example of its use within a term. After you study the entire list, self-check your understanding by writing the meaning of each wordbyte on the line. As you continue through the *Learning Guide,* identify (and fill in) additional terms that contain the same wordbyte.

Wordbyte	Self-check	Meaning	Example(s)
a-	_____	without	*a*trophy
cyto-	_____	cell	*cyto*logist
-elle	_____	small	organ*elle*
homo-	_____	same	*homo*logous
hydro-	_____	water	*hydro*static
hyper-	_____	above	*hyper*tonic
hypo-	_____	below	*hypo*tonic
iso-	_____	equal	*iso*tonic
meta-	_____	beyond	*meta*stasis
neo-	_____	new	*neo*plasm
-oma	_____	tumor	carcin*oma*
-philic	_____	loving	hydro*philic*
-phobic	_____	fearing	hydro*phobic*
-plasia, -plasm	_____	growth	dys*plasia*
-stasis, -static	_____	stand, stay	meta*stasis*
-tonic	_____	pressure	hyper*tonic*
-trophy	_____	nourish	hyper*trophy*

CHECKPOINTS

A. Generalized animal cell (page 34–35)

■ **A1.** List the four principal parts of a generalized animal cell.

A2. Contrast: *cytoplasm* with *cytosol*.

B. Movement of materials across plasma membranes (pages 55–66)

■ **B1.** Refer to Figure LG 3.1 to complete the following exercise about the fluid mosaic model of the plasma membrane. First color each part of the membrane listed below (a–e) with a different color; then use the same colors to fill in the color code ovals next to the letters below. Next label these parts on the figure. Now use the lines provided next to the name of each membrane part to write one or more functions for each part.

○ a. Phospholipid polar "head" _____

○ b. Phospholipid nonpolar "tail" _____

○ c. Integral protein _____

○ d. Glycoprotein _____

○ e. Cholesterol _____

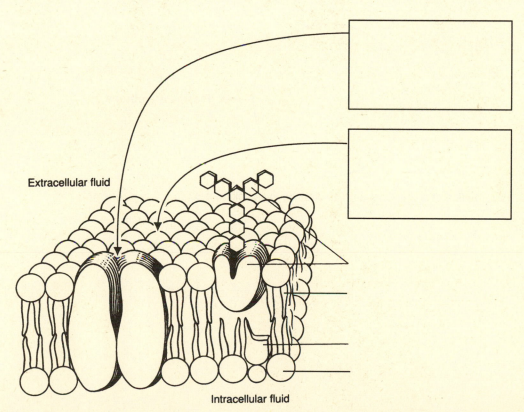

Extracellular fluid

Intracellular fluid

Figure LG 3.1 Plasma membrane, according to the fluid mosaic model. Color as directed in Checkpoint B1. Fill in answers in boxes according to Checkpoint B4.

■ **B2.** *For extra review.* Complete this exercise about the plasma membrane.

a. Plasma membranes are made up of about _____% proteins and _____% lipids by weight. About 20% of the lipids are *(cholesterol? glycolipids? phospholipids?)*.

b. About 75% of the lipids are *(cholesterol? glycolipids? phospholipids?)*. Because phos-

   pholipids have polar and nonpolar parts, they are said to be _____.
   The polar part of a phospholipid molecule faces toward the *(interior? surface?)* of the membrane.

c. Glycolipids appear only in the layer that faces *(extracellular? intracellular?)* fluid.

   This location is logical because these lipids may be important in _____.

d. Which type of membrane molecule is larger? *(Lipid? Protein?)* In different types of membranes, there is more variety in *(lipids? proteins?)*. For example, some integral

   proteins serve as receptors for ligands such as _____,

   _____, or _____. Some *(integral? peripheral?)*
   proteins serve as cytoskeletal anchors. Other membrane proteins known as

   _____ form catalysts for chemical reactions.

■ **B3.** Complete this activity about the electrochemical gradient.

a. Refer to Figure LG 3.2. Write inside the plasma membrane the symbols for the major ions or electrolytes concentrated in intracellular fluid (ICF). Write outside the plasma membrane the symbols for major ions or electrolytes concentrated in extracellular fluid (ECF). Use these symbols:

$$K^+ \qquad Na^+ \qquad Cl^- \qquad PO_4^{3-} \qquad Protein^-$$

— Plasma membrane

Figure LG 3.2 Outline of a cell. Complete as directed in Checkpoint B3.

b. This difference between chemicals inside and outside the plasma membrane is known as a(n) *(chemical? electrical?)* gradient. A difference in charge across the membrane accounts for a(n) *(chemical? electrical?)* gradient.

c. The inside of the plasma membrane usually has a more *(positive? negative?)* charge.

Therefore the membrane maintains a voltage called the _____,

which is measured in _____. This voltage is a measure of *(kinetic? potential?)* energy available to accomplish cell work, such as nerve transmission.

■ **B4.** Complete this activity about selective permeability of membranes.

a. Plasma membranes are more permeable to *(nonpolar chemicals such as O_2 and CO_2? most polar molecules and ions?)* because these molecules dissolve readily in the *(protein? phospholipid?)* portion of the membrane.

b. Charged chemicals (such as Na^+, K^+, or Ca^{2+}) and most polar chemicals pass through the *(protein? phospholipid?)* portions of the membrane. Two such proteins are

_____ and _____.

c. Refer again to Figure LG 3.1. Write the names of the following chemicals in the correct boxes on the figure to differentiate chemicals that pass through the phospholipid bilayer from those that pass through protein channels:

Cl^- CO_2 H_2O K^+ Na^+

O_2 Urea Vitamin D

B5. Explain why substances must move across your plasma membranes in order for you to survive and maintain homeostasis. (What types of chemicals must be allowed into cells or kept out of cells?)

B6. Complete parts a–d of Table LG 3.1

Table LG 3.1 Summary of mechanisms for movement of material across membrane.

Name of Process	What Moves (Particles or Water)?	Where (Direction, eg., High to Low Concentration)?	Membrane Necessary (Yes or No)?	Carrier Necessary (Yes or No)?	Active or Passive?
a. Simple diffusion					
b.	Water	From high to low concentration of water across semipermeable membrane			
c.		From high-pressure area to low-pressure area			
d. Facilitated diffusion					
e. Active transport		From low to high concentration of solute			
f. Phagocytosis			Yes		
g.	Water	From outside cell to inside cell			
h. Exocytosis					Active

■ **B7.** Select from the following list of terms to identify passive transport processes described below.

Fac.	Facilitated diffusion	O.	Osmosis
Fil.	Filtration	SD.	Simple diffusion

_____ a. Net movement of any substance (such as cocoa powder in hot milk) from region of higher concentration to region of lower concentration; membrane not required

_____ b. Same as (a) except movement across a semipermeable membrane with help of a carrier; ATP not required

_____ c. Net movement of water from region of high water concentration (such as 2%

NaCl) to region of low water concentration (such as 10% NaCl) across semipermeable membrane; important in maintenance of normal cell size and shape

_____ d. Movement of molecules from high pressure zone to low pressure zone, for example, in response to force of blood pressure

■ **B8.** Complete the following exercise about osmosis in blood.

a. Human red blood cells (RBCs) contain intracellular fluid that is osmotically similar to

_____ NaCl, which is known as normal saline.

   A. 2.0%   B. 0.90%   C. 0% (pure water)

b. A solution that is hypertonic to RBCs contains (more? fewer?) solute particles and (more? fewer?) water molecules than blood.

c. Which of these solutions is hypertonic to RBCs?

   A. 2.0% NaCl   B. 0.90% NaCl

d. If RBCs are surrounded by hypertonic solution, water will tend to move (into? out of?) them, so they will (crenate? hemolyze?).

e. A solution that is _____-tonic to RBCs will maintain the shape and

size of the RBC. An example of such a solution is _____.

f. Which solution will cause RBCs to hemolyze?

   A. 2.0% NaCl   B. 0.90% NaCl   C. Pure water

g. Which solution has the highest osmotic pressure?

   A. 2.0% NaCl   B. 0.90% NaCl   C. Pure water

■ **B9.** A large part of the human diet is starch. When starch is digested, it is broken down to glucose. Answer these questions related to movement of glucose into body cells.

a. Explain why glucose cannot readily pass through plasma membranes by simple diffusion.

b. By what process does glucose cross plasma membranes as it is transported into many

body cells? _____

c. To start this process, glucose must first attach to a chemical known as a *(transporter such as GluT1? kinase? phosphate?)*, which then changes shape and moves glucose into the cell.

d. Three factors that are likely to speed up facilitated diffusion of glucose are a *(high? low?)* concentration gradient of glucose across the membrane, a *(large? small?)* number of transporters available, and presence of the hormone _____.

■ **B10.** Describe active transport in the following exercise.

a. List three categories of substances that must enter or leave cells by active processes.

b. There are two categories of active processes: _____ transport and

_____ transport. Active transport may be either primary or secondary. The sodium (or Na^+/K^+) pump is the most abundant form of *(primary? secondary?)* transport in cells. The effect of it is to concentrate sodium (Na^+) *(inside? outside?)* of cells. (See Figure LG 3.2.) The Na^+/K^+ pump requires energy from ATP

directly by splitting ATP with an enzyme known as _____-ase. In

fact, a typical cell may use up to _____% of its ATP for primary active transport.

c. Two forms of secondary active transport are known. Symport (cotransport) involves movement of a substance (such as dietary glucose) in *(the same? opposite?)* direction as sodium (Na^+). Antiport (countertransport) involves movement of a substance (such as Ca^{2+} or H^+) in *(the same? opposite?)* direction as sodium (Na^+).

d. Bulk transport includes _____-cytosis and _____-cytosis.

B11. Outline the steps of receptor-mediated endocytosis. Be sure to include these terms: *ligand, receptor, endocytic vesicle, endosome, recycle, lysosome.*

B12. State one advantage of the use of liposomes with antibiotic or cancer drug therapy.

■ **B13.** Complete parts e–h of Table LG 3.1.

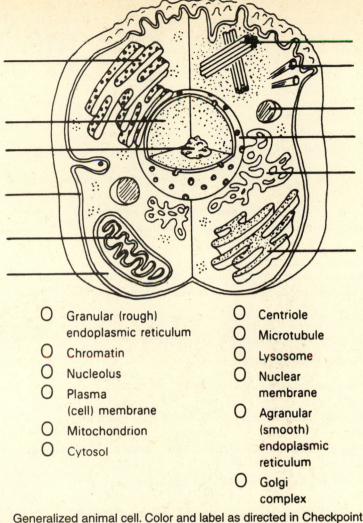

- ◯ Granular (rough) endoplasmic reticulum
- ◯ Chromatin
- ◯ Nucleolus
- ◯ Plasma (cell) membrane
- ◯ Mitochondrion
- ◯ Cytosol

- ◯ Centriole
- ◯ Microtubule
- ◯ Lysosome
- ◯ Nuclear membrane
- ◯ Agranular (smooth) endoplasmic reticulum
- ◯ Golgi complex

Figure LG 3.3 Generalized animal cell. Color and label as directed in Checkpoint C3.

C. Cytosol, organelles (pages 66–75)

■ **C1.** Contrast cytosol and cytoplasm in this activity.

a. Cytosol consists of about _____ to _____% water plus solid components. Name two

types of chemicals that are colloids suspended in cytosol: _____

and _____.

b. Cytoplasm consists of cytosol plus the cyto- _____, as well as all of

the _____, except for the nucleus.

C2. The cell is compartmentalized by the presence of organelles. Of what advantage is this to the cell?

■ **C3.** Color and then label all of the parts of Figure LG 3.3 listed with color code ovals.

■ **C4.** Describe parts of the nucleus by completing this exercise.

    a. The nuclear membrane has pores that are *(larger? smaller?)* than channels in the plasma membrane. Of what significance is this?

    b. Ribosomes are assembled at sites known as _____, where protein, DNA, and RNA are stored.

    c. Chromosomes consist of beadlike subunits known as _____,

    which are made of double-stranded _____ wrapped around core

    proteins called _____. Nucleosome "beads" are held together, necklace-fashion, by *(histones? linker DNA?)*. Nucleosomes are then held into

    chromatin fibers and folded into _____.

    d. Before cell division, DNA duplicates and forms compact, coiled conformations known as *(chromatin? chromatids?)* that may be more easily moved around the cell during the division process.

■ **C5.** In the blanks below, write *rough ER* or *smooth ER,* as appropriate.

    _____ a. This form of ER is also known as agranular ER.

    _____ b. Synthesis of fats, such as fatty acids, phospholipids, and steroids, takes place here.

    _____ c. In muscle cells, calcium ions (Ca^{2+}) released from this form of ER trigger muscle contraction.

    _____ d. Enzymes within these membranes detoxify alcohol and other toxic substances within liver cells.

    _____ e. Protein is synthesized and stored in ribosomes located here.

■ **C6.** Describe the sequence of events in synthesis, secretion, and discharge of protein from the cell by numbering these five steps in correct order.

    _____ A. Proteins accumulate and are modified, sorted, and packaged in cisternae of Golgi complex.

    _____ B. Proteins pass within a transport vesicle to *cis* cisternae of Golgi complex.

    _____ C. Secretory granules move toward cell surface, where they are discharged.

    _____ D. Proteins are synthesized at ribosomes on rough ER.

    _____ E. Vesicles containing protein pinch off of Golgi complex *trans* cisternae to form secretory vesicles.

■ **C7.** Match the organelles listed in the box with the best descriptions of their functions. Use each answer once.

Cen.	Centrioles	I.	Intermediate filaments	Mit.	Mitochondria
Cil.	Cilia	L.	Lysosomes	N.	Nucleus
ER.	Endoplasmic reticulum	Mf.	Microfilaments	P.	Peroxisomes
F.	Flagella	Mt.	Microtubules	R.	Ribosomes
G.	Golgi complex				

_____ a. Site of direction of cellular activities by means of DNA located here

_____ b. Cytoplasmic sites of protein synthesis; may occur attached to ER or scattered freely in cytoplasm

_____ c. System of membranous channels providing pathways for transport within the cell and surface areas for chemical reactions; may be granular or agranular

_____ d. Stacks of cisternae with vesicular ends; involved in packaging and secretion of proteins and lipids and synthesis of glycoproteins

_____ e. Vesicles that contain powerful enzymes that can digest bacteria, organelles, or entire cells by processes of phagocytosis, autophagy, or autolysis

_____ f. Similar to lysosomes, but smaller; contain the enzyme catalase that uses hydro-

gen peroxide (H_2O_2) to detoxify harmful chemicals in cells of liver and kidney

_____ g. Cristae-containing structures, called "powerhouses of the cell" because ATP production occurs here

_____ h. Form part of cytoskeleton; involved with cell movement and contraction

_____ i. Part of cytoskeleton, provides support and gives shape to cell; form flagellae, cilia, centrioles, and spindle fibers; made of protein tubulin

_____ j. Help organize mitotic spindle used in cell division and also help form flagella and cilia

_____ k. Long, hairlike structures that help move entire sperm cell

_____ l. Short, hairlike structures that move particles over cell surfaces

_____ m. Contribute to shape of cell; hold organelles (such as nucleus) in place

■ **C8.** _For extra review._ Considering the functions of organelles listed in the above exercise, choose the answers (organelles) that fit the following descriptions. Answers in that list may be used more than once.

_____ a. Abundant in white blood cells, which use these organelles to help in phagocytosis of bacteria

_____ b. Present in large numbers in muscle and liver cells, which require much energy

_____ c. Located on surface of cells of the respiratory tract; help to move mucus

_____ d. Absence of a single enzyme from these organelles causes Tay–Sachs, an inherited disease affecting nerve cells

_____ e. An acid pH must be present inside these organelles for enzymes to function effectively

_____ f. Sites of cellular respiration requiring oxygen (O_2); contain some DNA used for self-replication

_____ g. Function much like conveyer belts to move chemicals and organelles through cytosol; assist in movement of pseudopods of phagocytic cells such as white blood cells

_____ h. Described as having a "nine and two" structure based on arrangement of nine pairs of microtubules around a core of two single microtubules (two answers)

_____ i. Might be described as a "clean-up service" since enzymes released from these organelles help to remove debris within and around injured cells (two answers)

D. Cell inclusions   (page 75)

D1.   Contrast *organelle* with *inclusion*.

■ **D2.**   Contrast these three types of cell inclusions in this exercise.

| G. Glycogen | M. Melanin | T. Triglycerides |

_____ a. Stored in adipocytes; may be broken down to form ATP

_____ b. A complex carbohydrate stored in cells of the liver, skeletal muscle, uterus, and vagina

_____ c. A pigment that gives brown or black color to skin, hair, or eyes

E. Gene action: protein synthesis   (pages 75–79)

■ **E1.**   State the significance of protein synthesis in the cell.

■ **E2.**   Refer to Figure LG 3.4 and describe the process of protein synthesis in this exercise..

a. Write the names of the two major steps of protein synthesis on the two lines in Figure LG 3.4.

b. Transcription takes place in the *(nucleus? cytoplasm?)* with the help of the enzyme

called _____. In this process, a portion of one side of a double-

stranded DNA serves as a mold or _____ and is called the *(sense?*

antisense?) strand. Nucleotides of _____ line up in a complementary fashion next to DNA nucleotides (much as occurs between DNA nucleotides in
replication of DNA prior to mitosis). Select different colors and color the components of
 DNA nucleotides (P, D, and bases T, A, C, and G).

c. Complete the strand of mRNA by writing letters of complementary bases. Then color
the parts of mRNA (P, R, and bases U, A, C, and G).

d. The second step of protein synthesis is known as _____ because it
involves translation of one "language" (the nucleotide base code of

_____ RNA) into another "language" (the correct sequence of the

20 _____ to form a specific protein).

e. Translation occurs in the *(nucleus? cytoplasm?)*, specifically at a *(chromosome? lysosome? ribosome?)* to which mRNA attaches. The *(large? small?)* ribosomal subunit
binds to one end of the mRNA and finds the point where translation will begin, known
as *(ATP? the start codon?)*.

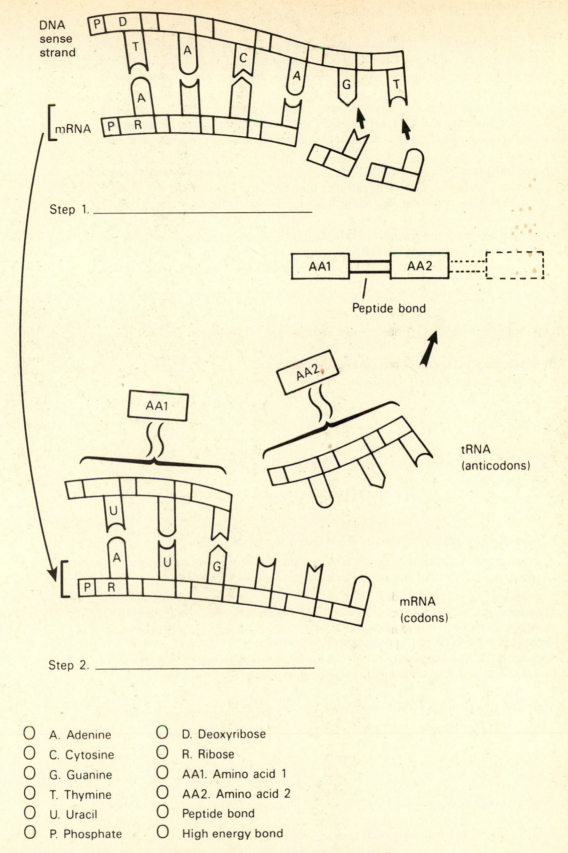

DNA
sense
strand

mRNA

Step 1. _____

Peptide bond

AA1 —— AA2 - - - - [ ]

tRNA
(anticodons)

AA1

AA2

mRNA
(codons)

Step 2. _____

○ A. Adenine   ○ D. Deoxyribose
○ C. Cytosine   ○ R. Ribose
○ G. Guanine   ○ AA1. Amino acid 1
○ T. Thymine   ○ AA2. Amino acid 2
○ U. Uracil   ○ Peptide bond
○ P. Phosphate   ○ High energy bond

Figure LG 3.4   Protein synthesis. Label and color as directed in Checkpoint E2.

f. Each amino acid is transported to this site by a particular _____
RNA characterized by a specific three-base unit or *(codon? anticodon?)*. This portion of
tRNA, for example, U A C, binds to a complementary portion of mRNA *(codon? anticodon?)*. In step 2 of Figure LG 3.4, complete the labeling of bases and color all components of tRNA and mRNA. Be sure you select colors that correspond to those used in
step 1.

g. Each amino acid is bonded to its specific amino acid by a *(high? low?)* energy bond.
The energy in this bond derives from *(ATP? ADP)*. Color such a bond on the figure.

h. As the ribosome moves along each mRNA codon, additional amino acids are trans-

ferred into place by _____ RNA. _____ bonds form between adja-
cent amino acids with the help of enzymes from the *(large? small?)* subunit of the ribosome. Energy for bonds *between* amino acids is provided by the high energy bonds

that were present between amino acids and _____. Now color the
amino acids and peptide bonds that join them.

i. As the ribosome moves on to the next codon on mRNA, what happens to the "empty"
tRNA?

j. When the protein is complete, synthesis is stopped by the _____
codon. As the protein is released, what happens to the ribosome?

k. When several ribosomes attach to the same strand of mRNA, a

_____ is formed, permitting repeated synthesis of the identical
protein.

■ **E3.** Complete this additional activity about protein synthesis.

a. Sections of DNA that do code for parts of proteins are known as *(exons? introns?)*.
whereas sections of DNA that do not code for parts of proteins are called

_____. The mRNA formed initially by transcription contains *(only
exons? only introns? both exons and introns?)*.

b. RNA splicing is the process of removal of *(exons? introns?)*, making up over 75% of the
initial mRNA. This process occurs *(before? after?)* mRNA moves into cytoplasm. In
fact, variations of RNA splicing result in changes in function of a single gene to

form different proteins as cells undergo the process of _____.

c. Define the term *gene*.

d. The exon portion of a typical gene contains about _____ to _____ nucleotides,

which code for a protein containing _____ to _____ amino acids.

E4. List six or more therapeutic substances that have been produced by bacteria containing recombinant DNA.

F. Normal cell division: somatic and reproductive (pages 80–87)

F1. Once you have reached adult size, do your cells continue to divide? *(Yes? No?)* Of what significance is this fact?

■ **F2.** Describe aspects of cell division by completing this exercise.

a. Division of any cell in the body consists of two processes: division of the

_____ and division of the _____.

b. In the formation of mature sperm and egg cells, nuclear division is known as

_____. In the formation of all other body cells, that is, *somatic*

cells, nuclear division is called _____.

c. Cytoplasmic division in both somatic and reproductive division is known as

_____. In *(meiosis? mitosis?)* the two daughter cells have the same hereditary material and genetic potential as the parent cell.

■ **F3.** Describe interphase in this activity.

a. Interphase occurs *(during cell division? between cell divisions?)*.

b. Chromosomes are replicated during phase *(S? G_1? G_2?)* of interphase. In fact "S"

stands for _____ of new DNA. During this time, the DNA helix partially uncoils and the double DNA strand separates at points where

_____ connect so that new complementary nucleotides can attach here. (See Figure LG 2.3, page 36.) In this manner the DNA of a single chromosome

is doubled to form two identical _____.

c. Two periods of growth occur during interphase, preparing the cell for cell division. The *(G_1? G_2?)* phase immediately precedes the S phase, whereas the *(G_1? G_2?)* phase follows the S portion of interphase. The "G" of these two phases also refers to "gaps"

because these periods are gaps in _____.

d. Most nerve cells *(are? are not?)* capable of mitosis after birth; their development is

permanently arrested in the _____ phase.

■ **F4.** Carefully study the phases of the cell cycle shown in Figure 3.26, page 82, in your text. Then check your understanding of the process by identifying major events in each phase.

> | A. Anaphase | P. Prophase |
> |---|---|
> | I. Interphase | T. Telophase |
> | M. Metaphase | |

_____ a. This phase immediately follows inter-phase.

_____ b. Chromatin condenses into distinct chromosomes (chromatids) each held together by a centromere complexed with a kinetochore.

_____ c. Nucleoli and nuclear envelope break up; the two centrosomes move to opposite poles of the cell, and formation of the mitotic spindle begins.

_____ d. Chromatids line up with their centromeres along the equatorial plate.

_____ e. Centromeres split; chromatids (now called daughter chromosomes) are dragged by microtubules to opposite poles of the cell. Cytokinesis begins.

_____ f. Events of this phase are essentially a reversal of prophase; cytokinesis is completed.

_____ g. This phase of cell division is the longest.

F5. Contrast roles of three types of microtubules in mitosis and identify which type shortens as daughter chromosomes are drawn toward opposite poles of the cell.

■ **F6.** Describe the time intervals involved in cell division by circling the correct answers below.

a. Typically, which phase is shortest?

$$G_1 \quad S \quad G_2 \quad \text{Mitosis and cytokinesis}$$

b. Which phase is most variable in duration (from virtually nonexistent to years)?

$$G_1 \quad S \quad G_2 \quad \text{Mitosis and cytokinesis}$$

■ **F7.** Do the following exercise about reproductive cell division.

a. Meiosis is a special type of nuclear division that occurs only in

_____ and is one process in the production of sex cells or

_____. Fusion of gametes during the process of

_____ results in formation of a _____.

b. Cells that begin the process of meiosis are ovarian or testicular cells that have been

undergoing mitosis up to this point. Each cell initially contains _____ chromosomes, which is the *(diploid or 2n? haploid or n?)* number for humans. These

chromosomes consist of _____ homologous pairs.

c. One pair is known as the *(autosomes? sex chromosomes?)*. In females, this pair consists of *(two X chromosomes? one X and one Y chromosome?)*. The remaining 22 pairs

   are known as _____.

d. By the time that cells complete meiosis, they will contain *(46? 23?)* chromosomes, the *(2n? n?)* number.

e. Meiosis differs from mitosis in that meiosis involves *(one? two? three?)* divisions with replication of DNA only *(once? twice?)*. The first division is known as the

   _____ division, whereas the second division is the

   _____ division.

f. Replication of DNA occurs *(before? after?)* reduction division, resulting in doubling

   of chromosome number in humans from 46 chromosomes to _____ "chromatids." (Chromatid is a name given to a chromosome while the chromosome is att ached to its replicated counterpart, that is, in the doubled state.) Doubled or paired chro- matids are held together by means of a *(centromere? centrosome?)*.

g. A unique event that occurs in Prophase I is the lining up of chromosomes in homolo-

   gous pairs, a process called _____. Note that this process *(does? does not?)* occur during mitosis. The four chromatids of the homologous pair are

   known as a _____. The proximity of chromatids permits the

   exchange or _____ of DNA between maternal and paternal chro- matids. What is the advantage of crossing over to sexual reproduction?

h. Another factor that increases possibilities for variety among sperm or eggs you can pro- duce is the random (or chance) assortment of chromosomes as they line up on either side of the equator in Metaphase *(I? II?)*.

i. Division I reduces the chromosome number to _____ chromatids. Division II results in the separation of doubled chromatids by splitting of

   _____ so that resulting cells contain only _____ chromosomes.

j. Meiosis is only one part of the process of forming sperm or eggs. Each original diploid cell in a testis will lead to production of *(1? 2? 4?)* mature sperm, while each diploid ovary cell produces *(1? 2? 4?)* mature eggs (ova) plus three small cells known as

   _____.

F8. Write the full names and a brief description of the roles of the following three chemicals that regulate cell division.

a. MPF

b. cdc2 proteins

c. Cyclin

G. Cells and aging   (page 87)

G1. Describe changes that normally occur in each of the following during the aging process.

a. Number of cells in the body

b. Collagen fibers

c. Elastin

d. Glucose

G2. In a sentence or two, summarize the main points of each of these theories of aging.

a. An "aging gene," affecting rate of mitosis

b. Free radicals

c. Immune system

H. Disorders: cancer   (pages 88–89)

■ **H1.** Complete this exercise about cancer.

a. The term _____ refers to the study of cancer.

_____ and _____ nurses are health care providers
who specialize in work with clients who have cancer.

b. A term that means "tumor or abnormal growth" is _____.
A cancerous growth is a *(benign? malignant?)* neoplasm, whereas a noncancerous
tumor is a *(benign? malignant?)* neoplasm.

c. Tumors grow quickly by increasing their rate of cell division, a process known as

_____. Cancer cells also compete with normal body cells by

robbing those cells of _____. Tumors also favor their own growth

by synthesizing chemicals known as TAFs. TAFs refer to _____

_____ _____, indicating that these chemicals

stimulate growth of new _____ around tumor cells.

d. Malignant tumors are *(more? less?)* likely than benign tumors to spread and possibly

cause death. A term that means "spread" of cancer cells is _____.
By what routes do cancers reach distant parts of the body?

e. What may cause pain associated with cancer?

■ **H2.** Match terms in the box with the definitions below.

C. Carcinoma	O. Osteogenic sarcoma
L. Lymphoma	S. Sarcoma

_____ a. General term for malignant tumor
arising from any connective tissue

_____ b. Malignant tumor derived from bone
(a connective tissue)

_____ c. General term for malignant tumor of
epithelial tissue

_____ d. A malignancy of lymph tissue

■ **H3.** Identify the types of viruses associated with the following types of cancer:

EBV. Epstein–Barr virus	HPV. Human papilloma virus
HBV. Hepatitis B virus	HSV. Herpes simplex virus
HIV. Human immunodeficiency virus	HTLV–1. Human T-cell leukemia-lymphoma virus-1

_____ a. Cancers of white blood cells (leukemia and lymphoma)

_____ b. Kaposi's sarcoma (KS), a cancer of connective tissues and blood vessels occurring most commonly in AIDS patients, related to impairment of the immune system

_____ c. Hodgkin's disease (a cancer of the lymphatic system)

_____ d. Liver cancer

_____ e. Cervical cancer (two answers)

H4. Relate the following terms to cancer:

a. Carcinogens

b. Proto-oncogenes

c. Oncogenes

d. Growth factors

e. Mutation

f. P-glycoprotein

H5. *A clinical challenge.* Are all cancer cells alike within a single tumor? *(Yes? No?)* Of what significance is this in planning cancer treatment?

I. Medical terminology   (page 89)

■ **I1.** Match the terms in the box describing alterations in cells or tissues with the descriptions below.

> D. Dysplasia   HT. Hypertrophy
> HP. Hyperplasia   M. Metaplasia

_____ a. Increase in size of tissue or organ by increase in size (not number) of cells, such as growth of your biceps muscle with exercise

_____ b. Increase in size of tissue or organ due to increase in number of cells, such as a callus on your hand or growth of breast tissue during pregnancy

_____ c. Change of one cell type to another normal cell type, such as change from single row of tall (columnar) cells lining airways to multilayers of cells as response to constant irritation of smoking

_____ d. Abnormal change in cells in a tissue as due to irritation or inflammation. May revert to normal if irritant removed, or may progress to neoplasia

■ **I2.** Match the terms in the box with the definitions below.

> A. Atrophy   N. Necrosis
> B. Biopsy

_____ a. Death of a group of cells

_____ b. Decrease in size of cells with decrease in size of tissue or organ

_____ c. Removal and examination of tissue from the living body for diagnosis

A1. Plasma (cell) membrane, cytosol, organelles, inclusions.

B1. See Figure LG 3.1A. (a) Water-soluble and facing the membrane surface, so compatible with watery environment surrounding the membrane. (b) Forms a water-insoluble barrier between cell contents and extracellular fluid; nonpolar molecules such as lipids, oxygen, and carbon dioxide pass readily between "tails." (c) Channels for passage of water or ions, carriers. (d) Receptors. (e) Cholesterol molecules strengthen the membrane, but decrease its flexibility and permeability.

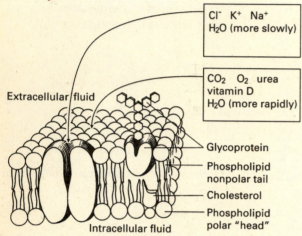

Figure LG 3.1A Plasma membrane.

B2. (a) 50, 50; cholesterol. (b) Phospholipids; amphipathic; surface. (c) Extracellular; cell adhesion, recognition, and/or communication. (d) Protein; proteins; hormones, neurotransmitters, or nutrients; peripheral; enzymes.

B3. (a) See Figure LG 3.2A (b) Chemical; electrical. (c) Negative; membrane potential, millivolts (mV); potential.

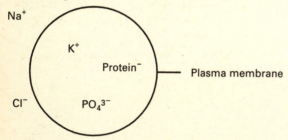

Figure LG 3.2A Major ions in intracellular fluid (ICF) and extracellular fluid (ECF).

B4. (a) Nonpolar chemicals such as O_2 and CO_2, phospholipid. (b) Protein; channels (pores), transporters (carriers). (c) See Figure LG 3.1A.

B6.
Table LG 3.1 (a–d)

Name of Process	What Moves (Particles or Water)?	Where (Direction, eg., High to Low Concentration)?	Membrane Necessary (Yes or No)?	Carrier Necessary (Yes or No)?	Active or Passive?
a. Simple diffusion	Particles or water	From high to low concentration of solute	No	No	Passive
b. Osmosis	Water	From high to low concentration of water across semi-permeable membrane	Yes	No	Passive
c. Filtration	Particles or water	From high-pressure area to low-pressure area	Yes (or other barrier such as filter paper)	No	Passive
d. Facilitated diffusion	Particle, such as glucose	From high to low concentration of solute	Yes	Yes	Passive

B7. (a) SD. (b) Fac. (c) O. (d) Fil.

B8. (a) B. (b) More, fewer. (c) A. (d) Out of, crenate. (e) Iso-, 0.90% NaCl (normal saline) or 5.0% glucose. (f) C. (g) A.

B9. (a) Glucose is too large. (b) Facilitated diffusion. (c) Transporter such as GluT1. (d) High, large, insulin.

B10. (a) Substances that are large (such as proteins and even whole bacteria or red blood cells), have the wrong charge, or must move into an area already highly concentrated with the same substance. (b) Active, bulk; primary; outside; ATP; 40. (c) The same; opposite. (d) Endo-, exo-.

B13.
Table LG 3.1 (e–h)

Name of Process	What Moves (Particles or Water)?	Where (Direction, eg., High to Low Concentration)?	Membrane Necessary (Yes or No)?	Carrier Necessary (Yes or No)?	Active or Passive?
e. Active transport	Solutes, such as glucose or ions (Na^2, K^+)	From low to high concentration of solute	Yes	Yes	Active
f. Phagocytosis	Large particles	From outside of cell to inside of cell	Yes	No	Active
g. Pinocytosis	Water	From outside cell to inside cell	Yes	No	Active
h. Exocytosis	Large molecules or particles	From inside of cell to outside of cell	Yes	No	Active

C1. (a) 75, 90; proteins, as well as polysaccharides such as glycogen. (b) Skeleton, organelles.

C3.

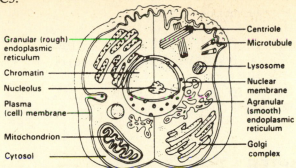

Figure LG 3.3A Generalized animal cell.

C4. (a) Larger; permit passage of large molecules such as RNA and proteins. (b) Nucleoli. (c) Nucleosomes, DNA, histones; linker DNA; loops. (d) Chromatids.

C5. (a–d) Smooth ER. (e) Rough ER.

C6. 3, 2, 5, 1, 4 (or D, B, A, E, C).

C7. (a) N. (b) R. (c) ER. (d) G. (e) L. (f) P. (g) Mit. (h) Mf. (i) Mt. (j) Cen. (k) F. (l) Cil. (m) I.

C8. (a) L. (b) Mit. (c) Cil. (d) L. (e) L. (f) Mit. (g) Mt. (h) Cil, F. (i) L, P.

D2. (a) T. (b) G. (c) M.

E1. Proteins serve a number of significant functions. Refer to Chapter 2, page 33 of the *Learning Guide,* Activity D8.

E2. (a) See Figure LG 3.4A. (b) Nucleus, RNA polymerase; template, sense; messenger RNA (mRNA) (see Figure LG 3.4A). (c) See Figure LG 3.4A. (d) Translation, m (messenger), amino acids. (e) Cytoplasm, ribosome; small, the start codon. (f) t (transfer), anticodon; codon (see Figure LG 3.4A). (g) High; ATP; see Figure LG 3.4A. (h) t (transfer); peptide, large; tRNA; see Figure LG 3.4A. (i) It is released and may be recycled to transfer another amino acid. (j) Stop; it splits into its large and small subunits. (k) Polyribosome.

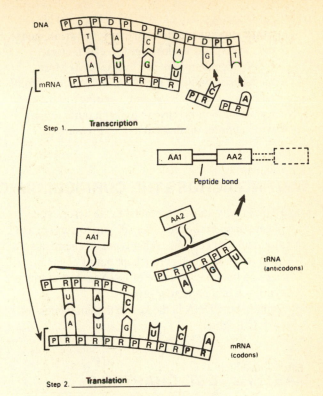

Figure LG 3.4A Protein synthesis.

E3. (a) Exons, introns; both exons and introns. (b) Introns; before; differentiation. (c) A group of nucleotides on a DNA molecule that codes for synthesis of a specific sequence of bases in mRNA (by transcription) and ultimately, a specific sequence of amino acids (by transcription). (d) 300, 3000, 100, 1000.

F2. (a) Nucleus, cytoplasm. (b) Meiosis; mitosis. (c) Cytokinesis; mitosis.

F3. (a) Between cell divisions. (b) S; synthesis; nitrogen bases; chromatids. (c) G_1; G_2; DNA synthesis. (d) Are not, G_1.

F4. (a) P. (b) P. (c) P. (d) M. (e) A. (f) T. (g) P.

F6. (a) Mitosis and cytokinesis (see Figure 3.24, page 80 of text). (b) G_1.

F7. (a) Gonads (ovaries or testes), gametes; fertilization, zygote. (b) 46, diploid or $2n$; 23. (c) Sex chromosomes; two X chromosomes; autosomes. (d) 23, n. (e) Two, once; reduction, equatorial. (f) Before, "46×2" (or 46 doubled); centromere. (g) Synapsis; does not; tetrad; crossing-over; permits exchange of DNA between chromatids and greatly increases possibilities for variety among sperm or eggs and thus offspring. (h) I. (i) "23 $\times$ 2" (or 23 doubled chromatids); centromeres, 23. (j) 4, 1, polar bodies.

H1.　(a) Oncology; oncologists, oncology. (b) Neoplasm; malignant, benign. (c) Hyperplasia; nutrients and space; tumor angiogenesis factors,
blood vessels (*angio* = vessel). (d) More;
metastasis; blood or lymph, or by detaching from
an organ, such as liver, and "seeding" into a cavity (as abdominal cavity) to reach other organs
which border on that cavity. (e) Pressure on
nerve, obstruction of passageway, loss of
function of a vital organ.

H2.　(a) S. (b) O. (c) C. (d) L.

H3.　(a) HTLV-1. (b) HIV. (c) EBV. (d) HBV.
(e) HPV, HSV (often a consequence of
HIV also).

I1.　(a) HT. (b) HP. (c) M. (d) D.

I2.　(a) N. (b) A. (c) B.

WRITING ACROSS THE CURRICULUM: CHAPTER 3

1. Explain why substances must move across your
plasma membranes in order for you to survive and
maintain homeostasis. What types of chemicals
must move into cells or be kept out of cells?

2. Explain how the structure of DNA is related to
synthesis of a particular protein.

3. Explain how DNA is the key to the uniqueness of
each individual.

4. Tell the "life history" of a digestive enzyme made
within a pancreas cell. Remember that most of any
enzyme consists of protein. Start with the "birth"
(synthesis) of this protein and outline its "travels"
from one part of the cell to another, including its
exit from the cell via exocytosis.

5. Describe several steps in the progression of carcinogenesis.

MASTERY TEST: CHAPTER 3

Questions 1–9: Circle the letter preceding the one best answer to each question.

1. The organelle that carries out the process of
autophagy in which old organelles are digested so
that their components can be recycled is:
   A. Peroxisome
   B. Mitochondrion
   C. Centrosome
   D. Lysosome
   E. Endoplasmic
   reticulum

2. Which statement about proteins is *false?*
   A. Proteins are synthesized on ribosomes.
   B. Proteins are so large that they normally will
   not pass out of blood that undergoes filtration
   in kidneys.
   C. Proteins are so large that they can be expected
   to create osmotic pressure important in maintaining fluid volume in blood.
   D. Proteins are small enough to pass through
   pores of typical plasma (cell) membranes.

3. Choose the *false* statement about prophase of
mitosis.
   A. It occurs just after metaphase.
   B. It is the longest phase of mitosis.
   C. Nucleoli disappear.
   D. Centrioles move apart and become connected
   by spindle fibers.

4. Choose the *false* statement about the Golgi
complex.
   A. It is usually located near the nucleus.
   B. It consists of stacked membranous sacs known
   as cisternae.
   C. It is involved with protein and lipid secretion.
   D. It pulls chromosomes toward poles of the cell
   during mitosis.

5. Choose the *false* statement about genes.
   A. Genes contain DNA.
   B. Genes contain information that controls
   heredity.
   C. Each cell in the human body has a total of
   46 genes.
   D. Genes are transcribed by messenger RNA
   during the first step of protein synthesis.

6. Choose the *false* statement about protein synthesis.
   A. Translation occurs in the cytoplasm.
   B. Messenger RNA picks up and transports an
   amino acid during protein synthesis.
   C. Messenger RNA travels to a ribosome in the
   cytoplasm.
   D. Transfer RNA is attracted to mRNA due to
   their complementary bases.

7. All of the following structures are part of the
nucleus *except:*
   A. Nucleolus
   B. Chromatin
   C. Chromosomes
   D. Centrosome

8. The fact that the aroma of cookies baking in the
kitchen reaches you in the living room is due to:
   A. Active transport
   B. Osmosis
   C. Diffusion
   D. Phagocytosis

9. Bone cancer is an example of what type of cancer?
   A. Sarcoma
   B. Lymphoma
   C. Carcinoma

Questions 10–20: Circle T (true) or F (false). If the statement is false, change the underlined word or phrase so that the statement is correct.

T F 10. Plasma membranes consist of a double layer of <u>carbohydrate molecules</u> with proteins embedded in the bilayer.

T F 11. Free ribosomes in the cytosol are concerned primarily with synthesis of <u>proteins for insertion in the plasma membrane or for export from the cell.</u>

T F 12. Glucose in foods you eat is absorbed into cells lining the intestine by the process of <u>facilitated transfusion.</u>

T F 13. <u>Cristae</u> are folds of membrane in mitochondria.

T F 14. Sperm move by means of lashing their long tails named <u>cilia.</u>

T F 15. Hypertrophy refers to increase in size of a tissue related to increase in the <u>number</u> of cells.

T F 16. <u>Diffusion, osmosis, and filtration</u> are all passive transport processes.

T F 17. A 5% glucose solution is <u>hypotonic</u> to a 10% glucose solution.

T F 18. <u>The free radical theory of aging</u> suggests that electrically charged, unstable, and highly reactive free radicals cause oxidative damage to chemical components of aging cells.

T F 19. The mitotic phase that follows metaphase is <u>anaphase.</u>

T F 20. <u>Cytokinesis</u> is another name for mitosis.

Questions 21–25: Fill-ins. Complete each sentence with the word which best fits.

_____ 21. The first meiotic division is known as the _____ division.

_____ 22. Active processes involved in movement of substances across cell membranes are those that use energy from the splitting of _____.

_____ 23. White blood cells engulf large solid particles by the process of _____.

_____ 24. _____ DNA consists of DNA combined from several sources, such as from human cells and from bacteria.

_____ 25. The sequence of bases of mRNA which would be complementary to DNA bases in the sequence A-T-T-C-A-C would be _____.

ANSWERS TO MASTERY TEST: ☆ CHAPTER 3

Multiple Choice

1. D
2. D
3. A
4. D
5. C
6. B
7. D
8. C
9. A

True–False

10. F. Phospholipid molecules
11. F. Proteins for use inside of the cell
12. F. Facilitated diffusion
13. T
14. F. Flagella
15. F. Size
16. T
17. T
18. T
19. T
20. F. Nuclear division (Cytokinesis is cytoplasmic division.)

Fill-ins

21. Reduction
22. ATP
23. Phagocytosis
24. Recombinant
25. U-A-A-G-U-G

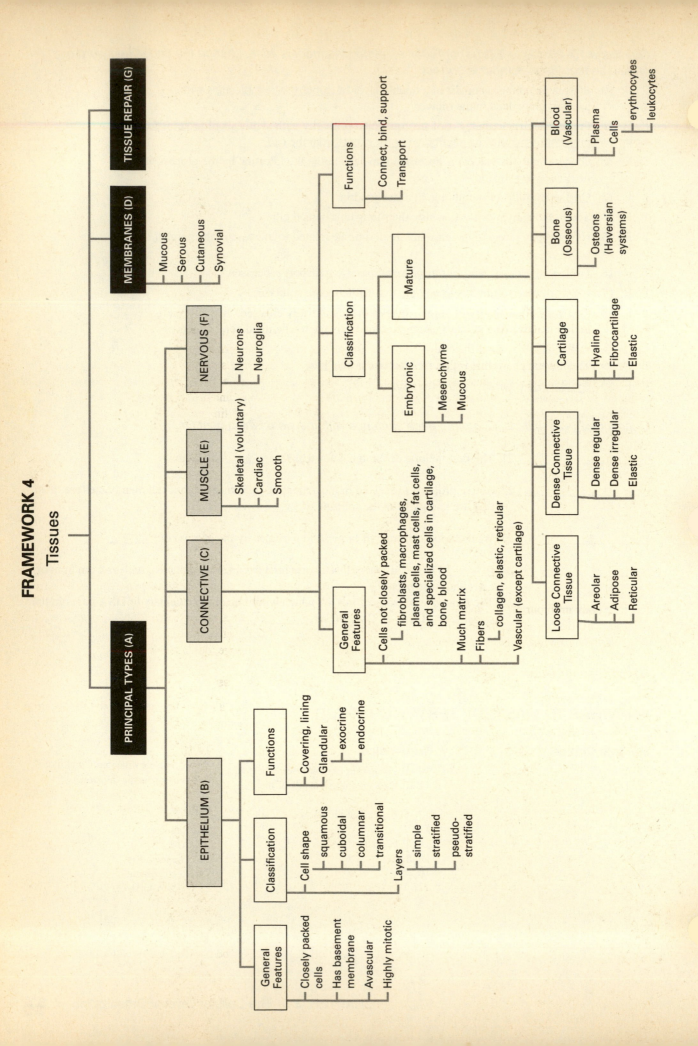

FRAMEWORK 4

Tissues

PRINCIPAL TYPES (A)

MEMBRANES (D)
- Mucous
- Serous
- Cutaneous
- Synovial

TISSUE REPAIR (G)

EPITHELIUM (B)

General Features
- Closely packed cells
- Has basement membrane
- Avascular
- Highly mitotic

Classification
- Cell shape
  - squamous
  - cuboidal
  - columnar
  - transitional
- Layers
  - simple
  - stratified
  - pseudo-stratified

Functions
- Covering, lining
- Glandular
  - exocrine
  - endocrine

CONNECTIVE (C)

MUSCLE (E)
- Skeletal (voluntary)
- Cardiac
- Smooth

NERVOUS (F)
- Neurons
- Neuroglia

General Features
- Cells not closely packed
  - fibroblasts, macrophages, plasma cells, mast cells, fat cells, and specialized cells in cartilage, bone, blood
- Much matrix
- Fibers
  - collagen, elastic, reticular
- Vascular (except cartilage)

Classification

Embryonic
- Mesenchyme
- Mucous

Mature

Functions
- Connect, bind, support
- Transport

Loose Connective Tissue
- Areolar
- Adipose
- Reticular

Dense Connective Tissue
- Dense regular
- Dense irregular
- Elastic

Cartilage
- Hyaline
- Fibrocartilage
- Elastic

Bone (Osseous)
- Osteons (Haversian systems)

Blood (Vascular)
- Plasma
- Cells
  - erythrocytes
  - leukocytes

The Tissue Level of Organization

CHAPTER 4

Tissues consist of cells and the intercellular (between cells) materials secreted by these cells. All of the organs and systems of the body consist of tissues that may be categorized into four classes or types: epithelium, connective, muscle, and nerve. Each tissue exhibits structural characteristics (anatomy) that determine the function (physiology) of that tissue. For example, epithelial tissues consist of tightly packed cells that form the secretory glands and also provide protective barriers, as in the membranes that cover the surface of the body and line passageways or cavities. Tissues are constantly stressed by daily wear and tear and sometimes by trauma or infection. Tissue repair is therefore an ongoing human maintenance project.

As you begin your study of tissues, carefully examine the Chapter 4 Framework and note relationships among concepts and key terms there. Also refer to the Topic Outline and Objectives; you may want to check off each objective as you complete it.

TOPIC OUTLINE AND OBJECTIVES

A. Types of tissues; cell junctions

- ☐ 1. Define a tissue and classify the tissues of the body into four major types.
- ☐ 2. Describe the structure and functions of the three principal types of cell junctions.

B. Epithelial tissue

- ☐ 3. Describe the general features of epithelial tissue.
- ☐ 4. List the structure, location, and function for the following types of epithelium: simple squamous, simple cuboidal, simple columnar (nonciliated and ciliated), stratified squamous, stratified cuboidal, stratified columnar, transitional, and pseudostratified columnar.
- ☐ 5. Define a gland and distinguish between exocrine and endocrine glands.

C. Connective tissue

- ☐ 6. Describe the general features of connective tissue.
- ☐ 7. Discuss the cells, ground substance, and fibers that compose connective tissue.

- ☐ 8. List the structure, function, and location of mesenchyme, mucous connective tissue, areolar connective tissue, adipose tissue, reticular connective tissue, dense regular and irregular connective tissue, elastic connective tissue, cartilage, bone, and blood.

D. Membranes

- ☐ 9. Define a membrane and describe the location and function of mucous, serous, cutaneous, and synovial membranes.

E. Muscle tissue

- ☐ 10. Contrast the three types of muscle tissue with regard to structure, location, and modes of control.

F. Nervous tissue

- ☐ 11. Describe the structural features and functions of nervous tissue.

G. Tissue repair

- ☐ 12. Describe tissue repair in restoring homeostasis.

WORDBYTES

Now become familiar with the language of this chapter by studying each wordbyte, its meaning, and an example of its use within a term. After you study the entire list, self-check your understanding by writing the meaning of each wordbyte on the line. As you continue through the Learning Guide, identify (and fill in) additional terms that contain the same wordbyte.

Wordbyte	Self-check	Meaning	Example(s)
a-	_____	without	*a*vascular
acin	_____	grape	*acin*ar gland
adip(o)-	_____	fat	*adip*ose
apex, apic-	_____	tip	*apic*al
areol-	_____	small space	*areol*ar
-crine	_____	to secrete	endo*crine*
endo-	_____	within	*endo*crine
exo-	_____	outside	*exo*crine
macro-	_____	large	*macro*phage
multi-	_____	many	*multi*cellular
pseudo-	_____	false	*pseudo*stratified
squam-	_____	thin plate	*squam*ous
strat-	_____	layer	*strat*ified
uni-	_____	one	*uni*cellular
vasc-	_____	blood vessel	a*vasc*ular

CHECKPOINTS

A. Types of tissues; cell junctions (pages 93–95)

    A1. Define the following terms.

    a. Tissue

    b. Histology

    c. Pathologist

    d. Biopsy

A2. Complete the table about the four main classes of tissues.

Tissue	General Functions
a. Epithelial tissue	
b. Connective tissue	
c.	Movement
d.	Initiates and transmits nerve impulses that help coordinate body activities

A3. Circle the two types of tissue in Checkpoint A1 that cannot undergo mitosis after birth. In other words, those tissues grow only by means of *(hyperplasia? hypertrophy?)*.

A4. *The Big Picture: Looking Ahead.* Refer to Exhibit 29.1 on page 965 of your text. Then do the following matching exercise on the three embryonic germ layers that form all body organs.

> Ecto. Ectoderm   Endo. Endoderm   Meso. Mesoderm

_____ a. Epithelial lining of the organs of diges-   _____ c. Muscles and bones
tion and of respiratory passageways

_____ b. Brain and nerves   _____ d. Outer layers of skin

A5. Complete this exercise about extracellular materials and connections between cells.

a. Two major types of extracellular fluid (ECF) are the _____ located between

cells in tissues, and the liquid portion of blood, known as _____. It may
be helpful to review Checkpoint D1 and Figure LG 1.1 on page 9.

b. Name one type of cell that typically moves around within extracellular areas _____.

c. Most body cells stay in one place. They are held in place by points of contact between

adjacent cell membranes; such points are known as cell _____. For example,
(anchoring gap? tight?) junctions are helpful in forming a tight seal between cells, such as that

required in the lining of the stomach or the_____.

d. *(Anchoring? Gap? Tight?)* junctions are common in areas subject to stretching or friction.
One example of a location of an anchoring junction is in the uterus. Write one or more

example.

e. A gap junction is consists of proteins called_____ that form tunnels between

cells. Passage of ions through these tunnels permits spread of action _____
within the nervous system or heart muscle. The lack of appropriate communication and coordinated
growth of cancer cells is related to the *(lack of? many?)* gap junctions between cancer cells.

B. Epithelial tissue (pages 95–103)

■ **B1.** Name the two subtypes of epithelial tissue based on location and function.

_____ and _____

B2. Complete the table by describing five or more locations and functions of epithelial tissue. One is done for you.

Location	Function
a. Eyes, nose, outer layer of skin	Sensory reception
b.	
c.	
d.	
e.	

■ **B3.** Describe features of epithelium in this exercise.
   a. Epithelium consists of *(closely packed cells? intercellular material with few cells?)*. *(Many? Few?)* cell junctions are present, securing these cells *(tightly? loosely?)* to each other.
   b. Blood vessels *(do? do not?)* typically penetrate epithelium. A term meaning "lack of blood vessels"

   is _____. State the significance of this feature of epithelium.

   c. Epithelium *(has? lacks?)* a nerve supply.
   d. Epithelium *(does? does not?)* have the capacity to undergo mitosis.

■ **B4.** Complete these sentences describing *basement membrane.*
   a. The *(apical? basal?)* surfaces of epithelial cells attach by a _____
      membrane to underlying connective tissue.
   b. This membrane consists of *(cells? extracellular material?)*. The part of the membrane adjacent to epithelium is secreted by *(epithelium? connective tissue?)*. It consists of the protein named *(collagen? reticulin?)* along with proteoglycans and is known as the *(reticular lamina? basal lamina?)*.
   c. The second, underlying layer of the membrane is secreted by *(epithelium? connective tissue?)*

      and is called the _____ lamina.

B5. Study diagrams of epithelial tissue types (A–F) in Figure LG 4.1 and then write the following information on the lines provided on the figure:
■ a. The name of each tissue
   b. One or more locations of each type of tissue
   c. One or more functions of each type of tissue

Now color the following structures on each diagram:
   ○ Nucleus   ○ Intercellular material
   ○ Cytoplasm   ○ Basement membrane

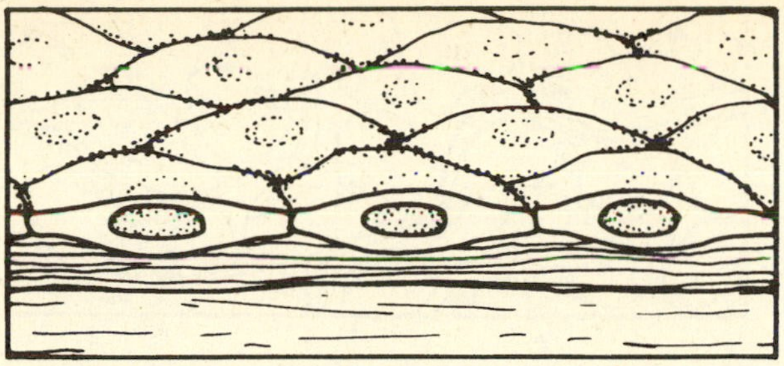

A Name _____

Location _____

Function _____

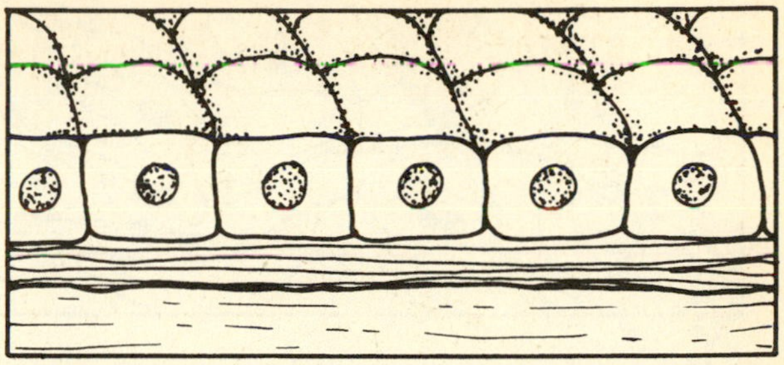

B Name _____

Location _____

Function _____

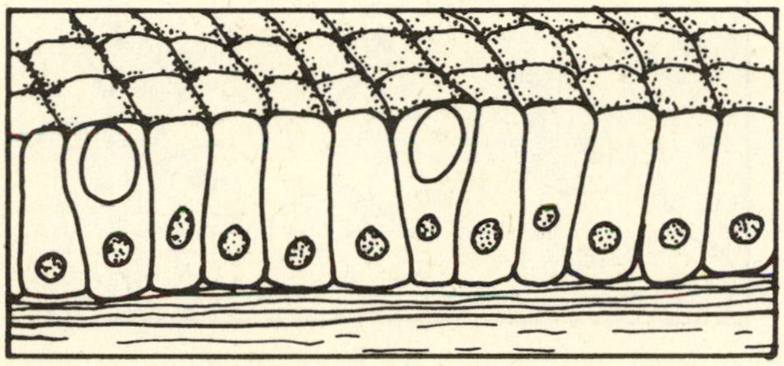

C Name _____

Location _____

Function _____

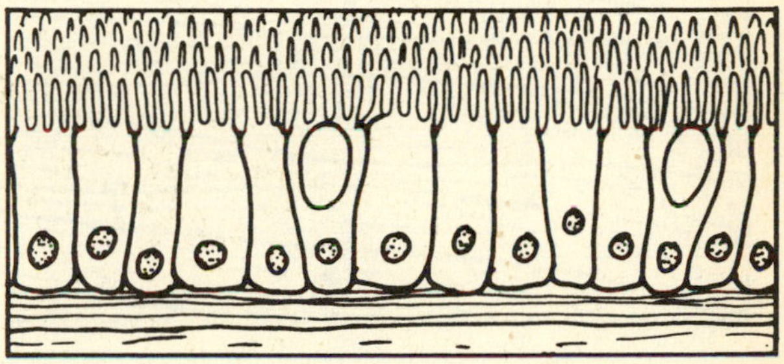

D Name _____

Location _____

Function _____

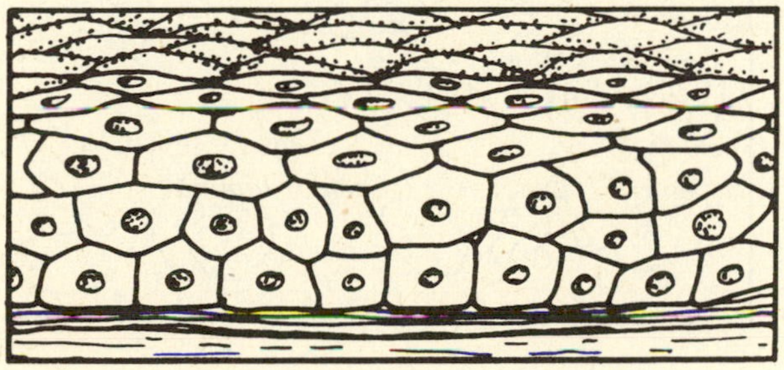

E Name _____

Location _____

Function _____

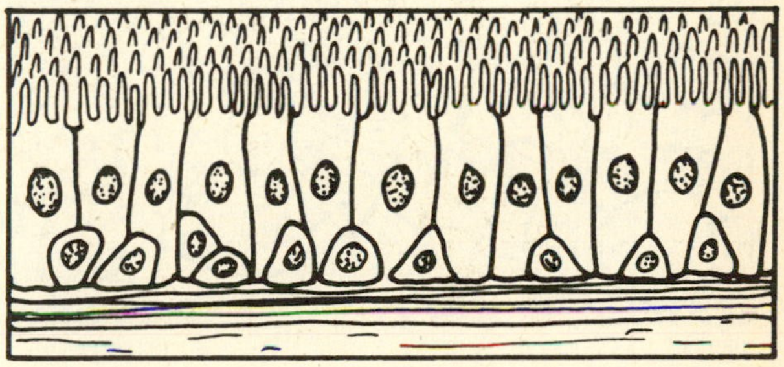

F Name _____

Location _____

Function _____

Figure LG 4.1 Diagrams of selected tissues types. Complete as directed.

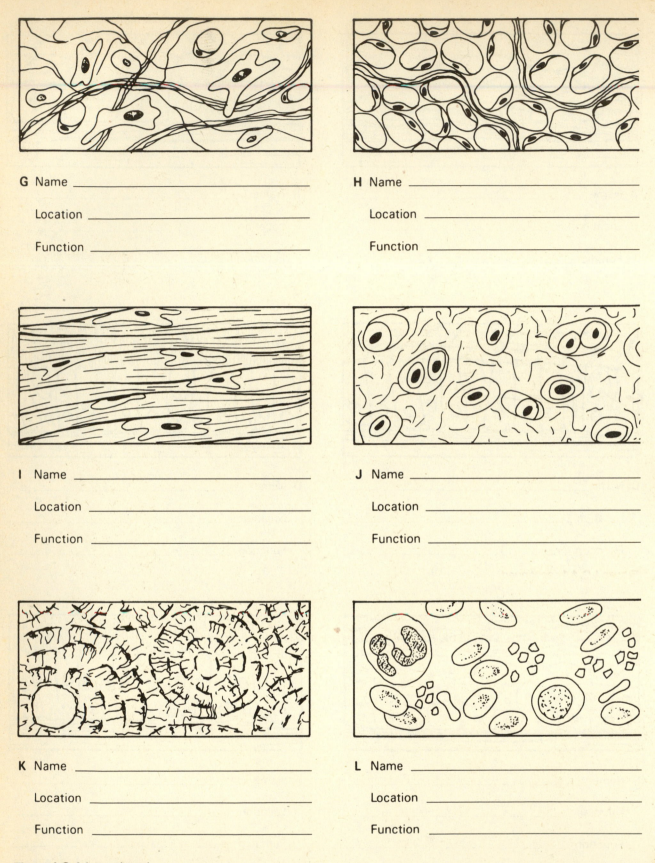

G Name _____

　　Location _____

　　Function _____

H Name _____

　　Location _____

　　Function _____

I Name _____

　　Location _____

　　Function _____

J Name _____

　　Location _____

　　Function _____

K Name _____

　　Location _____

　　Function _____

L Name _____

　　Location _____

　　Function _____

Figure LG 4.1 *continued*

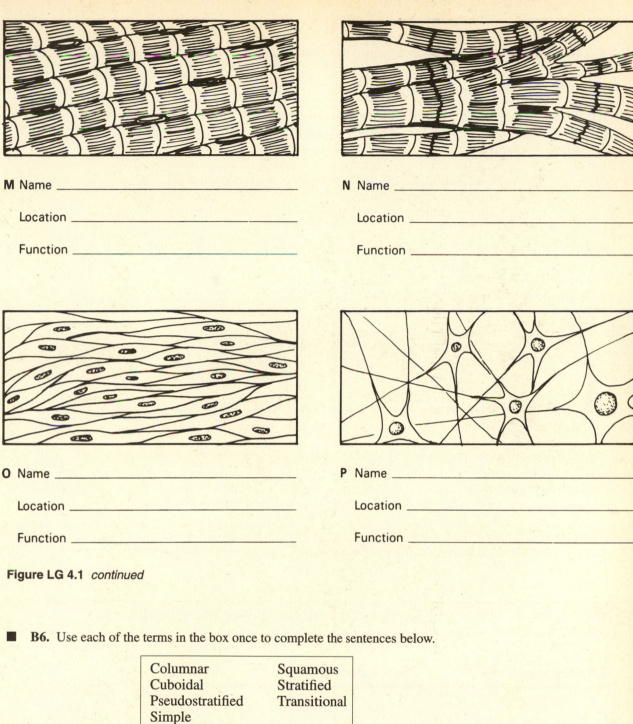

M Name _____

   Location _____

   Function _____

N Name _____

   Location _____

   Function _____

O Name _____

   Location _____

   Function _____

P Name _____

   Location _____

   Function _____

Figure LG 4.1 *continued*

■ **B6.** Use each of the terms in the box once to complete the sentences below.

Columnar	Squamous
Cuboidal	Stratified
Pseudostratified	Transitional
Simple	

a. Epithelial cells arranged in one layer are known as _____ epithelium.

   _____ epithelium is composed of many layers of cells.

b. _____ epithelium consists of cells so flat that diffusion readily occurs across them.

   Tall, cylindrical or rectangular cells are called_____epithelium. Intermediate

   between these two types of cells are the cells that form_____ epithelium.

c. _____ epithelium is not a true stratified tissue; all cells lie on a basement membrane, but some do not reach the surface.

d. The urinary bladder is lined with epithelium that must stretch, so the shape of cells must change.

This is _____ epithelium.

■ **B7.** *The Big Picture: Looking Ahead.* Check your understanding of the most common types of epithelium listed in the box by writing the name of the type after the phrase that describes it. To confirm your answers, refer to text pages (listed in parentheses) of upcoming chapters.

Pseudostratified ciliated columnar	Simple cuboidal
Simple columnar, ciliated	Simple squamous
Simple columnar, nonciliated	Stratified squamous

a. Lines the inner surface of the stomach (Figure 24.12, pages 769–770) and intestine

(Figure 24.23, page 785): _____.

b. Forms the outer surface of skin (Figures 5.1, 5.3, 5.5, and 5.6,

pages 125–134): _____.

c. Single layer of cube-shaped cells; found in some glands, such as hormone-producing follicles of the thyroid gland (Figure 18.13, page 521), as well as ducts of

glands: _____.

d. Endothelium, as in capillaries (Figure 21.1, page 613) as well as mesothelium:

_____.

e. Lines airways and facilitates removal of dust and other particles before they enter lungs

(Figure 23.6 page 716): _____.

f. Lines air sacs of lungs where thin cells are required for diffusion of gases into blood

(Figure 23.11, page 722): _____.

g. Forms the delicate Bowman's capsule surrounding each filtering apparatus (glomerulus)

(Figure 26.7, page 856): _____.

B8. Complete the table describing distinguishing characteristics of specialized epithelial cells. Also fill in sites where such modified cells would be likely to be located.

Modification	Description	Location
a. Cilia		
b. Microvilli		Cells lining digestive tract
c. Keratin production	Waterproofing protein	
d. Goblet cells		

B9. Define glands? Why are glands studied in this section on epithelium?

■ **B10.** Write EXO before descriptions of *exocrine* glands, and ENDO before descriptions of *endocrine* glands. (Endocrine glands will be studied further in Chapter 18.)

_____ a. Their products are secreted into ducts that lead either directly or indirectly to the outside of the body.

_____ b. Their products are secreted into ECF and then the blood and so stay within the body; they are ductless glands.

_____ c. Examples are glands that secrete sweat, oil, mucus, and digestive enzymes.

_____ d. Examples are glands that secrete hormones

■ **B11.** Do this activity contrasting classes of glands. Select the answer in the box that best fits each description. Use each answer only once. (Lines after descriptions will be used in Checkpoint B12.)

Apo.	Apocrine	Multi.	Multicellular
Holo.	Holocrine	Uni.	Unicellular
Mero.	Merocrine		

_____ a. The apical region of gland cells—containing the secretion—pinches off to release the secretion

along with that part of the cell. _____

_____ b. Cells accumulate secretions and then cells die. Glandular secretions contain both dead cells

and their contents._____

_____ c. Secretions are discharged from these gland cells by exocytosis; the gland cell stays intact._____

_____

_____ d. This is defined as a many-celled gland. **Sweat, oil, or salivary gland.**

_____ e. A goblet cell is this type of gland._____

■ **B12.** Now identify the class of the types of glands listed below. On the lines that follow the descriptions in Checkpoint B11, write names of specific glands listed here. One is done for you.

Oil glands	Mammary glands
Glands lining digestive, respiratory, and reproductive tracts	Salivary and pancreatic glands
	Sweat, oil, or salivary glands

C. Connective tissue (pages 103-115)

■ **C1.** Write *connective tissue* or *epithelial tissue* following the description below that correctly describes that tissue.

a. Consists of many cells with little intercellular substance (matrix and ground substance): _____

b. Penetrated by blood vessels (vascular): _____

c. Does not cover body surfaces or line passageways and cavities but is more internally located; binds, supports, protects: _____

■ **C2.** List four or more categories of connective tissue matrix. (Note that *inter*cellular means "between cells" or "extracellular.") What is the source of these materials?

■ **C3.** Do this activity about connective tissue cells.

a. Cells in connective tissues derive from *(endo? meso? ecto?)*-derm cells in the embryo. These cells are

called _____ cells.

b. Most matrix is secreted by *(immature? mature?)* connective tissue cells. Which type of cells have greater capacity for mitosis? *(Mature? Immature?)*

c. Names of immature cells end in *(-blast? -cyte?),* whereas names of mature cells end in *(-blast? -cyte?).* To check your knowledge further, write correct names for the following

cells: immature bone cell: _____; immature cartilage cell:

_____; mature bone cell: _____.

■ **C4.** Identify characteristics of each connective tissue cell type by writing the name of the correct cell type after its description. (Note that several of these cell types are shown in diagrams G, H, and I of Figure LG 4.1.)

Adipocyte	Mast cell
Fibroblast	Plasma cell
Macrophage	

a. Derived from B lymphocyte; gives rise to antibodies; helpful in defense:

_____

b. Phagocytic cell that engulfs bacteria and cleans up debris; important during infection; formed

from monocyte: _____

c. Fat cell: _____

d. Forms collagenous and elastic fibers in

injured tissue: _____

e. Abundant along walls of blood vessels; contains heparin and produces histamine, which

dilates blood vessels: _____

■ **C5.** The ground substance of connective tissue consists of *(large? small?)* molecules that are combinations of polysaccharides and *(lipids? proteins?).* Identify components of ground substance by selecting the answers listed in the box that best fit the descriptions below.

AP.	Adhesion proteins	HA.	Hyaluronic acid
CS.	Chondroitin sulfate	KS.	Keratan sulfate
DS.	Dermatan sulfate		

a. _____ Chemicals that stabilize position of cells; includes proteins such as fibronectin, collagen, and fibrinogen

b. _____ Viscous, slippery substance that binds cells together and lubricates joints

c. _____ Chemical component of bone, cartilage, and the cornea of the eye

d. _____ Jellylike; provides support and adhesiveness in bone, skin, blood vessels, and cartilage

e. _____ Found in bone, skin, blood vessels, and heart valves

■ **C6.** Match the correct type of fiber in connective tissue matrix with the descriptions below. Use the answers in the box.

> C. Collagen   E. Elastic   R. Reticular

a. _____ Formed of collagen and glycoproteins; form branching networks that provide stroma for soft organs such as spleen and networks around fat, nerve, and muscle cells

b. _____ Tough fibers in bundles that provide great strength, as needed in bone and

tendons; formed of the most abundant protein in the body

c. _____ Can be greatly stretched without breaking, an important quality of connective tissues forming skin, blood vessels, and lungs

■ **C7.** As described earlier, the embryonic connective tissue from which all of the connective

tissues arise is called _____. *(Much? Little?)* of this type of tissue is found in the body after birth. Another embryonic tissue, known as mucous connective tissue or

_____ jelly, gives support to the umbilical cord. Tissues such as areolar connective tissue, cartilage, or bone are known as *(embryonic? mature?)* connective tissues.

C8. After you study each mature connective tissue type in the text, refer to diagrams of connective tissues on Figure LG 4.1, G–L. Write the following information on the lines provided on the figure.

■ a. The name of each tissue
b. One or more locations of the tissue
c. One or more functions of the tissue

Now color the following structures on each diagram, using the same colors as you did for diagrams A–F:
  ○ Nucleus   ○ Matrix
  ○ Cytoplasm   ○ Fat globule (a cellular inclusion), diagram H

■ From this activity, you can conclude that connective tissues appear to consist mostly of *(cells? intercellular material [matrix]?)*.

C9. Label fibers in Figure LG 4.1, diagrams G, I, and J.
(Note that fibers *are* present in bone but not visible in diagram K.)

■ **C10.** Which two types of tissue form subcutaneous tissue (superficial fascia)?

_____, _____

C11. *A clinical challenge.* Answer the following questions on clinical topics.
a. Explain how the enzyme *hyaluronidase* may facilitate absorption of injected drugs.

b. Describe the typical stature of a person with *Marfan syndrome.* List three systems likely to be affected by this connective tissue disorder.

■ **C12.** Answer these questions about adipose tissue.

a. Explain what accounts for the "signet ring" appearance (like a class ring with a stone) of adipocytes, as in Figure LG 4.1H.

b. List three or more functions of fat in the body.

c. Explain what gives the brown color to "brown fat," and state how this type of fat may be helpful to newborns.

■ **C13.** *The Big Picture: Looking Ahead*. Refer to figures in your textbook to help you identify the dense connective tissues described below. Use the answers in the box. Use each answer once..

A. Aponeurosis	P. Periosteum
C. Capsule	SL. Suspensory ligament
EWA. Elastic walls of arteries	T. Tendon
F. Fascia	TVC. True vocal cords
L. Ligament	V. Valves

_____ a. Figure 6.3 (page146) shows the dense covering over bone which affords a pathway for blood vessels to penetrate bone and also serves as a site of osteoblast cells that build new bone. _____

_____ b. Strong but flexible bands that attach bone to bone are shown in Figure 9.7 (pages 227–228). _____

_____ c. In Figure 9.8 (page 229) a strong, fibrous band is shown attaching the quadriceps femoris muscle across the patella and eventually inserting it into bone (tibia). Similar fibrous bands attach forearm muscles into bones of the hand (Figure 11.17, page 307). _____

_____ d. Figure 11.10 (page 291) shows broad, sheetlike attachments of external and internal oblique muscles. _____

_____ e. A broad sheet of fibrous tissue that provides support to the floor of the pelvic cavity (urogenital diaphragm) is shown in Figure 11.12 (page 294). _____

_____ f. Figure 20.4 (page 585) demonstrates the strong connective tissue leaflets that direct the flow of blood within the heart. _____

_____ g Figure 21.1 (page 613) shows the tissue that provides blood vessels walls with the elasticity needed as the heart regularly pumps blood into these vessels. _____

_____ h. Figure 23.5 (page 714) demonstrates the structures that require strength yet flexibility as their continual alteration of tension changes vocal quality. _____

_____ i. Connective tissue surrounds, supports, and protects each kidney (Figure 26.4, page 852). _____

_____ j. Figure 28.1 (page 909) shows the tissue that suspends and supports the penis. _____

■ **C14.** Now identify the category to which each of the connective tissues listed in Activity C13 belongs. On the lines following each description above, write DR for dense regular, DI for dense irregular, or E for elastic.

■ **C15.** Do this activity about cartilage.

a. In general, cartilage can endure *(more? less?)* stress than other connective tissues studied so far.

b. The strength of cartilage is due to its *(collagen? elastic? reticular?)* fibers, and its resilience is related to the presence of *(chondroitin? dermatan? keratan?)* sulfate.

c. Cartilage heals *(more? less?)* rapidly than bone. Explain why this is so.

d. Growth of cartilage occurs by two mechanisms. In childhood and adolescent years, cartilage increases in size by *(appositional? interstitial?)* growth. This process is also called *(exogenous? endogenous?)* because it is growth by enlargement of existing cells within the cartilage.

e. Post adolescence, cartilage increases in thickness by *(appositional? interstitial?)* growth.

This process involves cells in the _____ that covers cartilage. Here cells known as

(chondroblasts? chondrocytes? fibroblasts?) differentiate into_____, which lay down matrix and eventually become mature chondrocytes.

■ **C16.** Match the types of cartilage with the descriptions given.

| E. Elastic   F. Fibrous   H. Hyaline |

_____ a. Found where strength and rigidity are needed, as in discs between vertebrae and in symphysis pubis

_____ b. White, glossy cartilage covering ends of bones (articular), covering ends of ribs (costal), and giving strength to . nose, larynx, and trachea

_____ c. Provides strength and flexibility, as in external part of ear.

_____ d. Forms menisci (cartilage pads) of the knees

_____ e. Forms most of the skeleton of the human embryo

■ **C17.** Complete this exercise about bone tissue.

a. Bone tissue is also known as _____ tissue. Compact bone consists of concentric rings or *(lamellae? canaliculi?)* with bone cells, called *(chondrocytes? osteocytes?)*, located in tiny spaces called

_____. Nutrients in blood reach osteocytes by vessels located in the _____

canal and then via minute, radiating canals called _____ that extend out to lacunae.

The basic unit of structure and function of compact bone is known as an _____ or Haversian system.

b. Spongy bone *(is? is not?)* arranged in osteons. Spongy bone consists of plates of bone called

_____.

c. List three or more functions of bone tissue.

■ **C18.** Fill in the blanks about blood tissue.

Blood, or _____ tissue, consists of a fluid called _____ containing three

types of formed elements. Red blood cells, or _____, transport oxygen and carbon dioxide;

white cells, also known as _____, provide defense for the body; thrombocytes assist in the

function of blood _____.

C19. Now label the following cell types found in diagrams J, K, or L of Figure LG 4.1: *chondrocytes, osteocytes, erythrocytes, leukocytes, thrombocytes.*

D1. Complete the table about the four types of membranes in the body.

Type of Membrane	Location	Example of Specific Location	Function(s)
a.	Lines body cavities leading to exterior		
b. Serous			Allows organs to glide easily over each other
c.		Lines knee and hip joints	
d.	Covers body surfaces	Skin	

■ **D2.** *The Big Picture: Looking Ahead.* Refer to figures in your textbook to help you identify the types of membranes described below. Use the answers in the box.

Cutaneous	Pleural
Mucous	Peritoneal
Pericardial	Synovial

a. A _____ membrane is another name for skin, shown in Figure 5.1, page 125.

b. Figure 9.1 (page 218) shows a _____ membrane that secretes a lubricating fluid that reduces friction at joints. This type of membrane contains connective (rather than epithelial) tissue.

c. On Figure 20.2 (page 580), the serous membrane that covers the heart is shown. This is the

_____ membrane. The same figure shows the _____ membrane covering the lungs and lining the chest.

d. Figure 24.2 (page 755) contrasts the membrane lining the inside of digestive organs, known as a

_____ membrane, with the _____ membrane that covers the outside of such organs. Figure 26.23 (page 879) shows similar membranes of the urinary bladder:

the pink _____ membrane with a rippled pattern (rugae) inside the organ and the

_____ membrane over the outside of the bladder.

■ **D3.** Describe specific layers of serous and mucous membranes in this exercise.

    a. The layer of a serous membrane that adheres to an organ, such as the liver, is the *(parietal? visceral?)* layer, whereas the portion lining the body cavity wall is the *(parietal? visceral?)* layer.

    b. The epithelial layer of mucous membranes varies, with *(simple columnar? pseudostratified columnar? stratified squamous?)* lining the mouth, _____ lining the small intestine,

    and _____ lining the large airways to the lungs.

    c. The connective tissue of a mucous membrane is known as the lamina _____
    Unlike the surface epithelial layer, the lamina propria *(is? is not?)* vascular.

■ **D4.** *A clinical challenge.* Serous membranes are normally relatively free of microorganisms, whereas mucous and cutaneous membranes have many microorganisms on them. Based on this information, state a rationale for the administration of antibiotics to patients undergoing stomach or intestinal surgery.

E. Muscle tissue (pages 116–118)

E1. After you study each muscle tissue type in the text, refer to diagrams of muscle tissues in Figure LG 4.1, M–O. Write the following information on the lines provided on the figure.

■   a. The name of each muscle tissue
    b. One or more locations of the tissue
    c. One or more functions of the tissue

Now color the following structures on each diagram, using the same colors as you did for diagrams A–L.

◯ Nucleus   ◯ Cytoplasm   ◯ Intercellular material

■ From this activity, you can conclude that muscle tissues appear to consist mostly of *(cells? intercellular material [matrix]?)*.

■ **E2.** Check your understanding of the three muscle types listed in the box by selecting types that best fit descriptions below.

C. Cardiac   Sk. Skeletal   Sm. Smooth

_____ a. Tissue forming most of the wall
            of the heart

_____ b. Attached to bones

_____ c. Spindle-shaped cells with ends
            tapering to points

_____ d. Contains intercalated discs

_____ e. Found in walls of intestine, urinary
            bladder, and blood vessels

_____ f. Voluntary muscle

F. Nervous tissue (page 118)

F1. Refer to diagram P in Figure LG 4.1. Write the following information on the lines provided on the figure.
■   a. The name of the tissue
    b. One or more locations of the tissue
    c. One or more functions of the tissue

■ **F2.** *(Axons? Dendrites?)* carry nerve impulses toward the cell body of a neuron, whereas

_____ transmit nerve impulses away from the cell body.

G. Tissue repair (pages 118–119)

■ **G1.** Circle all of the tissue types that maintain the capacity to undergo mitosis after birth.

a. Blood

b. Bone

c. Cardiac muscle

d. Epithelium lining the mouth

e. Epithelium on the skin surface

f. Nervous tissue (neurons)

g. Skeletal muscle

■ **G2.** Choose the correct term to complete each sentence. Write P for parenchymal or S for stromal.

a. The part of an organ that consists of functioning cells, such as secreting epithelial cells lining the intestine, is composed of _____ tissue.

b. The connective tissue cells that support the functional cells of the organ are called _____ cells.

c. If only the _____ cells are involved in the repair process, the repair will be close to perfect.

d. If _____ cells are involved in the repair, they will lay down fibrous tissue known as a fibrosis or a scar.

G3. Define each of the following terms related to tissue repair.

a. Regeneration

b. Granulation tissue

c. Adhesions

■ **G4.** You have most likely experienced an injury that led to formation of a *scab.* What causes scabs to form?

■ **G5.** List three factors that enhance tissue repair.

■ **G6.** *A clinical challenge.* Wound healing is likely to be impaired in persons with inadequate nutrition. Identify the vitamins—A, B, C, D, E, or K—that perform the following functions in the process of wound healing:

_____ a. This vitamin is necessary for healing fractures because it enhances calcium absorption from foods in the intestine.

_____ b. This is important in repair of connective tissue and walls of blood vessels.

_____ c. Thiamine, riboflavin, and nicotinic acid are categorized as vitamins of this group. They enhance metabolic processes and the cell division necessary for wound repair.

_____ d. This vitamin is helpful in replacement of epithelial tissues, for example, in the lining of the respiratory tract.

_____ e. Believed to promote healing of injured tissues, this vitamin may also help to prevent scarring.

_____ f. This vitamin aids in blood clotting and so helps to prevent excessive blood loss from wounds.

A3. Nerve tissue and most muscle tissue; hypertrophy, which is tissue growth by enlargement of existing cells (see page 89 in your text).

A4. (a) Endo. (b) Ecto. (c) Meso. (d) Ecto.

A5. (a) Interstitial fluid, plasma. (b) Phagocyte. (c) Junctions; tight, urinary bladder. (d) Anchoring; epidermis of skin, cardiac muscle, lining of the GI tract. (e) Connexons; potentials; lack of.

B1. Lining, glandular.

B3. (a) Closely packed cells; many, tightly. (b) Do not; avascular; epithelial tissue is protective because it does not bleed when injured; however, the avascular epithelial cells must draw nutrients from adjacent tissues. (c) Has. (d) Does.

B4. (a) Basal, basement. (b) Extracellular material; epithelium; collagen, basal lamina. (c) Connective tissue, reticular.

B5. (A) Simple squamous epithelium. (B) Simple cuboidal epithelium. (C) Simple columnar epithelium (unciliated). (D) Simple columnar epithelium (ciliated). (E) Stratified squamous epithelium. (F) Pseudostratified columnar epithelium, ciliated.

B6. (a) Simple; stratified. (b) Squamous; columnar; cuboidal. (c) Pseudostratified. (d) Transitional.

B7. (a) Simple columnar, nonciliated. (b) Stratified squamous. (c) Simple cuboidal. (d) Simple squamous. (e) Pseudostratified ciliated columnar. (f, g) Simple squamous.

B10. (a) EXO. (b) ENDO. (c) EXO. (d) ENDO.

B11, 12. (a) Apo (mammary glands). (b) Holo: remember that "whole" cell dies; (oil glands). (c) Mero (salivary and pancreatic glands). (d) Multi. (e) Uni (glands lining digestive, respiratory, and reproductive tracts).

C1. (a) Epithelial tissue. (b) Connective tissue (except cartilage). (c) Connective tissue.

C2. Fluid, semifluid, gelatinous, fibrous, or calcified; secreted by connective tissue (or adjacent) cells.

C3. (a) Meso-; mesenchymal. (b) Immature; immature. (c) -blast, -cyte; osteoblast, chondroblast, osteocyte.

C4. (a) Plasma cell. (b) Macrophage. (c) Adipocyte. (d) Fibroblast. (e) Mast cell.

C5. Large, proteins. (a) AP. (b) HA. (c) KS. (d) CS. (e) DS.

C6. (a) R. (b) C. (c) E.

C7. (a) Mesenchyme; little; Wharton's, mature.

C8. G) Loose (areolar) connective tissue. (H) Adipose tissue. (I) Dense connective tissue. (J) Cartilage. (K) Osseous (bone) tissue. (L) Vascular (blood) tissue. Intercellular material (matrix).

C10. Loose (areolar) connective tissue; adipose tissue.

C12. (a) Most of the cell consists of a globule of stored triglycerides (fat); the nucleus and the rest of the cytoplasm are located in a narrow ring around the fat globule. (b) Fat insulates the body, supports and protects organs, and serves as a major source of energy. (c) This type of fat has an abundance of blood vessels as well as mitochondria (which appear brown due to the presence of copper-containing cytochromes). Brown fat may help to generate heat to maintain proper body temperature in newborns.

C13, 14. (a) P, DI. (b) L, DR. (c) T, DR. (d) A, DR. (e) F, DI. (f) V, DI. (g) EWA, E. (h) TVC, E. (i) C, DI. (j) SL, E.

C15. (a) More. (b) Collagen, chondroitin. (c) Less; cartilage is avascular, so chemicals needed for repair must reach cartilage by diffusion from the perichondrium. (d) Interstitial; endogenous. (e) Appositional; perichondrium; fibroblasts, chondroblasts.

C16. (a) F. (b) H. (c) E. (d) F. (e) H.

C17. (a) Osseous; lamellae, osteocytes, lacunae; central (Haversian), canaliculi; osteon. (b) Is not; trabeculae. (c) Supports and protects, plays a role in movement, stores mineral salts, houses red and yellow marrow, including lipids.

C18. Vascular, plasma; erythrocytes, leukocytes, clotting.

D2. (a) Cutaneous. (b) Synovial. (c) Pericardial; pleural. (d) Mucous, peritoneal (or serous); mucous, peritoneal (or serous).

D3. (a) Visceral, parietal. (b) Stratified squamous, simple columnar, pseudostratified columnar. (c) Propria; is.

D4. Microorganisms from skin or from within the stomach or intestine may be introduced into the peritoneal cavity during surgery; the result may be an inflammation of the serous membrane (peritonitis).

E1. (M) Skeletal muscle. (N) Cardiac muscle. (O) Smooth muscle. Cells [known as muscle fibers].

E2. (a) C. (b) Sk. (c) Sm. (d) C. (e) Sm. (f) Sk.

F1. Nervous tissue (neurons).

F2. Dendrites, axons (Hint: Remember A for Away from cell body.)
G1. Blood, bone, epithelium lining the mouth, epithelium on the skin surface.
G2. (a) P. (b) S. (c) P. (d) S.
G4. Fibrin clots form to limit blood loss at wounds. As the clot dries out, the two sides of the wound are drawn together. This process (clot retraction) forms a temporary scab from the dried out fibrin.
G5. Nutrition, adequate blood circulation, and (younger) age.
G6. (a) D. (b) C. (c) B. (d) A. (e) E. (f) K.

WRITING ACROSS THE CURRICULUM: CHAPTER 4

1. The outer part (epidermis) of skin is composed of epithelium. Describe what makes epithelium suitable for this location.
2. Which tissue is likely to heal better: bone or cartilage? Explain why.
3. Contrast an embryo with a fetus. What is the role of mesenchyme in the embryo?
4. Serous membranes are normally relatively free of microorganisms, whereas mucous and cutaneous membranes have many microorganisms on them. Based on this information, why may antibiotics be administered to patients undergoing stomach or intestinal surgery?
5. Explain how special characteristics of the shark skeleton may help in the fight against cancer.
6. Describe structural features of mucous membranes that offer defense mechanisms and others that bind the membrane to underlying tissue and provide a rich blood supply.

MASTERY TEST: CHAPTER 4

Questions 1–6: Circle the letter preceding the one best answer to each question.

1. A surgeon performing abdominal surgery will pass through the skin, and then loose connective tissue (subcutaneous), then fascia covering muscle and muscle tissue itself to reach the _____ membrane lining the inside wall of the abdomen.
   A. Parietal pleura
   B. Parietal pericardium
   C. Parietal peritoneum
   D. Visceral pleura
   E. Visceral pericardium
   F. Visceral peritoneum

2. The type of tissue that covers body surfaces, lines body cavities, and forms glands is:
   A. Nervous
   B. Muscular
   C. Connective
   D. Epithelial

3. Which of the four types of tissue is the most abundant and most widely distributed tissue in the body?
   A. Nervous tissue
   B. Muscle tissue
   C. Connective tissue
   D. Epithelial tissue

4. Which statement about connective tissue is *false*?
   A. Cells are very closely packed together.
   B. Most connective tissues have an abundant blood supply.
   C. Intercellular substance is present in large amounts.
   D. It is the most abundant tissue in the body.

5. Modified columnar cells that are unicellular glands secreting mucus are known as:
   A. Cilia
   B. Microvilli
   C. Goblet cells
   D. Branched tubular glands

6. Which tissues are avascular?
   A. Skeletal muscle and cardiac muscle
   B. Bone and dense connective tissue
   C. Cartilage and epithelium
   D. Areolar and adipose tissues.

Questions 7–12: Match tissue types with correct descriptions

Adipose	Simple columnar epithelium
Cartilage	Simple squamous epithelium
Dense connective tissue	Stratified squamous epithelium

7. Contains lacunae and chondrocytes:

_____

8. Forms fascia, tendons, and deeper layers of dermis

of skin:_____

9. Forms thick surface layer of skin on hands and feet,

providing extra protection:_____

10. Single layer of flat, scalelike cells:

_____

11. Lines stomach and intestine:

_____

12. Fat tissue: _____

Questions 13–20: Circle T (true) or F (false). If the statement is false, change the underlined word or phrase so that the statement is correct.

T F 13. Stratified transitional epithelium lines the <u>urinary bladder.</u>

T F 14. Epithelial tissue has <u>many</u> cell junctions.

T F 15. Hormones are classified as <u>exocrine</u> secretions.

T F 16. Elastic connective tissue, because it provides both stretch and strength, is found in <u>walls of elastic arteries, in the vocal cords, and in some ligaments.</u>

T F 17. The surface attachment between epithelium and connective tissue is called <u>basement membrane.</u>

T F 18. <u>Simple</u> squamous epithelium is most likely to line the areas of the body that are subject to wear and tear.

T F 19. Marfan syndrome is a <u>muscle tissue</u> disorder that results in persons being <u>short with short extremities.</u>

T F 20. Appositional growth of cartilage <u>starts later in life than interstitial growth and continues throughout life.</u>

Questions 21–25: Fill-ins. Complete each sentence with the word that best fits.

_____ 21. A _____ is a sample of living tissue for microscopic examination, for example, for diagnosis.

_____ 22. _____ is a dense, irregular connective tissue covering over bone.

_____ 23. The kind of tissue that lines alveoli (air sacs) of lungs is _____.

_____ 24. Another term for intercellular is _____.

_____ 25. All glands are formed of the tissue type named _____.

ANSWERS TO MASTERY TEST: ☆ CHAPTER 4

Multiple Choice
1. C   4. A
2. D   5. C
3. C   6. C

Matching
7. Cartilage
8. Dense connective tissue
9. Stratified squamous epithelium
10. Simple squamous epithelium
11. Simple columnar epithelium
12. Adipose

True–False
13. T
14. T
15. F. Endocrine
16. T
17. T
18. F. Stratified
19. F. Connective tissue; tall with long extremities
20. T

Fill-ins
21. Biopsy
22. Periosteum
23. Simple squamous epithelium
24. Extracellular or interstitial
25. Epithelium

FRAMEWORK 5

Integument

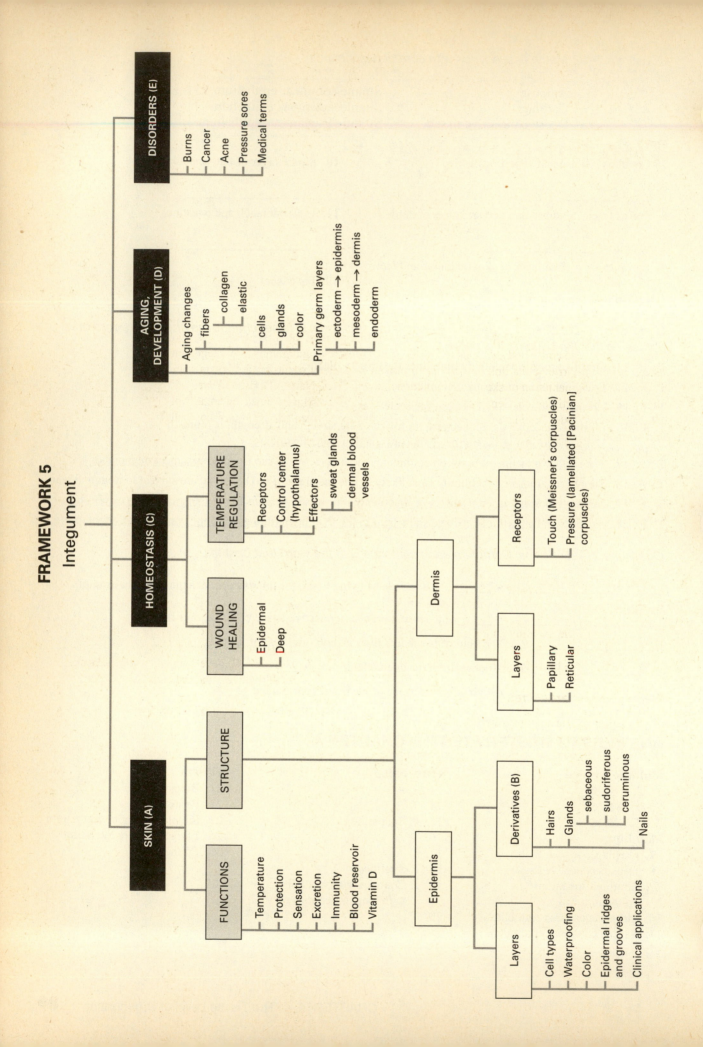

SKIN (A)

- **FUNCTIONS**
  - Temperature
  - Protection
  - Sensation
  - Excretion
  - Immunity
  - Blood reservoir
  - Vitamin D

- **STRUCTURE**
  - Epidermis
    - Layers
      - Cell types
      - Waterproofing
      - Color
      - Epidermal ridges and grooves
      - Clinical applications
    - Derivatives (B)
      - Hairs
      - Glands
        - sebaceous
        - sudoriferous
        - ceruminous
      - Nails
  - Dermis
    - Layers
      - Papillary
      - Reticular
    - Receptors
      - Touch (Meissner's corpuscles)
      - Pressure (lamellated [Pacinian] corpuscles)

HOMEOSTASIS (C)

- **WOUND HEALING**
  - Epidermal
  - Deep

- **TEMPERATURE REGULATION**
  - Receptors
  - Control center (hypothalamus)
  - Effectors
    - sweat glands
    - dermal blood vessels

AGING, DEVELOPMENT (D)

- Aging changes
  - fibers
    - collagen
    - elastic
  - cells
  - glands
  - color
- Primary germ layers
  - ectoderm → epidermis
  - mesoderm → dermis
  - endoderm

DISORDERS (E)

- Burns
- Cancer
- Acne
- Pressure sores
- Medical terms

The Integumentary System

In your human body tour you have now arrived at your first system, the integumentary system, which envelops the entire body. Composed of the skin, hair, nails, and glands, the integument serves as both a barrier and a link with the environment. The many waterproofed layers of cells, as well as the pigmentation of skin, afford protection to the body. Nails, glands, and even hairs offer additional fortification. Sense receptors and blood vessels in skin increase awareness of conditions around the body and facilitate appropriate responses. Even so, this microscopically thin surface cover may be traumatized, or invaded, or it may simply succumb to the passage of time. The integument then demands repair and healing for continued maintenance of homeostasis.

   As you begin your study of the integumentary system, carefully examine the Chapter 5 Framework and note relationships among concepts and key terms there. Also refer to the Topic Outline and Objectives; you may want to check off each objective as you complete it.

TOPIC OUTLINE AND OBJECTIVES

A. Skin

☐ 1. Describe the anatomy and seven functions of the skin.
☐ 2. Explain the basis for skin color.

B. Epidermal derivatives

☐ 3. Compare the anatomy, distribution, and physiology of hair, sebaceous (oil) glands, sudoriferous (sweat) glands, and ceruminous glands.

C. The skin and homeostasis

☐ 4. Outline the steps involved in epidermal wound healing and deep wound healing.
☐ 5. Explain how the skin helps maintain normal body temperature.

D. Effects of aging, developmental anatomy

☐ 6. Describe the effects of aging on the integumentary system.
☐ 7. Describe the development of the epidermis, its derivatives, and the dermis.

E. Disorders; medical terminology

☐ 8. Describe the causes and effects for the following skin disorders: burns, skin cancer, acne, and pressure sores.
☐ 9. Define medical terminology associated with the integumentary system.

WORDBYTES

Now become familiar with the language of this chapter by studying each wordbyte, its meaning, and an example of its use within a term. After you study the entire list, self-check your understanding by writing the meaning of each wordbyte on the line. As you continue through the *Learning Guide,* identify (and fill in) additional terms that contain the same wordbyte.

Wordbyte	Self-check	Meaning	Example(s)
cera-	_____	wax	*cer*umen
cut-	_____	skin	*cut*aneous
derm-	_____	skin	*derm*atologist
epi-	_____	over	*epi*dermis
fer-	_____	carry	sudori*fer*ous
melan-	_____	black	*melan*oma
seb-	_____	fat	*seb*aceous
sub-	_____	under	*sub*cutaneous
sudor-	_____	sweat	*sudor*iferous

CHECKPOINTS

A. Skin (pages 124–129)

■ **A1.** Name the structures included in the integumentary system.

A2. *A clinical challenge.* Skin may provide clues about the health of the body; in this case at least, you *can* tell a lot about a book by its cover. Write three examples of how a physical or emotional health state may be detected by inspection of skin.

■ **A3.** Answer these questions about the two portions of skin. Label them (bracketed regions) on Figure LG 5.1.

a. The outer layer is named the _____. It is composed of (*connective tissue? epithelium?*).

b. The inner portion of skin, called the _____, is made of (*connective tissue? epithelium?*). The dermis is (*thicker? thinner?*) than the epidermis.

■ **A4.** The tissue underlying skin is called *subcutaneous,* meaning _____.

This layer is also called _____. It consists of two types of tissue,

_____ and _____. What functions does subcutaneous tissue serve?

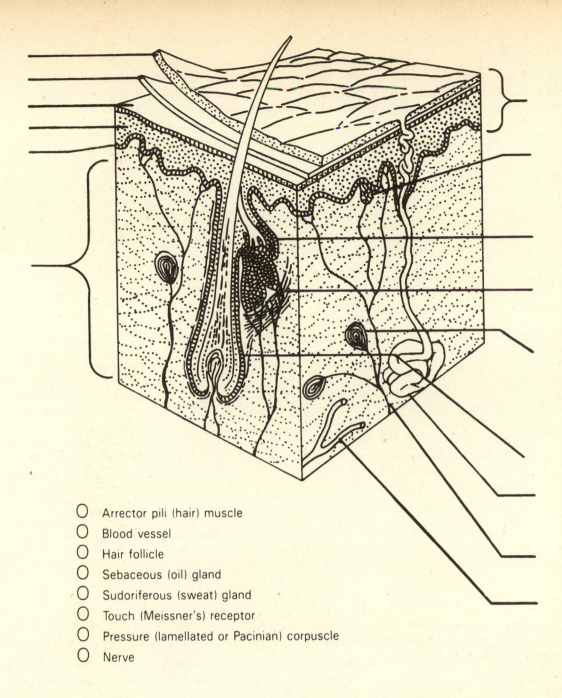

O   Arrector pili (hair) muscle
O   Blood vessel
O   Hair follicle
O   Sebaceous (oil) gland
O   Sudoriferous (sweat) gland
O   Touch (Meissner's) receptor
O   Pressure (lamellated or Pacinian) corpuscle
O   Nerve

Figure LG 5.1 Structure of the skin. Labels as directed in Checkpoints A3, A7, and A10.

A5. Skin may be one of the most underestimated organs in the body. What functions does your skin perform while it is "just lying there" covering your body? List seven functions on the lines provided.

_____   _____

_____   _____

_____   _____

_____

■ **A6.** Epidermis contains four distinct cell types. Fill in the name of the cell type that fits each description.

a. Most numerous cell type, this cell produces keratin which helps to waterproof skin:

_____.

b. This type of cell produces the pigments which give skin its color:

_____.

c. These cells, derived from macrophages, function in immunity: _____.

d. Located in the deepest layer of the epidermis, these cells may contribute to the sensation

of touch: _____.

■ **A7.** Label each of the five layers on the left side of Figure LG 5.1.

■ **A8.** Match the names of the epidermal layers (strata) with the descriptions below.

> B. Basale   L. Lucidum
> C. Corneum   S. Spinosum
> G. Granulosum

_____ a. Also known as the stratum germinativum,   _____ d. A clear layer that is normally found
      because new cells arise here   only in thick layers of palms and soles

_____ b. Stratum immediately superficial   _____ e. Uppermost layer, consisting of
      to the stratum basale; cells look prickly   25 to 30 rows of flat, dead cells filled
      under microscopic examination   with keratin

_____ c. Development of keratohyalin,
      precursor of keratin, occurs in these cells

A9. Briefly describe the relationship of epidermal growth factor (EGF) to skin cancer.

■ **A10.** Most sensory receptors, nerves, blood vessels, and glands are embedded in the *(epidermis? dermis?)*. Fill in all label lines on the right side of Figure LG 5.1. Then color those structures and their related color code ovals.

■ **A11.** Describe the dermis in this exercise.

a. The outer one-fifth of the dermis is known as the *(papillary? reticular?)* region. It consists of *(loose? dense?)* connective tissue. Present in fingerlike projections known as dermal

_____ are Meissner's corpuscles, sense receptors sensitive to *(pressure? touch?)*.

b. The remainder of the dermis is known as the _____ region, composed of *(loose? dense?)* connective tissue. Skin is strengthened by *(elastic? collagenous?)* fibers in the reticular layer. Skin is

extensible and elastic due to _____ fibers.

c. Immediately deep to the reticular region is the _____ layer, known as the *(deep? superficial?)* fascia. Receptors located in this layer are sensitive to *(touch? pressure?)*.

■ **A12.** Circle areas of the body in which skin is very thick:

Eyelids   Palms   Soles   Penis   Scrotum

Dorsal surface of the body   Ventral surface of the body

■ **A13.** Explain what accounts for skin color by doing this exercise.
  a. Dark skin is due primarily to *(a larger number of melanocytes? greater melanin production per melanocyte?).* Melanocytes are in greatest abundance in the stratum _____.

  b. Melanin is derived from the amino acid _____. The enzyme that converts tyrosine to melanin is _____. This enzyme is activated by _____ light. In the inherited condition known as _____, melanocytes cannot synthesize tyrosinase, so skin, hair, and eyes lack melanin.

  c. What are freckles and liver spots?

  Do they tend to become cancerous? _____
  d. The yellowish color of skin of persons of Asian origin is due to the combination of the pigments _____ and _____.
  e. What accounts for the pinker color of skin during blushing and acts as a cooling mechanism during exercise?

  This condition is known as *(cyanosis? erythema?).*

  f. Jaundiced skin has a more *(blue? red? yellow?)* hue, often due to _____ problems. Cyanotic skin appears more *(blue? red? yellow?)* due to lack of oxygen and excessive carbon dioxide in the blood vessels of the skin.

■ **A14.** *A clinical challenge.* Identify types of skin grafts that might be used for Tim, a severely burned patient. Choose from these answers; not all will be used.

Autograft	Homograft
Autologous skin transplantation	Isograft
Heterograft	

  a. A graft taken from an unburned region of Tim's skin _____

  b. A section of skin donated by Tim's identical twin, Tom _____

  c. A portion of Tim's own epidermis grown in the laboratory for several weeks and then transplanted back to the burn site _____

  d. A section of skin donated by Tim's friend Nick while Tim awaits the graft described in (c); this donated skin will protect the site from fluid loss _____

B. Epidermal derivatives (pages 129–132)

■ **B1.** What is the main function of hair? _____

■ **B2.** Complete this exercise about the structure of a hair and its follicle.

    a. Arrange the parts of a hair from superficial to deep: _____ _____ _____
       A. Shaft   B. Bulb   C. Root

    b. Arrange the layers of a hair from outermost to innermost: _____ _____ _____
       A. Medulla   B. Cuticle   C. Cortex

    c. A hair is composed of *(cells? no cells, but only secretions of cells?)*. What accounts for the fact that hairs are waterproofed?

    d. Surrounding the root of a hair is the hair _____. The follicle consists of two parts. The *(external? internal?)* root sheath is an extension of deeper layers of the *(epidermis? dermis?)*.

      The internal root sheath consists of cells derived from the _____.

    e. The _____ is the part of a hair follicle where cells undergo mitosis permitting growth of a new hair. What is the function of the papilla of the hair?

B3. Contrast the *growth stage* and the *resting stage* in the growth cycle of a hair.

B4. Describe the different chemicals in hairs that account for these colors:

    a. Brown or black

    b. Blonde or red

    c. Graying

    d. White

B5. Describe the relationship between the following pairs of terms related to hairs:

a. *Arrector pili muscle/"goose bump"*

b. Roles of androgens in *hirsutism/male-pattern baldness*

■ **B6.** Answer these questions about sebaceous glands.

a. In which parts of the body are these glands largest?

b. In which body parts are sebaceous glands absent?

c. What functions are attributed to these glands?

d. What is a "blackhead"? Is the blackness of a blackhead due to accumulated dirt?

■ **B7.** Contrast types of sudoriferous (sweat) glands in this activity. Use answers in the box.

| Apo. Apocrine   Ecc. Eccrine   Mam. Mammary |

_____ a. Specialized sudoriferous glands found in breasts; secrete milk

_____ b. Of apocrine and eccrine, the more common type of gland

_____ c. Responsible for "sweaty palms" because these glands are found in large numbers in palms

_____ d. Associated with foot odor because these glands are found in large numbers on soles

_____ e. Produce most of the sweat in armpits (axillae) and in pubic areas

_____ f. Of apocrine and eccrine glands, secrete more viscous (less watery) fluid

■ **B8.** Write one major function and one minor function of sweat (perspiration).

B9. Where is *cerumen* produced and what is its function?

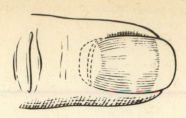

Figure LG 5.2 Structure of nails. Label according to Checkpoint B10.

B10. Refer to Figure LG 5.2. Now look at one of your own fingernails. Label the following parts of the nail on the figure: *free edge, nail body, nail root, lunula,* and *eponychium (cuticle).* Check your accuracy by consulting Figure 5.4, page 133 in your text.

B11. Answer these questions about nails and hair.
   a. Are nails formed of cells? *(Yes? No?)* Do nails contain keratin? *(Yes? No?)* What type of tissue forms nails? *(Dermis? Epidermis?)*
   b. How is the function of the *nail matrix* similar to that of the *matrix of a hair?*

   c. Why does the nail body appear pink, yet the lunula and free edge appear white?

C. The skin and homeostasis (pages 132–135)

C1. Describe the process of epidermal wound healing in this exercise.
   a. State two examples of epidermal wounds.

   b. Usually the deepest part of the wound is the *(central? peripheral?)* region.
   c. In the process of repair, epidermal cells of the stratum *(corneum? basale?)* break contact from the basement membrane. These are cells at the *(center? periphery?)* of the wound.
   d. These basal cells migrate toward the center of the wound, stopping when they meet other similar advancing cells. This cessation of migration is an example of the phenomenon known as

   _____. Cancer cells *(do? do not?)* exhibit this characteristic.

   e. Both the migrated cells and the remaining epithelial cells at the periphery undergo _____ to fill in the epithelium up to a normal (or close to normal) level.

C2. The process of deep wound healing involves four phases. List the three or four major events that occur in each phase.

a. Inflammatory

b. Migratory

c. Proliferative

d. Maturation

■ **C3.** *For extra review* of deep wound healing, match the phases in the box with descriptions below.

I. Inflammation	Mig. Migration
Mat. Maturation	P. Proliferation

_____ a. Blood clot temporarily unites edges of wound; blood vessels dilate so neutrophils enter to clean up area.

_____ b. Clot forms a scab; epithelial cells migrate into scab; fibroblasts also migrate to start scar tissue; pink granulation tissue contains delicate new blood vessels.

_____ c. Epithelium and blood vessels grow; fibroblasts lay down many fibers.

_____ d. Scab sloughs off; epidermis grows to normal thickness; collagenous fibers give added strength to healing tissue; blood vessels are more normal.

C4. Contrast these two terms: *hypertrophied scar/keloid.*

C5. Humans are *warm-blooded animals.* What is the meaning of that term?

■ **C6.** Complete this exercise about temperature regulation.

a. Human body temperature is maintained at about *(37°C? 38°C? 39°C?).*

b. The temperature control center of the brain is the *(cerebrum? medulla? hypothalamus?).*

c. When the body temperature is too hot, as during vigorous exercise, nerve messages from the brain inform sweat glands to *(in? de?)*-crease sweat production. In addition, blood vessels in the dermis are directed to *(dilate? constrict?).*

d. Temperature regulation is an example of a *(negative? positive?)* feedback system because the response (cooling) is *(the same as? opposite to?)* the stimulus (heating).

D. Effects of aging; developmental anatomy (pages 135–136)

■ **D1.** Complete the table relating observable changes in aging of the integument to their causes.

Changes	Causes
a. Wrinkles; skin springs back less when gently pinched	
b.	Macrophages become less efficient; decrease in number of Langerhans cells
c.	Loss of subcutaneous fat
d. Dry, easily broken skin	
e.	Decrease in number and size of melanocytes

■ **D2.** List and briefly describe two or more factors that can help keep your skin healthy throughout your lifetime.

■ **D3.** Do this activity on the effects of light rays and chemicals upon skin.

a. Sunburn is caused by overexposure to _____ light rays.

UV rays cause damage called _____-damage or _____-aging to skin.
Write two examples of health practices that may limit such damage to skin.

b. Tretinoin (retin-A) has been used as a treatment for conditions such as

_____, _____, and _____.
Does it reverse aging changes and decrease risk of skin cancer? *(Yes? No?)*

■ **D4.** *The Big Picture: Looking Ahead.* Do this exercise by referring to specific figures and exhibits in Chapter 29 in your text.
a. Identify and list the three primary germ layers (Figure 29.5b, page 964) developed from a fertilized egg:

_____-derm, _____-derm, and _____-derm.

b. The epidermis of skin develops from the _____-derm
(Exhibit 29.1, page 965). All epidermal layers are formed by the *(second? fourth?)* month of the
nine-month human gestation period (Exhibit 29.2, page 969).
c. Nails form from the *(dermis? epidermis?)* and reach the ends of fingertips by the end of the
(fourth? ninth?) month of gestation (Exhibit 29.2, page 969).

d. Delicate fetal hair called _____ covers the body during the middle trimester of fetal development. It is normally *(present at? shed before?)* birth (Exhibit 29.2, page 969), so its presence on a newborn is an indicator of prematurity.

e. The dermis and its associated blood vessels derive from the _____-derm (Exhibit 29.1, page 965).

f. Subcutaneous fat deposited during months *(6 and 7? 8 and 9?)* decreases the wrinkled appearance of the skin (Exhibit 29.2, page 969). This timing helps explain the small and wrinkled appearance of premature infants.

E. Disorders; medical terminology (pages 137-139)

E1. Contrast systemic effects with local effects of burns.

■ **E2.** Identify characteristics of the three different classes of burns by completing this exercise.
a. In a first-degree burn, only the superficial layers of the *(dermis? epidermis?)* are involved.

The tissue appears _____ in color. Blisters *(do? do not?)* form. Give one example of a first-degree burn.

b. Which parts of the skin are injured in a second-degree burn?

Blisters usually *(do? do not?)* form. Epidermal derivatives, such as hair follicles and glands, *(are? are not?)* injured. Healing usually occurs in about three to four *(days? weeks?)*.

c. Third-degree burns are called *(partial? full?)*-thickness burns. Such skin appears *(red and blistered? white, brown, or black and dry?)*. Such burned areas are usually *(painful? numb?)* because nerve endings are destroyed. Regeneration is usually *(rapid? slow?)*.

■ **E3.** *A clinical challenge.* Answer these questions about the Lund–Browder method.
a. What does the Lund–Browder method estimate? *(Depth of burn? Amount of body surface area burned?)* This method is based upon differences in body *(size? proportions?)* of age groups.
b. If an adult and a one-year-old each experience burns over the entire anterior surface of both legs, who is more burned? *(Adult? One-year-old?)* If both are burned over the entire anterior surface of the head, who is more burned? *(Adult? One-year-old?)*

■ **E4.** Circle the risk factor for cancer in each pair.
a. Skin type:
   A. Fair with red hair
   B. Tans well with dark hair
b. Sun exposure:
   A. Lives in Canada at sea level
   B. Lives in high-altitude mountainous area of South America
c. Age:
   A. 30 years old
   B. 75 years old
d. Immunological status:
   A. Taking immunosuppressive drugs for cancer or following organ transplant
   B. Normal immune system

■ **E5.** Match the name of the disorder with the description given.

A. Acne	M. Malignant melanoma
B. Basal cell carcinoma	N. Nevus
D. Decubitus ulcers	P. Pruritus
I. Impetigo	

_____ a. Pressure sores

_____ b. Mole

_____ c. Staphylococcal or streptococcal infection
which may become epidemic in nurseries

_____ d. Inflammation of sebaceous glands
especially in chin area; occurs under
hormonal influence

_____ e. Rapidly metastasizing form of cancer

_____ f. The most common form of skin cancer

_____ g. Itching

E6. Contrast the following pairs of terms.
a. *Topical/intradermal*

b. *Corn/wart*

A1. Skin, its derivatives (hair, nails, and glands), and nerve endings.

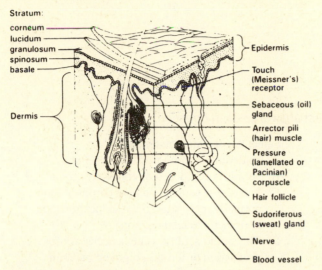

Stratum:
corneum
lucidum
granulosum
spinosum
basale

Epidermis

Dermis

Touch (Meissner's) receptor

Sebaceous (oil) gland

Arrector pili (hair) muscle

Pressure (lamellated or Pacinian) corpuscle

Hair follicle

Sudoriferous (sweat) gland

Nerve

Blood vessel

Figure LG 5.1A Structure of the skin.

A3. See Figure LG 5.1 above. (a) Epidermis; epithelium. (b) Dermis, connective tissue; thicker.

A4. Under the skin; superficial fascia or hypodermis; areolar, adipose; anchors skin to underlying tissues and organs.

A6. (a) Keratinocyte. (b) Melanocyte. (c) Langerhans cell. (d) Merkel cell.

A7. See Figure LG 5.1A above.

A8. (a) B. (b) S. (c) G. (d) L. (e) C.

A10. Dermis. See Figure LG 5.1A.

A11. (a) Papillary; loose; papillae, touch. (b) Reticular, dense; collagenous; elastic. (c) Subcutaneous, superficial; pressure.

A12. Palms, soles, dorsal surface of the body.

A13. (a) Greater melanin production per melanocyte; basale. (b) Tyrosine; tyrosinase; ultraviolet (UV); albinism. (c) Clusters of melanocytes; no. (d) Carotene, melanin. (e) Widening (dilation) of blood vessels; erythema. (f) Yellow, liver; blue.

A14. (a) Autograft. (b) Isograft. (c) Autologous skin transplantation. (d) Homograft.

B1. Protection.

B2. (a) A C B. (b) B C A. (c) Cells; keratin (especially in cuticle). (d) Follicle; external, epidermis; matrix. (e) Matrix (of bulb of hair follicle); contains blood vessels that nourish the hair.

B6. (a) In skin of the breast, face, neck, and upper chest. (b) Palms and soles. (c) Keep hair and skin from drying out, lower risk of dehydration and infection of skin. (d) Enlarged sebaceous gland with accumulated sebum; no, it is due to the presence of melanin and oxidation of the oil within sebum.

B7. (a) Mam. (b–d) Ecc. (e–f) Apo.

B8. Major: regulation of body temperature; minor: eliminate wastes such as urea, uric acid, ammonia, lactic acid, and excess salts.

B11. (a) Yes; yes; epidermis. (b) They bring about growth of nails and hairs, respectively. (c) The pink color is related to visibility of blood vessels deep to the nail body; the free edge has no tissue (so no blood vessels) deep to it, whereas the thickened stratum basale obscures blood vessels deep to the lunula.

C1. (a) Skinned knee, first- or second-degree burn. (b) Central. (c) Basale; periphery. (d) Contact inhibition; do not. (e) Mitosis.

C3. (a) I. (b) Mig. (c) P. (d) Mat.

C6. (a) 37°C. (b) Hypothalamus. (c) In; dilate. (d) Negative, opposite to.

D1.

Changes	Causes
a. Wrinkles; skin springs back less when gently pinched	**Elastic fibers thicken into clumps and fray**
b. **Increased susceptibility to skin infections and skin breakdown**	Macrophages become less efficient; decrease in number of Langerhans cells
c. **Loss of body heat; increase likelihood of skin breakdown, as in decubitus ulcers**	Loss of subcutaneous fat
d. Dry, easily broken skin	**Decreased secretion of sebum by sebaceous gland; decreased sweat production**
e. **Gray or white hair; atypical skin pigmentation**	Decrease in number and size of melanocytes

D2. Good nutrition, decreased stress, balance of rest and exercise, not smoking, protection from the sun.

D3. (a) Ultraviolet; photo, photo; using a sunblock with SPF of at least 15 and avoiding exposure to direct sunlight midday, such as by wearing a hat or staying out of sun. (b) Acne, wrinkles, liver spots; no.

D4. (a) Endo, meso, ecto. (b) Ecto; fourth (c) Epidermis; ninth. (d) Lanugo; shed before. (e) Meso. (f) 8 and 9.

E2. (a) Epidermis; redder; do not; typical sunburn. (b) All of epidermis and possibly upper regions of dermis; do; are not; weeks. (c) Full; white, brown, or black and dry; numb; slow.

E3. (a) Amount of body surface burned; proportions. (b) Adult; one-year-old.

E4. (a) A. (b) B. (c) B. (d) A.

E5. (a) D. (b) N. (c) I. (d) A. (e) M. (f) B. (g) P.

WRITING ACROSS THE CURRICULUM: CHAPTER 5

1. Write a paragraph suggesting health tips that may prevent vitamin D deficiency.
2. Explain what causes different hair colors such as red or blond(e), black or brown, and gray or white.
3. Describe five normal aging changes of skin, including a scientific rationale for each.

MASTERY TEST: CHAPTER 5

Questions 1–5: Circle the letter preceding the one best answer to each question.

1. "Goose bumps" occur as a result of:
    A. Contraction of arrector pili muscles
    B. Secretion of sebum
    C. Contraction of elastic fibers in the bulb of the hair follicle
    D. Contraction of papillae

2. Select the one *false* statement about the stratum basale.
    A. It is the one layer of cells that can undergo cell division.
    B. It consists of a single layer of squamous epithelial cells.
    C. It is the stratum germinativum.
    D. It is the deepest layer of the epidermis.

3. Select the one *false* statement.
    A. Epidermis is composed of epithelium.
    B. Dermis is composed of connective tissue.
    C. Pressure-sensitive Pacinian corpuscles are normally more superficial in location than Meissner's touch receptors.
    D. The amino acid tyrosine is necessary for production of the skin pigment melanin.

4. At about what age do epidermal ridges which cause fingerprints develop?
    A. Months 3–4 in fetal development
    B. Age 3–4 months
    C. Age 3–4 years
    D. Age 30–40 years

5. Which is derived from mesoderm?
    A. Epidermis
    B. Dermis

Questions 6–10: Arrange the answers in correct sequence.

_____ _____ _____
6. From most serious to least serious type of burn:
    A. First-degree
    B. Second-degree
    C. Third-degree

_____ _____ _____
7. Deep wound healing involves four phases. List in order the phases following the inflammatory phase.
    A. Migration
    B. Maturation
    C. Proliferation

_____ _____ _____
8. From outside of hair to inside of hair:
    A. Medulla
    B. Cortex
    C. Cuticle

_____ _____ _____
9. From most superficial to deepest:
    A. Dermis
    B. Epidermis
    C. Superficial fascia

_____ _____ _____
10. From most superficial to deepest:
    A. Stratum lucidum
    B. Stratum corneum
    C. Stratum germinativum

Questions 11–20: Circle T (true) or F (false). If the statement is false, change the underlined word or phrase so that the statement is correct.

T F 11. Hairs are <u>noncellular structures composed entirely of nonliving substances secreted by follicle cells.</u>

T F 12. The color of skin is due primarily to a pigment named <u>keratin.</u>

T F 13. The <u>outermost layers of epidermis</u> are composed of dead cells.

T F 14. Eccrine sweat glands are <u>more</u> numerous than apocrine sweat glands and are especially dense on <u>palms and soles.</u>

T F 15. The dermis consists of two regions; <u>the papillary region is more superficial, and the reticular region is deeper.</u>

T F 16. Temperature regulation is a <u>positive</u> feedback system.

T F 17. The <u>internal root sheath</u> is a downward continuation of the epidermis.

T F 18. <u>Both epidermis and dermis contain</u> blood vessels (<u>are vascular</u>).

T F 19. Hair, glands, and nails are all derived from the <u>dermis.</u>

T F 20. Pacinian corpuscles are <u>pressure</u>-sensitive nerve endings most abundant in the <u>subcutaneous tissue,</u> rather than in the <u>epidermis.</u>

Questions 21–25: Fill-ins. Complete each sentence with the word or phrase that best fits.

_____ 21. Skin contains a chemical which, under the influence of ultraviolet radiation, leads to formation of vitamin _____.

_____ 22. The cells that are sloughed off as skin cells age and undergo keratinization are those of the stratum _____.

_____ 23. Fingerprints are the result of a series of grooves called _____.

_____ 24. The oily glandular secretion that keeps skin and hairs from drying is called _____.

_____ 25. When the temperature of the body increases, nerve messages from brain to skin will decrease body temperature by _____.

ANSWERS TO MASTERY TEST: ✮ CHAPTER 5

Multiple Choice

1. A   4. A
2. B   5. B
3. C

Arrange

6. C B A   9. B A C
7. A C B   10. B A C
8. C B A

True–False

11. F. Composed of different kinds of cells
12. F. Melanin
13. T
14. T
15. T
16. F. Negative
17. F. External root sheath
18. F. Only the dermis contains, is vascular
19. F. Epidermis
20. T

Fill-ins

21. D
22. Corneum
23. Epidermal ridges or grooves
24. Sebum
25. Stimulating sweat glands to secrete and blood vessels to dilate

Principles of Support and Movement

FRAMEWORK 6
Skeletal Tissue

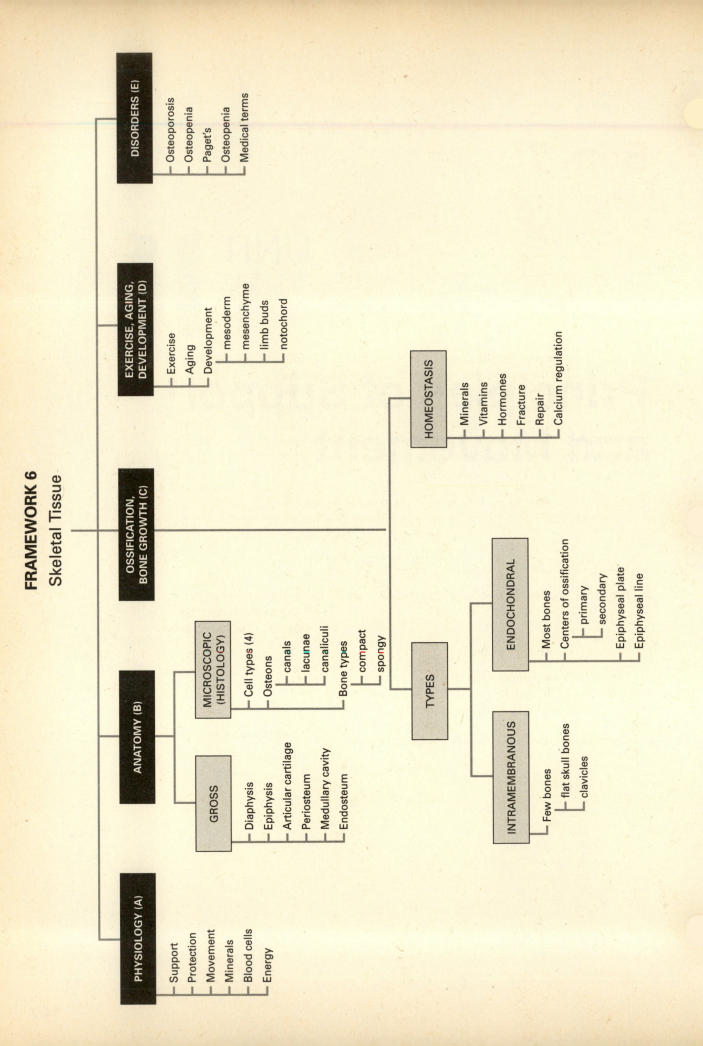

PHYSIOLOGY (A)
- Support
- Protection
- Movement
- Minerals
- Blood cells
- Energy

ANATOMY (B)

GROSS
- Diaphysis
- Epiphysis
- Articular cartilage
- Periosteum
- Medullary cavity
- Endosteum

MICROSCOPIC (HISTOLOGY)
- Cell types (4)
- Osteons
  - canals
  - lacunae
  - canaliculi
- Bone types
  - compact
  - spongy

TYPES

INTRAMEMBRANOUS
- Few bones
  - flat skull bones
  - clavicles

ENDOCHONDRAL
- Most bones
- Centers of ossification
  - primary
  - secondary
- Epiphyseal plate
- Epiphyseal line

OSSIFICATION, BONE GROWTH (C)

HOMEOSTASIS
- Minerals
- Vitamins
- Hormones
- Fracture
- Repair
- Calcium regulation

EXERCISE, AGING, DEVELOPMENT (D)
- Exercise
- Aging
- Development
  - mesoderm
  - mesenchyme
  - limb buds
  - notochord

DISORDERS (E)
- Osteoporosis
- Osteopenia
- Paget's
- Osteopenia
- Medical terms

Bone Tissue

CHAPTER

6

The skeletal system provides the framework for the entire body, affording strength, support, and firm anchorage for the muscles that move the body. In this initial chapter in Unit II you will explore tissue that forms the skeleton. Bones provide rigid levers covered with connective tissue designed for joint formation and muscle attachment. Microscopically, bones present a variety of cell types intricately arranged in an osseous sea of calcified intercellular material. Bone growth or ossification occurs throughout life. Essential nutrients, hormones, and exercise regulate the growth and maintenance of the skeleton. Fractures and other disorders may result from a lack of such normal regulation.

As you begin your study of bone tissue, carefully examine the Chapter 6 Framework and note relationships among concepts and key terms there. Also refer to the Topic Outline and Objectives; you may want to check off each objective as you complete it.

TOPIC OUTLINE AND OBJECTIVES

A. Physiology: functions of skeletal tissue

☐ 1. Discuss the functions of bone.

B. Anatomy, histology of bone

☐ 2. Identify the parts of a long bone.
☐ 3. Describe the histological features of compact and spongy bone tissue.

C. Ossification, bone growth, homeostasis

☐ 4. Contrast the steps involved in intramembranous and endochondral ossification.
☐ 5. Describe the processes involved in bone remodeling.
☐ 6. Define a fracture, describe several common kinds of fractures, and describe the sequence of events involved in fracture repair.
☐ 7. Describe the role of bone in calcium homeostasis.

D. Effects of exercise and aging on the skeletal system; developmental anatomy of the skeletal system

☐ 8. Explain the effects of exercise and aging on the skeletal system.
☐ 9. Describe the development of the skeletal system.

E. Disorders, medical terminology

☐ 10. Contrast the causes and clinical symptoms associated with osteoporosis and Paget's disease.
☐ 11. Define medical terminology associated with bone tissue.

WORDBYTES

Now become familiar with the language of this chapter by studying each wordbyte, its meaning, and an example of its use within a term. After you study the entire list, self-check your understanding by writing the meaning of each wordbyte on the line. As you continue through the *Learning Guide,* identify (and fill in) additional terms that contain the same wordbyte.

Wordbyte	Self-check	Meaning	Example(s)
-blast	_____	germ, to form	osteo*blast*
chondr-	_____	cartilage	peri*chondr*ium
-clast	_____	to break	osteo*clast*
os-, osteo-	_____	bone	*os*sification, *osteo*cyte
peri-	_____	around	*peri*osteum

CHECKPOINTS

A. Functions of skeletal tissue (page 143)

■ **A1.** List six functions of the skeletal system.

_____   _____

_____   _____

_____   _____

■ **A2.** What other systems of the body depend on a healthy skeletal system? Explain why in each case.

A3. Define *osteology.*

B. Anatomy, histology of bone (pages 143–147)

■ **B1.** The skeletal system consists of four types of connective tissue. Name these.

■ **B2.** On Figure LG 6.1, label the *diaphysis, epiphysis, medullary (marrow) cavity,* and *nutrient foramen.* Then color the parts of a long bone using color code ovals on the figure.

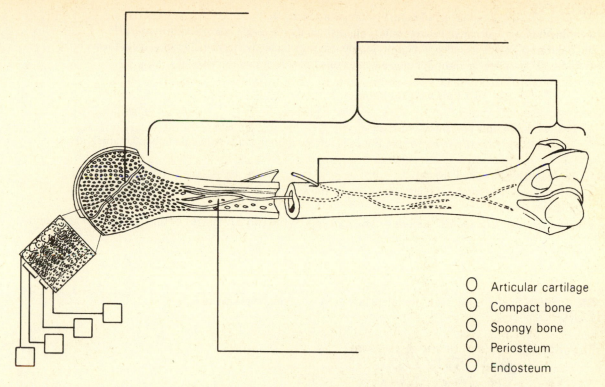

○ Articular cartilage
○ Compact bone
○ Spongy bone
○ Periosteum
○ Endosteum

Figure LG 6.1 Diagram of a long bone that has been partially sectioned lengthwise. Color and label as directed in Checkpoints B2 and C4.

■ **B3.** Match the names of parts of a long bone listed in the box with the descriptions below.

A. Articular cartilage	M. Metaphysis
E. Endosteum	O. Osteogenic periosteum
F. Fibrous periosteum	

_____ a. Thin layer of hyaline cartilage at end of long bone

_____ b. Region of mature bone where diaphysis joins epiphysis

_____ c. Outer layer of covering over bone into which ligaments and tendons attach

_____ d. Inner layer of covering over bone; osteoblasts here permit increase in diameter of bone

_____ e. Layer of bone cells lining the marrow cavity

■ **B4.** In the following exercise, bone cells are described in the sequence in which they form and later destroy bone. Write the name of the correct bone type after its description.

a. Derived from mesenchyme, these cells can undergo mitosis and differentiate into osteoblasts: _____
↓

b. Form bone initially by secreting mineral salts and fibers; not mitotic: _____
↓

c. Maintain bone tissue; not mitotic: _____
↓

d. Resorb (degrade) bone in bone repair, remodeling, and aging: _____

■ **B5.** Describe the components of bone by doing this exercise.
  a. Typical of all connective tissues, bone consists mainly of *(cells? intercellular material?)*.
  b. The intercellular substance of bone is unique among connective tissues. Protein fibers form about *(25? 50?)* % of the weight of bone, whereas mineral salts account for about

  _____ % of the weight of bone.

  c. The two main salts present in bone are _____ and _____.
  d. The *(hardness? tensile strength?)* of bones is related to inorganic chemicals in bone, namely the

  mineral salts deposited around _____ fibers. Resistance to being stretched
  or torn apart, a property of bone known as *(hardness? tensile strength?)*, is afforded by *(mineral salts? collagen fibers?)*.
  e. Bone *(is completely solid? contains some spaces?)*. Write two advantages of this structural feature of bone.

■ **B6.** Refer to Figure LG 6.2 and complete this exercise about bone structure.

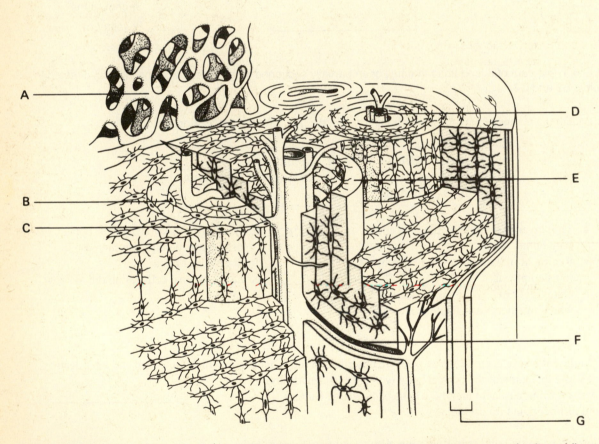

Figure LG 6.2 Osteons (Haversian systems) of compact bone. Identify lettered structures and color as directed in Checkpoint B6.

  a. Compact bone is arranged in concentric circle patterns known as

  _____. Each individual concentric layer of bone is known as a

  _____, labeled with letter _____ in the figure.

b. The osteon pattern of compact bone permits blood vessels and nerves to supply bone cells trapped in hard bone tissue. Blood vessels and nerves penetrate bone from the periosteum, labeled with letter

_____ in the figure. These structures then pass through horizontal canals, labeled _____ and known as perforating (Volkmann's) canals. These vessels and nerves finally pass into microscopic channels,

labeled_____, in the center of each osteon. Color blood vessels red and blue in the figure.

c. Mature bone cells, known as _____, are located relatively far apart in bone tissue.

These are present in "little lakes," or _____, labeled _____ in the figure. Color ten lacunae green.

d. Minute canals, known as canaliculi, are labeled with letter _____. What are the functions of these channels?

■ **B7.** Do this exercise about spongy bone.
   a. Spongy bone is arranged in *(osteons? trabeculae?)*, which may be defined as:

   _____

   b. Spongy bone makes up most of the *(diaphyses? epiphyses?)* of long bones. Spongy bone *(is not? is also?)* located within bones that are short, flat, and irregular in shape.
   c. *A clinical challenge.* Red blood cell formation (hematopoiesis) normally takes place in *(all? only certain areas of?)* spongy bone tissue. Name four or more bones in which this process takes place.

   _____   _____

   _____   _____

   Of what clinical importance is this information?

B8. Briefly describe the blood and nerve supply of bones.

C. Ossification, bone growth, homeostasis (pages 147–156)

■ **C1.** Complete this overview of bone formation.

a. Bone formation is also called _____.

b. All bones develop from preexisting (*epithelial? connective? muscle?*) tissue,

specifically from _____. These immature cells grow and develop into

osteo-_____ cells.

c. Those cells can then develop into either _____-blasts or

_____-blasts. A factor that determines which type of "blast" cells

will form is the presence or absence of _____.

d. If blood vessels are absent in the developing structure, (*chondroblasts? osteoblasts?*)
will develop, and bone will develop by (*intramembranous? endochondral?*)
ossification. Where blood vessels are present, (*chondroblasts? osteoblasts?*) will develop.

e. Name bones that form by each of these methods of ossification:

endochondral _____

intramembranous _____

f. These two methods of bone formation described above (*do? do not?*) lead to different types
of bone structurally.

■ **C2.** Refer to Figure 7.6, page 170 in your text. Locate the parietal bone, a typical "flat bone"
forming in the fetal skull. Describe intramembranous formation of this bone by using each of
the terms in the box once.

Compact	Mesenchyme	Osteoprogenitor
Fontanel	Ossification	Protection
Growth	Osteoblast	Spongy
Marrow	Osteocyte	

a. Fibrous membranes surrounding the head of the developing baby contain embryonic

_____ cells. These cells cluster in areas known as centers of

_____, and these mark areas where bone tissue will form.

b. Differentiation of these cells leads to _____ cells and then into

_____ and _____ cells that surround themselves
with organic and inorganic matrix forming spongy bone. (For help, look back at Checkpoint B4, page 107.)

c. Eventually, mesenchyme on the outside of the bone develops into a periosteal covering over the skull bone.
This lays down some compact bone over spongy bone. The resulting typical flat bone is analogous to a

flattened jelly sandwich: two firm (flat) "slices of bread" comparable to _____ bone

with a "jelly" of _____ bone filled with red _____.

d. By birth, fetal skull bones are still not completed. Some fibrous membranes still remain

at points where skull bones meet, known as _____. Of what advantage are

these "soft spots"? They permit additional _____ of skull bones as well as

safer passage (_____) of the fetal head during birth.

■ **C3.** Endochondral ossification refers to bone formation from an initial model made of *(fibrous membranes? cartilage?)*. Increase your understanding of this process by building your own outline in this exercise.

    a. First identify the four major steps in endochondral ossification by filling in the four long lines (I–IV) below with the correct terms selected from those in the box below.

Development	Growth
Diaphysis and epiphyses	Primary ossification center

      I. _____ of the cartilage model   **E**___ _____

      II. _____ of the cartilage model _____ _____ _____ _____

      III. Development of the _____   _____ _____

      IV. Development of the _____   _____

    b. Now fill in details of the process by placing the following statements of events in the correct sequence. Write the letters on the short lines to the right of the correct phase (I–IV) above. One is done for you.

      A. The cartilage model grows in length as chondrocytes divide and secrete more cartilage matrix; it grows in thickness as new chondroblasts develop within the perichondrium.

      B. Cartilage cells in the center of the diaphysis accumulate glycogen, enlarge, and burst, releasing chemicals that alter pH and trigger calcification of cartilage.

      C. Cartilage cells die because they are deprived of nutrients; in this way spaces are formed within the cartilage model.

      D. Blood vessels penetrate the perichondrium, stimulating perichondrial cells to form osteoblasts. A collar of bone forms and gradually thickens around the diaphysis. The membrane covering the developing bone is now called the periosteum.

      E. A hyaline cartilage model of future bone is laid down by differentiation of mesenchyme into chondroblasts.

      F. A perichondrium develops around the cartilage model.

      G. Secondary ossification centers develop in epiphyses, forming spongy bone there about the time of birth.Hyaline cartilage remains as the epiphyseal plate for as long as the bone grows.

      H. Capillaries grow into spaces, and osteoblasts deposit bone matrix over disintegrating calcified cartilage. In this way spongy bone is forming within the diaphysis at the primary ossification center.

      I. Osteoclasts break down the newly formed spongy bone in the very center of the bone, thereby leaving the medullary (marrow) cavity.

■ **C4.** On Figure LG 6.1, label the epiphyseal plate. This plate of growing bone consists of four zones. First identify each region by placing the letter of the name of that zone next to the correct description. Next place these same four letters in the boxes in the inset on Figure LG 6.1 to indicate the locations of these cartilage zones within the epiphyseal plate of the developing bone.

C. Calcified	P. Proliferating
H. Hypertrophic	R. Resting

_____ a. Zone of cartilage which is not involved in bone growth but anchors the epiphyseal plate (site of bone growth) to the bone of the epiphysis

_____ b. New cartilage cells form here by mitosis and are arranged like stacks of coins

_____ c. Cartilage cells mature and die here as matrix around them calcifies

_____ d. Osteoblasts and capillaries from the diaphysis invade this region to lay down bone upon the calcified cartilage remnants here

■ **C5.** Complete this summary statement about bone growth at the epiphyseal plate.

Cartilage cells multiply on the *(epiphysis? diaphysis?)* side of the epiphyseal plate, providing temporary new tissue. Cartilage cells then die and are replaced by bone cells on the *(epiphysis? diaphysis?)* side of the epiphyseal plate. If the epiphyseal plate of a growing bone is damaged, for example by fracture, the bone is likely to be *(shorter than normal? of normal length? longer than normal?)*.

■ **C6.** Complete this activity about maturation of the skeleton.
   a. The epiphyseal *(line? plate?)* is the bony region in bones of adults that marks the original cartilaginous epiphyseal *(line? plate?)*.
   b. Ossification of most bones is finished by age *(18? 25?)*. Completion of this process usually occurs earlier in *(females? males?)*.
   c. Bones must grow in diameter to match growth in length. *(Osteoblasts? Osteoclasts?)* are cells that destroy bone to enlarge the medullary cavity, whereas *(osteoblasts? osteoclasts?)* are cells from the periosteum that increase the thickness of compact bone.

■ **C7.** Fill in the names of hormones involved in bone growth in this exercise. Use the answers in the box.

hGH.	Human growth hormone	INS. Insulin
IGFs.	Insulinlike growth factor	SH. Sex hormones

_____ a. Hormone made by the pituitary. In excessive amounts, results in giantism; if deficient, leads to dwarfism.

_____ b. Produced by bone cells and also by the liver, these chemicals stimulate protein synthesis and growth of bones

_____ c. Made by the pancreas; promotes normal bone growth

_____ d. Produced by ovaries and testes, these hormones stimulate osteoblast activity and lead to growth spurt and changes in skeletal structure at puberty.

C8. Defend or dispute this statement: "Once a bone, such as your thighbone, is formed, the bone tissue is never replaced unless the bone is broken."

■ **C9.** Explain the roles of lysosomes and acids in osteoclastic activity.

■ **C10.** List factors that are necessary for normal bone growth, remodeling, and repair of fractured bone.

   a. Five minerals: _____ _____ _____ _____ _____

   b Four vitamins: _____ _____ _____ _____

   c. Weight-bearing _____

C11. Contrast the terms related to types of fractures in each pair.

a. *Simple/compound*

b. *Partial/complete*

c. *Pott's/Colles'*

d. *Comminuted/greenstick*

C12. *A clinical challenge.* What is the major difference between these two methods of setting a fracture: *closed reduction and open reduction?*

■ **C13.** Describe the four main steps in repair of a fracture by filling in terms from this list in the spaces below. One answer will be used twice.

Cell Type	Process or Stage	Time Period
Chondroblasts	Fracture hematoma	Hours
Fibroblasts	Inflammation and cell death	Months
Osteoblasts	Procallus	Weeks
Osteoclasts	Remodeling	
Phagocytes		

a. Blood flows from torn blood vessels, forming a clot known as a _____

within 6 to 8 _____ of the injury. Because of the trauma and lack of

circulation, _____ occurs in the area. As a result, cells known as

_____ and _____ enter the area and remove debris

over a period of several _____.

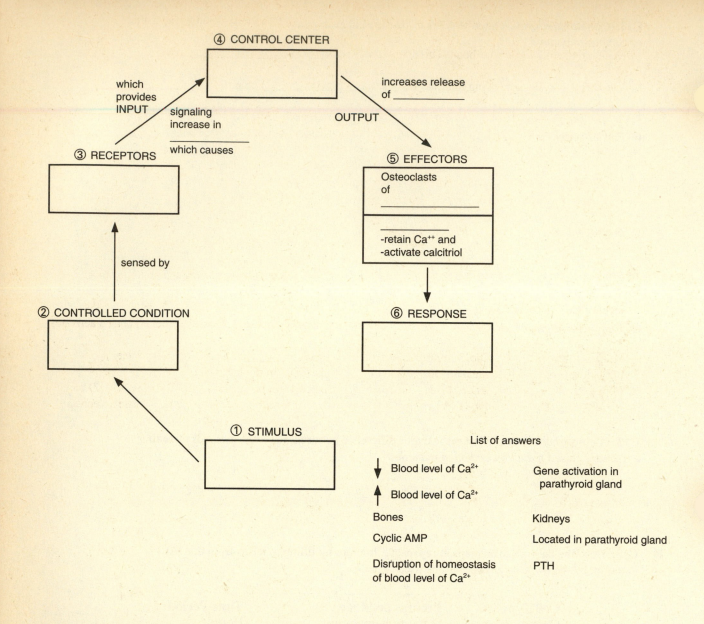

④ CONTROL CENTER

which provides INPUT

signaling increase in _____

which causes

increases release of _____

OUTPUT

③ RECEPTORS

⑤ EFFECTORS

Osteoclasts of _____

_____
-retain Ca⁺⁺ and
-activate calcitriol

sensed by

② CONTROLLED CONDITION

⑥ RESPONSE

① STIMULUS

List of answers

↓ Blood level of Ca^{2+}

↑ Blood level of Ca^{2+}

Bones

Cyclic AMP

Disruption of homeostasis of blood level of Ca^{2+}

Gene activation in parathyroid gland

Kidneys

Located in parathyroid gland

PTH

Figure LG 6.3 Negative feedback mechanism for regulation of blood calcium (Ca^{2+}) concentration. Complete according to directions in Checkpoint C14.

b. Next, granulation tissue known as a _____ forms. Then cells called

_____ produce collagen fibers which connect broken ends of the bone.

Other cells, the _____, form a soft callus at a distance from the remaining

healthy, vascular bone. This stage lasts about 3 _____.

c. The hard callus stage occurs next as _____ cells form bone adjacent to the

   remaining healthy, vascular bone. This stage lasts 3 to 4 _____.

d. The callus is reshaped during the final (or _____) phase.

■ **C14.** *The Big Picture: Looking Ahead.* Refer to Figures 6.10 (page 155) and 18.18 (page 527) in your text, and answer these questions about regulation of calcium homeostasis.
   a. Figure 1.4 (page 10) of your text introduced the concept of negative feedback systems. Using that reference and Figure 6.10 (page 155) of your text, complete Figure LG 6.3. Fill in the six boxes and lines (with * and **) using the list of answers provided on the figure.
   b. Figure LG 6.3 shows that parathyroid gland cells respond to a lowered blood level of Ca^{2+} by signaling

      increased production of cyclic _____. This INPUT then *("turns on"? "turns off"?)* the PTH cell gene in parathyroid glands. The increased blood level of PTH acts as OUTPUT to "tell" effectors to respond.

      Effectors include kidneys as well as _____; these organs *(raise? lower?)* the blood

      level of Ca^{2+} in several ways.
   c. PTH signals kidneys to *(eliminate? retain?)* calcium in blood so that less is lost in urine. PTH also activates

      kidneys to produce *(calcitriol? calcitonin?)*, which is vitamin _____. This vitamin also raises blood levels of Ca^{2+} by increasing absorption of Ca^{2+} from foods in the intestine.
   d. PTH activates osteo-*(blasts? clasts?)* of bone cells so that calcium is *(deposited in? resorbed from?)* bone.
   e. Regulation of blood calcium (Ca^{2+}) exemplifies a *(positive? negative?)* feedback cycle because a decrease

      in blood Ca^{2+} ultimately causes PTH to _____-crease blood Ca^{2+}.
   f. The effect of PTH is *(the same as? opposite to?)* that of calcitonin (Figure 18.18, page 527). Calcitonin causes calcium to be *(deposited in? resorbed from?)* bone. In other words, calcitonin, which is released

      by the *(thyroid? parathyroid?)* gland, causes the blood level of Ca^{2+} to _____-crease.
   g. Chapter 27 (page 897) points out that *(hypo? hyper?)*-parathyroidism refers to a deficiency of PTH. This condition is likely to lead to *(high? low?)* blood calcium levels, known as *(hypo? hyper?)*-calcemia. Symptoms of this disorder include increased *(hardness? fragility?)* of bones because bones have *(lost? gained?)* calcium and of *(hyper? hypo?)*-active muscles because low blood levels of calcium increase activity of nerves.
   h. List two or more other functions of the human body affected by the blood concentration of calcium.

D. Effects of exercise and aging on the skeletal system; developmental anatomy of the skeletal system (pages 156–158)

■ **D1.** Circle the correct answers about the effects of exercise upon bones.
   a. Exercise *(strengthens? weakens?)* bones. A fractured bone that is not exercised is likely to become *(stronger? weaker?)* during the period that it is immobilized.
   b. The pull of muscles upon bones, as well as the tension on bones as they support body weight during exercise, causes *(increased? decreased?)* production of the protein collagen.
   c. The stress of exercise also stimulates *(osteoblasts? osteoclasts?)* and increases production of the hormone calcitonin, which inhibits *(osteoblasts? osteoclasts?)*.

■ **D2.** Complete this exercise about skeletal changes that occur in the normal aging process.

a. The amount of calcium in bones *(de? in?)*-creases with age. As a result, bones of the elderly are likely to be *(stronger? weaker?)* than bones of younger persons. This change occurs at a younger age in *(men? women?)*.

b. Another component of bones that decreases with age is _____.
   What is the significance of this change?

c. Write two or more health practices that can minimize skeletal changes associated with aging.

■ **D3.** Complete this learning activity about development of the skeleton.

a. Bones and cartilage are formed from *(ecto? meso? endo?)*-derm which later differentiates into the

   embryonic connective tissue called _____. Some of these cells become

   chondroblasts, which eventually form *(bone? cartilage?)*, whereas _____-blasts
   develop into bone.

b. Limb buds appear during the *(fifth? seventh? tenth?)* week of development. At this point the skeleton in

   these buds consists of *(bone? cartilage?)*. Bone begins to form during week _____.

c. By week *(six? seven? eight?)* the upper extremity is evident, with defined shoulder, arm, elbow, forearm, wrist, and hand. The lower extremity is also developing but at a slightly *(faster? slower?)* pace.

d. The notochord develops in the region of the *(head? vertebral column? pelvis?)*. Most of the

   notochord eventually *(forms vertebrae? disappears?)*. Parts of it persist in the _____.

E. Disorders, medical terminology (page 159)

■ **E1.** Describe osteoporosis in this exercise.

a. Osteoporotic bones exhibit _____-creased bone mass and _____-creased susceptibility to fractures.

   This disorder is associated with a loss of the hormone _____. This hormone

   stimulates osteo-_____ activity.

b. Circle the category of persons at higher risk for osteoporosis in each pair:
   Younger persons/older persons   Men/women
   Short, thin persons/tall, large-build persons   Black persons/white persons

c. List three factors in daily life that will help to prevent osteoporosis.
   (*Hint:* Focus on diet and activities.)

■ **E2.** Check your understanding of bone disorders by matching the terms in the box with related descriptions.

Osteoarthritis	Osteopenia
Osteomyelitis	Osteosarcoma

a. Decreased bone mass: _____

b. Malignant bone tumor: _____

c. Bone infection, for example, caused by *Staphylococcus:* _____

d. Degenerative joint disease caused by destruction of articular cartilage: _____

ANSWERS TO SELECTED CHECKPOINTS: CHAPTER 6

A1. Support, protection, movement, storage of minerals such as calcium, storage of energy as in the fat in bone marrow, and hemopoiesis.

A2. Essentially all do. For example, muscles need intact bones for movement to occur; bones are site of blood formation; bones provide protection for viscera of nervous, digestive, urinary, reproductive, cardiovascular, respiratory, and endocrine systems; broken bones can injure integument.

B1. Cartilage, bone (osseous) tissue, bone marrow, and periosteum.

B2.

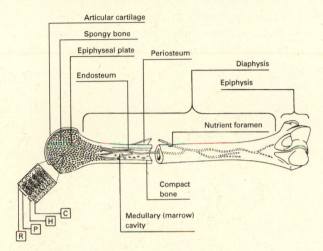

Figure LG 6.1A Diagram of a developing long bone that has been partially sectioned lengthwise.

B3. (a) A. (b) M. (c) F. (d) O. (e) E.

B4. (a) Osteoprogenitor. (b) Osteoblast. (c) Osteocyte. (d) Osteoclast.

B5. (a) Intercellular material. (b) 25, 50. (c) Tricalcium phosphate, calcium carbonate. (d) Hardness, collagen; tensile strength, collagen fibers. (e) Contains some spaces; provides channels for blood vessels and makes bones lighter weight.

B6. (a) Osteons (Haversian systems); lamella, E. (b) G; F; D. (c) Osteocytes; lacunae, B. (d) C; contain extensions of osteocytes bathed in extracellular fluid (ECF); permit communication between osteocytes at gap junctions; provide routes for nutrients and oxygen to reach osteocytes and for wastes to diffuse away.

B7. (a) Trabeculae, irregular latticework of plates of bone tissue containing osteocytes surrounded by red marrow and blood vessels. (b) Epiphyses; is also. (c) Only certain areas of; hipbones, ribs, sternum, vertebrae, skull bones, and epiphyses of long bones such as femurs. These areas (such as the sternum) may be biopsied to examine for aplastic anemia or for response of bone marrow to antianemia medications. These sites (such as hipbones) may be utilized for marrow transplant

(for example, for patients with severe anemia or cancers involving bone marrow).

C1. (a) Ossification. (b) Connective, mesenchyme; progenitor. (c) Chondro or osteo (either order); capillaries (blood vessels). (d) Chondroblasts, endochondral; osteoblasts. (e) Endochondral: almost all bones of the body; intramembranous: flat bones of the cranium, mandible (lower jaw bone), and clavicles (collarbones). (f) Do not.

C2. (a) Mesenchyme; ossification. (b) Osteoprogenitor, osteoblast, osteocyte. (c) Compact, spongy, marrow. (d) Fontanels; growth, protection.

C3. Cartilage. (I) Development: E F. (II) Growth: A B C D. (III) Primary ossification center: H I. (IV) Diaphysis and epiphyses: G.

C4. See Figure LG 6.1A. (a) R. (b) P. (c) H. (d) C.

C5. Epiphysis; diaphysis; shorter than normal.

C6. (a) Line, plate. (b) 25; females. (c) Osteoclasts, osteoblasts.

C7. (a) hGH. (b) IGFs. (c) INS. (d) SH.

C9. Lysosomes release enzymes that may digest the protein collagen, whereas acids may cause minerals of bones to dissolve. Both contribute to bone resorption.

C10. (a) Calcium, phosphorus, magnesium, boron, manganese. (b) A, B_{12}, C, D. (c) Exercise.

C13. (a) Fracture hematoma, hours; inflammation and cell death; osteoclasts, phagocytes, weeks. (b) Procallus; fibroblasts; chondroblasts; weeks. (c) Osteoblasts; months. (d) Remodeling.

C14. (a) See Figure LG 6.3A. (b) AMP; "turns on"; bones, raise. (c) Retain; calcitriol, D. (d) Clasts, resorbed from. (e) Negative, in. (f) Opposite to; deposited in; thyroid, de. (g) Hypo; low, hypo; fragility, lost, hyper. (h) Blood clotting, heart activity, breathing, enzyme function.

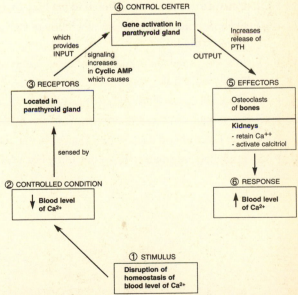

Figure LG 6.3A Negative feedback regulation of calcium.

D1. (a) Strengthens; weaker. (b) Increased.
    (c) Osteoblasts, osteoclasts.
D2. (a) De; weaker; women. (b) Protein; bones are
    more brittle and vulnerable to fracture. (c)
    Weight-bearing exercise; good diet with adequate
    calcium, vitamin D, and protein; not smoking
    because smoking constricts blood vessels.
D3. (a) Meso-, mesenchyme; cartilage, osteo.
    (b) Fifth; cartilage; six or seven. (c) Eight; slower.
    (d) Vertebral column; disappears; intervertebral
    discs.

E1. (a) De, in; estrogen; blast. (b) Older; women;
    short, thin persons; white persons. (c) Adequate
    intake of calcium, vitamin D, weight-bearing exer-
    cise, and not smoking; estrogen replacement ther-
    apy (HRT) may be advised for some women.
E2. (a) Osteopenia. (b) Osteosarcoma.
    (c) Osteomyelitis. (d) Osteoarthritis.

WRITING ACROSS THE CURRICULUM: CHAPTER 6

1. Explain how the skeletal system is absolutely
   vital for basic functions that keep you alive, for
   example, for breathing, eating, movement, and
   elimination of wastes.
2. Do you think bone is dynamic or "dead"?
   State your rationale.
3. Contrast compact bone with spongy bone
   according to structure and locations.
4. Describe mesenchyme and osteoprogenitor roles
   in formation, growth, and maintenance of skeletal
   tissues.

MASTERY TEST: CHAPTER 6

Questions 1–15: Circle T (true) or F (false). If the statement is false, change the underlined word or phrase so that the statement is correct.

T F 1. Greenstick fractures occur only in <u>adults.</u>
T F 2. <u>Appositional</u> growth of cartilage or bone means growth in thickness due to action of cells from the perichondrium or periosteum.
T F 3. Calcitriol is an <u>active form of calcium.</u>
T F 4. The epiphyseal plate appears <u>earlier</u> in life than the epiphyseal line.
T F 5. Osteons (Haversian systems) are found in <u>compact bone, but not in spongy bone.</u>
T F 6. Another name for the epiphysis is the <u>shaft of the bone.</u>
T F 7. A compound fracture is defined as one in which the bone <u>is broken into many pieces.</u>
T F 8. Haversian canals run <u>longitudinally (lengthwise) through bone, but perforating (Volkmann's) canals run horizontally across bone.</u>
T F 9. Canaliculi are tiny canals containing <u>blood which nourishes bone cells in lacunae.</u>
T F 10. Compact bone that is of intramembranous origin differs <u>structurally</u> from compact bone developed from cartilage.
T F 11. In a long bone the primary ossification center is located in the <u>diaphysis, whereas the secondary center of ossification is in the epiphysis.</u>
T F 12. <u>Osteoblasts</u> are bone-destroying cells.
T F 13. Most bones start out in embryonic life as <u>hyaline cartilage.</u>
T F 14. The layer of compact bone is <u>thicker</u> in the diaphysis than in the epiphysis.
T F 15. <u>Lamellae</u> are small spaces containing bone cells.

_____ _____ _____ 16. Steps in marrow formation during endochondral bone formation, from first to last:
  A. Cartilage cells burst, causing intercellular pH to become more alkaline.
  B. pH changes cause calcification, death of cartilage cells; eventually spaces left by dead cells are penetrated by blood vessels.
  C. Cartilage cells hypertrophy with accumulated glycogen.

_____ _____ _____ 17. From most superficial to deepest:
  A. Endosteum
  B. Periosteum
  C. Compact bone

_____ _____ _____ 18. Phases in repair of a fracture, in chronological order:
  A. Fracture hematoma formation
  B. Remodeling
  C. Callus formation

_____ _____ _____ _____ 19. Phases in formation of bone in chronological order:
  A. Mesenchyme cells
  B. Osteocytes
  C. Osteoblasts
  D. Osteoprogenitor cells

_____ _____ _____ _____ 20. Portions of the epiphyseal plate, from closest to epiphysis to closest to diaphysis:
  A. Zone of resting cartilage
  B. Zone of proliferating cartilage
  C. Zone of calcified cartilage
  D. Zone of hypertrophic cartilage

Questions 21–25: Fill-ins. Write the word or phrase that best completes the statement.

_____ 21. _____ is a term that refers to the shaft of the bone.

_____ 22. _____ is a disorder primarily of older women associated with decreased estrogen level; it is characterized by weakened bones.

_____ 23. _____ is the connective tissue covering over cartilage in adults and also over embryonic cartilaginous skeleton.

_____ 24. The majority of bones formed by intramembranous ossification are located in the _____ .

_____ 25. Vitamin _____ is a vitamin that is necessary for absorption of calcium from the gastrointestinal tract, and so it is important for bone growth and maintenance.

ANSWERS TO MASTERY TEST: ☆ CHAPTER 6

True–False

1. F. Children
2. T
3. F. Active form of vitamin D
4. T
5. T
6. F. End of the bone (often bulbous)
7. F. Protrudes through the skin
8. T

9. F. Parts of osteocytes and fluid from blood vessels in Haversian canals, but not blood itself
10. F. Only in origin
11. T
12. F. Osteoclasts
13. T
14. T
15. F. Lacunae

Arrange

16. C A B
17. B C A
18. A C B
19. A D C B
20. A B D C

Fill-ins

21. Diaphysis
22. Osteoporosis
23. Perichondrium
24. Skull or cranium; also the clavicles
25. D (or D_3)

FRAMEWORK 7
Axial Skeleton

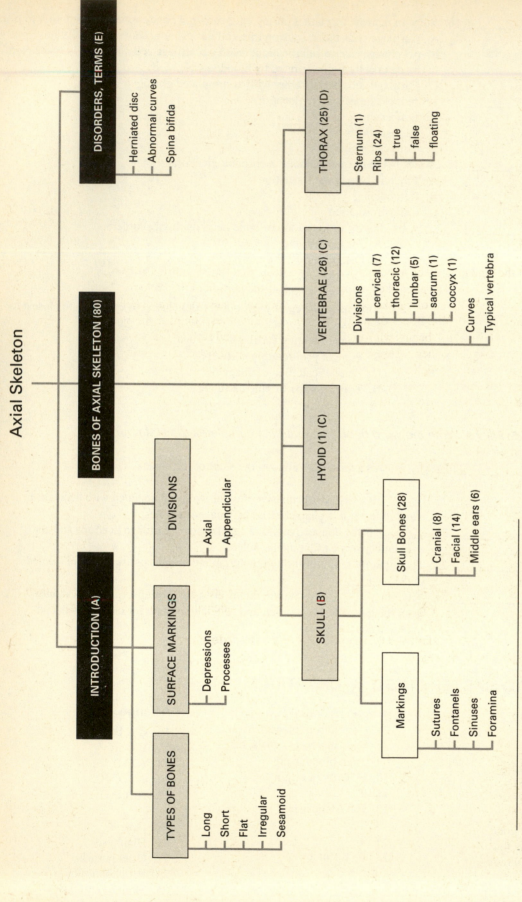

INTRODUCTION (A)

TYPES OF BONES
- Long
- Short
- Flat
- Irregular
- Sesamoid

SURFACE MARKINGS
- Depressions
- Processes

DIVISIONS
- Axial
- Appendicular

BONES OF AXIAL SKELETON (80)

SKULL (B)

Markings
- Sutures
- Fontanels
- Sinuses
- Foramina

Skull Bones (28)
- Cranial (8)
- Facial (14)
- Middle ears (6)

HYOID (1) (C)

VERTEBRAE (26) (C)

Divisions
- cervical (7)
- thoracic (12)
- lumbar (5)
- sacrum (1)
- coccyx (1)

Curves

Typical vertebra

THORAX (25) (D)
- Sternum (1)
- Ribs (24)
  - true
  - false
  - floating

DISORDERS, TERMS (E)
- Herniated disc
- Abnormal curves
- Spina bifida

Note: numbers in ( ) refer to number of each bone in the body.

The Skeletal System: The Axial Skeleton

CHAPTER 7

The 206 bones of the human skeleton are classified into two divisions—axial and appendicular—based on their locations. The axial skeleton is composed of the portion of the skeleton immediately surrounding the axis of the skeleton, primarily the skull bones, vertebral column, and bones surrounding the thorax. Bones of the appendages, or extremities, comprise the appendicular skeleton. Chapter 7 begins with an introduction to types and surface characteristics (markings) of all bones and then focuses on the 80 bones of the axial skeleton. Disorders of the vertebral column are also included.

As you begin your study of the axial skeleton, carefully examine the Chapter 7 Framework and note relationships among concepts and key terms there. Also refer to the Topic Outline and Objectives; you may want to check off each objective as you complete it.

TOPIC OUTLINE AND OBJECTIVES

A. Introduction: types of bones, surface markings, divisions of the skeletal system

☐ 1. Classify the principal types of bones on the basis of shape and location.
☐ 2. Describe the various surface markings on bones.

B. Skull

☐ 3. Identify the bones of the skull and the major markings associated with each.
☐ 4. Identify the principal sutures, fontanels, paranasal sinuses, and foramina of the skull.

C. Hyoid bone and vertebral column

☐ 5. Identify the bones of the vertebral column and their principal markings.

D. Thorax

☐ 6. Identify the bones of the thorax and their principal markings.

E. Disorders

☐ 7. Contrast herniated (slipped) disc, abnormal curves, and spina bifida as disorders associated with the skeletal system.

WORDBYTES

Now become familiar with the language of this chapter by studying each wordbyte, its meaning, and an example of its use within a term. After you study the entire list, self-check your understanding by writing the meaning of each wordbyte on the line. As you continue through the *Learning Guide,* identify (and fill in) additional terms that contain the same wordbyte.

Wordbyte	Self-check	Meaning	Example(s)
annulus	_____	ring	*annulus* fibrosus
costa-	_____	rib	*costal* cartilage
cribr-	_____	sieve	*cribr*iform plate
crist-	_____	crest	*crista* galli
ethm-	_____	sieve	*ethm*oid bone
lambd-	_____	L-shaped	*lambd*oidal suture
pulp	_____	soft, flesh	nucleus *pulp*osus

CHECKPOINTS

A. Introduction: types of bones, surface markings, divisions of the skeletal system (pages 163–166)

A1. Explain how the skeletal system is absolutely necessary for vital life functions of eating, breathing, and movement.

A2. Complete the table about the four major and two minor types of bones.

Type of Bone	Structural Features	Examples
a. Long	Slightly curved to absorb stress better	
b.		Wrist, ankle bones
c.	Composed of two thin plates of bone	

d. Irregular		
e. Sutural (or Wormian)		
f. Small bones in tendons		

A3. In general, what are the functions of surface markings of bones?

A4. Contrast the bone markings in each of the following pairs.
a. *Tubercle/tuberosity*

b. *Crest/line*

c. *Fossa/foramen*

d. *Condyle/epicondyle*

■ **A5.** Write the correct answer from the following list of specific bone markings on the line next to its description. The first one has been done for you.

Articular *facet* on vertebra	Maxillary *sinus*
External auditory *meatus*	Optic *foramen*
Greater *trochanter*	Styloid *process*
Head of humerus	Superior orbital *fissure*

a. Air-filled cavity within a bone, connected to

   nasal cavity   **Maxillary sinus**

b. Narrow, cleftlike opening between adjacent parts of bone; passageway for blood vessels

   and nerves:_____

c. Rounded hole, passageway for blood vessels

   and nerves:_____

d. Tubelike passageway through bone:

   _____

e. Large, rounded projection above constricted

   neck: _____

f. Large, blunt projection on the femur:

   _____

g. Sharp, slender projection:

   _____

h. Smooth, flat surface:

   _____

■ **A6.** Describe the bones in the two principal divisions of the skeletal system by completing this exercise. It may help to refer to Figure LG 8.1, page 145.

a. Bones that lie along the axis of the body are included in the *(axial? appendicular?)* skeleton.

b. The axial skeleton includes the following groups of bones. Indicate how many bones are in each category.

_____ Skull (cranium, face)   _____ Vertebrae

_____ Earbones (ossicles)   _____ Sternum

_____ Hyoid   _____ Ribs

c. The total number of bones in the axial skeleton is _____.

d. The appendicular skeleton consists of bones in which parts of the body?

e. Write the number of bones in each category. Note that you are counting bones on one side of the body only.

_____ Left shoulder girdle   _____ Left hipbone

_____ Left upper extremity (arm, forearm,   _____ Left lower extremity (thigh, kneecap,
          wrist, hand)   leg, foot)

f. There are _____ bones in the appendicular skeleton.

g. In the entire human body there are _____ bones.

B. Skull (pages 166–181)

■ **B1.** Answer these questions about the cranium.

a. What is the main function of the cranium?

b. Which bones make up the cranium (rather than the face)?

c. Define *sutures*.

Between which two bones is the sagittal suture located?_____
On Figure LG 7.1a, label the following sutures: *coronal, squamous, lambdoid.*

d. "Soft spots" of a newborn baby's head are known as _____.

What is the location of the largest one?_____

○ Ethmoid bone
○ Frontal bone
○ Lacrimal bone
○ Mandible bone
○ Maxilla bone
○ Nasal bone
○ Occipital bone
○ Parietal bone
○ Sphenoid bone
○ Temporal bone
○ Zygomatic bone

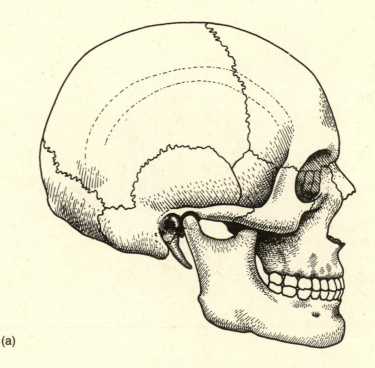

(a)

Figure LG 7.1 Skull bones. (a) Skull viewed from right side. Color and label as directed in Checkpoints B1, B3, and B6.

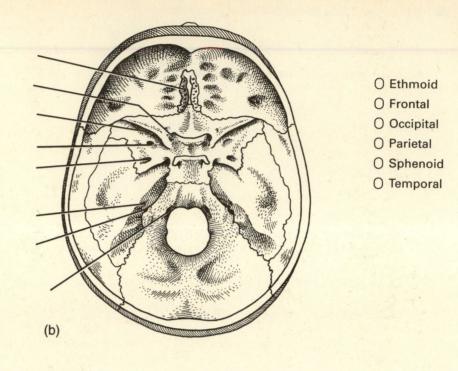

○ Ethmoid
○ Frontal
○ Occipital
○ Parietal
○ Sphenoid
○ Temporal

(b)

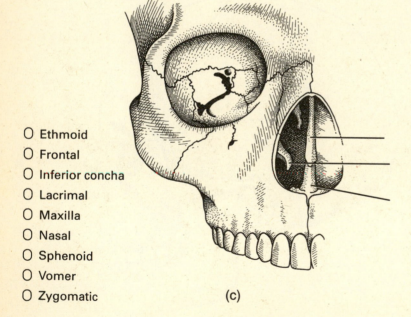

○ Ethmoid
○ Frontal
○ Inferior concha
○ Lacrimal
○ Maxilla
○ Nasal
○ Sphenoid
○ Vomer
○ Zygomatic

(c)

Figure LG 7.1 Skull bones. (b) Floor of the cranial cavity. (c) Anterior view of bones that form the right orbit and the nose. Color and label as directed in Checkpoints B1, B3, B6, B7, B10, and B12.

B2. List three functions of the *fontanels*.

■ **B3.** Color the skull bones on Figure LG 7.1a, b, and c. Be sure to color the corresponding color code oval for each bone listed on the figure.

■ **B4.** Check your understanding of locations and functions of skull bones by matching names of bones in the box with their descriptions below. Use each bone only once.

Ethmoid	Maxilla	Sphenoid
Frontal	Nasal	Temporal
Inferior concha	Occipital	Vomer
Lacrimal	Palatine	Zygomatic (malar)
Mandible	Parietal	

_____ a. This bone forms the lower jaw, including the chin.

_____ b. These are the cheek bones; they also form lateral walls of the orbit of the eye.

_____ c. Tears pass through tiny foramina in these bones; they are the smallest bones in the face.

_____ d. The bridge of the nose is formed by these bones.

_____ e. Organs of hearing (internal part of ears) and mastoid air cells are located in and protected by these bones.

_____ f. This bone sits directly over the spinal column; it contains the foramen through which the spinal cord connects to the brain.

_____ g. The name means "wall." The bones form most of the roof and much of the side walls of the skull.

_____ h. These bones form most of the roof of the mouth (hard palate) and contain the sockets into which upper teeth are set.

_____ i. L-shaped bones form the posterior parts of the hard palate and nose.

_____ j. Commonly called the forehead, it provides protection for the anterior portion of the brain.

_____ k. A fragile bone, it forms much of the roof and internal structure of the nose.

_____ l. It serves as a "keystone" because it binds together many of the other bones of the skull. It is shaped like a bat, with the wings forming part of the sides of the skull and the legs at the back of the nose.

_____ m. This bone forms the inferior part of the septum dividing the nose into two nostrils.

_____ n. Two delicate bones form the lower parts of the side walls of the nose.

■ **B5.** Complete the table describing major markings of the skull.

Marking	Bone	Function
a. Greater wings	Sphenoid	
b.		Forms superior portion of septum of nose
c.		Site of pituitary gland
d. Petrous portion		
e.		Largest hole in skull; passageway for spinal cord
f.	(2)	Bony sockets for teeth
g.		Passageway for sound waves to enter ear
h. Condylar process	Mandible	

■ **B6.** Now label each of the markings listed in the above table on Figure LG 7.1.

■ **B7.** *The Big Picture: Looking Ahead.* Refer to figures on text pages listed in parentheses as you do this activity. The 12 pairs of nerves attached to the brain are called cranial nerves (Figure 14.5, page 398 of your text). Holes in the cranium permit passage of these nerves to and from the brain. These nerves are numbered according to the order in which they attach to the brain (and leave the cranium) from I (most anterior) to XII (most posterior). To help you visualize their sequence, label the foramina for cranial nerves in order on the left side of Figure LG 7.1b. Complete the table summarizing these foramina. The first one is done for you.

Number and Name of Cranial Nerve	Location of Opening for Nerve
a. I Olfactory (Figure 16.1, p. 454)	Cribriform plate of ethmoid bone
b. II Optic (Figure 11.6a, p. 283)	
c. III Oculomotor (Figure 17.1, p. 490) IV Trochlear V Trigeminal (ophthalmic branch) VI Abducens	
d. IV Trigeminal (maxillary branch)	
e. V Trigeminal (mandibular branch)	
f. VII Facial (Figure 17.1, p. 490) VIII Vestibulocochlear (Figure 16.16a, p. 472)	
g. IX Glossopharyngeal (Figure 17.1, p. 490) X Vagus (Figure 17.1, p. 490) XI Accessory	
h. XII Hypoglossal	

■ **B8.** *For extra review.* Write the names of the markings listed in the box next to the bones in which those markings are located. Write the name of one marking on each line.

Carotid canal	Mental foramen	Superior nasal conchae
Condylar process	Olfactory foramina	Superior nuchal line
Crista galli	Palatine process	Superior orbital fissure
Inferior orbital fissure	Perpendicular plate	Supraorbital margin
Lesser wings	Pterygoid processes	Zygomatic process
Mastoid air cells		

a. Ethmoid: _____ _____

_____ _____

b. Frontal: _____

c. Mandible: _____ _____

d. Maxilla: _____ _____

e. Occipital: _____

f. Sphenoid: _____ _____

_____

g. Temporal: _____ _____

_____

■ **B9.** *The Big Picture: Looking Ahead* at details of markings of the skull, circle the correct answers in each statement. Refer to figures on text pages listed in parentheses as you complete this Checkpoint.

a. The zygomatic arch (Figure 11.5, pages 281–282) is formed of two bones; these are the *(parietal? sphenoid? temporal? zygomatic?)* bones.

b. The styloid process (Figure 11.7, page 285) is a penlike projection from the *(parietal? sphenoid? temporal?)* bone.

c. The tear duct (Figure 16.4b, page 458) passes through a canal in the *(lacrimal? mental?)* foramen.

d. The frontal sinus (Figure 23.2b, page 710) is located *(superior? inferior?)* to the nasal cavity.

e. The nasal conchae (Figure 23.2b, page 710) are located *(superior to? within?)* the nasal cavity.

f. The hard palate (Figure 23.2b, page 710) is formed anteriorly of the *(hyoid? mandible? maxilla? palatine?)* bone and posteriorly of the *(mandible? maxilla? palatine?)* bone.

■ **B10.** Do this exercise about bony structures related to the nose.
a. Name four bones that contain paranasal sinuses.

Practice identifying their locations the next time you have a cold!
b. List three functions of paranasal sinuses.

c. Most internal portions of the nose are formed by a bone that includes these markings: paranasal sinuses, conchae, and part of the nasal septum. Name the bone.

d. What functions do nasal conchae perform?

e. On Figure LG 7.1c, label the bone that forms the inferior portion of the nasal septum.

The anterior portion of the nasal septum consists of flexible _____.

■ **B11.** *A clinical challenge. TMJ syndrome* refers to disorders involving the only movable joint

in the skull, the joint between the _____ and _____ bones.
List three signs or symptoms of this problem.

■ **B12.** A good way to test your ability to visualize locations of important skull bones is to try to identify bones that form the orbit of the eye. On Figure LG 7.1c, locate six bones that form the orbit. (*Note:* The tiny portion of the seventh bone, the superior tip of the palatine bone, is not visible in the figure.)

C. Hyoid bone and vertebral column (pages 181–189)

C1. Identify the location of the hyoid bone on yourself. Place your hand on your throat and swallow. Feel your larynx move upward? The hyoid sits just *(superior? inferior?)* to the larynx at the level of the mandible.

■ **C2.** In what way is the hyoid unique among all the bones of the body?

■ **C3.** The five regions of the vertebral column are grouped (A–E) in Figure LG 7.2. Color vertebrae in each region and be sure to select the same color for the corresponding color code oval. Now write on lines next to A–E the number of vertebrae in each region. One is done for you. Also label the first two cervical vertebrae.

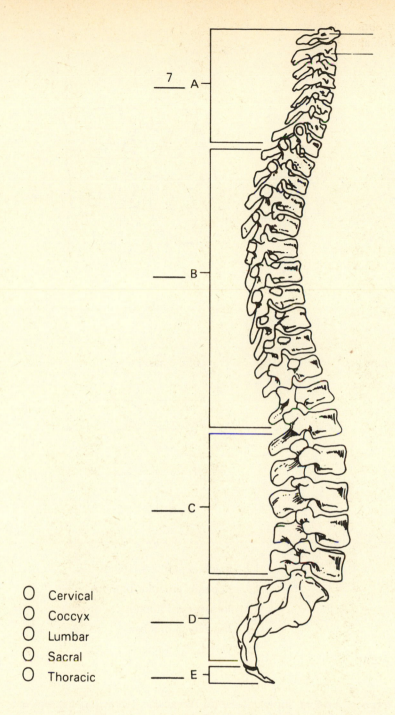

O Cervical
O Coccyx
O Lumbar
O Sacral
O Thoracic

Figure LG 7.2 Right lateral view of the vertebral column. Color and label as directed in Checkpoint C3.

■ **C4.** Note which regions of the vertebral column in Figure 7.2 normally retain an anteriorly concave

curvature in the adult: _____ and _____. These are considered
(primary? secondary?) curvatures. This classification is based upon the fact that these curves
(were present originally during fetal life? are more important?).

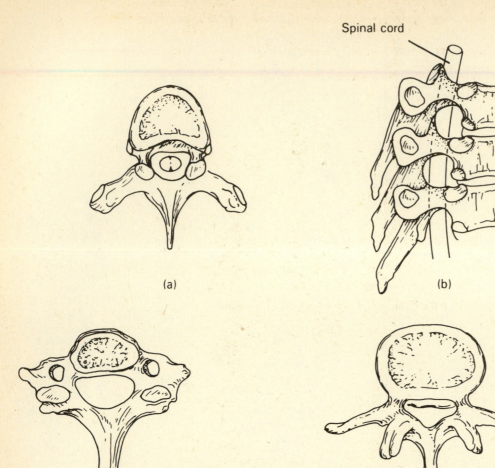

Spinal cord

(a)

(b)

(c)

(d)

○ Body
○ Facets for articulation with rib
○ Intervertebral disc
○ Intervertebral foramen
○ Lamina
○ Pedicle

○ Spinous process (spine)
○ Superior articular process
○ Transverse foramen
○ Transverse process
○ Vertebral foramen

Figure LG 7.3 Typical vertebrae. (a) Thoracic vertebra, superior view. (b) Thoracic vertebrae, right lateral view. (c) Cervical vertebra, superior view. (d) Lumbar vertebra, superior view. Color and label as directed in Checkpoint C5.

■ **C5.** On Figure LG 7.3, color the parts of vertebrae and corresponding color code ovals. As you do this, notice differences in size and shape among the three vertebral types in the figure.

■ **C6.** Identify distinctive features of vertebrae in each region.

C. Cervical	S. Sacral
L. Lumbar	T. Thoracic

_____ a. Small body, foramina for vertebral blood vessels in transverse processes

_____ b. Only vertebrae that articulate with ribs

_____ c. Massive body, blunt spinous process and articular processes directed medially or laterally

_____ d. Long spinous processes that point inferiorly

_____ e. Articulate with the two hipbones

■ **C7.** *A clinical challenge.* State the clinical significance of these markings.
a. Sacral hiatus

b. Sacral promontory

c. Odontoid process

D. Thorax (pages 189–192)

D1. Name the structures that compose the thoracic cage. (Why is it called a "cage"?)

■ **D2.** On Figure LG 7.4, color the three parts of the sternum.

■ **D3.** *A clinical challenge.* State one reason for performing a sternal puncture.

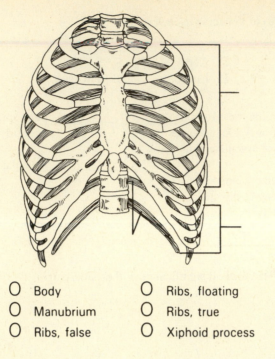

○ Body	○ Ribs, floating
○ Manubrium	○ Ribs, true
○ Ribs, false	○ Xiphoid process

Figure LG 7.4 Anterior view of the thorax. Color as directed in Checkpoints D2 and D4

■ **D4.** Complete this exercise about ribs. Refer again to Figure LG 7.4.

a. There is a total of _____ ribs (_____ pairs) in the human skeleton.

b. Ribs slant in such a way that the anterior portion of the rib is *(superior? inferior?)* to the posterior end of the rib.

c. Posteriorly, all ribs articulate with _____. Ribs also pass *(anterior? posterior?)* to and articulate with transverse processes of vertebrae. What is the functional advantage of such an arrangement?

d. Anteriorly, ribs numbered _____ to _____ attach to the sternum directly by means of strips

of hyaline cartilage, called _____ cartilage. These ribs are called *(true? false?)* ribs. Color these ribs on Figure LG 7.4, leaving the costal cartilages white.

e. Ribs 8 to 12 are called _____. Color these ribs a different color, again leaving the costal cartilages white. Do these ribs attach to the sternum? *(Yes? No?)* If so, in what manner?

f. Ribs _____ and _____ are called "floating ribs." Use a third color for these ribs. Why are they so named?

g. What function is served by the costal groove?

h. What occupies intercostal spaces?

■ **D5.** At what point are ribs most commonly fractured?

E. Disorders (page 193)

■ **E1.** Complete this exercise about slipped discs.
   a. The normal intervertebral disc consists of two parts: an outer ring of *(hyaline? elastic? fibro-?)*

   cartilage called _____ and a soft, elastic, inner portion called the _____.

   b. Ligaments normally keep discs in alignment with vertebral bodies. What may happen if these
      ligaments weaken?

   c. Why might pain result from a slipped disc?

   d. In what part of the vertebral column are slipped discs most common? What symptoms may
       result from a slipped disc in this region?

■ **E2.** Match types of abnormal curvatures of the vertebral column with descriptions below.

| K. Kyphosis | L. Lordosis | S. Scoliosis |

_____ a. Exaggerated lumbar curvature;   _____ c. S- or C-shaped lateral bending
            "swayback"
_____ b. Exaggerated thoracic curvature;
            "hunchback"

E3. Imperfect union of the vertebral arches at the midline is the condition known as _____.
Why is it crucial that the vertebral foramen be completely surrounded by bone? What problems may result from
incomplete closure?

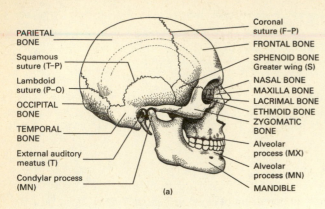

Labels for (a):
PARIETAL BONE
Squamous suture (T-P)
Lambdoid suture (P-O)
OCCIPITAL BONE
TEMPORAL BONE
External auditory meatus (T)
Condylar process (MN)

Coronal suture (F-P)
FRONTAL BONE
SPHENOID BONE Greater wing (S)
NASAL BONE
MAXILLA BONE
LACRIMAL BONE
ETHMOID BONE
ZYGOMATIC BONE
Alveolar process (MX)
Alveolar process (MN)
MANDIBLE

(a)

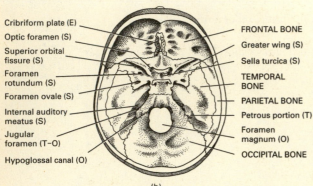

Labels for (b):
Cribriform plate (E)
Optic foramen (S)
Superior orbital fissure (S)
Foramen rotundum (S)
Foramen ovale (S)
Internal auditory meatus (S)
Jugular foramen (T-O)
Hypoglossal canal (O)

FRONTAL BONE
Greater wing (S)
Sella turcica (S)
TEMPORAL BONE
PARIETAL BONE
Petrous portion (T)
Foramen magnum (O)
OCCIPITAL BONE

(b)

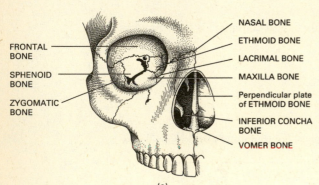

Labels for (c):
FRONTAL BONE
SPHENOID BONE
ZYGOMATIC BONE

NASAL BONE
ETHMOID BONE
LACRIMAL BONE
MAXILLA BONE
Perpendicular plate of ETHMOID BONE
INFERIOR CONCHA BONE
VOMER BONE

(c)

Figure LG 7.1A Skull bones. (a) Skull viewed from right side. (b) Floor of the cranial cavity. (c) Anterior view of bones that form the right orbit and the nose. Bone labels are capitalized; markings are in lowercase with related bone initial in parentheses.

A5. (b) Superior orbital *fissure*. (c) Optic *foramen*. (d) External auditory *meatus*. (e) *Head* of humerus. (f) Greater *trochanter*. (g) Styloid *process*. (h) Articular *facet* on vertebra.

A6. (a) Axial. (b) 22 skull, 6 earbones (studied in Chapter 16), 1 hyoid, 26 vertebrae, 1 sternum, 24 ribs. (c) 80. (d) Shoulder girdles, upper extremities, hipbones, lower extremities. (e) 2 left shoulder girdle, 30 left upper extremity, 1 left hipbone, 30 left lower extremity. (f) 126. (g) 206.

B1. (a) Protects the brain. (b) Frontal, parietals, occipital, temporals, ethmoid, sphenoid. (c) Immovable, fibrous joints between skull bones; between parietal bones; see Figure LG 7.1A. (d) Fontanels; between frontal and parietal bones.

B3. See Figure LG 7.1A.

B4. (a) Mandible. (b) Zygomatic (malar). (c) Lacrimal. (d) Nasal. (e) Temporal. (f) Occipital. (g) Parietal. (h) Maxilla. (i) Palatine. (j) Frontal. (k) Ethmoid. (l) Sphenoid. (m) Vomer. (n) Inferior concha.

B5.

Marking	Bone	Function
a. Greater wings	Sphenoid	**Form part of side walls of skull**
b. **Perpendicular plate**	**Ethmoid**	Forms superior portion of septum of nose
c. **Sella turcica**	**Sphenoid**	Site of pituitary gland
d. Petrous portion	**Temporal**	**Houses middle ear and inner ear**
e. **Foramen magnum**	**Occipital**	Largest hole in skull; passageway for spinal cord
f. **Alveolar processes**	**(2) Maxillae and mandible**	Bony sockets for teeth
g. **External auditory meatus**	**Temporal**	Passageway for sound waves to enter ear
h. Condylar process	Mandible	**Articulates with temporal bone (in TMJ)**

B6. See Figure LG 7.1A

B7. See Figure LG 7.1A

Number and Name of Cranial Nerve	Location of Opening for Nerve
a. I Olfactory (Figure 16.1, p. 454)	Cribriform plate of ethmoid bone
b. II Optic (Figure 11.6a, p. 283)	**Optic foramen of sphenoid bone**
c. III Oculomotor (Figure 17.1, p. 490) IV Trochlear V Trigeminal (ophthalmic branch) VI Abducens	**Superior orbital fissure of sphenoid bone**
d. IV Trigeminal (maxillary branch)	**Foramen rotundum of sphenoid bone**
e. V Trigeminal (mandibular branch)	**Foramen ovale of sphenoid bone**
f. VII Facial (Figure 17.1, p. 490) VIII Vestibulocochlear (Figure 16.16a, p. 472)	**Internal auditory meatus**
g. IX Glossopharyngeal (Figure 17.1, p. 490) X Vagus (Figure 17.1, p. 490) XI Accessory	**Jugular foramen**
h. XII Hypoglossal	**Hypoglossal canal**

B8. (a) Crista galli, olfactory foramina, perpendicular plate, superior nasal concha. (b) Supraorbital margin. (c) Condylar process, mental foramen. (d) Inferior orbital fissure, palatine process. (e) Superior nuchal line. (f) Lesser wings, pterygoid processes, superior orbital fissure. (g) Carotid canal, mastoid air cells, zygomatic process.

B9. (a) Temporal, zygomatic. (b) Temporal. (c) Lacrimal. (d) Superior. (e) Within. (f) Maxilla, palatine.

B10. (a) Frontal, ethmoid, sphenoid, maxilla. (b) Warm and humidify air because air sinuses are lined with mucous membrane; serve as resonant chambers for speech and other sounds; make the skull lighter weight. (c) Ethmoid. (d) Like paranasal sinuses, they are covered with mucous membrane, which warms, humidifies, and cleanses air entering the nose. (e) Vomer; cartilage.

B11. Temporal, mandible; pain, clicking noise, or limitation of movement associated with the joint.

B12. See on Figure LG 7.1cA; frontal, zygomatic, maxilla, lacrimal, ethmoid, sphenoid.

C1. Superior.

C2. It articulates (forms a joint) with no other bones.

C3. A, cervical (7); B, thoracic (12); C, lumbar (5); D, sacrum (1); E, coccyx (1); C1, atlas; C2, axis.

C4. B (thoracic) and D (sacral); primary; were present originally during fetal life.

C5. See Figure LG 7.3A

C6. (a) C. (b) T. (c) L. (d) T. (e) S.

C7. (a) Site used for administration of epidural anesthesia. (b) Obstetrical landmark for measurement of size of the pelvis. (c) In whiplash, it may injure the brainstem.

D2. See Figure LG 7.4A

D3. Biopsy red bone marrow.

D4. (a) 24, 12. (b) Inferior. (c) Bodies of thoracic vertebrae; anterior; prevents ribs from slipping posteriorward. (d) 1, 7, costal; true. (e) False; yes, ribs 8–10 attach to sternum indirectly via seventh costal cartilage. (f) 11, 12; they have no anterior attachment to sternum. (g) Provides protective channel for intercostal nerve, artery, and vein. (h) Intercostal muscles. Refer also to Figure LG 7.4A.

D5. Just anterior to the costal angle, especially involving ribs 3 to 10.

E1. (a) Fibro-, annulus fibrosus, nucleus pulposus. (b) Annulus fibrosus may rupture, allowing the nucleus pulposus to protrude (herniate), causing a "slipped disc." (c) Disc tissue may press on spinal nerves. (d) L4 to L5 or L5 to sacrum; pain in posterior of thigh and leg(s) due to pressure on sciatic nerve, which attaches to spinal cord at L4 to S3.

E2. (a) L. (b) K. (c) S.

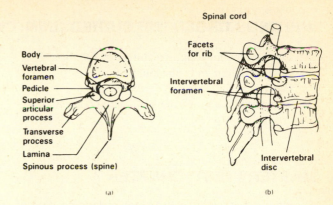

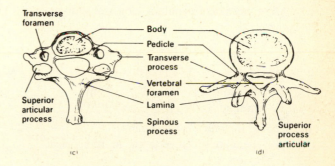

Figure LG 7.3A Typical vertebrae. (a) Thoracic vertebra, superior view. (b) Thoracic vertebrae, right lateral view. (c) Cervical vertebra, superior view. (d) Lumbar vertebra, superior view.

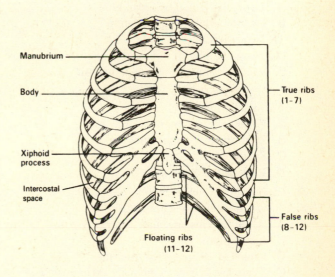

Figure LG 7.4A Anterior view of the thorax.

WRITING ACROSS THE CURRICULUM: CHAPTER 7

1. Contrast the location, structure, and function of the ethmoid and sphenoid bones.
2. Discuss advantages and disadvantages of having 26 separate bones forming the vertebral column or "spine."
3. Contrast the location, structure, and function of cervical, thoracic, and lumbar vertebrae.
4. Contrast the location, structure, and function of the atlas and axis bones.

MASTERY TEST: CHAPTER 7

Questions 1–10: Circle the letter preceding the one best answer to each question.

1. All of these bones contain paranasal sinuses *except:*
   A. Frontal
   B. Maxilla
   C. Nasal
   D. Sphenoid
   E. Ethmoid

2. Choose the one *false* statement.
   A. There are seven vertebrae in the cervical region.
   B. The cervical region normally exhibits a curve that is slightly concave anteriorly.
   C. The lumbar vertebrae are superior to the sacrum.
   D. Intervertebral discs are located between bodiesof vertebrae.

3. The hard palate is composed of _____ bones.
   A. Two maxilla and two mandible
   B. Two maxilla and two palatine
   C. Two maxilla
   D. Two palatine
   E. Vomer, ethmoid, and two temporal

4. The lateral wall of the orbit is formed mostly by which two bones?
   A. Zygomatic and maxilla
   B. Zygomatic and sphenoid
   C. Sphenoid and ethmoid
   D. Lacrimal and ethmoid
   E. Zygomatic and ethmoid

5. Choose the one *true* statement.
   A. All of the ribs articulate anteriorly with the sternum.
   B. Ribs 8 to 10 are called true ribs.
   C. There are 23 ribs in the male skeleton and 24 in the female skeleton.
   D. Rib 7 is larger than rib 3.
   E. Cartilage discs between vertebrae are called costal cartilages.

6. Which is the largest fontanel?
   A. Frontal
   B. Occipital
   C. Sphenoid
   D. Mastoid

7. Immovable joints of the skull are called:
   A. Wormian bones
   B. Sutures
   C. Conchae
   D. Sinuses
   E. Fontanels

8. _____ articulate with every bone of the face except the mandible.
   A. Lacrimal bones
   B. Zygomatic bones
   C. Maxillae
   D. Sphenoid bones
   E. Ethmoid bones

9. All of these markings are parts of the sphenoid bone *except:*
   A. Lesser wings
   B. Optic foramen
   C. Crista galli
   D. Sella turcica
   E. Pterygoid processes

10. All of these bones are included in the axial skeleton *except:*
    A. Rib
    B. Sternum
    C. Clavicle
    D. Hyoid
    E. Ethmoid

Questions 11–15: Arrange the answers in correct sequence.

_____ _____ _____ 11. From anterior to posterior:
    A. Ethmoid bone
    B. Sphenoid bone
    C. Occipital bone

_____ _____ _____ 12. Parts of vertebra from anterior to posterior:
    A. Vertebral foramen
    B. Body
    C. Lamina

_____ _____ _____ 13. Vertebral regions, from superior to inferior:
    A. Lumbar
    B. Thoracic
    C. Cervical

_____ _____ _____ 14. From superior to inferior:
    A. Atlas
    B. Axis
    C. Occipital bone

_____ _____ _____ 15. From superior to inferior:
    A. Atlas
    B. Manubrium of sternum
    C. Hyoid

Questions 16–20: Circle T (true) or F (false). If the statement is false, change the underlined word or phrase so that the statement is correct.

T F 16. The space between two ribs is called the <u>costal groove.</u>

T F 17. The thoracic and sacral curves are called <u>primary</u> curves, meaning that they retain the original curve of the fetal vertebral column.

T F 18. The jugular vein passes through the same foramen as cranial nerves <u>V, VI, and VII.</u>

T F 19. In general, <u>foramina, meati, and fissures</u> serve as openings in the skull for nerves and blood vessels.

T F 20. The annulus fibrosus portion of an intervertebral disc is a <u>firm ring of fibrocartilage</u> surrounding the nucleus pulposus.

Questions 21–25: Fill-ins. Write the word or phrase that best completes the statement.

_____ 21. A finger- or toothlike projection called the dens is part of the _____ bone.

_____ 22. The ramus, angle, mental foramen, and alveolar processes are all markings on the _____ bone.

_____ 23. The squamous, petrous, and zygomatic portions are markings on the _____ bone.

_____ 24. The perpendicular plate, crista galli, and superior and middle conchae are markings found on the _____ bone.

_____ 25. The sternum is often used for a marrow biopsy because _____ .

ANSWERS TO MASTERY TEST: ☆ CHAPTER 7

Multiple Choice

1. C
2. B
3. B
4. B
5. D
6. A
7. B
8. C
9. C
10. C

Arrange

11. A B C
12. B A C
13. C B A
14. C A B
15. A C B

True–False

16. F. Intercostal space
17. T
18. F. IX, X, and XI
19. T
20. T

Fill-ins

21. Axis (second cervical vertebra)
22. Mandible
23. Temporal
24. Ethmoid
25. It contains red bone marrow and it is readily accessible.

FRAMEWORK 8

The Appendicular Skeleton

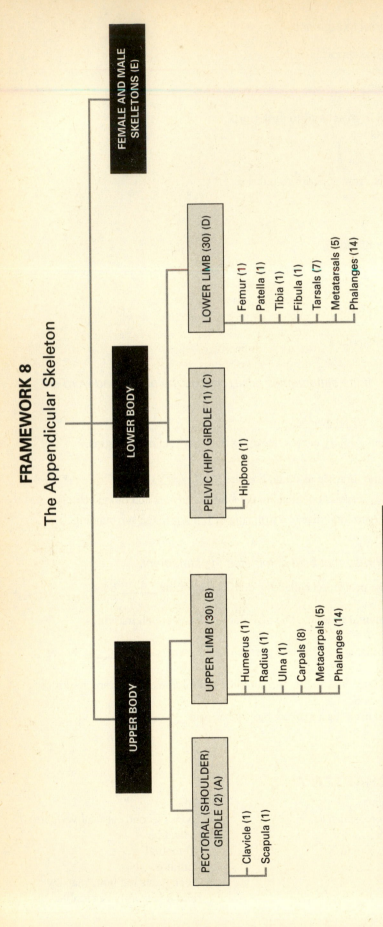

FEMALE AND MALE SKELETONS (E)

LOWER BODY

LOWER LIMB (30) (D)
- Femur (1)
- Patella (1)
- Tibia (1)
- Fibula (1)
- Tarsals (7)
- Metatarsals (5)
- Phalanges (14)

PELVIC (HIP) GIRDLE (1) (C)
- Hipbone (1)

UPPER BODY

UPPER LIMB (30) (B)
- Humerus (1)
- Radius (1)
- Ulna (1)
- Carpals (8)
- Metacarpals (5)
- Phalanges (14)

PECTORAL (SHOULDER) GIRDLE (2) (A)
- Clavicle (1)
- Scapula (1)

Note: numbers in ( ) refer to number of each bone on *one* side of the body.

The Skeletal System: The Appendicular Skeleton

The appendicular skeleton includes the bones of the limbs, or extremities, as well as the supportive bones of the shoulder and hip girdles. Male and female skeletons exhibit some differences in skeletal structure, noted particularly in bones of the pelvis.

As you begin your study of the appendicular skeleton, carefully examine the Chapter 8 Topic Outline and Objectives and note relationships among concepts and key terms in the Framework.

TOPIC OUTLINE AND OBJECTIVES

A. Pectoral (shoulder) girdle

☐ 1. Identify the bones of the pectoral (shoulder) girdle and their principal markings.

B. Upper limb

☐ 2. Identify the upper limb, its component bones, and their principal markings.

C. Pelvic (hip) girdle

☐ 3. Identify the components of the pelvic (hip) girdle and their principal markings.

D. Lower limb

☐ 4. Identify the lower limb, its component bones, and their principal markings.

☐ 5. Define the structural features and importance of the arches of the foot.

E. Female and male skeletons

☐ 6. Compare the principal structural differences between female and male skeletons, especially those that pertain to the pelvis.

WORDBYTES

Now become familiar with the language of this chapter by studying each wordbyte, its meaning, and an example of its use within a term. After you study the entire list, self-check your understanding by writing the meaning of each wordbyte on the line. As you continue through the *Learning Guide*, identify (and fill in) additional terms that contain the same wordbyte.

Wordbyte	Self-check	Meaning	Example(s)
acro-	_____	tip	*acro*mion process
cap-	_____	head	*cap*itulum
meta-	_____	beyond	*meta*tarsal
-physis	_____	to grow	pubic sym*physis*
semi-	_____	half	*semi*lunar notch
sym-	_____	together	pubic *sym*physis

CHECKPOINTS

A. Pectoral (shoulder) girdle (pages 196–197)

■ **A1.** Do this exercise about the pectoral girdle.
   a. Which bones form the pectoral girdle?

   b. Do these bones articulate (form a joint) with vertebrae or ribs? _____
   c. The pectoral girdle is part of the *(axial? appendicular?)* skeleton. Identify the point (marked by *) on Figure LG 8.1 at which the shoulder girdle articulates with the axial skeleton. Name the two bones forming that joint. Palpate (press and feel) the bones at this joint on yourself.

■ **A2.** Write the name of the bone that articulates with each of these markings on the scapula.

   a. Acromion process: _____

   b. Glenoid cavity: _____

   c. Coracoid process: _____

■ **A3.** Study a scapula carefully, using a skeleton or Figure 8.3, page 198 in your text. Then match the markings in the box with the descriptions given.

A. Axillary border	M. Medial border
I. Infraspinatus fossa	S. Spine

_____ a. Sharp ridge on the posterior surface   _____ c. Edge closest to the vertebral column

_____ b. Depression inferior to the spine;   _____ d. Thick edge closest to the arm
            location of infraspinatus muscle

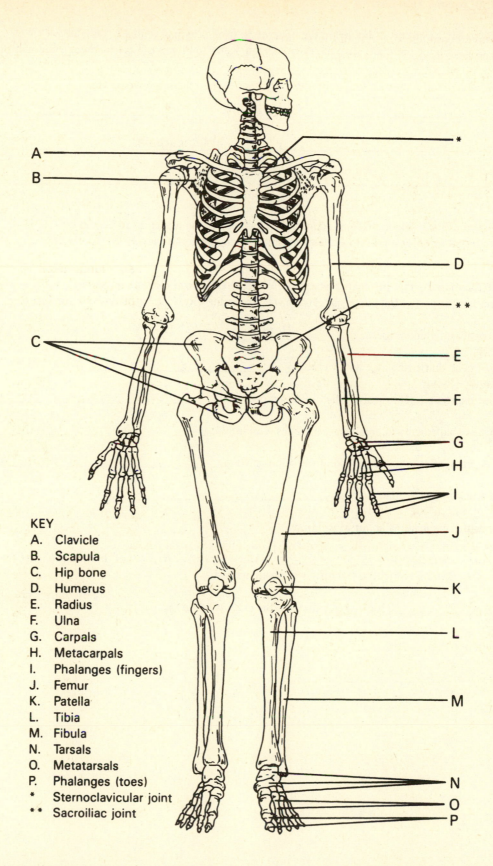

KEY
A. Clavicle
B. Scapula
C. Hip bone
D. Humerus
E. Radius
F. Ulna
G. Carpals
H. Metacarpals
I. Phalanges (fingers)
J. Femur
K. Patella
L. Tibia
M. Fibula
N. Tarsals
O. Metatarsals
P. Phalanges (toes)
* Sternoclavicular joint
** Sacroiliac joint

Figure LG 8.1 Anterior view of the skeleton. Color, label, and answer questions as directed in Checkpoints A1, B1, C1, D1, D2, and D7.

B. Upper limb (pages 197–202)

■ **B1.** List the bone (or groups of bones) in the upper limb from proximal to distal. Indicate how many of each bone there are. Two are done for you. Refer to Figure LG 8.1 to check your answers.

a. **Humerus** _____ **(1)**   d. _____ **( )**

b. _____ **( )**   e. **Metacarpals** _____ **(5)**

c. _____ **( )**   f. _____ **( )**

■ **B2.** On Figure LG 8.2, select different colors and color each of the markings indicated by ○. Where possible, color markings on both the anterior and posterior views.

■ **B3.** Name the marking that fits each description. Write H, U, or R to indicate whether the marking is part of the humerus (H), ulna (U), or radius (R). One has been done for you.

a. Articulates with glenoid fossa: _____ **Head** _____ (**H**)

b. Rounded head that articulates with radius: _____ (_____)

c. Posterior depression that receives the olecranon process: _____ (_____)

d. Half-moon-shaped curved area that articulates with trochlea: _____ (_____)

e. Slight depression in which the head of the radius pivots: _____ (_____)

f. Biceps brachii muscle attaches here: _____ (_____)

■ **B4.** Answer these questions about wrist bones.

a. Wrist bones are called _____. There are *(5? 7? 8? 14?)* of them in each wrist.
b. The wrist bone most subject to fracture is the bone named

_____, located just distal to the *(radius? ulna?)*.

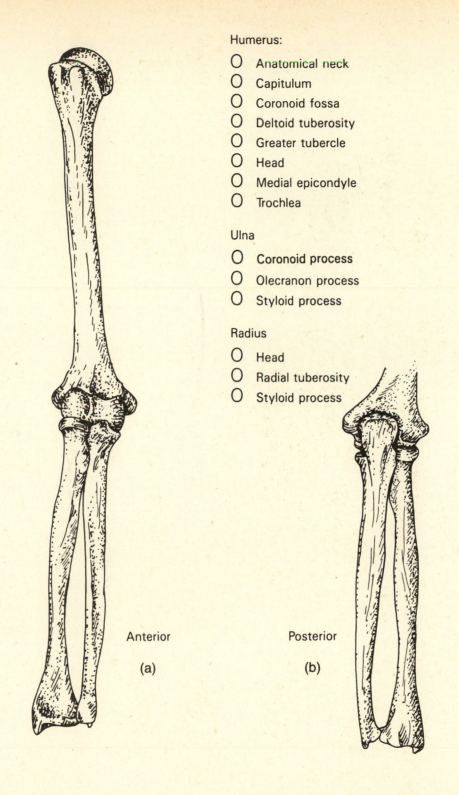

Humerus:

O Anatomical neck
O Capitulum
O Coronoid fossa
O Deltoid tuberosity
O Greater tubercle
O Head
O Medial epicondyle
O Trochlea

Ulna

O Coronoid process
O Olecranon process
O Styloid process

Radius

O Head
O Radial tuberosity
O Styloid process

Anterior

(a)

Posterior

(b)

Figure LG 8.2   Right upper limb. (a) Anterior view. (b) Posterior view. Color and label as directed in Checkpoint B2.

■ **B5.** Trace an outline of your hand. Draw in and label all bones.

C. Pelvic (hip) girdle (pages 202–206)

■ **C1.** Describe the pelvic bones in this exercise.

    a. Name the bones that form the pelvic girdle. _____

       Which bones form the pelvis? _____
    b. Which of these bones is/are part of the axial skeleton?

    c. Locate the point (**) on Figure LG 8.1 at which the pelvic girdle portion of the appendicular skeleton articulates with the axial skeleton. Name the bones involved in that joint.

       _____ and _____

■ **C2.** Answer the following questions about coxal bones.
    a. Each coxal (hip) bone originates as three bones that fuse early in life. These bones are the

       _____, _____, and _____.

       At what location do the bones fuse? _____

    b. The largest of the three bones is the _____.
       A ridge along the superior border is called the iliac crest. Locate this on yourself.

    c. The iliac crest ends anteriorly as the _____ spine. This marking causes a dimpling of the skin just lateral to the sacrum, which can be used as a landmark for administering hip injections accurately.

■ **C3.** Complete the table about markings of the coxal bones.

Marking	Location on Coxal Bone	Function
a. Greater sciatic notch		
b.		Supports most of body weight in sitting position
c.		Fibrocartilaginous joint between two coxal bones
d.		Socket for head of femur
e. Obturator oramen	Large foramen surrounded by pubic and ischial rami and acetabulum	

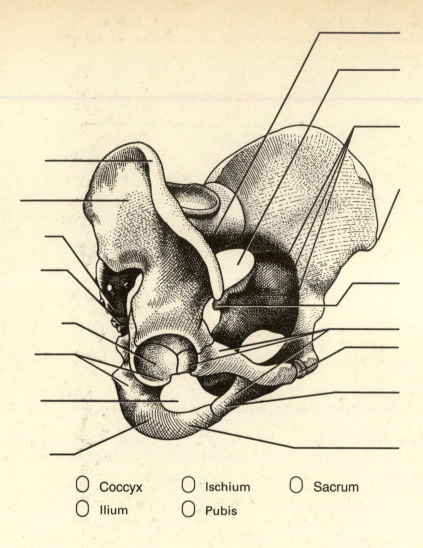

○ Coccyx   ○ Ischium   ○ Sacrum

○ Ilium   ○ Pubis

Figure LG 8.3   Right anterolateral view of the pelvis. Color, label, and answer questions as directed in Checkpoints C4 and C6.

■ **C4.** Refer to Figure LG 8.3 and answer these questions about pelvic markings.
  a. Trace with your pencil, and then label, the *brim of the pelvis* on the figure. It is a relatively *(smooth? irregular?)* line that demarcates the *(superior? inferior?)* border of the lesser (true) pelvis. The pelvic brim encircles the pelvic *(inlet? outlet?)*.
  b. The bony border of the pelvic outlet is *(smooth? irregular?)*. Why is the pelvic outlet so named?

C5. Contrast the two principal parts of the pelvis. Describe their locations and name the structures that compose them.
  a. Greater (false) pelvis

b. Lesser (true) pelvis

■ **C6.** Color the pelvic structures indicated by color code ovals on Figure LG 8.3. Also label the following markings: *acetabulum, anterior superior iliac spine, greater sciatic notch, iliac crest, inferior pubic ramus, ischial spine, ischial tuberosity, obturator foramen, sacral promontory,* and *symphysis pubis.*

D. Lower limb (pages 206–212)

■ **D1.** Refer to Figure LG 8.1 and list the bones (or groups of bones) in the lower limb from proximal to distal. Indicate how many of each bone there are. One is done for you.

a. **Femur** **(1)**   e. _____ ( )

b. _____ ( )   f. _____ ( )

c. _____ ( )   g. _____ ( )

d. _____ ( )

D2. Contrast the size, location, and names of the bones of the upper and lower limbs by coloring the bones on Figure LG 8.1 as follows. Color on one side of the figure only.

Humerus and femur (red)   Carpals and tarsals (blue)

Patella (brown)   Metacarpals and metatarsals (orange)

Ulna and tibia (green)   Phalanges (purple)

Radius and fibula (yellow)

■ **D3.** Circle the term that correctly indicates the location of these parts of the lower limb.
  a. The head is the *(proximal? distal?)* epiphysis of the femur.
  b. The greater trochanter is *(lateral? medial?)* to the lesser trochanter.
  c. The intercondylar fossa is on the *(anterior? posterior?)* surface of the femur.
  d. The tibial condyles are more *(concave? convex?)* than the femoral condyles.
  e. The lateral condyle of the femur articulates with the *(fibula? lateral condyle of the tibia?)*.
  f. The tibial tuberosity is *(superior? inferior?)* to the patella.
  g. The tibia is *(medial? lateral?)* to the fibula.
  h. The outer portion of the ankle is the *(lateral? medial?)* malleolus, which is part of the *(tibia? fibula?)*.

■ **D4.** The tarsal bone that is most superior in location (and that articulates with the

  tibia and fibula) is the _____. The largest and strongest of the tarsals is the

  _____.

D5. Answer these questions about the arch of the foot.

a. How is the foot maintained in an arched position?

b. Locate each of these arches on your own foot. Refer to Figure 8.14, page 212 of your text. (*For extra review:* List the bones that form each arch.)
Longitudinal: medial side

Longitudinal: lateral side

Transverse

c. What causes flatfoot?

■ **D6.** Now that you have seen all of the bones of the appendicular skeleton, complete this table relating common and anatomical names of bones.

Common Name	Anatomical Name
a. Shoulder blade	
b.	Pollex
c. Collarbone	
d. Heel bone	
e.	Olecranon process
f. Kneecap	
g.	Tibial crest
h. Toes	
i. Palm of hand	
j. Wrist bones	

D7. *For extra review.* Label all bones marked with label lines on Figure LG 8.1.

E. Female and male skeletons   (pages 212–213)

E1. State several characteristics of the female pelvis that make it more suitable for childbirth than the male pelvis.

■ **E2.** Identify specific differences in pelvic structure in the two sexes by placing M before characteristics of the male pelvis and F before structural descriptions of the female pelvis.

_____ a. Shallow greater pelvis

_____ b. Heart-shaped inlet

_____ c. Pubic arch greater than 90° angle

_____ d. Acetabulum small

_____ e. Pelvic inlet comparatively small

■ **E3.** *The Big Picture: Looking Ahead.* Refer to text pages in upcoming chapters as you check your understanding of the appendicular skeleton in this Checkpoint.
  a. Contrast Figures 11.12 (page 294) and 11.13 (page 296). In each figure, find the coccyx and the ischial tuberosities. Which pelvis has a relatively wider distance between ischial tuberosities? *(Female? Male?)* Also notice that *(bone? muscle?)* forms most of the floor of the human pelvis.
  b. The ischial tuberosities are points of attachment of several large muscles (Figure 11.19c and d, pages 319–320, and Exhibit 11.21, page 321). Name two or more muscles attached to these bony markings of the hip bone.

  c. Which muscles cover the posterior surface of the scapula, padding this thin, hard bone (Figure 11.15c, page 302)?

  The _____-spinatus muscle lies superior to the spine of the scapula,

  whereas the _____-spinatus and teres _____ are located inferior to the spine.
  d. Name two muscles that are anchored into the comma-shaped coracoid process of the scapula

  (Figure 11.14a, page 298): the _____ brachii and the pectoralis *(major? minor?).* These two muscles both lie closer to the *(anterior? posterior?)* surface of the body.
  e. Figure 11.16a (page 304) shows that the distal tendon of the biceps brachii attaches to the anterior of the *(humerus? radius? ulna?).*

ANSWERS TO SELECTED CHECKPOINTS: CHAPTER 8

A1. (a) Two clavicles and two scapulas. (b) No.
(c) Appendicular; clavicles, manubrium of
sternum.

A2. (a) Clavicle. (b) Humerus. (c) None; muscles
and ligaments attach here.

A3. (a) S. (b) I. (c) M. (d) A.

B1. (a) Humerus, 1. (b) Ulna, 1. (c) Radius, 1.
(d) Carpals, 8. (e) Metacarpals, 5.
(f) Phalanges, 14.

B2.

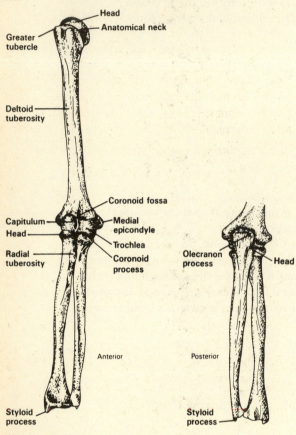

Figure LG 8.2A Right upper limb. (a) Anterior view.
(b) Posterior view.

B3. (b) Capitulum (H). (c) Olecranon fossa (H).
(d) Semilunar (trochlear) notch (U). (e) Radial
notch (U). (f) Radial tuberosity (R).

B4. (a) Carpals, 8. (b) Scaphoid, radius.

C1. (a) Two coxal (hip) bones; hip bones plus sacrum
and coccyx. (b) Sacrum and coccyx. (c) Sacrum,
iliac portion of hip bone (sacroiliac).

C2. (a) Ilium, ischium, pubis; acetabulum. (b) Ilium.
(c) Anterior superior iliac; posterior superior
iliac spine.

C3.

Marking	Location on Coxal Bone	Function
a. Greater sciatic notch	Inferior to posterior inferior iliac spine	Sciatic nerve passes inferior to notch
b. Ischial tuberosity	Posterior and inferior to obturator foramen	Supports most of body weight in sitting position
c. Symphysis pubis	Most anterior portion of pelvis	Fibrocartilaginous joint between two coxal bones
d. Acetabulum	At junction of ilium, ischium, and pubis	Socket for head of femur
e. Obturator oramen	Large foramen surrounded by pubic and ischial rami and acetabulum	Blood vessels and nerves pass through

C4.

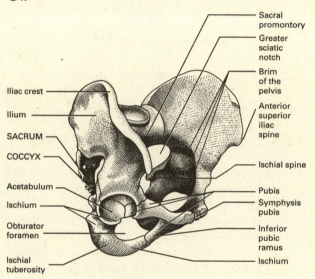

Figure LG 8.3A Right anterolateral view of the pelvis.
Bone labels are capitalized; markings are in
lowercase.

See Figure LG 8.3A. (a) Smooth, superior, inlet.
(b) Irregular; feces, urine, semen, mentstrual flow,
and baby (at birth) exit via openings in the
muscular floor attached to bony outlet.

C6. See Figure LG 8.3A.

D1. (a) Femur, 1. (b) Patella, 1. (c) Tibia, 1.
(d) Fibula, 1. (e) Tarsals, 7. (f) Metatarsals, 5.
(g) Phalanges, 14.

D3. (a) Proximal. (b) Lateral. (c) Posterior.
(d) Concave. (e) Lateral condyle of the tibia.
(f) Inferior. (g) Medial. (h) Lateral, fibula.

D4. Talus; calcaneus.

D6. (a) Scapula. (b) Thumb. (c) Clavicle. (d) Calcaneus. (e) Elbow. (f) Patella. (g) Shinbone.
(h) Phalanges. (i) Metacarpals. (j) Carpals.

E2. (a) F. (b) M. (c) F. (d) M. (e) M.

E3. (a) Female; muscle. (b) The three heads of the
hamstrings (semimembranosus, semitendinosus,
and biceps femoris), as well as the adductor
magnus, quadratus femoris, and inferior gemellus
(Exhibit 11.20). (c) Supra, infra, minor.
(d) Biceps, minor; anterior. (e) Radius.

WRITING ACROSS THE CURRICULUM: CHAPTER 8

1. Contrast the structure and function of the pectoral (shoulder) girdle with that of the pelvic (hip) girdle.

2. Contrast the structure and function of the upper limb (arm) with that of the lower limb (leg).

3. Contrast the location, structure, and function of the carpal and tarsal bones.

MASTERY TEST: CHAPTER 8

Questions 1–6: Circle the letter preceding the one best answer to each question.

1. The point at which the upper part of the appendicular skeleton is joined to (articulates with) the axial skeleton is at the joint between:
   A. Sternum and ribs
   B. Humerus and clavicle
   C. Scapula and clavicle
   D. Scapula and humerus
   E. Sternum and clavicle

2. All of the following are markings on the femur *except:*
   A. Acetabulum
   B. Head
   C. Condyles
   D. Greater trochanter
   E. Intercondylar fossa

3. All of the following are bones in the lower limb *except:*
   A. Talus
   B. Tibia
   C. Calcaneus
   D. Ulna
   E. Fibula

4. Which structures are on the posterior surface of the upper limb (in anatomical position)?
   A. Radial tuberosity and lesser tubercle
   B. Trochlea and capitulum
   C. Coronoid process and coronoid fossa
   D. Olecranon process and olecranon fossa

5. The humerus articulates with all of these bones *except:*
   A. Ulna
   B. Radius
   C. Clavicle
   D. Scapula

6. Choose the *false* statement.
   A. The capitulum articulates with the head of the radius.
   B. The medial and lateral epicondyles are located at the distal ends of the tibia and fibula.
   C. The coronoid fossa articulates with the ulna when the forearm is flexed.
   D. The trochlea articulates with the trochlear notch of the ulna

Questions 7–11: Arrange the answers in correct sequence.

_____ _____ _____ 7. According to size of the bones, from largest to smallest:
  A. Femur
  B. Ulna
  C. Humerus

_____ _____ _____ 8. Parts of the humerus, from proximal to distal:
  A. Anatomical neck
  B. Surgical neck
  C. Head

_____ _____ _____ 9. From proximal to distal:
  A. Phalanges
  B. Metacarpals
  C. Carpals

_____ _____ _____ 10. From superior to inferior:
  A. Lesser pelvis
  B. Greater pelvis
  C. Pelvic brim

_____ _____ _____ 11. Markings on hipbones in anatomical position, from superior to inferior:
  A. Acetabulum
  B. Ischial tuberosity
  C. Iliac crest

Questions 12–20: Circle T (true) or F (false). If the statement is false, change the underlined word or phrase so that the statement is correct.

T F 12. Another name for the true pelvis is the greater pelvis.

T F 13. The scapulae do articulate with the vertebrae.

T F 14. The olecranon process is a marking on the ulna, and the olecranon fossa is a marking on the humerus.

T F 15. The female pelvis is deeper and more heart shaped than the male pelvis.

T F 16. The greater tubercle of the humerus is lateral to the lesser tubercle.

T F 17. There are 14 phalanges in each hand and also in each foot.

T F 18. The fibula articulates with the femur, tibia, talus, and calcaneus.

T F 19. The organs contained within the right iliac, hypogastric, and left iliac portions (ninths) of the abdomen are located in the true pelvis.

T F 20. The total number of bones in one upper limb (including arm, forearm, wrist, hand, and fingers, but excluding shoulder girdle) is 29.

Questions 21–25: Fill-ins. Write the word or phrase that best completes the statement.

_____ 21. In about three-fourths of all carpal bone fractures, only the _____ bone is involved.

_____ 22. A fracture of the distal end of the fibula with injury to the tibial articulation is known as a _____ fracture.

_____ 23. The _____ is the thinnest bone in the body compared to its length.

_____ 24. The point of fusion of the three bones forming the hip bone is the _____.

_____ 25. The kneecap is the common name for the _____.

ANSWERS TO MASTERY TEST: ★ CHAPTER 8

Multiple Choice

1. E
2. A
3. D
4. D
5. C
6. B

Arrange

7. A C B
8. C A B
9. C B A
10. B C A
11. C A B

True–False

12. F. Lesser
13. F. Do not
14. T
15. F. Shallower and more oval
16. T
17. T
18. F. Only tibia and talus
19. F. False
20. F. 30

Fill-ins

21. Scaphoid
22. Pott's
23. Fibula
24. Acetabulum
25. Patella

FRAMEWORK 9
Articulations

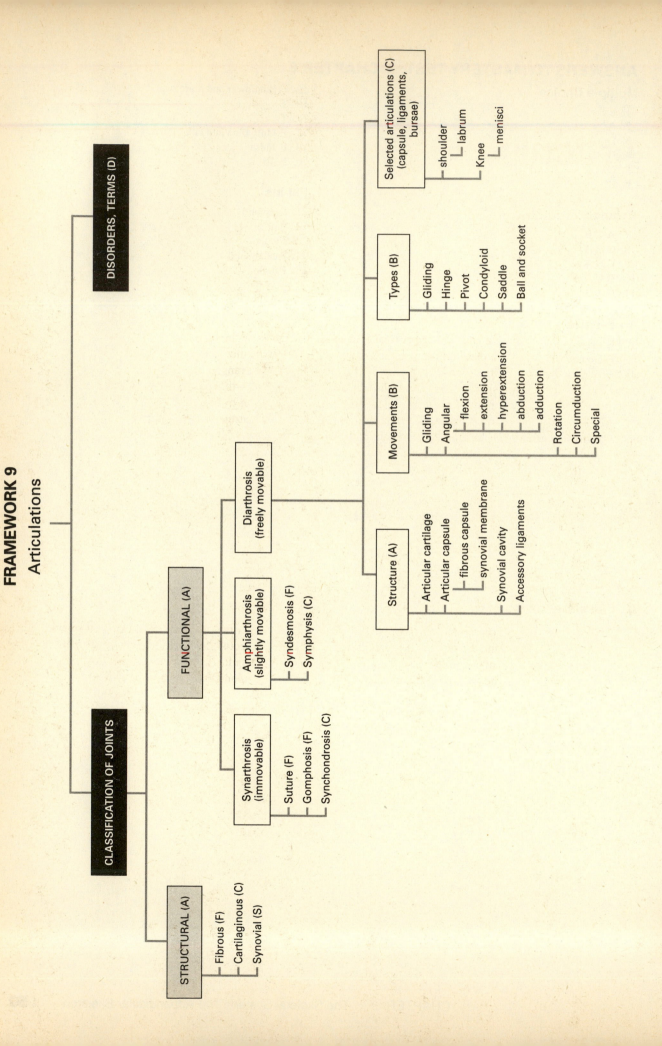

DISORDERS, TERMS (D)

CLASSIFICATION OF JOINTS

STRUCTURAL (A)
- Fibrous (F)
- Cartilaginous (C)
- Synovial (S)

FUNCTIONAL (A)

Synarthrosis (immovable)
- Suture (F)
- Gomphosis (F)
- Synchondrosis (C)

Amphiarthrosis (slightly movable)
- Syndesmosis (F)
- Symphysis (C)

Diarthrosis (freely movable)

Structure (A)
- Articular cartilage
- Articular capsule
  - fibrous capsule
  - synovial membrane
- Synovial cavity
- Accessory ligaments

Movements (B)
- Gliding
- Angular
  - flexion
  - extension
  - hyperextension
  - abduction
  - adduction
- Rotation
- Circumduction
- Special

Types (B)
- Gliding
- Hinge
- Pivot
- Condyloid
- Saddle
- Ball and socket

Selected articulations (C) (capsule, ligaments, bursae)
- shoulder
  - labrum
- Knee
  - menisci

Articulations

In the last three chapters you have learned a great deal about the 206 bones in the body. Separated, or disarticulated, these bones would constitute a pile as disorganized as the rubble of a ravaged city deprived of its structural integrity. Fortunately, bones are arranged in precise order and held together in specific conformations—articulations or joints—that permit bones to function effectively. Joints may be classified by function (how movable) or by structure (the type of tissue forming the joint). Synovial joints are most common; their structure, movements, and types will be discussed. A detailed examination of two important synovial joints (shoulder and knee) is also included. Joint disorders such as arthritis are usually not life threatening but plague much of the population, particularly the elderly. An introduction to joint disorders completes this chapter.

As you begin your study of articulations, carefully examine the Chapter 9 Topic Outline and Objectives and note relationships among concepts and key terms in the Framework.

TOPIC OUTLINE AND OBJECTIVES

A. Classification of joints

☐ 1. Define an articulation (joint) and identify the factors that determine the types and degree (range) of movement at a joint.
☐ 2. Classify joints on the basis of structure and function.
☐ 3. Contrast the structure, kind of movement, and location of immovable, slightly movable, and freely movable joints.

B. Movements and types of synovial (diarthrotic) joints

☐ 4. Describe the structure, types, and movements of freely movable joints.

C. Details of the shoulder and knee joints

☐ 5. Describe the shoulder and knee joints with respect to the bones that enter into their formation, structural classification, articular components, and movements.

D. Disorders, medical terminology

☐ 6. Describe the causes and symptoms of common joint disorders, including rheumatism, rheumatoid arthritis (RA), osteoarthritis (OA), gouty arthritis, Lyme disease, bursitis, ankylosing spondylitis, sprain, and strain.
☐ 7. Define medical terminology associated with articulations.

Now become familiar with the language of this chapter by studying each wordbyte, its meaning, and an example of its use within a term. After you study the entire list, self-check your understanding by writing the meaning of each wordbyte on the line. As you continue through the *Learning Guide*, identify (and fill in) additional terms that contain the same wordbyte.

Wordbyte	Self-check	Meaning	Example(s)
amphi-	_____	both	*amphi*arthrotic
arthr-	_____	joint	osteo*arthritis*
articulat-	_____	joint	*articulat*ion
cruci-	_____	cross	*cruci*ate
-itis	_____	inflammation	arth*ritis*
-osis	_____	condition of	syndesm*osis*
rheum-	_____	watery discharge	*rheum*atoid arthritis
syn-	_____	together	*syn*arthrotic

CHECKPOINTS

A. Classification of joints (pages 216–219)

■ **A1.** Define the term *articulation (joint)*.

List three structural features that affect movement at a joint.

■ **A2.** Name three classes of joints based on structure.

■ **A3.** Fill in the blanks below to name three classes of joints according to the amount of movement they permit.

a. Synarthrosis: _____

b. _____: slightly movable

c. _____: freely movable

■ **A4.** Describe synarthrotic (immovable) joints by completing this exercise.

a. Fibrous joints *(have? lack?)* a joint cavity. They are held together by

_____ connective tissue.

b. One type of fibrous joint is a _____ found between skull bones.

Such joints are *(freely? slightly? im-?)* movable, or _____-arthrotic.
Replacement of a fibrous suture with bony fusion is known as a *(synchondrosis? synostosis?)*.

c. Which of the following is a site of a *gomphosis?*
   A. Joint at distal ends of tibia and fibula
   B. Epiphyseal plate
   C. Attachment of tooth by periodontal ligament to tooth socket in maxilla or mandible

d. Synchondroses involve *(hyaline? fibrous?)* cartilage between regions of bone. An

example is the _____ cartilage between diaphysis and epiphysis
of a growing bone. This cartilage *(persists through life? is replaced by bone during adult life?)*. Synchondroses are *(somewhat movable? immovable?)*.

■ **A5.** Describe amphiarthrotic joints by completing this exercise.

a. The distal end of the tibia/fibula joint is a fibrous joint. It is *(more? less?)* mobile than

a suture and is therefore _____-arthrotic. It is called a

_____.

b. Fibrocartilage is present in the type of joint known as a _____.

These joints permit some movement and so are called _____-arthrotic.

Two locations of symphyses are _____ and _____.

■ **A6.** What structural features of synovial joints make them more freely movable than fibrous or cartilaginous joints?

■ **A7.** On Figure LG 9.1, color the indicated structures.

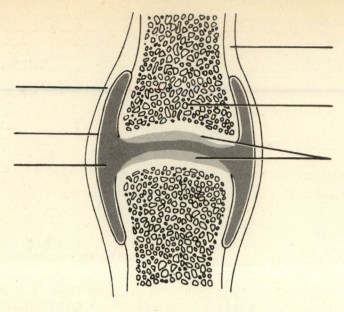

O	Articular cartilage	O	Periosteum
O	Articulating bone	O	Synovial (joint) cavity
O	Fibrous capsule	O	Synovial membrane

Figure LG 9.1 Structure of a generalized synovial joint. Color as directed in Checkpoint A7.

■ **A8.** Select the parts of a synovial joint (listed in the box) that fit the descriptions below.

> A. Articular cartilage   SF. Synovial fluid
> F. Fibrous capsule   SM. Synovial membrane
> L. Ligaments

_____ a. Hyaline cartilage that covers ends of articulating bones but does not bind them together

_____ b. With the consistency of uncooked egg white or oil, it lubricates the joint and nourishes the avascular articular cartilage

_____ c. Connective tissue membrane that lines synovial cavity and secretes synovial fluid

_____ d. Parallel fibers in some fibrous capsules; bind bones together

_____ e. Together these form the articular capsule (two answers)

_____ f. Provides both flexibility and tensile strength

_____ g. Reduces friction at the joint

_____ h. Contains phagocytes that remove debris from joint

A9. Describe the structure, function, and location of the following structures that are associated with synovial joints.

a. Intracapsular ligaments

b. Extracapsular ligaments

c. Bursae

A10. Do this activity about "torn cartilage" of the knee.

a. In this condition, the injured tissues are *(articular cartilages? menisci?)*.
b. What functions are normally served by these cartilages?

c. How may these cartilages be repaired with minimal destruction to other tissues?

B. Movements and types of synovial joints   (pages 219–277)

■ **B1.** *The Big Picture: Looking Back and Looking Ahead.* The design of synovial joints permits free movement of bones. However, if bones moved too freely, they would potentially move right out of their joint cavities (dislocation). A number of factors modify movement at synovial joints. Identify these factors as you refer to figures and text in your textbook. Use answers listed in the box to fill in lines with *.

| Bones | Hormones | Muscle | Ligaments | Soft parts |

a. Shape of articulating *_____ affects movement at joints, for example, at shoulder (Figure 8.4, page 199) and hip joints (Figure 8.10, page 207). Which joint has a deeper socket, contributing to increased stability and decreased mobility? *(Shoulder? Hip?)*

b. Position and tautness of *_____, such as the anterior cruciate (Figure 9.8e, page 230), strengthen the joint and limit excessive knee movement.

In about _____% of all serious knee injuries, this ligament is damaged. It normally

forms a cross ("cruci") with the _____ cruciate ligament. Both of these are *(intra? extra?)*-capsular ligaments.

c. Figure 11.16 (page 304) shows *_____ located on the anterior and posterior of the arm. The *(biceps? triceps?)* lies on the anterior of the arm. When this muscle contracts, the arm moves forward (flexion). The triceps, located on the *(anterior? posterior?)* of the arm, must stretch during this movement; because it has a limited ability to stretch, the triceps *(enhances? limits and opposes?)* flexion of the arm. In fact, the triceps is said to be an *antagonist* to the biceps (page 273).

d. Bend (flex) your own forearm (Figure 11.1, page 271). Try to place your wrist directly on your shoulder. If this is not possible for you, it is probably due to the meeting

(apposition) of *_____ of your forearm against those covering your arm. Muscles are "soft parts." Name several others.

e. One other factor affecting movement at joints is *_____ such as relaxin

(Exhibit 18.9, page 539). Made by ovaries and the _____ toward the end of pregnancy, this hormone helps to relax the anterior joint between hipbones, known as

the _____ _____ (Figure 29.10, page 972).

■ **B2.** Complete Table LG 9.1 on subtypes of diarthrotic joints. Select answers from lists in the box. The first one is done for you.

Subtype of diarthrotic joint	Planes of movement	Types of movement	Examples of joints
Ball-and-socket	Biaxial	ABD. Abduction	List name of joint,
Condyloid (ellipsoidal)	Monoaxial	ADD. Adduction	articulating bones
Gliding	Triaxial	CIR. Circumduction	with specific
Hinge		E. Extension	markings at points
Pivot		F. Flexion	of articulation.
Saddle		ROT. Rotation	

Table LG 9.1 Subtypes of diarthrotic joints.

Subtype of Diarthrotic Joint	Planes of Movement	Types of Movement	Examples
a. Ball-and-socket	Triaxial	F, E, ADD, ABD, CIR, ROT	1. Shoulder: scapula (glenoid cavity)–humerus (head) 2. Hip: hip bone (acetabulum)–femur (head)
b. Condyloid (ellipsoidal)			Wrist (radius–carpals)
c.			Carpal (trapezium)–first metacarpal (base)
d. Hinge		F, E	1. Knee: femur (condyles)–tibia (condyles)
e.	Monoaxial	ROT	
f.	(Omit)	Gliding	

■ **B3.** *For extra review* of types of synovial joints, choose the type of joint that fits the description. (Answers may be used more than once.)

B. Ball-and-socket	H. Hinge
C. Condyloid	P. Pivot
G. Gliding	S. Saddle

_____ a. Monaxial joint; only rotation possible

_____ b. Examples include atlas–axis joint and joint between head of radius and radial notch at proximal end of ulna

_____ c. Triaxial joint, allowing movement in all three planes

_____ d. Hip and shoulder joints

_____ e. Spoollike (convex) surface articulated with concave surface, for example, elbow, ankle, and joints between phalanges

_____ f. One type of monoaxial joint in which only flexion and extension are possible

_____ g. Found in joints at sternoclavicular and claviculoscapular joints

_____ h. Thumb joint located between metacarpal of thumb and carpal bone (trapezium)

_____ i. Biaxial joints (two answers)

■ **B4.** From the terms listed in the box, choose the one that fits the type of movement in each case. Not all answers will be used.

Abd. Abduction	E. Extension	I. Inversion
Add. Adduction	F. Flexion	P. Plantar flexion
C. Circumduction	G. Gliding	R. Rotation
D. Dorsiflexion		

_____ a. Decrease in angle between anterior surfaces of bones (or between posterior surfaces at knee and toe joints)

_____ b. Simplest kind of movement that can occur at a joint; no angular or rotary motion involved; example: ribs moving against vertebrae

_____ c. State of entire body when it is in anatomical position

_____ d. Movement away from the midline of the body

_____ e. Movement of a bone around its own axis

_____ f. Position of foot when heel is on the floor and rest of foot is raised

■ **B5.** Perform the action described. Then write in the name of the type of movement.

a. Describe a cone with your arm, as if you were winding up to pitch a ball.

The movement at your shoulder joint is called _____.

b. Stand in anatomical position (palms forward). Turn your palms backward. This action

is called _____.

c. Move your fingers from "fingers together" to "fingers apart" position. This action is

_____ of fingers.

d. Raise your shoulders, as if to shrug them. This movement is called

_____ of the shoulders.

e. Stand on your toes. This action at the ankle joint is called _____.

f. Grasp a ball in your hand. Your fingers are performing the type of movement called

_____.

g. Sit with the soles of your feet pressed against each other. In this position, your feet are

performing the action called _____.

h. Thrust your jaw outward (gently!). This action is _____ of the mandible.

■ **B6.** Identify the kinds of movements shown in Figure LG 9.2. Write the name of the movement below each figure. Use the following terms: *abduction, adduction, extension, flexion,* and *hyperextension*.

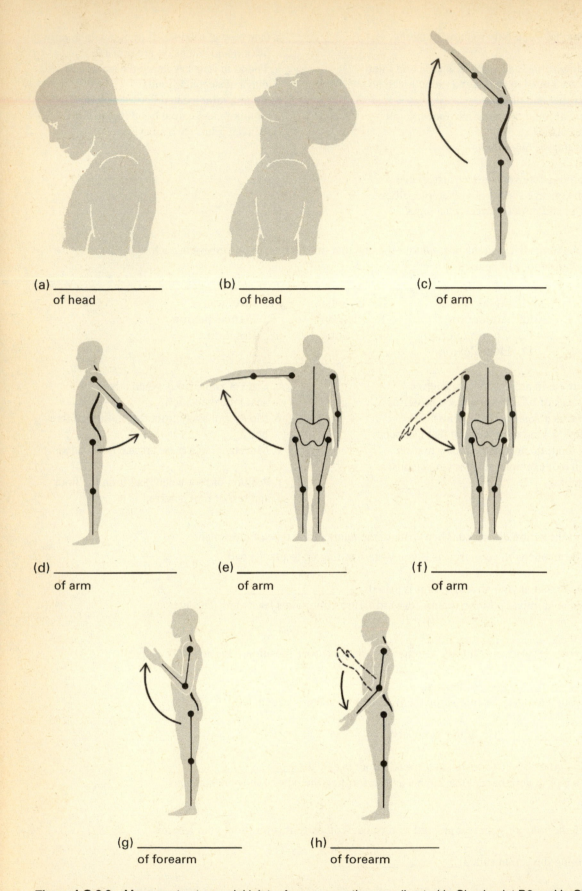

(a) _____
of head

(b) _____
of head

(c) _____
of arm

(d) _____
of arm

(e) _____
of arm

(f) _____
of arm

(g) _____
of forearm

(h) _____
of forearm

Figure LG 9.2 Movements at synovial joints. Answer questions as directed in Checkpoint B6 and in Chapter 11.

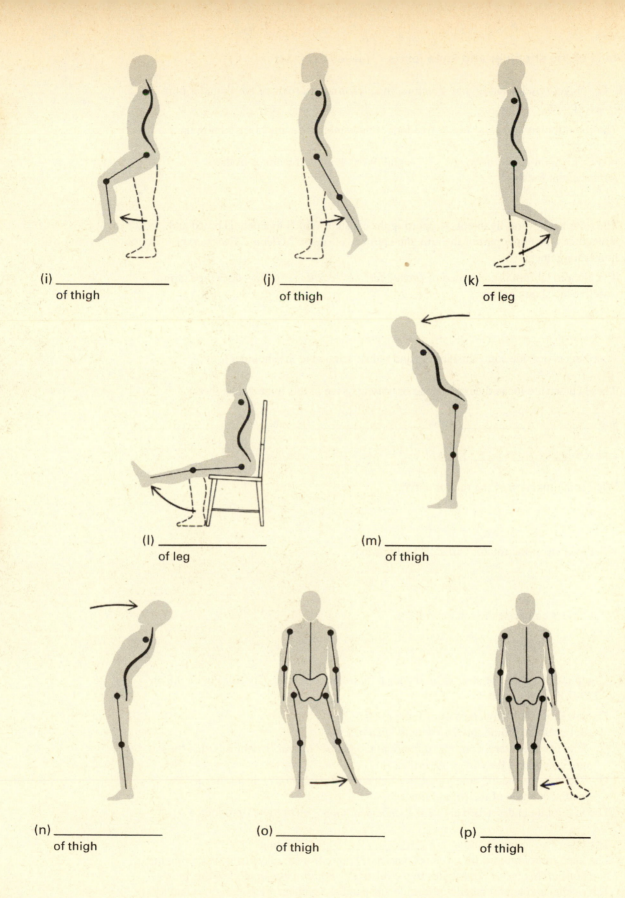

(i) _____
    of thigh

(j) _____
    of thigh

(k) _____
    of leg

(l) _____
    of leg

(m) _____
    of thigh

(n) _____
    of thigh

(o) _____
    of thigh

(p) _____
    of thigh

C. Details of the shoulder and knee joints   (pages 227–234)

■ **C1.** Check your understanding of structure, function, and disorders of the shoulder joint in this Checkpoint.

a. The shoulder has *(greater? less?)* freedom of movement than any other joint in the

body. This joint is said to be _____-axial. What structural features of the joint account for this?

b. The joint has several ligaments supporting the capsule. Which ligament is broad and strengthens the upper part of the articular capsule? *(Coraco-? Gleno-? Transverse)* humeral ligament

c. The glenoid labrum is composed of *(bone? fat? fibrocartilage?)*. How does the labrum stabilize the shoulder joint?

d. Four bursae provide some cushioning and reduce friction at the shoulder joint.

These bursae are located deep to bone or muscle tissue in this joint. Name them:

sub-_____, sub-_____, sub-_____,

and sub-_____.

e. What is the function of the rotator cuff?

What happens if the rotator cuff fails to do its job?

What tissues comprise the rotator cuff?

■ **C2.** Complete this exercise describing the knee (tibiofemoral) joint. Consult Figure 9.8 on pages 229–230 in your text.

a.   The knee joint actually consists of three joints:
   1. Between the femur and the *(patella? fibula?)*
   2. Between the lateral condyle of the femur, *(medial? lateral?)* meniscus, and the lateral condyle of the *(tibia? fibula?)*
   3. Between the medial condyle of the femur, *(medial? lateral?)* meniscus, and the medial condyle of the *(tibia? fibula?)*

b.   The first joint (1) described above is a *(modified hinge? pivot? gliding?)* joint.

   The other two joints (2 and 3) are _____ joints.

c.   The knee (tibiofemoral) joint *(does? does not?)* include a complete capsule uniting the two bones. This factor contributes to the relative *(strength? weakness?)* of this joint.

d.   The medial and lateral patella retinacula and patellar ligament are both tissues that serve as insertions of the *(hamstrings? quadriceps femoris?)* muscles.

e. A number of other ligaments provide strength and support. Identify whether the following ligaments are extracapsular (E) or intracapsular (I).

    1. Arcuate popliteal ligament: _____

    2. Anterior and posterior cruciate ligaments: _____

    3. Tibial (medial) and fibular (lateral) collateral ligaments: _____

f. Which of the ligaments listed in (e) is damaged in most serious knee injuries?

g. Two cartilages, much like two C-shaped stadiums facing one another, each making an incomplete circle, are located between condyles of the femur and tibia. These articular

discs are called _____. What is their function?

h. There are *(no? several?)* bursae associated with the knee joint. Inflammation of bursae is called _____.

i. *A clinical challenge.* The most common knee injuries of football players are likely to involve three structures beginning with the letter C. List the "three C's":

C3. Briefly describe each term associated with the knee joint.

a. Swollen knee and "water on the knee"

b. Dislocated knee

c. Arthroscopy

d. Arthroplasty

■ **C4.** Match the names of joints with descriptions provided.

A. Ankle joint	S. Shoulder
Ao. Atlanto-occipital joint	T. Temporomandibular joint
E. Elbow joint	W. Wrist
H. Hip (coxal) joint	

_____ a. Talocrural joint; capable of plantar flexion and dorsiflexion

_____ b. Glenohumoral joint

_____ c. A synovial joint with hinge and gliding actions; moves the lower (but not upper) jaw bone within the mandibular fossa of the temporal bone

_____ d. Radiocarpal joint

_____ e. Joint between the skull and the first cervical vertebra; condyloid joint

_____ f. Joint between the humerus (trochlea) and the trochlear (semilunar) notch of the ulna as well as the head of the radius

_____ g. Joint between the head of the femur and acetabulum of the coxal bone

D. Disorders, medical terminology (pages 224–235)

■ **D1.** How are *arthritis* and *rheumatism* related? Choose the correct answer.

    A. Arthritis is a form of rheumatism.
    B. Rheumatism is a form of arthritis.

■ **D2.** List several forms of arthritis. Name one symptom common to all forms of this ailment.

■ **D3.** Contrast *rheumatoid arthritis* with *osteoarthritis*. Write *Yes, No, Larger,* or *Smaller* for answers.

Table LG 9.2 Comparison of major types of arthritis.

	Rheumatoid Arthritis	Osteoarthritis
a. Is known as "wear-and-tear" arthritis		
b. Which types of joints are affected?		
c. Is this an inflammatory autoimmune condition?		
d. Is synovial membrane affected?		
e. Does articular cartilage degenerate?		
f. Does fibrous tissue join bone ends?		
g. Is movement limited?		

■ **D4.** Complete this exercise describing disorders involving articulations.

    a. Gouty arthritis is a condition due to an excess of _____ in the

    blood leading to deposit of _____ in joints. Gout can also involve

    damage to joints or to organs such as the _____ because crystals
    maybe deposited there also. This condition is more common among (*females? males?*).

    b. An acute chronic inflammation of a bursa is called _____.

    c. *Luxation,* or _____, is displacement of a bone from its joint
    with tearing of ligaments.

    d. Pain in a joint is known as _____.

e. _____ is a cluster of conditions named after a town in Connecticut. It is caused by *(bacteria? fungi? viruses?)* known as *Borrelia burgdorferi,* which are transmitted

by *(fleas? mosquitoes? ticks?)*. Signs and symptoms include skin _____, then possibly cardiac problems and/or facial paralysis, and then arthritis of *(larger? smaller?)* joints.
f. Ankylosing spondylitis affects primarily *(ankle and wrist? intervertebral and sacroiliac?)* joints.
g. A *(sprain? strain?)* is an overstretching of a muscle. A *(sprain? strain?)* involves more serious injury to joint structures. The joint most often sprained is the *(ankle? elbow?)* joint.

ANSWERS TO SELECTED CHECKPOINTS: CHAPTER 9

A1. Point of contact between bones, between bone and cartilage, or between teeth and bone; shape and fit of articulating bones, tension and position of connecting tissues (such as tendons, ligaments, and muscles).

A2. Fibrous, cartilage, and synovial.

A3. (a) Immovable. (b) Amphiarthrosis. (c) Diarthrosis.

A4. (a) Lack; fibrous. (b) Suture; im-, syn; syntostosis. (c) C. (d) Hyaline; epiphyseal; is replaced by bone during adult life; immovable.

A5. (a) More, amphi; syndesmosis. (b) Symphysis; amphi-; discs between vertebrae, symphysis pubis between hipbones.

A6. The space (synovial cavity) between the articulating bones and the absence of tissue between those bones (which might restrict movement) make the joints more freely movable. The articular capsule and ligaments also contribute to free movement.

A7.

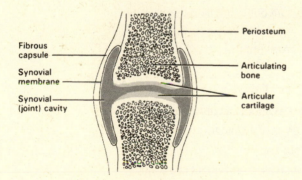

Figure LG 9.1A Structure of a generalized synovial joint.

A8. (a) A. (b) SF. (c) SM. (d) L. (e) SM and F. (f) F (and L, somewhat). (g) SF. (h) SF.

B1. (a) Bones; hip. (b) Ligaments; 70; posterior; intra. (c) Muscles; biceps; posterior; limits and opposes. (d) Soft parts; skin, fat, bursae, tendons, ligaments, and blood vessels. (e) Hormones; placenta, pubic symphysis.

B2.

Table LG 9.1A Subtypes of diarthrotic joints.

Subtype of Diarthrotic Joint	Planes of Movement	Types of Movement	Examples
b. Condyloid (ellipsoidal)	**Biaxial**	**F, E, ADD, ABD, CIR**	Wrist (radius–carpals)
c. **Saddle**	**Biaxial**	**F, E, ABD, ADD, CIR**	Carpal (trapezium)–first metacarpal (base)
d. Hinge	**Monoaxial**	F, E	1. Knee: femur (condyles)–tibia (condyles) 2. **Elbow: humerus (trochlea)–ulna (trochlear [semilunar] notch).** 3. **Ankle: Tibia and fibula–talus.** 4. **Phalanges** 5. **Atlanto–occipital: atlas–occipital bone (condyles)**
e. **Pivot**	Monoaxial	ROT	1. Atlas–axis (dens) 2. Radius (head)–ulna (radial notch)
f. Gliding	—	Gliding	1. Ribs (heads, tubercles)–vertebrae (bodies, transverse processes) 2. Clavicle–sternum (manubrium), clavicle–scapula (acromion)

B3. (a) P. (b) P. (c) B. (d) B. (e) H. (f) H. (g) G. (h) S. (i) C, S.

B4. (a) F. (b) G. (c) E. (d) Abd. (e) R. (f) D.

B5. (a) Circumduction. (b) Pronation. (c) Abduction. (d) Elevation. (e) Plantar flexion. (f) Flexion. (g) Inversion. (h) Protraction.

B6. (a) Flexion. (b) Hyperextension. (c) Flexion. (d) Hyperextension. (e) Abduction. (f) Adduction. (g) Flexion. (h) Extension. (i) Flexion, while leg also slightly flexed. (j) Hyperextension, while leg extended. (k) Flexion, while thigh extended. (l) Extension, with thigh flexed. (m) Flexion. (n) Hyperextension. (o) Abduction. (p) Adduction.

C1. (a) Greater; tri; loose articular capsule and large, shallow glenoid cavity/socket for the head of the humerus. (b) Coraco-. (c) Fibrocartilage; increases the depth of the glenoid cavity. (d) Acromion, coracoid, deltoid, scapular. (e) Helps to hold the humerus in the glenoid cavity; subluxation (dislocation) of the humerus out of the shoulder socket; four muscles and their tendons: supraspinatus, infraspinatus, teres minor, and subscapularis.

C2. (a) (1) Patella, (2) lateral, tibia, (3) medial, tibia. (Note that neither the femur nor the patella articulates with the fibula.) (b) Gliding; modified hinge. (c) Does not; weakness. (d) Quadriceps femoris. (e) (1) E, (2) I, (3) E. (f) Anterior cruciate. (g) Menisci; provide some stability to an otherwise unstable joint. (h) Several; bursitis. (i) (Tibial) collateral ligament, (anterior) cruciate ligament, and (medial meniscus) cartilage.
C4. (a) A. (b) S. (c) T. (d) W. (e) Ao. (f) E. (g) H.
D1. A.
D2. Rheumatoid arthritis (RA), osteoarthritis (OA), and gouty arthritis; pain.

D3.
Table LG 9.2A   Types of arthritis.

	Rheumatoid Arthritis	Osteoarthritis
a. Which is known as "wear-and-tear" arthritis?	No	Yes
b. Which types of joints are affected?	Smaller	Larger
c. Is this an inflammatory autoimmune condition?	Yes	No
d. Is synovial membrane affected?	Yes	no
e. Does articular cartilage degenerate?	Yes	Yes
f. Does fibrous tissue join bone ends?	Yes	No
g. Is movement limited?	Yes	Yes

D4. (a) Uric acid, sodium urate; kidneys; males. (b) Bursitis. (c) Dislocation. (d) Arthralgia. (e) Lyme disease; bacteria, ticks; rash, larger. (f) Intervertebral and sacroiliac. (g) Strain; sprain; ankle.

WRITING ACROSS THE CURRICULUM: CHAPTER 9

1. Describe arthroscopy and explain why this procedure may be advantageous for practitioners working with knee injuries.
2. Briefly describe arthroplasty of the hip joint.
3. Contrast the structure and function of fibrous joints and synovial joints.
4. Describe structures that strengthen the shoulder (scapulohumeral) joint.

MASTERY TEST: CHAPTER 9

Questions 1–14: Circle T (true) or F (false). If the statement is false, change the underlined word or phrase so that the statement is correct.

T F 1. A fibrous joint is one in which there is <u>no joint cavity and bones are held together by fibrous connective tissue.</u>

T F 2. <u>Sutures, syndesmoses, and symphyses</u> are kinds of fibrous joints.

T F 3. A sprain of a joint is <u>more</u> serious than a strain.

T F 4. <u>Ball-and-socket, gliding, pivot, and ellipsoidal joints</u> are all diarthrotic joints.

T F 5. All fibrous joints <u>are synarthrotic and all cartilaginous joints are amphiarthrotic.</u>

T F 6. In synovial joints synovial membranes <u>cover the surfaces of articular cartilages.</u>

T F 7. Bursae are <u>saclike structures that reduce friction</u> at joints.

T F 8. Synovial fluid becomes <u>more</u> viscous when there is increased movement at a joint.

T F 9. When your arm is in the supine position, your radius and ulna are <u>parallel (not crossed).</u>

T F 10. When you touch your toes, the major action you perform at your hip joint is called <u>hyperextension.</u>

T F 11. The <u>only type of joint that is triaxial</u> is the ball-and-socket.

T F 12. Abduction is movement <u>away from</u> the midline of the body.

T F 13. The elbow, knee, and ankle joints are all <u>hinge</u> joints.

T F 14. Joints that are relatively stable (such as hip joints) tend to have <u>more</u> mobility than joints that are less stable (such as shoulder joints).

Questions 15–16: Arrange the answers in correct sequence.

_____ _____ _____ 15. From most mobile to least mobile:
        A. Amphiarthrotic
        B. Diarthrotic
        C. Synarthrotic

_____ _____ _____ 16. Stages in rheumatoid arthritis, in chronological order:
        A. Articular cartilage is destroyed and fibrous tissue joins exposed bone.
        B. The synovial membrane produces pannus which adheres to articular cartilage.
        C. Synovial membrane becomes inflamed and thickened, and synovial fluid accumulates.

Questions 17–20: Circle the letter preceding the one best answer to each question.

17. Which structure is extracapsular in location?
    A. Meniscus
    B. Posterior cruciate ligament
    C. Synovial membrane
    D. Tibial collateral ligament

18. Which joint is amphiarthrotic and cartilaginous?
    A. Symphysis
    B. Synchondrosis
    C. Syndesmosis
    D. Synostosis

19. All of these structures are associated with the knee joint *except:*
    A. Glenoid labrum
    B. Patellar ligament
    C. Infrapatellar bursa
    D. Medial meniscus
    E. Fibular collateral ligament

20. A suture is found between:
    A. The two pubic bones
    B. The two parietal bones
    C. Radius and ulna
    D. Diaphysis and epiphysis
    E. Tibia and fibula (distal ends)

Questions 21–25: Fill-ins. Write the word or phrase that best completes the statement.

_____ 21. The term that means total hip replacement is a hip _____

_____ 22. _____ is the forcible wrenching or twisting of a joint with partial rupture of it, but without dislocation.

_____ 23. The action of pulling the jaw back from a thrust-out position so that it becomes in line with the upper jaw is the movement called _____.

_____ 24. Another name for a freely movable joint is _____.

_____ 25. The type of joint between the atlas and axis and also between proximal ends of the radius and ulna is a _____ joint.

ANSWERS TO MASTERY TEST: ☆ CHAPTER 9

True–False
1. T
2. F. Sutures, syndesmoses, and gomphoses
3. T
4. T
5. F. Either synarthrotic or amphiarthrotic, and the same is true of cartilaginous
6. F. Do not cover surfaces of articular cartilages, which may be visualized as floor and ceiling of a room; but syn-ovial membranes do line the rest of the inside of the joint cavity (much like wallpaper covering the four walls of the room).
7. T
8. F. Less
9. T
10. F. Flexion
11. T
12. T
13. T. (Or modified hinge joints)
14. F. Less

Arrange
15. B A C
16. C B A

Multiple Choice
17. D
18. A
19. A
20. B

Fill-ins
21. Arthroplasty
22. Sprain
23. Retraction
24. Diarthrotic
25. Pivot (or synovial or diarthrotic)

FRAMEWORK 10
Muscle Tissue

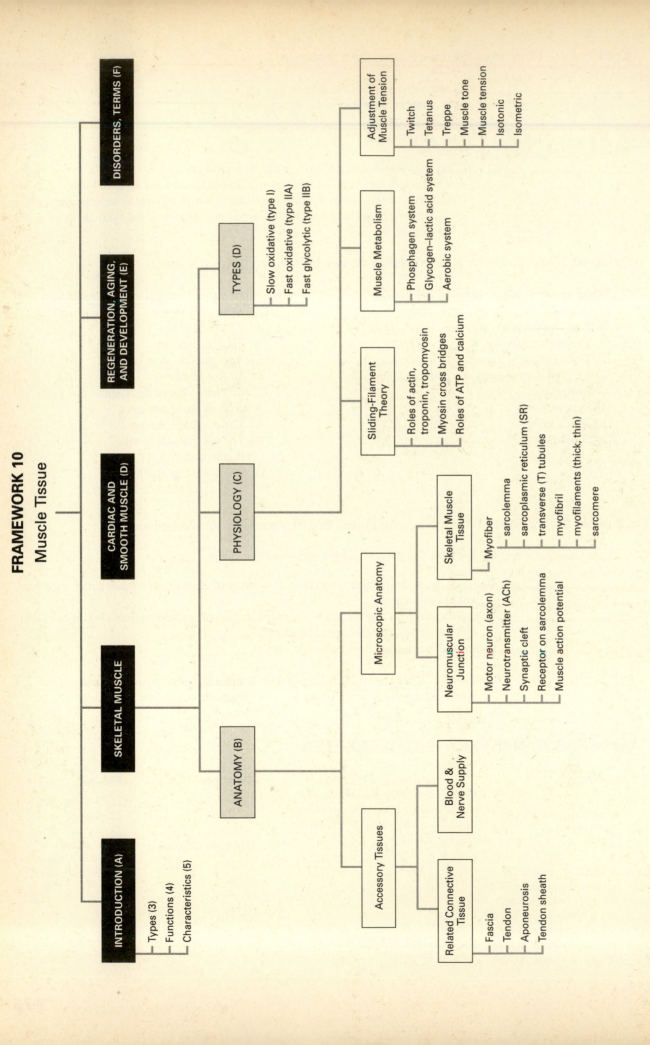

Muscle Tissue

Muscles are making it possible for you to read this paragraph: facilitating movement of your eyes and head, maintaining your posture, producing heat to keep you comfortable. Muscles do not work in isolation, however. Connective tissue binds muscle cells into bundles and attaches muscle to bones. Blood delivers the oxygen and nutrients for muscle work, and nerves initiate a cascade of events that culminate in muscle contraction.

A tour of the mechanisms of muscle action finds the visitor caught in the web of muscle protein filaments sliding back and forth as the muscle contracts and relaxes. Neurotransmitters, as well as calcium and power-packing ATP, serve as chief regulators. Muscles are observed contracting in a multitude of modes, from twitch to treppe and isotonic to isometric. It may be surprising to hear a tour guide's announcement that most muscle cells present now were actually around at birth. Very few muscle cells can multiply, but fortunately muscle cells can greatly enlarge, as with exercise. With aging, cell numbers decline and so do muscle strength and endurance. Certain disorders decrease muscle function and debilitate even in very early years of life.

As you begin your study of muscle tissue, carefully examine the Chapter 10 Topic Outline and Objectives; check off each one as you complete it. To organize your study of muscle tissue, glance over the Chapter 10 Framework now. Be sure to refer to the Framework frequently and note relationships among key terms in each section.

TOPIC OUTLINE AND OBJECTIVES

A. Types, functions, characteristics of muscle tissue

- [ ] 1. List the characteristics and functions of muscle tissue.

B. Skeletal muscle anatomy

- [ ] 2. Describe the structure and importance of a neuromuscular junction and a motor unit.

C. Skeletal muscle physiology: contraction, tension, metabolism

- [ ] 3. Describe the principal events associated with the sliding-filament mechanism of muscle contraction.

- [ ] 4. Explain the roles played by muscle in homeostasis of body temperature.
- [ ] 5. Explain how muscle tension can be varied.
- [ ] 6. Identify the sources of energy used during muscular contraction.

D. Types of skeletal muscle fibers; cardiac and smooth muscle

- [ ] 7. Describe the different types of skeletal muscle fibers and compare them to cardiac and smooth muscle fibers.
- [ ] 8. Compare the location, microscopic appearance, nervous control, functions, and regenerative capacities of the three kinds of muscle tissue.

E. Regeneration, aging, and development of muscle tissue

☐ 9. Explain the effects of aging on muscle tissue.
☐ 10. Describe the development of the muscular system.

F. Disorders, medical terminology

☐ 11. Define such common muscular disorders as fibromyalgia, muscular dystrophies, myasthenia gravis, spasm, cramps, tremor, fasciculation, fibrillation, and tic.
☐ 12. Define medical terminology associated with the muscular system.

WORDBYTES

Now become familiar with the language of this chapter by studying each wordbyte, its meaning, and an example of its use within a term. After you study the entire list, self-check your understanding by writing the meaning of each wordbyte on the line. As you continue through the *Learning Guide,* identify (and fill in) additional terms that contain the same wordbyte.

Wordbyte	Self-check	Meaning	Example(s)
a-	_____	not	*a*trophy
-algia	_____	pain	fibromy*algia*
apo-	_____	from	*apo*neurosis
dys-	_____	bad, difficult	muscular *dys*trophy
endo-	_____	within	*endo*mysium
-gen	_____	to produce	phospha*gen* system
-graph	_____	to write	electromyo*graphy*
-lemma	_____	rind, skin	sarco*lemma*
myo-	_____	muscle	*myo*fiber, *myo*sin
mys-	_____	muscle	epi*mys*ium
peri-	_____	around	*peri*mysium
sarco-	_____	flesh, muscle	*sarco*lemma
troph-	_____	nourishment	muscular dys*troph*y

CHECKPOINTS

A. Types, functions, characteristics of muscle tissue (pages 239–240)

■ **A1.** Match the muscle types listed in the box with descriptions below.

C. Cardiac   Sk. Skeletal   Sm. Smooth

_____ a. Involuntary muscle found in blood vessels and intestine

_____ b. Involuntary striated muscle

_____ c. Striated muscle attached to bones

_____ d. The only type of muscle that is voluntary

_____ e. Involuntary muscle in arrector pili muscles that causes "goose bumps."

■ **A2.** Describe four functions of muscles that are important for maintenance of homeostasis.

   a. _____ due to action of muscles pulling on bones.

   b. Movement of substances within the body, such as:

      1. _____ within blood vessels due to pumping of heart muscle,

      2. _____ within the digestive system,

      3. _____ through the urinary system, and

      4. _____ within the reproductive system.

   c. _____ of body position, and maintenance of stored

     fluids within organs such as _____.

   d. Thermogenesis, which is generation of _____ by muscles; an example

     is the warming effect of _____ during cold weather.

■ **A3.** Match the characteristics of muscles listed in the box with descriptions below.

Conductivity	Excitability
Contractility	Extensibility
Elasticity	

a. Ability of muscle to stretch without damaging the tissue _____

b. Tendency of stretched or contracted muscle to return to its original shape _____

c. Ability of muscle cells and nerve cells (neurons) to respond to stimuli by producing action potentials (impulses); irritability

_____

d. Ability of muscle (or nerve) cell to carry an action potential along the length of the cell _____

e. Ability of muscle tissue to shorten and thicken in response to action potential

_____

B. Skeletal muscle: anatomy (pages 240–247)

■ **B1.** Contrast two kinds of fascia by indicating which of the following are characteristics of superficial fascia (S) or deep fascia (D).

_____ a. Located immediately under the skin (subcutaneous)

_____ b. Composed of dense connective tissue that extends inward to surround and compartmentalize muscles

_____ c. A route for nerves and blood vessels to enter muscles; contains much fat, so provides insulation and protection

■ **B2.** Arrange the following terms (connective tissue) in correct sequence according to the amount of muscle surrounded: *endomysium, epimysium, perimysium.*

_____ → _____ → _____

   (entire muscle)   (bundle [fascicle] of muscle fibers)   (individual muscle fiber)

For extra review. Color these three connective tissues on Figure LG 10.1a. Label a *fascicle.*

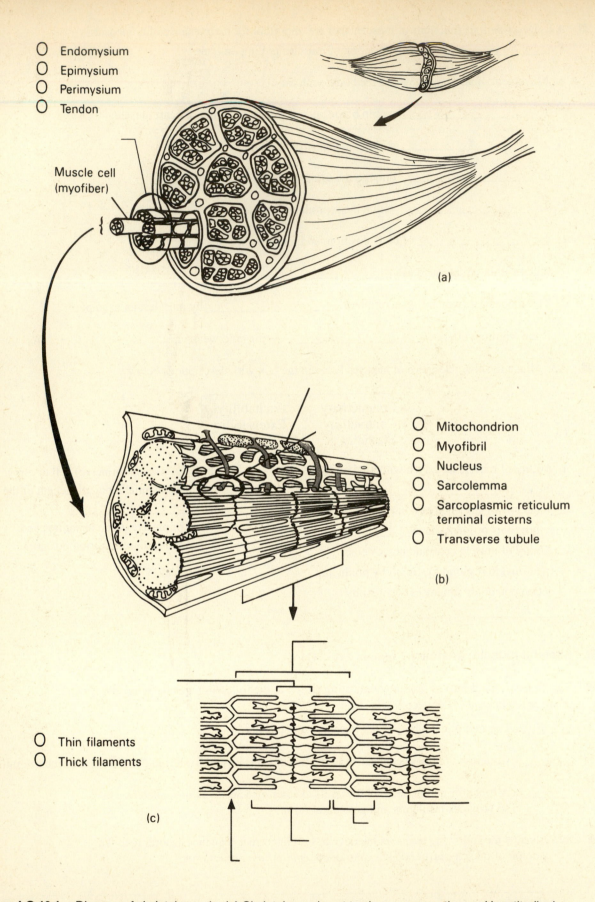

○ Endomysium
○ Epimysium
○ Perimysium
○ Tendon

Muscle cell
(myofiber)

(a)

○ Mitochondrion
○ Myofibril
○ Nucleus
○ Sarcolemma
○ Sarcoplasmic reticulum
  terminal cisterns
○ Transverse tubule

(b)

○ Thin filaments
○ Thick filaments

(c)

Figure LG 10.1   Diagram of skeletal muscle. (a) Skeletal muscle cut to show cross section and longtitudinal section with connective tissues. (b) Section of one muscle cell (myofiber). (c) Detal of sarcomere of muscle cell. Color and label as indicated in Checkpoints B2 and B7.

B3. Contrast the terms in the following pairs:

a. *Tendon/aponeurosis*

b. *Tendon/tendon sheath*

■ **B4.** Skeletal muscle tissue *(is? is not?)* vascular tissue. State two or more functions of blood vessels that supply muscles.

■ **B5.** Do this activity about the nerve supply of skeletal muscles.

a. In most cases a single nerve cell (neuron) innervates *(one muscle fiber [cell]? an average of 150 muscle fibers [cells]?)*. The combination of a neuron plus the muscle

fibers (cells) it innervates is called a _____. An example of a motor unit that is likely to consist of just two or three muscle fibers (cells) precisely innervated by one neuron is *(laryngeal muscles controlling speech? calf [gastrocnemius] muscles controlling walking?)*.

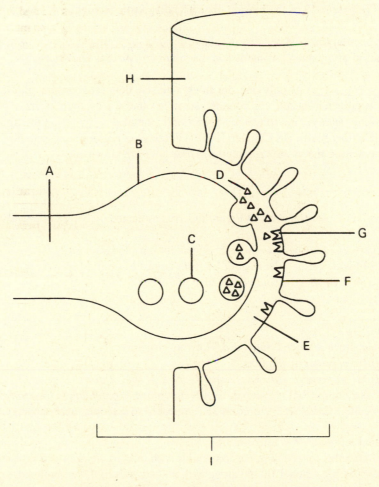

○ Motor neuron (axon with terminal)   ○ Neurotransmitter
○ Synaptic vesicle   ○ Skeletal muscle myofiber (cell)

Figure LG 10.2   Diagram of a neuromuscular junction (motor end plate). Refer to Checkpoints B5 and B6.

b. In other words, motor neurons form at least two, and possibly thousands of, branches

called _____, each of which supplies an individual skeletal
muscle fiber (cell). When the motor neuron "fires," *(just one muscle fiber supplied by
one axon? all muscle fibers within that motor unit?)* will be stimulated to contract.

c. Refer to Figure LG 10.2, in which we zoom in on the portion of a motor unit at which a
branch of one neuron stimulates a single muscle fiber (cell). A nerve impulse travels

along an axon terminal (at letter _____) toward one muscle fiber (letter _____).

The axon is enlarged at its end into a synaptic _____ (at letter B).

d. The nerve impulse causes synaptic vesicles (letter _____) to fuse with the
plasma membrane of the axon. Next the vesicles release the neurotransmitter (letter

_____) named _____ into the synaptic cleft (letter _____).

e. The region of the muscle fiber membrane (sarcolemma) close to the axon terminals is
called a *(motor end plate? neuromuscular junction [NMJ]?)*. This site contains specific

_____ (letter G) that recognize and bind to ACh. A typical motor end
plate contains about *(30–40? 3000–4000? 30–40 million?)* ACh receptors.

f. The effect of ACh is to cause Na⁺ channels in the sarcolemma to *(open? close?)* so that
Na⁺ enters the muscle fiber. As a result, an action potential is initiated, leading to
(contraction? relaxation?) of the muscle fiber. The diagnostic technique of recording

such electrical activity in muscle cells is called _____.

g. The combination of the axon terminals and the motor end plate is known as a

_____. At only one site along a muscle fiber (at about its middle)
does a nerve approach a muscle cell to innervate it. In other words, there is/are
usually *(only one? many?)* neuromuscular junction(s), labeled _____ on the figure,
for each muscle fiber (cell). (In Chapter 12 we will consider regions similar to NMJ's,
but where one neuron meets another neuron. These sites are known as

_____.)

B6. *For extra review.* Color the parts of Figure LG 10.2 indicated by color code ovals.

■ **B7.** Refer to Figure LG 10.1, and do this exercise about muscle structure.
a. Arrange the following terms in correct order from largest to smallest in size:
myofilaments (thick or thin), myofibrils, muscle fiber (cell).

_____ → _____ → _____
        (largest)   (smallest)

b. Using the leader lines provided, label the following structures:
Figure LG 10.1b: *sarcomere, nucleus*
Figure LG 10.1c: *sarcomere, A band, I band, H zone, M line, Z disc*

c. Sarcoplasmic reticulum (SR) is comparable to *(endoplasmic reticulum? ribosomes? mitochondria?)*
in nonmuscle cells. *(Ca²⁺? K⁺?)* stored in SR is released to sarcoplasm as the trigger

for muscle contraction. Ca²⁺ is released from SR by _____ channels.

d. Transverse (T) tubules *(are? are not?)* continuous with the sarcolemma and lie *(parallel?
perpendicular?)* to SR. Each T tubule along with dilated ends of SR on both sides

of it is known as a _____. Label the triad circled on Figure LG 10.1a.

e. Color all structures indicated by color code ovals in Figure LG 10.1.

■ **B8.** *For extra review.* Match the correct term from the list in the box with its description below.

A band	Sarcoplasm
I band	Triad
M line	Z disc (line)
Sarcomere	

a. Cytoplasm of muscle cell _____

b. A transverse tubule, along with terminal cisternae of sarcoplasmic reticulum on either

   side _____

c. Extends from Z disc to Z disc _____

d. Dark area in striated muscle; contains thick

   and thin filaments _____

e. Light area on either side of Z disc; location of

   thin filaments only _____

f. Located in the center of the H zone, it consists of protein molecules connecting adjacent thick

   filaments _____

g. Located in the center of the I band, it serves as an anchoring point for actin molecules. Its "zigzag" appearance may give a clue to its

   name. _____

■ **B9.** After you study Figure 10.6, page 246 in your text, contrast three types of filaments in skeletal muscle fibers.

a. Thin filaments are anchored at *(M lines? Z discs?)* and *(do? do not?)* extend into H zones. These filaments are composed mostly of bean-shaped *(actin? myosin?)* molecules that are twisted into a helix. Thin filaments also contain two other proteins. The protein that covers myosin-binding sites in relaxed muscle is called *(tropomyosin? troponin?).*

b. Each *(thick? thin?)* filament is composed of about 200 myosin molecules. Each molecule is shaped like two intertwined *(footballs? golf clubs?)*. The ends of the "handles" point toward

   the *(M lines? Z discs?)*. The rounded head of the "club" is called a _____ and it attaches to *(actin? troponin?)* as muscles begin contraction.

c. The third type of filament, called a(n) _____ filament, is composed

   of the protein named _____. Its name comes from the fact that this

   protein is _____.

d. Write C next to muscle proteins that are contractile and write R next to proteins that are regulatory.

   Actin: _____   Myosin: _____   Tropomyosin: _____   Troponin: _____

■ **B10.** *A clinical challenge.* Describe possible effects of intensive exercise upon muscles by filling in the blanks in this activity.

a. The acronym DOMS refers to D_____ O_____

   M_____ S_____.

b. Three muscle structures that may be damaged with DOMS are

   _____, _____, and _____.

c. _____ and _____ are two enzymes normally present within healthy muscle cells. Increased blood levels of these enzymes may be correlated to the degree of muscle cell damage, as these chemicals move out of injured cells and into blood.

d. DOMS are likely to occur within _____ to _____ hours after strenuous exercise.

C. Skeletal muscle physiology: contraction, tension, metabolism (pages 247–257)

■ **C1.** Summarize one theory of muscle contraction in this Checkpoint.

a. In order to effect muscle shortening (or contraction), heads (cross bridges) of

_____ myofilaments act like oars pulling on _____
molecules of thin filaments.

b. As a result, *(thick? thin?)* myofilaments move toward the center of the sarcomere.
Because the thick and thin myofilaments *(shorten? slide?)* to decrease the length of the

sarcomere, this theory of muscle contraction is known as the _____ theory.

■ **C2.** Complete this exercise describing the principal events that occur during muscle
contraction and relaxation.

a. In Checkpoint B5 we discussed stimulation of a muscle fiber at a neuromuscular junction.
The effect of release of the neurotransmitter *(ACh? AChE?)* upon the muscle is spread of
an action potential (impulse) from the sarcolemma via *(myosin? T tubules?)* to SR.

b. In a relaxed muscle the concentration of calcium ions (Ca^{2+}) in sarcoplasm is *(high? low?)*
due to action of Ca^{2+} active transport pumps located in *(sarcoplasm? SR membrane?).*
The effect of nerve stimulation of the muscle fiber is the opening of Ca^{2+}

_____ channels in the SR membrane. As a result, Ca^{2+} level

_____-creases in the sarcoplasm surrounding thick and thin myofilaments.

c. In relaxed muscle, myosin cross bridges *(are? are not?)* attached to actin in thin

filaments and _____ is bound to myosin cross bridges,

while the tropomyosin–_____ complex blocks binding sites on actin.

d. The released calcium ions attach to *(myosin? troponin?),* causing a structural change
which leads to exposure of binding sites on *(myosin? actin?).*

e. Breakdown of ATP (on myosin cross bridges) occurs via action of the enzyme named

_____ (also from the myosin cross bridge). Energy derived from ATP

activates myosin heads to bind to _____, swivel, and then slide actin

in an oarlike movement called a _____ stroke.

f. More ATP is then needed to bind to the myosin head and detach it from

_____. Splitting of ATP by ATPase located on *(actin? myosin?)*
again activates myosin and returns it to its original position, where it is ready to bind to

another _____-binding site on an actin further along the thin filament.

g. Repeated power strokes slide actin filaments *(toward or even across? away from?)*
the H zone and M line and so shorten the sarcomere (and entire muscle). If this process
is compared to running on a treadmill, *(actin? myosin?)* plays the role of the runner
staying in one place, whereas the treadmill that *slides* backward (toward H zones) is
comparable to the *(thick? thin?)* sliding filament.

h. Relaxation of a muscle occurs when a synaptic cleft enzyme named _____ destroys ACh. This terminates impulse conduction over the muscle. Calcium ions are then

sequestered from sarcoplasm back into _____ by *(active transport pumps?*

diffusion?) and via a calcium-binding protein named _____. These mechanisms are highly effective, causing the Ca^{2+} level in sarcoplasm of a relaxed muscle to be *(10? 10,000?)* times lower than that inside the SR.

i. With such a low level of Ca^{2+} now in the sarcoplasm surrounding myofilaments, the troponin–tropomyosin complex once again blocks binding sites on

_____. As a result, thick and thin filaments detach, slip back into

normal position, and the muscle is said to _____.

j. The movement of Ca^{2+} back into sarcoplasmic reticulum (C2h) is *(an active? a passive?)* process. After death, a supply of ATP *(is? is not?)* available, so an active transport process cannot occur. Explain why the condition of rigor mortis results.

■ **C3.** *For extra review* of roles of different chemicals within muscle during contraction and relaxation, match the names of chemicals in the box with descriptions below.

ACh	ATP	Calsequestrin
AChE	ATPase	Myosin
Actin	Ca^{2+}	Troponin–tropomyosin

_____ a. Neurotransmitter released by a nerve at a myoneural junction

_____ b. Enzyme that destroys ACh within the synaptic cleft

_____ c. Double golf-club-shaped molecule that forms thick filaments

_____ d. Main protein in thin filaments

_____ e. Called regulatory proteins because their attachment (or removal) from myosin-binding sites on actin determines whether muscle is in relaxed (or contracted) state

_____ f. Chemical that binds to and changes shape of troponin–tropomyosin complex, exposing myosin-binding site of actin

_____ g. Provides energy for swivel action of myosin upon actin in oarlike "power stroke"

_____ h. Enzyme located on myosin; breaks down (hydrolyzes) ATP

_____ i. Calcium-binding protein that effectively sequesters (hides) Ca^{2+} within SR

■ **C4.** Do this activity on the roles of muscle in regulation of your body temperature.

a. As you exercise, your body temperature *(in? de?)*-creases. What percentage of energy used during muscle contraction is likely to go to fuel your muscle contractions?

_____% What percentage is used to increase your body temperature? _____%

b. When your body is cold, an involuntary increase in muscle tone called

_____ can raise your body temperature. This process is initiated by

the "thermostat" portion of your brain, known as your _____.

■ **C5.** State the all-or-none principle.

This principle applies to *(individual motor units? an entire muscle such as the biceps?)*.

■ **C6.** List three or more factors that can decrease the strength of a muscle contraction.

■ **C7.** Match the terms in the box with definitions below. One answer will be used twice; one answer will not be used.

C. Contraction period	Ref. Refractory period
L. Latent period	Rel. Relaxation period
M. Myogram	T. Twitch

_____ a. Rapid, jerky response to a single action potential by a motor neuron

_____ b. Recording of a muscle contraction

_____ c. Period between application of a stimulus and start of a contraction; Ca^{2+} is being released from the SR during this time

_____ d. Period when a muscle is not responsive to a stimulus

_____ e. Active transport of Ca^{2+} back into the SR is occurring

_____ f. A short period (about 0.005 second) in skeletal muscle and a long one (about 0.30 second) in cardiac muscle

■ **C8.** Choose the type of contraction that fits each descriptive phrase.

Trepp	Tetanus

a. Sustained contraction due to stimulation at a rate of 20–100 stimuli per second:

_____

b. More forceful contraction of skeletal muscle in response to same strength stimuli after muscle has contracted several times:

_____

c. Phenomenon that is the principle behind athletic warmups: _____

d. Most voluntary contraction of muscles, such as biceps: _____

e. Due to a progressive buildup of Ca^{2+} in the sarcoplasm (two answers):

_____

■ **C9.** Do this activity about changes in the force of muscle contraction.

a. When a muscle is stretched to its optimal length, there is *(much? little or no?)* overlap of myosin cross bridges on thick filaments with actin on thin filaments. In this case, a muscle demonstrates *(minimal? maximal?)* force of contraction. When a skeletal muscle is stretched excessively (such as to 175% of its optimal length), then *(many? no?)* myosin cross bridges can bind to actin.

b. As a general rule, the more a muscle is stretched (within limits), the *(stronger? weaker?)* the contraction.

C10. Explain how motor unit *recruitment* is related to production of smooth movements.

■ **C11.** *A clinical challenge.* Select terms from the box that are related to the descriptions below.

Hypertonic	Muscle tone
Hypotonic	

a. Spasticity or rigidity of muscles:

_____

b. Flaccid muscles:

_____

c. Sustained, small contractions that cause firmness in relaxed muscle:

_____

C12. Contrast *active tension* with *passive tension* of muscles. Be sure to state factors contributing to each type of tension.

■ **C13.** Contrast isometric and isotonic contractions by doing this exercise.
  a. A contraction in which a muscle shortens while tension (tone) of the muscle remains constant is known as an *(isometric? isotonic?)* contraction.
  b. In an isometric contraction the muscle length *(shortens? stays about the same?)* and tension of the muscle *(increases? stays the same?)*.

C14. Contrast terms in each pair below.
a. *Muscular atrophy/muscular hypertrophy*

b. *Disuse atrophy/denervation atrophy*

■ **C15.** Complete this exercise about energy sources for muscle contraction.

a. Breakdown of *(ADP? ATP?)* provides the energy muscles use for contraction. Recall from Checkpoint C2e that ATP is attached to *(actin? myosin?)* cross bridges and so is available to energize the power stroke. Complete the chemical reaction showing ATP breakdown.

ATP →

b. ATP must be regenerated constantly. One method involves use of ADP and energy from food sources. Complete that chemical reaction.

ADP +

In essence, ADP is serving as a transport vehicle that can pick up and drop off energy stored in an extra high energy phosphate bond (~ P).

c. But ATP is used for other cell activities such as _____. To assure adequate energy for muscle work, muscle cells contain an additional molecule for

transporting high energy phosphate; this is _____. Complete the reaction in Figure LG 10.3 showing how the phosphocreatine (PC) and ADP transport "vehicles" can meet and transfer the high energy phosphate "trailer" so that more ATP is formed for muscle work.

d. How is phosphocreatine (PC) regenerated during time when muscles are at rest? Show this on Figure LG 10.3.

e. ATP and PC, together called the _____ system, provide only enough energy to power muscle activity for about *(an hour? 10 minutes? 15 seconds?)*. After that, muscles turn first to *(aerobic? anaerobic?)* pathways and later to *(aerobic? anaerobic?)* pathways. Complete Figure 10.4 to show how muscles get energy via these pathways.

f. During strenuous exercise an adequate supply of oxygen may not be available to muscles.

They must convert pyruvic acid to _____ acid by an *(aerobic? anaerobic?)* process. Excessive amounts of lactic acid in muscle tissue contribute to some muscle

_____. Name several types of cells that can use lactic acid to form ATP.

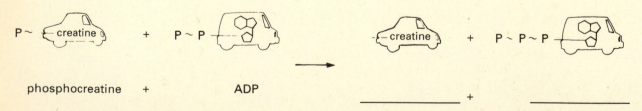

P~ [creatine]   +   P~P [   ]   -[creatine]   +   P~P~P [   ]

phosphocreatine   +   ADP   _____   +   _____

Figure LG 10.3 High energy molecules of the phosphagen system: the (~P) "trailer" tradeoff. Complete figure as indicated in Checkpoint C15c, d.

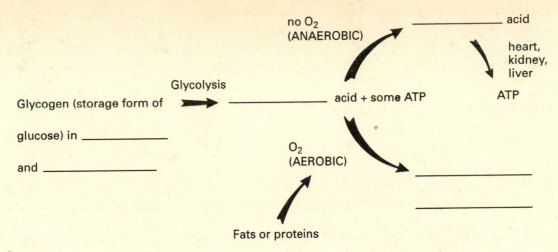

Figure LG 10.4 Anaerobic and aerobic energy sources for muscles. Complete figure as indicated in Checkpoint C15e.

g. One substance in muscle that stores oxygen until oxygen is needed by mitochondria is

_____. This protein is structurally somewhat like the _____

_____ -globin molecule in blood that also binds to and stores oxygen. Both of these molecules have a

_____ color that accounts for the color of blood and also of red muscle.

h. During exercise lasting more than 10 minutes, more than 90% of ATP is provided by the *(anaerobic? aerobic?)* breakdown of pyruvic acid (See Figure LG 10.4.). Athletic training *(de? in?)*-creases the maximal rate at which mitochondria can use oxygen for aerobic

catabolism. This rate is known as maximal _____ _____.

■ **C16.** *For extra review.* Complete this Checkpoint to identify metabolic processes used to supply energy for maximal exercise of varying durations. Select answers from those in the box.

> A. Aerobic metabolism
> G–L. Glycogen–lactic acid system
> P. Phosphagen system

a. _____ a. 15–30 sec (100-meter dash)

b. _____ b. 30–40 sec (300-meter race); carbohydrate loading prior to an athletic event helps prepare muscles by maximizing glycogen storage there; does not require O_2

c. _____ c. Longer than 10 minutes (2 miles to 26-mile marathon); "fuel" sources include glucose, fatty acids, and amino acids; does require O_2.

■ **C17.** Contrast effects of different types of training in this Checkpoint.

a. Repeated isometric contractions (such as weight lifting) rely on *(aerobic? anaerobic?)* ATP production. Repeated short bouts of exercise such as sprints also depend on *(aerobic? anaerobic?)* ATP production.

b. Repeated isotonic contractions (such as jogging or dancing) rely on *(aerobic? anaerobic?)* ATP production. *(Aerobic? Anaerobic?)* workouts as in endurance training or prolonged jogging are more likely to increase blood vessels to skeletal and heart muscle to provide needed oxygen.

C18. Describe these two potential effects of exercise.

a. Oxygen debt

b. Muscle fatigue

D. Types of skeletal muscle fibers; cardiac and smooth muscle (pages 257–263)

■ **D1.** Refer to the table below and contrast the three types of skeletal muscle fibers by doing this activity. You may find it helpful to refer back to Checkpoints C16 and C17 also.

Type and Rate of Contraction	Color	Myoglobin Concentration	Mitochondria and Blood Vessels	Source of ATP	Fatigue Easily?
I. Slow oxidative	Red	High	Many	Aerobic	No
IIA. Fast oxidative	Red	Very high	Very high	Aerobic	Moderate
IIB. Fast glycolytic	White	Low	Few	Anaerobic	Yes

a. Postural muscles (as in neck and back) are used *(constantly? mostly for short bursts of energy?)*. Therefore it is appropriate that they contract at a *(fast? slow?)* rate. Such muscle tissue consists mostly of slow red fibers that *(do? do not?)* fatigue easily. These appear red because they have *(much? little?)* myoglobin and *(many? few?)* blood vessels. They have *(many? few?)* mitochondria and therefore can depend on *(aerobic? anaerobic?)* metabolism for energy. These muscles are classified as type *(I? IIA? IIB?)*.

b. Muscles of the arms are used *(constantly? mostly for short bursts of energy?)* as in lifting and throwing. Therefore they must contract at a *(fast? slow?)* rate. The arms consist mainly of fast twitch white fibers. These respond rapidly to nerve impulses and contract *(fast? slowly?)*. They *(do? do not?)* fatigue easily because they are designed for the relatively inefficient processes of *(aerobic? anaerobic?)* metabolism. These muscles are classified as *(I? IIA? IIB?)*.

c. Endurance exercises tend to enhance development of the more efficient (aerobic)

_____ fibers, whereas weight lifting, which requires short bursts of

energy, tends to develop _____ fibers.

d. Most skeletal muscles, such as the biceps brachii, are *(all of one type? mixtures of types I, IIA, and IIB?)*. The muscles of any one motor unit are *(all of one type? mixtures of types I, IIA, and IIB?)*.

e. Arrange the three types of fibers in the sequence in which they are "called to action":

1. _____ → 2. _____ → 3. _____
   (weak contraction)   (stronger contraction)   (maximal contraction)

f. Endurance training (such as running or swimming) tends to cause transformation of some _____.

   A. Fast oxidative (IIA) fibers into fast glycolytic (IIB) fibers

   B. Fast glycolytic (IIB) fibers into fast oxidative (IIA) fibers

D2. State the desired effects as well as disadvantages of uses of anabolic steroids.

D3. Compare the structure and function of skeletal and cardiac muscle by completing this table. State the significance of characteristics that have asterisks (*).

Characteristic	Skeletal Muscle	Cardiac Muscle
a. Number of nuclei per myofiber	Several (multinucleate)	
b. Number of mitochondria per myofiber (More or Fewer)*		
c. Striated appearance due to alternated actin and myosin myofilaments (Yes or No)	Yes	
d. Arrangement of muscle fibers (Parallel or Branching)		
e. Nerve stimulation required for contraction (Yes or No)*		
f. Length of refractory period (Long or Short)*		
g. Duration of muscle contraction (Long or Short)*		

D4. State the significance of intercalated discs in the function of cardiac muscle fibers.

D5. Compare the two types of smooth muscle.

Muscle Type	Structure	Spread of Stimulus	Locations
a. Visceral		Impulse spreads and causes contraction of adjacent fibers	
b.			Blood vessels, iris of eye, large airways, arrector pili

■ **D6.** Contrast smooth muscle with the muscle tissue you have already studied by circling the correct answers in this paragraph.

Smooth muscle fibers are *(cylinder shaped? tapered at each end?)* with *(several nuclei? one nucleus?)* per cell. They *(do? do not?)* contain actin and myosin. However, due to the irregular arrangement of these filaments, smooth muscle tissue appears *(striated? nonstriated or "smooth"?)*. In general, smooth muscle contracts and relaxes more *(rapidly? slowly?)* than skeletal muscle does, and smooth muscle holds the contraction for a *(shorter? longer?)* period of time than skeletal muscle does. This difference in smooth muscle is due at least partly to the fact that smooth muscle has *(no? many?)* transverse tubules to transmit calcium into the muscle fiber and has a *(large? small?)* amount of SR.

■ **D7.** Smooth muscle *(is? is not?)* normally under voluntary control. List three chemicals released in the body or other factors that can also lead to smooth muscle contraction or relaxation.

■ **D8.** *For extra review* of the three muscle types, select the correct muscle type from the box that matches the description below.

C. Cardiac	Sk. Skeletal	Sm. Smooth

_____ a. Starts most slowly and lasts much longer

_____ b. Has no transverse tubules and only scanty sarcoplasmic reticulum

_____ c. Has no striations because thick and thin filaments are not arranged in a regular pattern

_____ d. Has intermediate filaments attached to dense bodies (comparable to Z discs)

_____ e. Calmodulin, the regulatory protein in this muscle, activates an enzyme called myosin lightchain kinase

_____ f. These muscles demonstrate the stress-relaxation response

_____ g. Spindle-shaped fiber with one nucleus per cell

_____ h. Unbranched, cylindrical cells; multinucleate

_____ i. Cell diameter may be very large; cells may reach lengths up to 30 cm (over 10 inches)

_____ j. Has two T tubules per sarcomere (at A band–I band junctions)

_____ k. Has fastest speed of contraction

_____ l. Stimulated only by the neurotransmitter acetylcholine (ACh)

_____ m. Includes three types: I, IIA, and IIB

_____ n. Branched, cylindrical cells; usually have one nucleus per cell

_____ o. Found only in the heart

_____ p. Striated, involuntary muscle

_____ q. Contains intercalated discs containing desmosomes and gap junctions

_____ r. Has one T tubule per sarcomere (at Z disc)

D9. *The Big Picture: Looking Back.* For additional review of the three muscle types, look back at Chapter 4, Checkpoints E1–E3, and Figure LG 4.1, M–O, pages 83 and 75.

E. Regeneration, aging, and development of muscle tissue   (pages 264–265)

■ **E1.** Arrange the three types of your muscle tissue in correct sequence according to ability (from most to least) to regenerate during your lifetime.

_____ _____ _____   A. Heart muscle
B. Biceps muscle
C. Muscle of an artery or intestine

■ **E2.** Both skeletal and cardiac muscle tissue increases in size by *(hyperplasia? hypertrophy?)*, which means increase in *(number? size?)* of cells. If either of these tissues is damaged, fibrosis,

which means replacement of muscle tissue with _____ tissue, is likely to occur.

E3. Describe the roles of the following cells in regeneration of muscle tissue:

a. Satellite cells

b. Pericytes

E4. List three changes in skeletal muscle that are likely to occur with normal aging.

■ **E5.** Complete this exercise about the development of muscles.

a. Which of the three types of muscle tissue develop from mesoderm? *(Skeletal? Cardiac? Smooth?)*

b. Part of the mesoderm forms columns on either side of the developing nervous system.

This tissue segments into blocks of tissue called _____. The first pair of somites forms on day *(10? 20? 30?)* of gestation. By day 30 a total of

_____ pairs of somites are present.

c. Which part of a somite develops into vertebrae? *(Myo-? Derma-? Sclero-?)* derm. The

dermis of the skin and other connective tissues is formed from _____

-tomes, while most skeletal muscles develop from _____-tomes.

F. Disorders, medical terminology (pages 265–266)

■ **F1.** Write the correct medical term related to muscles after its description.

a. Muscle or tendon pain and stiffness, such as "charley horse":

_____

b. Loss or impairment of motor or muscular function due to nerve or muscle disorder:

_____

c. Inherited, muscle-destroying disease causing atrophy of muscles:

_____

d. A muscle tumor: _____

e. Any disease of muscle: _____

■ **F2.** Describe myasthenia gravis in this exercise.

a. In order for skeletal muscle to contract, a nerve must release the chemical

_____ at the myoneural junction. Normally ACh binds to

_____ on the muscle fiber membrane.

b. It is believed that a person with myasthenia gravis produces _____
that bind to these receptors, making them unavailable for ACh binding. Therefore, ACh
(can? cannot?) stimulate the muscle, and it is weakened.

c. One treatment for this condition employs _____ drugs, which
enhance muscle contraction by permitting the ACh molecules that *do* bind to act longer
(and not be destroyed by AChE).

d. Other treatments include _____-suppressants, which decrease

the patient's antibody production, or _____, which segregates
the patient's harmful antibodies.

F3. Define each of these types of abnormal muscle contractions.

a. Spasm

b. Fibrillation

c. Tic

A1. (a) Sm. (b) C. (c) Sk. (d) Sk. (e) Sm.

A2. (a) Motion or movement. (b1) Blood, (b2) food, bile, enzymes, or wastes, (b3) urine, (b4) ova and sperm or eggs. (c) Stabilization, stomach or bladder. (d) Heat, shivering.

A3. (a) Extensibility. (b) Elasticity. (c) Excitability. (d) Conductivity. (e) Contractility.

B1. (a) S. (b) D. (c) S.

B2. Epimysium → perimysium → endomysium. See Figure LG 10.1A

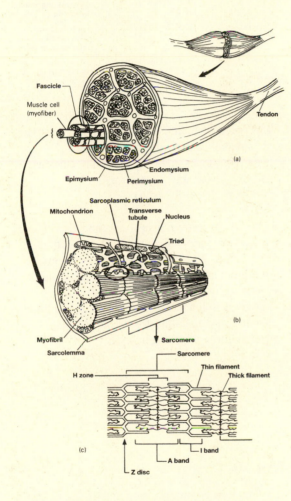

Figure LG 10.1A Diagram of skeletal muscle. (a) Skeletal muscle cut to show cross section and longitudinal section with connective tissues. (b) Section of one muscle cell (myofiber). (c) Detail of sarcomere of muscle cell.

B4. Is; (1) provide nutrients and oxygen for generation of ATP, (2) remove wastes.

B5. (a) An average of 150 muscle fibers [cells]; motor unit; laryngeal muscles controlling speech. (b) Axons; all muscle fibers within that motor unit. (c) A, H; end bulb. (d) C; D, acetylcholine (ACh), E. (e) Motor end plate; receptors; 30–40 million. (f) Open; contraction; electromyography (EMG). (g) Neuromuscular junction (NMJ); only one, I; synapses.

B7. (a) Muscle fiber (cell) → myofibrils → myofilaments (thick or thin). (b) See Figure LG 10.1A. (c) Endoplasmic reticulum; Ca^{2+}; Ca^{2+} release. (d) Are, perpendicular; triad; see Figure LG 10.1A. (e) See Figure LG 10.1A.

B8. (a) Sarcoplasm. (b) Triad. (c) Sarcomere. (d) A band. (e) I band. (f) M line. (g) Z disc (line).

B9. (a) Z discs, do not; actin; tropomyosin. (b) Thick; golf clubs; M lines; cross bridge, actin. (c) Elastic, titin (or connectin); a very large (titanic) protein, and it connects. (d) Actin and myosin: C; troponin and tropomyosin: R.

B10. (a) Delayed onset muscle soreness. (b) Sarcolemmas, myofibrils, Z discs. (c) Lactic dehydrogenase (LDH), creatine phosphokinase (CPK). (d) 12 to 48.

C1. (a) Myosin, actin. (b) Thin; slide, sliding-filament.

C2. (a) ACh, T tubules. (b) Low, SR membrane; release; in. (c) Are not, ATP, troponin. (d) Troponin, actin. (e) ATPase; actin, power. (f) Actin; myosin, myosin. (g) Toward or even across; myosin, thin. (h) AChE; SR, active transport pumps, calsequestrin; 10,000. (i) Actin; relax. (j). An active; is not; myosin cross bridges stay attached to actin and the muscles remain in a state of partial contraction (rigor mortis) for about a day.

C3. (a) ACh. (b) AChE. (c) Myosin. (d) Actin. (e) Troponin–tropomyosin. (f) Ca^{2+}. (g) ATP. (h) ATPase. (i) Calsequestrin.

C4. (a) In; 15; 85. (b) Shivering; hypothalamus.

C5. A single action potential (nerve impulse) elicits a single contraction in all of the muscle fibers of its motor unit; muscle fibers do not partially contract. Individual motor units.

C6. Decrease in frequency of stimulation by neurons, shorter length muscle fibers just prior to contraction, fewer and smaller motor units recruited, lack of nutrients, lack of oxygen, and muscle fatigue.

C7. (a) T. (b) M. (c) L. (d) Ref. (e) Rel. (f) Ref.

C8. (a) Tetanus. (b) Treppe. (c) Treppe. (d) Tetanus. (e) Treppe, tetanus.

C9. (a) Much; maximal; no. (b) Stronger.

C11. (a) Hypertonic. (b) Hypotonic. (c) Muscle tone.
C13. (a) Isotonic. (b) Stays about the same, increases.

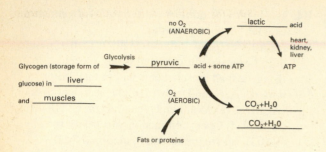

Figure LG 10.4A Anaerobic and aerobic energy sources for muscles.

C15. (a) ATP; myosin; ATP $\rightarrow$ ADP + P + energy.
(b) ADP + P + energy (from foods) $\rightarrow$ ATP.
(c) Active transport; creatine (that combines with phosphate to form creatine phosphate or phosphocreatine [PC]); phosphocreatine (PC) + ADP $\rightarrow$ creatine + ATP. (d) PC + ADP $\leftarrow$ C + ATP. (Draw arrow to LEFT in Figure LG 10.3.) (e) Phosphagen, 15 seconds; anaerobic, aerobic. See Figure LG 10.4A. (f) Lactic, anaerobic; fatigue; heart, liver, kidney; see Figure LG 10.4A. (g) Myoglobin; hemo; red. (h) Aerobic; in; oxygen uptake.
C16. (a) P. (b) G-L. (c) A.
C17. (a) Anaerobic; anaerobic. (b) Aerobic; aerobic.
D1. (a) Constantly; slow; do not; much, many; many, aerobic; I. (b) Mostly for short bursts of energy; fast; fast; do, anaerobic; IIB. (c) IIA (fast oxidative), IIB (fast glycolytic). (d) Mixtures of types I, IIA, and IIB; all of one type. (e1) Slow oxidative (type 1), (e2) fast glycolytic (type IIB), (e3) fast oxidative (type IIA). (f) B.

D3.

Characteristic	Skeletal Muscle	Cardiac Muscle
a. Number of nuclei per myofiber	Several (multinucleate)	One
b. Number of mitochondria per myofiber (More or Fewer)*	Fewer	More, because heart muscle requires constant generation of energy
c. Striated appearance due to alternated actin and myosin myofilaments (Yes or No)	Yes	Yes
d. Arrangement of muscle fibers (Parallel or Branching)	Parallel	Branching
e. Nerve stimulation required for contraction (Yes or No)*	Yes	No, so heart can contract without nerve stimulation, but nerves can increase or decrease heart rate
f. Length of refractory period (Long or Short)*	Short (.005 sec)	Long (0.30 sec) to allow the heart time to relax between beats
g. Duration of muscle contraction (Long or Short)*	Short (0.01 sec.– 0.100 sec)	10–15 times longer so heart can pump blood out of heart

D6. Tapered at each end, one nucleus; do; nonstriated or "smooth"; slowly, longer; no, small.
D7. Is not; hormones, pH or temperature changes, O_2 and CO_2 levels, and certain ions.
D8. (a–g) Sm. (h–m) Sk. (n–r) C.
E1. C B A
E2. Hypertrophy, size; fibrous (or connective or scar).
E5. (a) All three: skeletal, cardiac, and smooth. (b) Somites; 20; 44. (c) Sclero-; derma-, myo-.
F1. (a) Fibromyalgia. (b) Paralysis. (c) Muscular dystrophy. (d) Myoma. (e) Myopathy.
F2. (a) Acetylcholine (ACh); receptors. (b) Antibodies; cannot. (c) Anticholinesterase (antiAChE). (d) Immuno-, plasmapheresis.

WRITING ACROSS THE CURRICULUM: CHAPTER 10

1. Describe changes that occur in muscles and other body parts during exercise and immediately afterward. Be sure to explain the meaning of "recovery oxygen consumption."
2. Describe how muscles grow and change with age and training. Be sure to include effects of hormones.
3. Explain how communication occurs among cardiac muscle cells and among smooth muscle cells.
4. Describe several reasons why smooth muscle contracts more slowly than other muscle types.
5. Discuss the pros and cons of anabolic steroids as you would with a friend who is considering using these chemicals.

MASTERY TEST: CHAPTER 10

Questions 1–10: Circle T (true) or F (false). If the statement is false, change the underlined word or phrase so that the statement is correct.

T F 1. Tendons are <u>cords of connective tissue, whereas aponeuroses are broad, flat bands</u> of connective tissue.

T F 2. During contraction of muscle <u>both A bands and I bands</u> get shorter.

T F 3. Fascia is <u>one type of skeletal muscle tissue.</u>

T F 4. <u>Elasticity</u> is the ability of a muscle to be stretched or extended.

T F 5. Myasthenia gravis is caused by <u>an excess of acetylcholine production at the myoneural junction.</u>

T F 6. Muscle fibers remain relaxed if there are <u>few calcium ions in the sarcoplasm.</u>

T F 7. <u>Atrophy</u> means decrease in muscle mass.

T F 8. Muscles that are used mostly for quick bursts of energy (such as those in the arms) contain large numbers of <u>fast glycolytic (type IIB) fibers.</u>

T F 9. Most of the energy released during muscle contraction is used for <u>heat production.</u>

T F 10. In a relaxed muscle fiber <u>thin and thick myofilaments overlap to form the A band.</u>

Questions 11–13: Arrange the answers in correct sequence.

_____ _____ _____   11. Number of transverse tubules from greatest to least per muscle cell in these types of muscle:
    A. Smooth
    B. Skeletal
    C. Cardiac

_____ _____ _____   12. According to the amount of muscle tissue they surround, from most to least:
    A. Perimysium
    B. Endomysium
    C. Epimysium

_____ _____ _____   13. From largest to smallest:
    A. Myofibril
    B. Myofilament
    C. Muscle fiber (myofiber or cell)

14. All of the following molecules are parts of thin filaments *except:*
    A. Actin
    B. Myosin
    C. Tropomyosin
    D. Troponin

15. Choose the one statement that is *false:*
    A. A band refers to the anisotropic band.
    B. The A band is darker than the I band.
    C. Thick myofilaments reach the Z line in relaxed muscle.
    D. The H zone contains thick myofilaments, but not thin ones.
    E. Thick myofilaments are made of myosin.

16. All of the following terms are correctly matched with descriptions *except:*
    A. Denervation atrophy—wasting of a muscle to one-quarter of its size within two years of loss of nerve supply to a muscle
    B. Disuse atrophy—decrease of muscle mass in a person who is bedridden or who has a cast on
    C. Muscular hypertrophy—increase in size of a muscle by increase in the number of muscle cells
    D. Cramp—painful, spasmodic contraction

17. Which statement about muscle physiology in the relaxed state is *false?*
    A. Myosin cross bridges are bound to ATP.
    B. Calcium ions are stored in sarcoplasmic reticulum.
    C. Myosin cross bridges are bound to actin.
    D. Tropomyosin–troponin complex is bound to actin.

18. The staircase phenomenon refers to _____ contractions.
    A. Tetanic
    B. Treppe
    C. Tonic
    D. Isotonic

19. Choose the *false* statement about cardiac muscle.
    A. Cardiac muscle has a long refractory period.
    B. Cardiac muscle typically has one nucleus per fiber.
    C. Cardiac muscle cells are called cardiac muscle fibers.
    D. Cardiac fibers are separated by intercalated discs.
    E. Cardiac fibers are spindle shaped with no striations.

20. Most voluntary movements of the body are results of _____ contractions.
    A. Isometric
    B. Fibrillation
    C. Twitch
    D. Tetanic
    E. Spasm

_____ 21. _____ muscle cells are nonstriated and spindle shaped with one nucleus per cell.

_____ 22. Cardiac muscle remains contracted longer than skeletal muscle does because _____ is slower in cardiac than in skeletal muscle.

_____ 23. It is during the _____ period of a muscle contraction that calcium ions are released from sarcoplasmic reticulum and myosin cross bridge activity begins to occur.

_____ 24. _____ is the transmitter released from synaptic vesicles of axons supplying skeletal muscle.

_____ 25. ADP + phosphocreatine → _____ (Write the products.)

ANSWERS TO MASTERY TEST: ★ CHAPTER 10

True-False

1. T
2. F. A bands but not I bands
3. F. Dense connective tissue (which may surround skeletal muscle)
4. F. Extensibility
5. F. An autoimmune disorder in which antibodies are produced that bind onto receptors on the sarcolemma
6. T
7. T
8. T
9. T
10. T

Arrange

11. B C A
12. C A B
13. C A B

Multiple Choice

14. B
15. C
16. C
17. C
18. B
19. E
20. D

Fill-ins

21. Smooth
22. Passage of calcium ions from the extracellular fluid through sarcolemma
23. Latent
24. Acetylcholine
25. ATP + creatine

FRAMEWORK 11
The Muscular Systems

NAMING SKELETAL MUSCLES (A)
- Direction
- Location
- Size
- No. of origins
- Shape
- Site of origin, insertion
- Action

SKELETAL MUSCLES, MOVEMENT (B)
- Origin, insertion
- Lever, fulcrum
- Arrangements of fasciculi
- Group actions

PRINCIPAL SKELETAL MUSCLES

HEAD, NECK (C)
- Facial expression
- Mastication
- Eyeball movement
- Tongue movement
- Oral cavity
- Larynx
- Head movement

TRUNK MUSCLES (D)
- Abdominal wall
- Breathing
- Pelvic floor
- Perineum

UPPER BODY (E)
- Shoulder girdle
- Arm
- Forearm
- Wrist, hand, fingers

VERTEBRAL COLUMN (F)

LOWER BODY (G)
- Thigh
- Leg
- Foot & toes

INTRAMUSCULAR INJECTIONS, RUNNING INJURIES (H)

The Muscular System

CHAPTER
11

More than 600 different muscles attach to the bones of the skeleton. Muscles are arranged in groups that work much as a symphony, with certain muscles quiet while others are performing. The results are smooth, harmonious movements, rather than erratic, haphazard discord.

Much information can be gained by a careful initial examination of two aspects of each muscle: its name and location. A muscle name may offer such clues as size or shape, direction of fibers, or points of attachment of the muscle. A look at the precise location of the muscle and the joint it crosses—combined with logic—can usually lead to a correct understanding of action(s) of that muscle.

As you begin your study of the muscular system, carefully examine the Chapter 11 Topic Outline and Objectives; check off each one as you complete it. To organize your study of the muscular system, glance over the Chapter 11 Framework now. Be sure to refer to the Framework frequently and note relationships among key terms in each section.

TOPIC OUTLINE AND OBJECTIVES

A. Naming skeletal muscles

☐ 1. Define the criteria employed in naming skeletal muscles.

B. How skeletal muscles produce movement

☐ 2. Describe the relationship between bones and skeletal muscles in producing body movements.
☐ 3. Define a lever and fulcrum and compare the three classes of levers on the basis of placement of the fulcrum, effort, and resistance.
☐ 4. Identify the various arrangements of fascicles in a skeletal muscle and relate the arrangements to the strength of contraction and range of motion.
☐ 5. Discuss most body movements as activities of groups of muscles by explaining the roles of the prime mover, antagonist, synergist, and fixator.

C. Principal skeletal muscles of the head and neck

D. Principal skeletal muscles that act on the abdominal wall, muscles used in breathing, muscles of the pelvic floor and perineum

E. Principal skeletal muscles that move the shoulder girdle and upper limb

F. Principal skeletal muscles that move the vertebral column

G. Principal skeletal muscles that move the lower limb

☐ 6. Identify the principal skeletal muscles in different regions of the body by name, origin, insertion, action, and innervation.

H. Intramuscular (IM) injections; running injuries

☐ 7. Discuss the administration of drugs by intramuscular injections.
☐ 8. Describe several injuries related to running.

WORDBYTES

Now become familiar with the language of this chapter by studying each wordbyte, its meaning, and an example of its use within a term. After you study the entire list, self-check your understanding by writing the meaning of each wordbyte on the line. As you continue through the *Learning Guide,* identify (and fill in) additional terms that contain the same wordbyte.

Wordbyte	Self-check	Meaning	Example(s)
bi-	_____	two	*bi*ceps
brachi-	_____	arm	*brachi*alis
brev-	_____	short	flexor digitorum *brev*is
bucc-	_____	mouth, cheek	*bucc*inator
cap-, -ceps	_____	head	tri*ceps*
-cnem-	_____	leg	gastro*cnem*ius
delt-	_____	Greek D (D)	*del*toid
gastro-	_____	stomach, belly	*gastro* cnemius
genio-	_____	chin	*genio*glossus
glossus-	_____	tongue	*gloss*ary, stylo*glossus*
glute-	_____	buttock	*glute*us medius
grac-	_____	slender	*grac*ilis
-issimus	_____	the most	lat*issimus* dorsi
lat-	_____	broad	*lat*issimus dorsi
maxi-	_____	large	gluteus *maxi*mus
mini-	_____	small	gluteus *mini*mus
or-	_____	mouth	*or*bicularis oris
rect-	_____	straight	*rect*us femoris
sartor-	_____	tailor	*sartor*ius
serra-	_____	toothed, notched	*serra*tus anterior
teres-	_____	round	*teres* major
tri-	_____	three	*tri*ceps femoris
vast-	_____	large	*vast*us lateralis

CHECKPOINTS

A. Naming skeletal muscles   (page 270)

■ **A1.** Review the Wordbyte section above. Match the names of the following muscles with their meanings.

> A. Large muscle of the buttock region
> B. Belly–shaped muscle in leg
> C. Thigh muscle with four origins
> D. The broadest muscle of the back
> E. Large muscle in medial thigh area
> F. Muscle that raises the upper lip

_____ a. Latissimus dorsi   _____ d. Quadriceps femoris

_____ b. Vastus medialis   _____ e. Gastrocnemius

_____ c. Gluteus maximus   _____ f. Levator labii superioris

■ **A2.** As you study the names of muscles, you will find that most of them provide a good description of the muscle. For each of the following, indicate the type of clue that each part of the name gives. The first one is done for you.

> A. Action   N. Number of heads or origins
> D. Direction of fibers   P. Points of attachment of origin and insertion
> L. Location   S. Size or shape

_**D, L**_ a. Rectus abdominis   _____ d. Sternocleidomastoid

_____ b. Flexor carpi ulnaris   _____ e. Adductor longus

_____ c. Biceps brachii

B. How skeletal muscles produce movement   (pages 270–277)

■ **B1.** What structures constitute the *muscular system?*

■ **B2.** Refer to Figure LG 11.1 and consider flexion of your own forearm as you do this learning activity.

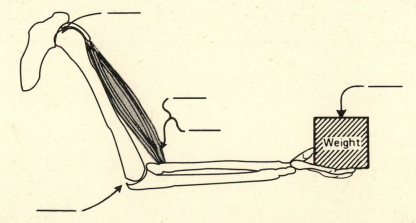

Figure LG 11.1   The lever–fulcrum principle is illustrated by flexion of the forearm. Complete the figure as directed in Checkpoint B2.

a. In flexion your forearm serves as a rigid rod, or _____, which moves

about a fixed point, called a _____ (your elbow joint, in this case).
b. Hold a weight in your hand as you flex your forearm. The weight plus your forearm
serve as the *(effort? fulcrum? resistance?)* during this movement.

c. The effort to move this resistance is provided by contraction of a _____ .
Note that if you held a heavy telephone book in your hand while your forearm was

flexed, much more _____ by your arm muscles would be required.
d. In Figure LG 11.1 identify the exact point at which the muscle causing flexion attaches
to the forearm. It is the *(proximal? distal?)* end of the *(humerus? radius? ulna?)*. Write
an E and an I on the two lines next to the arrow at that point in the figure. This indicates
that this is the site where the muscle exerts its effort (E) in the lever system, and it is
also the insertion (I) end of the muscle. (More about insertions in a minute.)
e. Each skeletal muscle is attached to at least two bones. As the muscle shortens, one bone
stays in place and so is called the *(origin? insertion?)* end of the muscle. What bone in

the figure appears to serve as the origin bone? _____ Write O
on that line at that point in the figure.
f. Now label the remaining arrows in Figure LG 11.1: F at fulcrum and R at resistance.
This is an example of a *(first? second? third?)* -class lever.

■ **B3.** Describe the roles of lever systems in the body by completing this exercise.

a. There are *(2? 3? 5? 7?)* classes of lever systems. Levers are categorized according to
relative positions of effort (E), fulcrum (F), and resistance (R).
b. Refer to Figure 11.2a, page 272 of your text. Now hyperextend your head as if to look at
the sky. The weight of your face and jaw serves as *(E? F? R?)*, while your neck muscles

provide *(E? F? R?)*. The fulcrum is the joint between the _____ and

the _____ bones. This is an example of a _____
class lever.
c. Define *leverage*.

d. Define *range of motion (ROM)*.

e. Consider muscles C and F, both with the same strength (force of contraction), but
muscle C is attached (inserted) closer to the joint and muscle F is attached (inserted)
farther from the joint. Which muscle is likely to be more powerful? *(C? F?)* Which
muscle is likely to produce greater ROM as well as greater speed? *(C? F?)* In other
words, power and ROM are related *(directly? inversely?)*, or as strength increases,

ROM _____-creases.
f. Refer to Figure 11.2b. The action of the calf (gastrocnemius) to lift the heel to stand on

the toes (_____ flexion) is an example of a *(first? second? third?)*
class lever. The gastrocnemius is a strong muscle to be able to lift so much body weight;
the ROM of this muscle is *(also? not?)* great. There are *(many? few?)* examples of this
type of lever in the body.

■ **B4.** Correlate fascicular arrangement with muscle power and range of motion of muscles.

a. A muscle with *(many? long?)* fibers will tend to have great strength. An example is the *(parallel? pennate?)* arrangement.

b. A muscle with *(many? long?)* fibers will tend to have great range of motion (but less power). An example is the *(parallel? pennate?)* arrangement.

■ **B5.** Refer again to Figure LG 11.1 and do this exercise about how muscles of the body work in groups.

a. The muscle that contracts to cause flexion of the forearm is called a _____.

An example of a prime mover in this action would be the _____ muscle.

b. The triceps brachii must relax as the biceps brachii flexes the forearm. The triceps is an extensor. Because its action is opposite to that of the biceps, the triceps is called *(a synergist? an agonist? an antagonist?)* of the biceps.

c. What would happen if the flexors of your forearm were functional, but not the antagonistic extensors?

d. What action would occur if both the flexors and extensors contracted simultaneously?

e. Muscles that assist or cooperate with the prime mover to cause a given action are known

as _____, whereas muscles that stabilize a bone (such as the scapula) so that prime movers and synergists can move another bone (such as the humerus) are

called _____.

C. Principal skeletal muscles of the head and neck   (pages 277–289)

■ **C1.** After studying Exhibit 11.3 (pages 277–278) in your text, check your understanding of the muscles of facial expression. Write the name of the muscle that answers each description. Locate muscles in Figure LG 11.2a O–R. Cover the key and write the name of each facial muscle next to its lettered leader line.

a. Allows you to show surprise by raising your eyebrows and forming horizontal forehead

wrinkle: _____

b. Muscle surrounding opening of your mouth; allows you to use your lips in kissing and

in speech: _____

c. Muscle for smiling and laughing because it draws the outer portion of the mouth upward

and outward: _____

d. Circular muscle around eye; closes eye: _____

■ **C2.** Essentially all of the muscles controlling facial expression receive nerve impulses via

the _____ nerve, which is cranial nerve *(III? V? VII?)*.

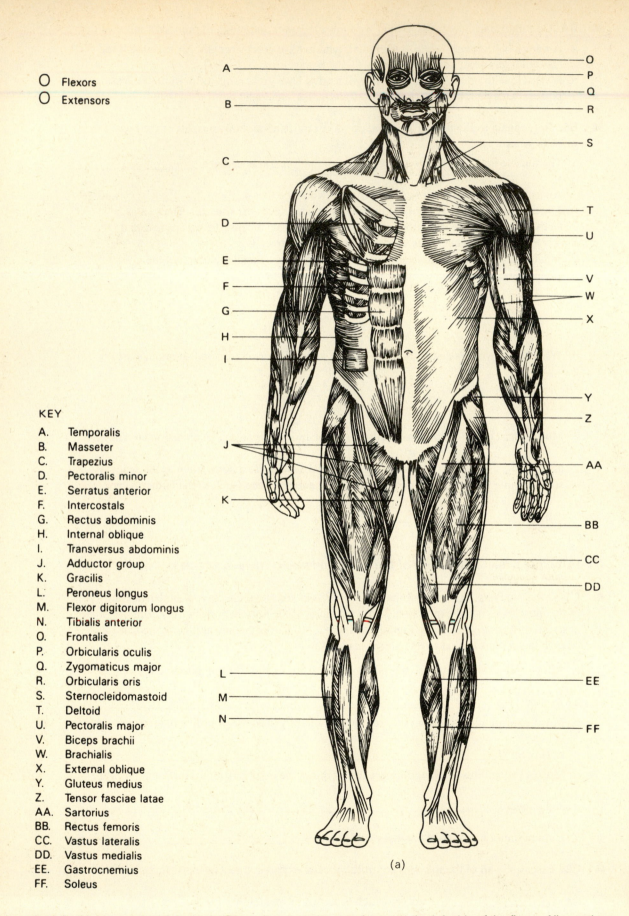

O Flexors
O Extensors

KEY

A. Temporalis
B. Masseter
C. Trapezius
D. Pectoralis minor
E. Serratus anterior
F. Intercostals
G. Rectus abdominis
H. Internal oblique
I. Transversus abdominis
J. Adductor group
K. Gracilis
L. Peroneus longus
M. Flexor digitorum longus
N. Tibialis anterior
O. Frontalis
P. Orbicularis oculis
Q. Zygomaticus major
R. Orbicularis oris
S. Sternocleidomastoid
T. Deltoid
U. Pectoralis major
V. Biceps brachii
W. Brachialis
X. External oblique
Y. Gluteus medius
Z. Tensor fasciae latae
AA. Sartorius
BB. Rectus femoris
CC. Vastus lateralis
DD. Vastus medialis
EE. Gastrocnemius
FF. Soleus

(a)

Figure LG 11.2 Major muscles of the body. Some deep muscles are shown on the left side of the figure. All muscles on the right side are superficial. Label and color as directed. (a) Anterior view. (b) Posterior view.

○ Flexors
○ Extensors

A
B
C
D
E
F
G
H
I
J
K

L
M
N
O
P
Q
R
S
T

(b)

KEY

A. Levator scapulae
B. Rhomboideus minor
C. Rhomboideus major
D. Supraspinatus
E. Infraspinatus
F. Teres major
G. Erector spinae
H. Adductors
I. Vastus lateralis
J. Biceps femoris
K. Semimembranosus and
    semitendinosus
L. Trapezius
M. Deltoid
N. Triceps brachii
O. Latissimus dorsi
P. External oblique
Q. Gluteus medius
R. Gluteus maximus
S. Gastrocnemius
T. Soleus

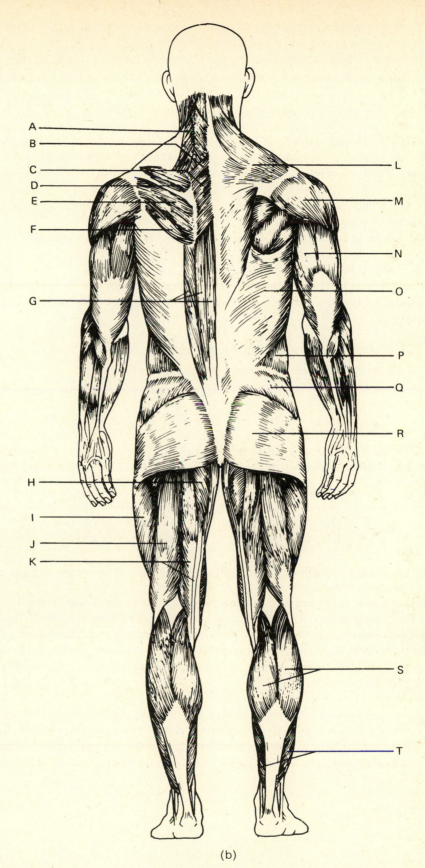

Figure LG 11.2 *Continued*

C3. Place your index finger and thumb on the origin and insertion of each of the muscles that move your lower jaw. (Refer to Exhibit 11.4 and Figure 11.5, pages 280–282 in the text, for help.) Then do this learning activity.

a. Two large muscles help you to close your mouth forcefully, as in chewing. Both of these act by *(lowering the maxilla? elevating the mandible?)*. The _____ covers your temple, and the _____ covers the ramus of the mandible.

b. Most of the muscles involved in chewing are ones that help you to *(open? close?)* your mouth. Think about this the next time you go to the dentist and try to hold your mouth open for a long time, with only your _____ muscles to force your mouth wide open.

c. The muscles that move the lower jaw and so aid in chewing are innervated by cranial nerve *(III? V? VII?)*, known as the _____ nerve.

d. Refer to Figure LG 11.2a. Muscles *A* and *B* are used primarily for *(facial expressions? chewing?)*, whereas muscles *O, P, Q,* and *R* are used mainly for *(facial expressions? chewing?)*.

C4. Complete this exercise about muscles that move the eyeballs.

a. Why are these muscles called *extrinsic* eyeball muscles?

b. Label Figure LG 11.3 with initials of the names of the six extrinsic eye muscles listed in the box to indicate *direction of movement* of each muscle. One has been done for you.

IO.	Inferior oblique	MR.	Medial rectus
IR.	Inferior rectus	SO.	Superior oblique
LR.	Lateral rectus	SR.	Superior rectus

c. Now write next to each muscle the number of the cranial nerve innervating that muscle. One has been done for you.

d. Note that most of the extrinsic eye muscles are innervated by cranial nerve *(III? IV? VI?)*, as indicated by the name of that nerve *(oculomotor)*. Cranial nerve VI, the *abducens,* supplies the muscles that cause most *(medial? lateral?)* movement, or abduction, of the eye. [*Hint:* Remember the chemical symbol for sulfate (SO_4) to remind yourself that the remaining muscle, superior oblique (SO), is supplied by the fourth cranial nerve.]

C5. *For extra review* of actions of eye muscles, work with a study partner. One person moves the eyes in a particular direction; the partner then names the eye muscles used for that action. *Note:* Will both eyes use muscles of the same name? For example, as you look to your right, will the lateral rectus muscles attached to both eyes contract?

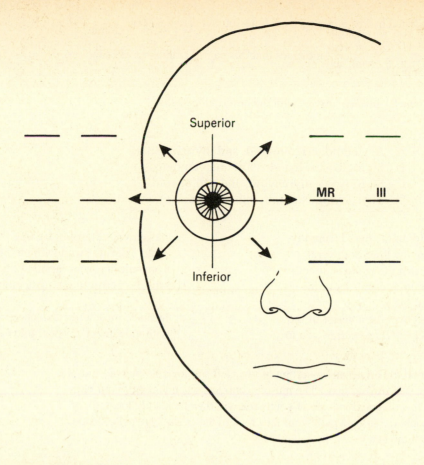

Superior

Inferior

MR   III

Figure LG 11.3   Right eyeball with arrows indicating directions of eye movements produced by the extrinsic eye muscles. Label as directed in Checkpoint C4.

■ **C6.** Actions such as opening your mouth for eating or speaking and also swallowing require integrated action of a number of muscles. You have already studied some that allow you to open your mouth (such as masseter and temporalis). We will now consider muscles of the tongue, floor of the oral cavity, pharynx, and larynx.

Identify locations of three bony structures relative to your tongue. Tell whether each is anterior or posterior and superior or inferior to your tongue. (For help refer to Figure 11.7, page 285 of the text.)

a. Styloid process of temporal bone _____

b. Hyoid bone _____

c. Chin (anterior portion)_____

■ **C7.** Remembering the locations of these bony points and that muscles move a structure (tongue) by pulling on it, determine the direction that the tongue is pulled by each of these muscles.

a. Styloglossus: _____

b. Hyoglossus: _____

c. Genioglossus: _____

■ **C8.** The three muscles listed in Checkpoint C7 are innervated by cranial nerve _____;

the name _____ nerve indicates that this pair of cranial nerves supplies the region "under the tongue."

■ **C9.** Match groups of muscles in each answer with the descriptions below.

> A. Arytenoid, cricothyroid, and thyroarytenoid
> B. Styloglossus, hyoglossus, and genioglossus
> C. Stylohyoid, mylohyoid, and geniohyoid
> D. Thyrohyoid, sternohyoid, and omohyoid

_____ a. Located superior to the hyoid (suprahyoid), these form the floor of the oral cavity. (Recall that the hyoid is directly posterior to the inferior border of the U-shaped mandible.)

_____ b. These muscles cover the anterior of the larynx and trachea. These permit you to elevate your thyroid cartilage ("Adam's apple") during swallowing to prevent food from entering your larynx. (Try it.)

_____ c. Muscles permitting tongue movements.

_____ d. Intrinsic muscles of the larynx associated with movement of vocal folds and speech.

■ **C10.** Using a mirror, find the origin and insertion of your left sternocleidomastoid muscle. (See Figure 11.9c, page 289 in your text.) The muscle contracts when you pull your chin down and to the right; this diagonal muscle of your neck will then be readily located. Note that the left sternocleidomastoid pulls your face toward the *(same? opposite?)* side. It also *(flexes? extends?)* the head.

■ **C11.** Alternately flex and extend vertebrae of your neck by looking at your toes and then toward the sky. As you look at the sky, you are *(flexing? extending and then hyperextending?)* your head and cervical vertebrae. On which surface of the neck would you expect to find extensors of the head and neck? *(Anterior? Posterior?)* Note that these muscles are the most superior muscles of the columns of extensors of the vertebrae. Find them on Figure 11.18 (page 313 in your text). Now name three of these muscles.

■ **C12.** *The Big Picture: Looking Ahead.* Refer to Figure 14.5, page 398 of your text, as well as Exhibit 14.4, pages 418–422, and summarize head and neck muscle innervation by cranial nerves. Write the correct cranial nerve numbers next to the names of muscles innervated. Select from answers in the box. All except one answer will be used.

> III. Oculomotor
> IV. Trochlear
> V (mand). Trigeminal (mandibular branch)
> VI. Abducens
> VII. Facial
>
> IX. Glossopharyngeal
> X (RL). Vagus (recurrent laryngeal branch)
> XI. Spinal accessory
> XII. Hypoglossal

_____ a. Most extrinsic eye muscles

_____ b. Superior oblique eye muscle

_____ c. Lateral rectus eye muscle

_____ d. Upper eyelid (raises it)

_____ e. Muscles of mastication (chewing), such as temporalis and masseter

_____ f. Muscles that move the tongue

_____ g. Muscles within the larynx, used for speech because they move vocal folds (cords)

_____ h. Muscles of facial expression, such as smiling, laughing, and frowning, as well as muscles used in speaking, pouting, or sucking

_____ i. Sternocleidomastoid muscle that causes flexion of the head

D. Principal skeletal muscles that act on the abdominal wall, muscles used in breathing, muscles of the pelvic floor and perineum (pages 290–296)

■ **D1.** Each half of the abdominal wall is composed of *(two? three? four?)* muscles. Describe these in the exercise below.

   a. Just lateral to the midline is the rectus abdominis muscle. Its fibers are *(vertical?*

   horizontal?), attached inferiorly to the _____ and superiorly to

   the _____. Contraction of this muscle permits *(flexion? extension?)* of the vertebral column.

   b. List the remaining abdominal muscles that form the sides of the abdominal wall from most superficial to deepest.

   _____ → _____ → _____
      most superficial   deepest

   c. Do all three of these muscles have fibers running in the same direction? *(Yes? No?)* Of what advantage is this?

■ **D2.** Answer these questions about muscles used for breathing.

   a. The diaphragm is _____-shaped. Its oval origin is located

   _____ _____. Its insertion is not into bone, but rather into dense connective tissue forming the roof of the diaphragm; this tissue is

   called the _____.

   b. Contraction of the diaphragm flattens the dome, causing the size of the thorax to *(increase? decrease?)*, as occurs during *(inspiration? expiration?)*.

   c. The name *intercostals* indicates that these muscles are located _____. Which set is used during expiration? *(Internal? External?)*

D3. *For extra review.* Cover the key to Figure LG 11.2a and write labels for muscles F, G, H, I, and X.

D4. Look at the inferior of the human pelvic bones on a skeleton (or refer to Figure 8.8, page 205 in your text). Note that a gaping hole (outlet) is present. Pelvic floor muscles attach to the bony pelvic outlet. Name these muscles and state functions of the pelvic floor.

■ **D5.** Fill in the blanks in this paragraph. *Diaphragm* means literally _____

   (dia-) _____ *(-phragm)*. You are familiar with the diaphragm that

   separates the _____ from the abdomen. The *pelvic diaphragm*

   consists of all of the muscles of the _____ floor plus their fasciae. It is a "wall" or barrier between the inside and outside of the body. Name the largest

   muscle of the pelvic floor. _____

E. Principal skeletal muscles that move the shoulder girdle and upper limb (pages 297–310)

■ **E1.** From the list in the box, select the names of all muscles that move the shoulder girdle as indicated.

LS. Levator scapulae	PM. Pectoralis minor
RMM. Rhomboideus major and minor	SA. Serratus anterior
	T. Trapezius

Superiorly (elevation)_____

Inferiorly (depression)_____

Toward vertebrae (adduction)_____

Away from vertebrae (abduction)_____

■ **E2.** Note on Figure LG 11.2a and b that the only two muscles listed in Checkpoint E1 that

are superficial are the _____ and a small portion of the

_____. The others are all deep muscles.

■ **E3.** On Figure LG 11.2a and b identify the pectoralis major, deltoid, and latissimus dorsi muscles. All three of these muscles are *(superficial? deep?)*. They are all directly involved with movement of the *(shoulder girdle? humerus? radius/ulna?)*.

Points of origin or insertion	Actions (of humerus)
C. Clavicle	Ab. Abducts
H. Humerus	Ad. Adducts
I. Ilium	EH. Extension, hyperextension
RC. Ribs or costal cartilages	F. Flexion
Sc. Scapula	
St. Sternum	
VS. Vertebrae and sacrum	

■ **E4.** Fill in letters for *all points of origin and insertion* and *all actions* that apply for each muscle.

Muscles	Points of Origin or Insertion	Actions (of Humerus)
a. Pectoralis major	_____	_____
b. Deltoid	_____	_____
c. Latissimus dorsi	_____	_____

■ **E5.** *The Big Picture: Looking Back.* Combine your knowledge of muscle actions with your knowledge of movements at joints from Chapter 9. Return to Figure LG 9.2, page 166. Write the name(s) of one or two muscles that produce each of the actions (a)–(h). Write muscle names next to each figure.

■ **E6.** Complete the table describing three muscles that move the forearm.

Muscle Name	Origin	Insertion	Action on Forearm
a.		Radial tuberosity	
b. Brachialis			
c.			Extension

■ **E7.** Complete this exercise about muscles that move the wrist and fingers.

a. Examine your own forearm, palm, and fingers. There is more muscle mass on the *(anterior? posterior?)* surface. You therefore have more muscles that can *(flex? extend?)* your wrist and fingers.

b. Locate the flexor carpi ulnaris muscle on Figure 11.17, page 307 in your text.

What action does it have other than flexion of the wrist? _____
What muscles would you expect to abduct the wrist?

c. What is the difference in location between flexor digitorum superficialis and flexor digitorum profundus?

d. What muscle helps you to point (extend) your index finger?

■ **E8.** For review of more details of the structures of the hand, do this exercise.

a. The *(extensor retinaculum? flexor retinaculum?)* is located over the palmar surface of the hand. This ligament secures the position of a number of long tendons of the wrist

and digits, as well as the _____ nerve. Inflammation of these tissues by repetitive wrist flexion may lead to carpal tunnel syndrome.

b. The flexor pollicis longus is a muscle that flexes the *(thumb? middle finger? little finger?)*.

c. Which action of the thumb is movement of the thumb medially across the palm? *(Hint:* refer to Exhibit 11.18.) *(Abduction? Adduction? Extension? Flexion? Opposition?)* Which action allows you to touch your thumb to the tip your of your little finger?

_____

d. Extensor digitorum muscles extend the *(wrist? fingers?)*.

e. Thenar and hypothenar muscles are *(extrinsic? intrinsic?)* muscles of the hand. Thenar muscles form the pronounced muscle mass of the hand that moves the *(thumb? little*

finger?), whereas hypothenar muscles move the _____.

■ **E9.** *For extra review* of muscles that move the upper limbs, write the name of one or more muscles that fit these descriptions.

a. Covers most of the posterior of the humerus:_____

b. Turns your hand from palm down to palm up position:_____

c. Originates from upper eight or nine ribs; inserts on scapula; moves scapula laterally:

_____

d. Used when a baseball is grasped:_____

e. Antagonist to serratus anterior: _____

f. Largest muscle of the chest region; used to throw a ball in the air (flex humerus)

and to adduct arm: _____

g. Raises or lowers scapula, depending on which portion of the muscle contracts:

_____

h. Controls action at the elbow for a movement such as the downstroke in hammering a nail

_____

i. Hyperextends the humerus, as in doing the "crawl" stroke in swimming or exerting a

downward blow; also adducts the humerus:_____

F. Principal skeletal muscles that move the vertebral column (pages 311–314)

■ **F1.** Describe the muscles that comprise the sacrospinalis. Locate and label on Figure LG 11.2b.

a. The sacrospinalis muscle is also called the _____.

b. The muscle consists of three groups: _____ (lateral),

_____ (intermediate), and _____ (medial).

c. In general, these muscles have attachments between _____.

d. They are *(flexors? extensors?)* of the vertebral column, and so are *(synergists? antagonists?)* of the rectus abdominis muscles.

■ **F2.** Explain why it is common for women in their final weeks of pregnancy to experience frequent back pains.

■ **F3.** Choose the correct origin and insertion of the scalene muscles:

A. Ribs—iliac crest

B. Cervical vertebrae—occipital and temporal bones

C. Cervical vertebrae—first two ribs

D. Thoracic vertebrae—sacrum

G. Principal skeletal muscles that move the lower limb   (pages 315–325)

G1. Cover the key in Figure LG 11.2a and b and identify by size, shape, and location the major muscles that move the lower limb.

■ **G2.** Now match muscle names in the box with their descriptions below.

Ad.	Adductor group	Ham.	Hamstrings
Gas.	Gastrocnemius	Il.	Iliopsoas
GMax.	Gluteus maximus	QF.	Quadriceps femoris
GMed.	Gluteus medius	Sar.	Sartorius

_____ a. Consists of four heads: rectus femoris and three vastus muscles (lateralis, medialis, and intermedius)

_____ b. This muscle mass lies in the posterior (flexor) compartment of the thigh; antagonist to quadriceps femoris

_____ c. Attached to lumbar vertebrae, anterior of ilium, and lesser trochanter, it crosses anterior to hip joint

_____ d. Large muscle mass of the buttocks; antagonist to the iliopsoas

_____ e. The only one of these muscles located in the leg (between knee and ankle), it forms the "calf;" attaches to calcaneus by "Achilles tendon"

_____ f. Forms the medial compartment of the thigh; moves femur medially

_____ g. Crossing the femur obliquely, it moves lower extremity into "tailor position"

_____ h. Located posterior to upper, outer portion of the ilium, it forms a preferred site for intramuscular (IM) injections

■ **G3.** Complete the table by marking an X below each action produced by contraction of the muscles listed. (Some muscles will have two answers.)

Key to actions in table: Ab. Abduct; Ad. Adduct; EH, extend or hyperextend; F, flex

	Movements of Thigh (Hip Joint)				Movement of Leg (Knee)	
	Ab	Ad	EH	F	EH	F
a. Iliopsoas						
b. Gluteus maximus and medius						
c. Adductor mass						
d. Tensor fasciae latae						
e. Quadriceps femoris						
f. Hamstrings						
g. Gracilis						
h. Sartorius						

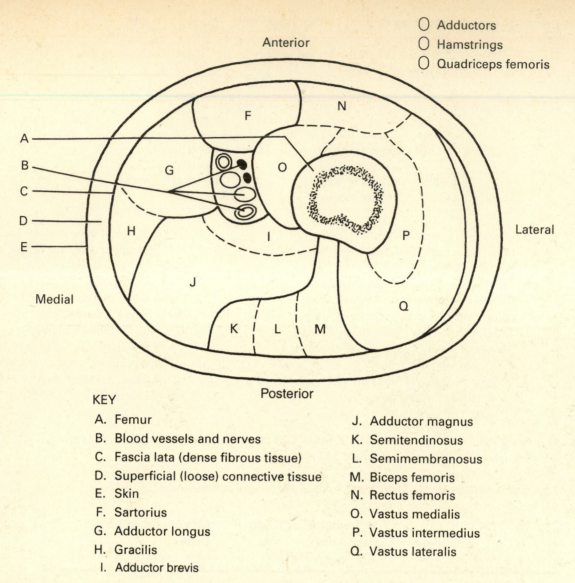

Anterior

○ Adductors
○ Hamstrings
○ Quadriceps femoris

Lateral

Medial

Posterior

KEY

A. Femur
B. Blood vessels and nerves
C. Fascia lata (dense fibrous tissue)
D. Superficial (loose) connective tissue
E. Skin
F. Sartorius
G. Adductor longus
H. Gracilis
I. Adductor brevis

J. Adductor magnus
K. Semitendinosus
L. Semimembranosus
M. Biceps femoris
N. Rectus femoris
O. Vastus medialis
P. Vastus intermedius
Q. Vastus lateralis

Figure LG 11.4 Cross section of the right thigh midway between hip and knee joints. Structures A to E are nonmuscle; F to Q are skeletal muscles of the thigh. Color as directed in Checkpoint G4.

■ **G4.** Refer to Figure LG 11.4.

a. Locate the major muscle groups of the right thigh in this cross section. Color the groups as indicated by color code ovals on the figure.

b. Cover the key to Figure LG 11.4 and identify each muscle. Relate position of muscles in this figure to views of muscles in Figure LG 11.2a and b.

■ **G5.** *The Big Picture: Looking Back.* What muscles cause the actions i–p shown in Figure LG 9.2 (page 167)? Write the muscle names next to each diagram.

G6. Perform these actions of your foot and toes. Feel which muscles are contracting. Then match names of actions with descriptions.

DF. Dorsiflex	F. Flex toes
Ev. Evert foot	In. Invert foot
Ex. Extend toes	PF. Plantar flex

_____ a. Jump, as if to touch ceiling

_____ b. Walk around on your heels

_____ c. Curl toes down

_____ d. Lift toes upward away from floor

_____ e. Move sole of foot medially

_____ f. Move sole of foot laterally

G7. To review details of leg muscles that move the foot, complete this table. Note that muscles within a compartment tend to have similar functions. Muscles with * flex or extend only the great toe (not all toes). Use the same key for foot and toe actions as in the box for Checkpoint G6.

	DF	PF	In	Ev	F	Ex
Posterior compartment: a. Gastrocnemius and soleus b. Tibialis posterior c. Flexor digitorum longus d. Flexor hallucis longus*						
Lateral compartment: e. Peroneus (longus and brevis)						
Anterior compartment: f. Tibialis anterior g. Extensor digitorum longus h. Extensor hallucis longus*						

G8. *For extra review* of all muscles, color flexors and extensors using color code ovals on Figure LG 11.2a and b. This activity will allow you to see on which sides of the body most muscles with those actions are located. Omit plantar flexors and dorsiflexors. Note that some muscles are flexors *and* extensors—at different joints. (See Mastery Test question 23 also.)

H. Intramuscular (IM) injections; running injuries   (pages 326–328)

H1. State three reasons why intramuscular (IM) injections may be the method of choice for administration of drugs.

H2. *A clinical challenge.* Why is the gluteus medius considered a safer site for intramuscular injections than the gluteus maximus?

H3. Two other muscles commonly used for intramuscular injections are the

_____ in the thigh and the _____ in the upper limb.

■ **H4.** Answer these questions about running injuries.

    a. The most common site of injury for runners is the *(calcaneal tendon? groin? hip? knee?).*

    b. *Patellofemoral stress syndrome* is a technical term for _____. Briefly describe this problem.

    c. *Shinsplint syndrome* refers to soreness along the *(patella? tibia? fibula?).*

    d. Write several suggestions you might make to a beginning runner to help to avoid runners' injuries.

    e. Initial treatment of sports injuries usually calls for *RICE* therapy. To what does RICE

    therapy refer? R_____ I_____

    C_____ E_____

ANSWERS TO SELECTED CHECKPOINTS: CHAPTER 11

A1. (a) D. (b) E. (c) A. (d) C. (e) B. (f) F.

A2. (b) A, P, L. (c) N, L. (d) P. (e) A, S.

B1. Skeletal muscle tissues and connective tissues.

B2. (a) Lever, fulcrum. (b) Resistance. (c) Muscle; effort. (d) Proximal, radius. (e) Origin, scapula. (f) See Figure LG 11.1A; third.

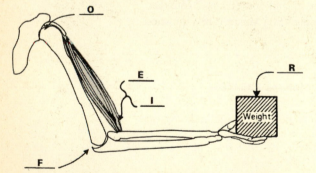

Figure LG 11.1A The lever–fulcrum principle is illustrated by flexion of the forearm.

B3. (a) 3. (b) *R, E;* atlas, occipital; first. (c) The mechanical advantage gained by a lever; it is largely responsible for the strength of a muscle and the range of motion. (d) Maximum ability to move bones of a joint through an arc. (e) C; F; inversely, de. (f) Plantar, second; not; few.

B4. (a) Many; pennate. (b) Long; parallel.

B5. (a) Prime mover (agonist); biceps brachii. (b) An antagonist. (c) Your forearm would stay in the flexed position. (d) None: each opposing muscle would negate the action of the other. (e) Synergists, fixators.

C1. (a) Frontalis portion of epicranius. (b) Orbicularis oris. (c) Zygomaticus major. (d) Orbicularis oculi.

C2. Facial, VII.

C3. (a) Elevating the mandible; temporalis, masseter. (b) Close; lateral pterygoid. (c) V, trigeminal. (d) Chewing, facial expression.

C4. (a) They are outside of the eyeballs, not intrinsic like the iris. (b) and (c) See Figure LG 11.3A. (d) III; lateral.

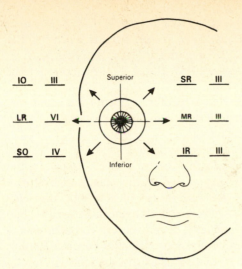

Figure LG 11.3A Right eyeball with arrows indicating directions of eye movements produced by the extrinsic eye muscles.

C5. No. The left eye contracts its medial rectus, while the right eye uses its lateral rectus and exerts some tension upon both oblique muscles.

C6. (a) Posterior and superior. (b) Inferior. (c) Anterior and inferior.

C7. Same answers as for C6.

C8. XII, hypoglossal.

C9. (a) C. (b) D. (c) B. (d) A.

C10. Opposite; flexes.

C11. Extending and then hyperextending; posterior; longissimus capitis, semispinalis capitis, and splenius capitis.

C12. (a) III. (b) IV: remember SO_4. (c) VI. (d) III. (e) V (mand). (f) XII. (g) X (RL). (h) VII. (i) XI.

D1. Four. (a) Vertical, pubic crest and symphysis pubis, ribs 5 to 7; flexion. (b) External oblique → internal oblique → transversus abdominis. (c) No; strength is provided by the three different directions.

D2. (a) Dome; around the bottom of the rib cage and on lumbar vertebrae; central tendon. (b) Increase, inspiration. (c) Between ribs (costa); internal.

D5. Across, wall; thorax; pelvic; levator ani (pubococcygeus and iliococcygeus).

E1. (a) LS, RMM, upper fibers of T. (b) PM, lower fibers of T. (c) RMM, T. (d) SA.

E2. Trapezius, serratus anterior.

E3. Superficial; humerus.

E4. (a) C, St, RC, H; Ad, F. (b) C, H, Sc; Ab, EH (posterior fibers), F (anterior fibers). (c) I, RC, VS, H; Ad, EH.

E5. (a) Sternocleidomastoid. (b) Three capitis muscles. (c) Pectoralis major, coracobrachialis, deltoid (anterior fibers), biceps brachii. (d) Latissimus dorsi, teres major, deltoid (posterior fibers), triceps brachii. (e) Deltoid and supraspinatus. (f) Pectoralis major, latissimus dorsi, teres major. (g) Biceps brachii, brachialis, brachioradialis. (h) Triceps brachii.

E6.

Muscle Name	Origin	Insertion	Action on Forearm
a. **Biceps brachii**	**Scapula (2 sites)**	Radial tuberosity (anterior)	**Flexion, supination** (*Note:* also flexes humerus)
b. Brachialis	**Anterior humerus**	Ulna (coronoid process)	**Flexion**
c. **Triceps brachii**	**Scapula and 2 sites on posterior of humerus**	**Posterior of ulna (olecranon)**	Extension (also extension of humerus)

E7. (a) Anterior; flex. (b) Adducts wrist; those lying over radius, such as flexor and extensor carpi radialis. (c) Superficialis is more superficial, and profundus lies deep. (d) Extensor indicis.

E8. (a) Flexor retinaculum; median. (b) Thumb. (c) Flexion; opposition. (d) Fingers (digits). (e) Intrinsic; thumb, little finger.

E9. (a) Triceps brachii. (b) Supinator and biceps brachii. (c) Serratus anterior. (d) Flexor digitorum superficialis and profundus. (e) Trapezius. (f) Pectoralis major. (g) Trapezius. (h) Triceps brachii. (i) Latissimus dorsi.

F1. (a) Erector spinae. (b) Iliocostalis, longissimus, spinalis. (c) Hipbone (ilium), ribs and vertebrae. (d) Extensors, antagonists.

F2. Extra weight of the abdomen demands extra support (contraction) by sacrospinalis muscles.

F3. C.

G2. (a) QF. (b) Ham. (c) Il. (d) GMax. (e) Gas. (f) Ad. (g) Sar. (h) Gmed.

G3.

	Movements of Thigh (Hip Joint)				Movement of Leg (Knee)	
	Ab	Ad	EH	F	EH	F
a. Iliopsoas				X		
b. Gluteus maximus and medius			X			
c. Adductor mass		X		X		
d. Tensor fasciae latae	X			X		
e. Quadriceps femoris				X	X	
f. Hamstrings			X			X
g. Gracilis		X				X
h. Sartorius				X		X

G4. (a) Adductors: G H I J; Hamstrings: K L M; Quadriceps: N O P Q. (b) See key to Figure LG 11.4.

G5. (i) Iliacus + psoas (iliopsoas), rectus femoris, adductors, sartorius; (j) Gluteus maximus, hamstrings, adductor magnus (posterior portion); (k) Hamstrings, gracilis, sartorius, gastrocnemius; (l) Quadriceps femoris; (m) Same as i but bilateral; (n) Same as j but bilateral; (o) Tensor fasciae latae, gluteus (medius and minimus), piriformis, superior and inferior gemellus, and obturator internus; (p) Adductors (longus, magnus, and brevis), pectineus, quadratus femoris, and gracilis.

G6. (a) PF. (b) DF. (c) F. (d) Ex. (e) In. (f) Ev.

G7.

	DF	PF	In	Ev	F	Ex
Posterior compartment:						
a. Gastrocnemius and soleus		X				
b. Tibialis posterior		X	X			
c. Flexor digitorum longus		X	X		X,	
d. Flexor hallucis longus*		X	X		X*	
Lateral compartment:						
e. Peroneus (longus and brevis)		X		X		
Anterior compartment:						
f. Tibialis anterior	X		X			
g. Extensor digitorum longus	X			X		X,
h. Extensor hallucis longus*	X		X			X*

G8. Figure LG 11.2a: flexors: G, J, K, S, T (anterior portion), U, V, W, AA, BB, EE; extensors: BB, CC, DD. Figure LG 11.2b: flexors: J, K, S; extensors: F, G, H, I, J, K, L, M (posterior fibers), N, O, R.

H2. The gluteus medius is superior and lateral to the sciatic nerve which runs deep to the gluteus maximus muscle.

H3. Vastus lateralis, deltoid.

H4. (a) Knee. (b) "Runner's knee"; the patella tracks (glides) laterally, causing pain. (c) Tibia. (d) Replace worn shoes with new, supportive ones; do stretching and strengthening exercises; build up gradually; get proper rest; consider other exercise if prone to leg injuries because most (70%) runners do experience some injuries. (e) Rest, ice, compression (such as by elastic bandage), and elevation (of the injured part).

WRITING ACROSS THE CURRICULUM: CHAPTER 11

1. Explain the kinds of information conveyed about muscles by examining muscle names. Give at least six examples of muscle names that provide clues to muscle locations, actions, size or shape, direction of fibers, numbers of origins, or points of attachment.

2. Identify types of movements required at shoulder, elbow, and wrist as you perform the action of tossing a tennis ball upward. Indicate groups of muscles that must contract and relax to accomplish this action.

3. Perform the actions of abducting and adducting your right thigh. Contrast the amount of muscle mass contracting to effect each of these movements. Name several muscles contracting as you carry out each action.

MASTERY TEST: CHAPTER 11

Questions 1–2: Arrange the answers in correct sequence.

_____ _____ _____ 1. Abdominal wall muscles, from superficial to deep:
    A. Transversus abdominis
    B. External oblique
    C. Internal oblique

_____ _____ _____ 2. From superior to inferior in location:
    A. Sternocleidomastoid
    B. Pelvic diaphragm
    C. Diaphragm and intercostal muscles

Questions 3–12: *Circle T (true) or F (false). If the statement is false, change the underlined word or phrase so that the statement is correct.*

T F   3. In extension of the thigh the hip joint serves as the fulcrum (F), while the hamstrings and gluteus maximus serve as the effort (E).

T F   4. The wheelbarrow and gastrocnemius are both examples of the action of second-class levers.

T F   5. The hamstrings are antagonists to the quadriceps femoris.

T F   6. The name deltoid is based on the action of that muscle.

T F   7. The most important muscle used for normal breathing is the diaphragm.

T F   8. The insertion end of a muscle is the attachment to the bone that does move.

T F   9. In general, adductors (of the arm and thigh) are located more on the medial than on the lateral surface of the body.

T F 10. Both the pectoralis major and latissimus dorsi muscles extend the humerus.

T F 11. The capitis muscles (such as splenius capitis) are extensors of the head and neck.

T F 12. The biceps brachii and biceps femoris are both muscles with two heads of origin located on the arm.

Questions 13–20: *Circle the letter preceding the one best answer to each question.*

13. All of these muscles are located in the lower limb *except:*
    A. Hamstrings
    B. Gracilis
    C. Tensor fasciae latae
    D. Deltoid
    E. Peroneus longus

14. All of these muscles are located on the anterior of the body *except:*
    A. Tibialis anterior
    B. Rectus femoris
    C. Sacrospinalis
    D. Pectoralis major
    E. Rectus abdominis

15. Contraction of all of the following muscles causes extension of the leg *except:*
    A. Rectus femoris
    B. Biceps femoris
    C. Vastus medialis
    D. Vastus intermedius
    E. Vastus lateralis

16. All of these muscles are attached to ribs *except:*
    A. Serratus anterior
    B. Intercostals
    C. Trapezius
    D. Iliocostalis
    E. External oblique

17. All of these muscles are directly involved with movement of the scapulae *except:*
    A. Levator scapulae
    B. Pectoralis major
    C. Pectoralis minor
    D. Rhomboideus major
    E. Serratus anterior

18. All of these muscles have attachments to the hip bones *except:*
    A. Adductor muscles (longus, magnus, brevis)
    B. Biceps femoris
    C. Rectus femoris
    D. Vastus medialis
    E. Latissimus dorsi

19. The masseter and temporalis muscles are used for:
    A. Chewing
    B. Pouting
    C. Frowning
    D. Depressing tongue
    E. Elevating tongue

20. All of these muscles are used for facial expression *except:*
    A. Zygomaticus major
    B. Orbicularis oculi
    C. Platysma
    D. Rectus abdominis
    E. Mentalis

Questions 21–25: Fill-ins. Refer to Figure LG 11.2a and b. Write the word or phrase or key letters of muscles that best complete the statement or answer the questions.

_____ 21. Muscles G, S, U, V, and W on Figure LG 11.2a all have in common the fact that they carry out the action of _____.

_____ 22. Muscles G, N, O, and R on Figure LG 11.2b all have in common the fact that they carry out the action _____.

_____ 23. If you colored all flexors red and all extensors blue on these two figures, the view of the _____ surface of the body would appear more blue.

_____ 24. Choose the letters of all of the muscles listed below that would contract as you raise your left arm straight in front of you, as if to point toward a distant mountain: Figure LG 11.2a: D T U V; Figure LG 11.2b: N O P.

_____ 25. Choose the letters of all of the muscles listed below that would contract as you raise your knee and extend your leg straight out in front of you, as if you are starting to march off to the distant mountain: Figure LG 11.2a: AA BB CC DD; Figure LG 11.2b; I J K R.

Arrange

1. B C A
2. A C B

True-False

3. T
4. T
5. T
6. F. Shape
7. T
8. T
9. T
10. F. The latissimus dorsi extends, but the pectoralis major flexes.
11. T
12. F. Two heads of origin; but origins of biceps brachii are on the scapula, and origins of biceps femoris are on ischium and femur.

Multiple Choice

13. D
14. C
15. B
16. C
17. B
18. D
19. A
20. D

Fill-ins

21. Flexion
22. Extension
23. Posterior
24. Figure LG 11.2a: T (anterior fibers) U V; Figure LG 11.2b: none.
25. Figure LG 11.2a: AA BB CC DD; Figure LG 11.2b: H I.

UNIT III

Control Systems of the Human Body

FRAMEWORK 12
Nervous Tissue

ORGANIZATION (A)

- CNS
  - Brain
  - Spinal cord
- PNS
  - Somatic
  - Autonomic
    - Sympathetic
    - Parasympathetic

ANATOMY (B)

- NEUROGLIA
  - Astrocytes
  - Oligodendrocytes
  - Microglia
  - Ependymal cells
  - Neurolemmocytes
  - Satellite cells
- NEURONS
  - Structure
    - Cell Body
      - Nucleus
      - Organelles
      - Lipofuschin
    - Cytoplasmic Processes
      - Dendrites
      - Axons
        - Hillock
        - Initial segment
        - Trigger zone
        - Terminals
        - Synaptic end bulbs
        - Synaptic vesicles
  - Classification
    - Structural
      - Multipolar
      - Bipolar
      - Unipolar
    - Functional
      - Afferent
        - Somatic
        - Visceral
      - Efferent
        - Somatic
        - Visceral

PHYSIOLOGY

- NERVE IMPULSE (C)
  - Resting membrane potential
  - Ion channels
  - Action potential
    - depolarization
    - repolarization
    - refractory period
  - Saltatory conduction
  - Speed of impulse propagation
- TRANSMISSION AT SYNAPSES (D)
  - Synapses
    - electrical
    - chemical
  - Excitatory (EPSP)
    - summation
  - Inhibitory (IPSP)
  - Neurotransmitters

REGENERATION AND REPAIRS, DISORDERS (E)

Nervous Tissue

Two systems—nervous and endocrine—are responsible for regulating our diverse body functions. Each system exerts its control with the help of specific chemicals, namely neurotransmitters (nervous system) and hormones (endocrine system). In Unit III both systems of regulation will be considered, starting with the tissue of the nervous system.

Nervous tissue consists of two types of cells: neurons and neuroglia. The name neuroglia (*glia* = glue) offers a clue as to function of these cells: they bind, support, and protect neurons. Neurons perform the work of transmitting nerve impulses. These long and microscopically slender cells sometimes convey information several feet along a single neuron. Their function relies on an intricate balance between ions (Na^+ and K^+) found in and around nerve cells. Neurons release neurotransmitters that bridge the gaps between adjacent neurons and at nerve–muscle or nerve–gland junctions. Analogous to a complex global telephone system, the nervous tissue of the human body boasts a design and organization that permits accurate communication, coordination, and integration of virtually all thoughts, sensations, and movements.

As you begin your study of nervous tissue, carefully examine the Chapter 12 Topic Outline and Objectives; check off each one as you complete it. To organize your study of nervous tissue, glance over the Chapter 12 Framework now. Be sure to refer to the Framework frequently and note relationships among key terms in each section.

TOPIC OUTLINE AND OBJECTIVES

A. Organization

☐ 1. Identify the three basic functions of the nervous system in maintaining homeostasis.
☐ 2. Classify the organs of the nervous system into central and peripheral divisions.

B. Anatomy

☐ 3. Contrast the histological characteristics and functions of neuroglia and neurons.
☐ 4. Describe the functions of neuroglia.
☐ 5. Describe the functions of neurons.
☐ 6. Define gray and white matter and give examples of each.

C. Neurophysiology

☐ 7. Describe the cellular properties that permit communication among neurons and muscle fibers.
☐ 8. Describe the factors that contribute to generation of a resting membrane potential.
☐ 9. List the sequence of events involved in generation of a nerve action potential (impulse).

D. Transmission at synapses

☐ 10. Explain the events of synaptic transmission.
☐ 11. Distinguish between spatial and temporal summation.

12. Give examples of excitatory and inhibitory neurotransmitters and describe how they may act.

13. List four ways that synaptic transmission may be enhanced or blocked.

14. Describe the various types of neuronal circuits.

E. Regeneration and repair of nervous tissue; disorders

15. Describe events of damage and repair of peripheral neurons.

16. Describe the symptoms and causes of epilepsy.

WORDBYTES

Now become familiar with the language of this chapter by studying each wordbyte, its meaning, and an example of its use within a term. After you study the entire list, self-check your understanding by writing the meaning of each wordbyte on the line. As you continue through the *Learning Guide,* identify (and fill in) additional terms that contain the same wordbyte.

Wordbyte	Self-check	Meaning	Example(s)
astro-	_____	star	*astro*cytes
dendr-	_____	tree	*dendr*ite
-glia	_____	glue	neuro*glia*
lemm-	_____	sheath	neuri*lemma*
neuro-	_____	nerve	*neuro*n
olig-	_____	few	*olig*odendrocytes
syn-	_____	together	*syn*apse

CHECKPOINTS

A. Organization   (page 332)

■ **A1.** List three principal functions of the nervous system.

■ **A2.** Check your understanding of the organization of the nervous system by selecting answers that best fit descriptions below.

Aff.	Afferent	Eff.	Efferent
ANS.	Autonomic nervous system	PNS.	Peripheral nervous system
CNS.	Central nervous system	SNS.	Somatic nervous system

_____ a. Brain and spinal cord

_____ b. Sensory nerves

_____ c. Carry information from CNS to skeletal muscles

_____ d. Consists of sympathetic and para-sympathetic divisions

_____ e. Nerves that convey impulses to smooth muscle, cardiac muscle, and glands

_____ f. Cranial nerves and spinal nerves

B. Anatomy   (pages 332–340)

■ **B1.** Write *neurons* or *neuroglia* after descriptions of these cells.

a. Conduct impulses from one part of the nervous system to another: _____

b. Provide support and protection for the nervous system: _____

c. Bind nervous tissue to blood vessels, form myelin, and serve phagocytic functions:

_____

d. Smaller in size, but more abundant in number: _____

■ **B2.** Check your understanding of neuroglia by selecting types of neuroglia in the box that fit descriptions below. Answers may be used once or more than once.

A. Astrocytes	N. Neurolemmocytes
E. Ependymal cells	O. Oligodendrocytes
M. Microglia	S. Satellite cells

_____ a. Neuroglia found in the PNS (two answers)

_____ b. Form myelin (two answers)

_____ c. Maintain proper balance of K+ needed for nerve impulses

_____ d. Help form the blood–brain barrier and connect neurons to blood vessels

_____ e. Star-shaped cells

_____ f. Protective because they have phagocytic functions

_____ g. Line the fluid-filled spaces (ventricles) within the brain

_____ h. Support neurons in clusters (ganglia) within the PNS

_____ i. Form neurolemma (sheath of Schwann) around axons of neurons in PNS

■ **B3.** Check your understanding of coverings over nerve fibers in this Checkpoint.

a. In the (*CNS? PNS?*), myelination requires neurolemmocytes. (*One? Up to 500?*) neurolemmocyte(s) is/are required for myelination of the long axons of the body.

b. Myelin is formed of _____.
   A. Up to 100 layers of neurolemmocyte membrane
   B. The outer nucleated cytoplasmic layer of the neurolemmocyte

c. Arrange these layers from outermost to innermost: _____ _____ _____
   A. Axon
   B. Neurolemma
   C. Myelin sheath

d. A neurolemma is found _____.
   A. Only around axons in the CNS
   B. Only around axons in the PNS
   C. Around axons in both the CNS and the PNS

e. A neurolemma is involved in the process of _____ of injured axons.

f. Neurofibral nodes (nodes of Ranvier) are sections of axons that lack

_____. Consequently, nerve impulses must "jump" from node to

node; this process is called _____ conduction. The ultimate effect is

that myelin arranged between nodes causes a(n) _____-crease in the rate of nerve
impulse transmission. (*Hint:* See page 346 in your text.)

g. Do any CNS axons have a neurolemma? (*Yes? No?*) State one consequence of this fact.
(*Hint:* See page 355 of your text.)

Do any CNS axons have a myelin sheath? (*Yes? No?*) What cells produce this?

_____

h. _____ is a condition in which myelin sheaths are destroyed and replaced
by hardened scars or plaques.
   A. Multiple sclerosis
   B. Muscular dystrophy
   C. Myasthenia gravis

■ **B4.** On Figure LG 12.1, label all structures with leader lines. Next draw arrows beside
the figure to indicate direction of nerve impulses. Then color structures to match color
code ovals.

B5. Provide "job descriptions" of:

a. Nerve growth factor (NGF)

b. Neurotropic factor (NT3)

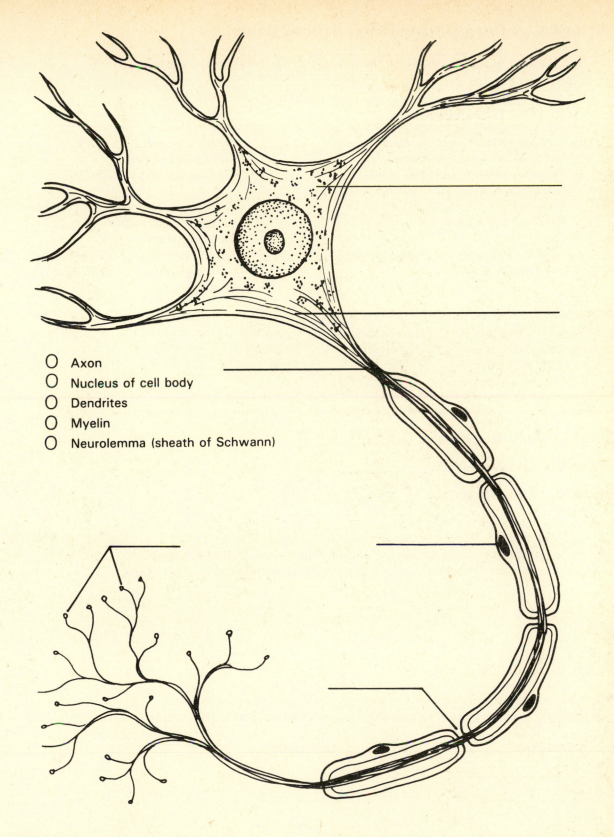

O Axon
O Nucleus of cell body
O Dendrites
O Myelin
O Neurolemma (sheath of Schwann)

Figure LG 12.1 Structure of a typical neuron as exemplified by an efferent (motor) neuron. Complete the figure as directed in Checkpoint B4.

■ **B6.** Match the parts of a neuron listed in the box with the descriptions below.

A. Axon	L. Lipofuscin
CB. Cell body (soma)	M. Mitochondria
CS. Chromatophilic substance (Nissl bodies)	NF. Neurofibrils
D. Dendrite	T. Trigger zone

_____ a. Contains nucleus; cannot regenerate because it lacks mitotic apparatus

_____ b. Yellowish brown pigment that increases with age; appears to be end product of lysosomes

_____ c. Provide energy for neurons

_____ d. Long, thin filaments that provide support and shape for the cell

_____ e. Orderly arrangement of rough ER; site of protein synthesis

_____ f. Conducts impulses toward cell body

_____ g. Conducts impulses away from cell body; has synaptic end bulbs that secrete neurotransmitter

_____ h. Located at junction of axon hillock and initial segment; nerve impulses arise here in many neurons

■ **B7.** Select terms from the box that match descriptions below.

Ganglion	Neuron
Nerve	Tract
Nerve fiber	

a. One single nerve cell, consisting of cell body, axon(s), and dendrites (in most cases): _____

b. One axon or one dendrite: _____

c. Bundle of nerve fibers in PNS: _____

d. Bundle of nerve fibers in CNS: _____

e. Cluster of neuron cell bodies in PNS: _____

B8. *A clinical challenge.* Contrast two types of *axonal transport* and tell which type is involved in the spread of the different microorganisms that cause herpes, rabies, and tetanus.

B9. Name and give a brief description of three types of neurons based on structural characteristics and three types of neurons classified according to functional differences.

Structural   **Description**

_____-polar _____

_____-polar _____

_____-polar _____

Functional   **Description**

_____-fferent _____

_____-fferent _____

_____ _____

■ **B10.** Refer to Checkpoint B9 above and place asterisks next to the most common types of neurons.

    a. Most common structural type of neuron in the brain and spinal cord

    b. Most common functional type of neuron in the body

■ **B11.** Using the list in the box, identify the type of nerve fiber that transmits each of the kinds of nerve impulses listed.

> GSA. General somatic afferent   GVE. General visceral efferent
> GSE. General somatic efferent   SA. Special afferent
> GVA. General visceral afferent

_____ a. Senses of vision, hearing, balance, taste, and smell are sensed via fibers of this type.

_____ b. Pain from a thorn in your skin is sensed via fibers of this type.

_____ c. Pain from a spasm of smooth muscle in the gallbladder is sensed by means of fibers of this type.

_____ d. With your eyes closed, you can tell the position of your skeletal muscles and joints due to nerve impulses that pass along this type of fiber.

_____ e. In order to increase your heart rate, impulses pass from your brain to your heart via this type of nerve fiber.

_____ f. These nerve fibers carry impulses from the CNS to muscles in your fingers used in writing.

■ **B12.** *The Big Picture: Looking Ahead.* Refer to figures in your text to complete this Checkpoint. Write G for gray matter or W for white matter next to related descriptions.

_____ a. Consists mostly of myelinated nerve fibers (axons or dendrites) (Figures 12.3, page 337, and 13.11b, page 376)

_____ b. Consists mostly of cell bodies and unmyelinated nerve fibers, such as in ganglia (Figure 17.2, page 491)

_____ c. Forms the H-shaped inner portion of the spinal cord as seen in cross section (Figure 13.3, page 366)

_____ d. Forms the outer portion (outside the "H") of the spinal cord (Figure 13.3, page 366)

_____ e. Forms the outer part of the brain known as the cerebral cortex (Figure 14.11, page 407)

_____ f. Forms nuclei (clusters of neuron cell bodies within the CNS) (Figures 14.9, page 404)

_____ g. Forms spinal nerves (Figure 13.7, page 371, and Exhibit 13.3, page 382) and cranial nerves (Figure 14.5, page 398) in the PNS

_____ h. Forms tracts in the CNS (Figures 13.4, page 367, 16.14, page 470, and 15.6, page 441). Note locations of tracts in the outer (white matter) portion of the spinal cord (Exhibit 15.1, pages 443–444).

■ **B13.** *For extra review,* complete the table contrasting structures in the nervous system.

Structure	Color (Gray or White)	Composition (Cell Bodies or Nerve Fibers)	Location (CNS or PNS)
a. Nerve	White		
b. Tract			CNS
c. Nucleus	Gray		
d. Ganglia		Cell bodies	

C. Neurophysiology   (pages 340–348)

■ **C1.** Complete this Checkpoint describing the properties of the plasma membrane of cells, such as those of muscles or glands, that may be stimulated (excited) by neurons.

 a. Excitable cells exhibit a membrane _____ along which an electric

 charge (or _____) may pass. In living cells, this current is usually carried by *(electrons? ions?)*.

 b. Two types of potential may be developed: _____ potential and

 _____ potential. These potentials can develop because of the

 presence of _____ channels that open or close to control ion flow. These channels provide the main path for flow of current across the membrane because the phospholipid bilayer normally permits *(little? much?)* passage of ions.

■ **C2.** Do this exercise about roles of ion channels in development of membrane potentials.

 a. There are two kinds of ion channels: *(gated? leakage?)* channels that are always open

 and _____ channels that require a stimulus to activate opening or closure. The plasma cell of a membrane is more permeable to *(K^+? Na^+?)* because there are more leakage channels for *(K^+? Na^+?)*.

 b. Identify the type of gated channel that is responsive in each case described here. Choose from these answers: *(voltage? chemically? mechanically? light?)* gated channels.

 Channels in photoreceptors (rods and cones) of the eyes: _____

 Channels responsive to the neurotransmitter acetylcholine: _____

 Channels in neurons responsive to touch (in skin): _____

■ **C3.** Contrast direct vs. indirect change in membrane permeability to ions in this Checkpoint.

 a. The neurotransmitter acetylcholine *(directly? indirectly?)* alters membrane permeability to ions such as Na^+, Ca^{2+}, or K^+.

 b. Name chemicals involved in an *indirect* method to change in membrane permeability.

■ **C4.** Refer to Figure LG 12.2a and describe the two principal factors that contribute to resting potential in this Checkpoint.

 a. One factor is the unequal distribution of _____ across the membrane. In the circles on the figure write the names of the major cation outside the neuron (in ECF) and the major cation inside the neuron (in ICF). In the rectangle write the name of the major anion in ICF. *(Hint:* refer to text Figure 27.4, page 895, and also to Figure LG 3.2A, page 64 in the *Learning Guide.)*

 b. Color the two arrows to indicate the second factor—the relative permeability of the

 membrane to Na^+ and K^+. The membrane is about _____ times more permeable to K^+ than to Na^+, so K^+ tends to diffuse out much more than Na^+ diffuses in.

 c. If K^+ did freely leak out of the membrane that is so permeable to it, the inside of the membrane would become increasingly *(negative? positive?)*, contributing to a *(positive? negative?)* membrane potential that attempts to balance the K^+ gradient.

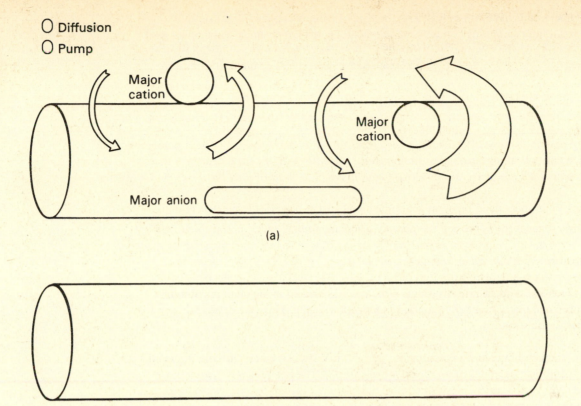

○ Diffusion
○ Pump

Major cation

Major cation

Major anion

(a)

(b)

Figure LG 12.2 Diagrams of a nerve cell. (a) Factors that contribute to membrane potential of a resting neuron. Label and color as directed in checkpoint C4. (b) Complete as directed in Checkpoint C8. ECF—Extracellular fluid; ICF—Intracellular fluid.

    d. Three other factors that contribute to this potential are inability of large negative ions

such as _____ to exit from the neuron, a slight leakage of _____ into cells, and the Na⁺/K⁺ pump, which pumps more cations out than it pumps in. Specifically, *(three? two?)* Na⁺ are pumped out for every *(three? two?)* K⁺ pumped in. Show this by coloring "pump" arrows on Figure LG 12.2a.

    e. As a result of factors described in (c) and (d), the membrane potential of a resting (nonactive) neuron is about _____ mV, or a range of _____ to _____ neurons. This means that the *(entire neuron? part of a neuron right next to the membrane?)* is much more *(positive? negative?)* than the fluid immediately outside the membrane.

■ **C5.** Describe *graded potentials* in this exercise.

    a. Graded potentials are so named because they vary in amplitude

(_____) depending on the strength of the _____. The size of the potentials varies *(directly? indirectly?)* with the number of gated ion channels open or closed as well as the amount of time they are open.

    b. These channels alter flow of _____ across the membrane. If the flow of ions makes the membrane more positive (such as −70 to −60 mV), it is said to be *(de? hyper?)*-polarized. If the potential becomes positive enough (such as −70 to −55 mV) to reach threshold (described below), an action potential *(is? is not?)* likely to occur. If the membrane potential becomes more negative (such as −70 to −80 mV), the

membrane is said to be _____-polarized, a point farther away from starting an action potential (nerve impulse).

c. Graded potentials have names based on the type of stimulus applied. For example, neurotransmitters affecting receptors of chemically gated ion channels at synapses lead

   to a post-_____ potential. Sensory receptors respond to stimuli by

   producing graded potentials called _____ potentials

   or _____ potentials.

d. Graded potentials most often occur in *(dendrites or cell bodies? axons?)*. These potentials produce currents along *(just a few micrometers? the entire length?)* of a neuron, whereas action potentials produce currents along *(short distances? short or long distances?)* of a neuron.

■ **C6.** Check your understanding of an action potential (nerve impulse) in this Checkpoint.

a. A stimulus causes the nerve cell membrane to become *(more? less?)* permeable to Na^+.

   Na^+ can then enter the cell as voltage-gated Na^+ _____ become activated and open. Notice from Figure 12.11, page 345, that Na^+ channels have two gates, much like a double door system. Which gate is located in the outer portion of the neuron cell membrane? *(Activation? Inactivation?)* gate. This gate opens in response to the stimulus.

b. At rest, the membrane had a potential of _____ mV. As Na^+ enters the cell, the inside of the membrane becomes more *(positive? negative?)*. The potential will tend to go toward *(−80? −55?)* mV. The process of *(polarization? depolarization?)* is occurring. This process causes structural changes in more Na^+ channels so that even more Na^+ enters. This is an example of a *(positive? negative?)* feedback mechanism. The result is

   a nerve impulse or nerve _____.

c. The membrane is completely depolarized at exactly *(−50? 0? +30?)* mV. Na^+ channels stay open until the inside of the membrane potential is *(reversed? repolarized?)* at +30 mV. Then inactivation gates close just milliseconds after the activation gates had opened, preventing more inward flow of Na^+.

d. After a fraction of a second, K^+ voltage-gated channels at the site of the original stimulus open. K^+ is more concentrated *(outside? inside?)* the cell (as you showed on Figure 12.2a); therefore K^+ diffuses *(in? out?)*. This causes the inside of the membrane to become more negative and

   return to its resting potential of _____ mV. The process is known as *(de? re?)*-polarization.

   In fact, outflow of K^+ may be so great that _____-polarization occurs in which membrane potential becomes closer to *(−50? −90?)* mV.

e. During depolarization, the nerve cannot be stimulated at all. This period is known as the *(absolute? relative?)* refractory period. *(Large? Small?)*-diameter axons have a longer absolute refractory period, with this about *(0.4? 4? 40?)* msec. In other words, slower impulses are likely to occur along *(large? small?)*-diameter neurons. Only a stronger-than-normal

   stimulus will result in an action potential during the _____ refractory period, which corresponds roughly with *(de? re?)*-polarization.

f. The nerve impulse is propagated (or _____) along the nerve, as adjacent areas are depolarized, causing more channels to be activated and more *(Na⁺? K⁺?)* to enter.

g. What effect do anesthetics such as procaine (Novocaine) have on action potentials?

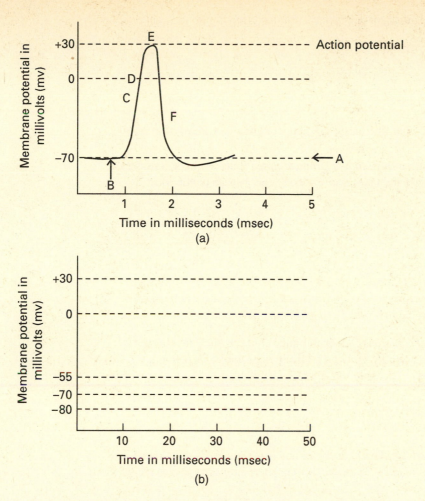

Figure LG 12.3 Diagrams for showing nerve action potentials. (a) Identify letter labels in Checkpoint C7. (b) Complete as directed in Checkpoint D2(b) and (d).

■ **C7.** Write the correct letter label from Figure LG 12.3a next to each description.

_____ a. The stimulus is applied at this point.

_____ b. Resting membrane potential is at this level.

_____ c. Membrane becomes so permeable to K^+ that K^+ diffuses rapidly out through K^+ channels.

_____ d. The membrane is becoming more positive inside as Na^+ enters; its potential is -30 mV. The process of depolarization is occurring.

_____ e. The membrane is completely depolarized at this point.

_____ f. The membrane is repolarizing at this point.

_____ g. Reversed polarization occurs; enough Na^+ has entered so that the region just inside the membrane at this part of the cell is more positive inside than outside.

C8. *For extra review.* Show the events that occur during initiation and propagation of a nerve impulse on Figure LG 12.2b. Compare your diagram to Figures 12.11 and 12.12a, pages 345 and 347 in the text.

C9. Explain how the all-or-none principle resembles a domino effect.

■ **C10.** Describe how myelination, fiber thickness, and temperature affect speed of impulse propagation.

a. Saltatory conduction occurs along *(myelinated? unmyelinated?)* nerve fibers. Saltatory transmission is *(faster? slower?)* and takes *(more? less?)* energy than continuous conduction. Explain why.

b. Type A fibers have the *(largest? smallest?)* diameter and *(are? are not?)* myelinated. These fibers conduct impulses *(rapidly? slowly?)*. Give two examples of A fibers.

c. Type C fibers have the *(largest? smallest?)* diameter and *(are? are not?)* myelinated. These fibers conduct impulses *(rapidly? slowly?)*. Give two examples of C fibers.

d. *A clinical challenge. (Warm? Cool?)* nerve fibers conduct impulses faster. How can this information be applied clinically?

■ **C11.** Contrast action potentials of nerve and muscle by completing this table.

Tissue	Typical Resting Membrane Potential	Duration of Nerve Impulse
a. Neuron		
b. Muscle		

C12. In Checkpoint C5 we considered graded potentials. Contrast *graded potentials* with *action potentials* by completing this table.

Characteristic	Graded Potential	Action Potential
a. Amplitude		All-or-none potential, usually about 100 mV
b. Duration	Long; may last several minutes	
c. Types of ion channels		Only voltage-gated
d. Location where potential arises	Usually dendrites or cell bodies	
e. Propagation		Long distances
f. Refractory period	None, so summation may occur	

D. Transmission at synapses   (pages 348–354)

■ **D1.** Do this activity about types of synapses.

a. In Chapter 10, you studied the point at which a neuron comes close to contacting a

muscle. This is known as a _____ junction, shown in Figure LG 10.2,

page 179. The minute space between two neurons is known as a _____.

b. There are two types of synapses. A(n) *(chemical? electrical?)* synapse is designed to allow spread of an ionic current from one neuron to the next, in fact, in both directions.

These synapses consist of _____ junctions made of hundreds of

proteinaceous tunnels called _____. Gap junctions are common in *(skeletal? smooth and cardiac?)* muscle where a group of muscle fibers works in synchrony.

c. In chemical synapses membranes of the two cells *(do? do not?)* touch. Does a nerve impulse actually "jump" across the synaptic cleft? Explain.

d. Now summarize the sequence of events at a chemical synapse by placing these events in

order from first to last: _____ → _____ → _____
    A. Electrical signal: postsynaptic potential (graded potential; if threshold level, action potential results)
    B. Electrical signal: presynaptic potential (nerve impulse)
    C. Chemical signal: release of neurotransmitter into synaptic cleft

e. Write these structures in order to summarize the anatomical pathway of a chemical synapse.

EB. End bulb of presynaptic neuron	SC. Synaptic cleft
N. Neurotransmitter	SV. Synaptic vesicle
NR. Neurotransmitter receptor	

_____ → _____ → _____ → _____ → _____
First   Last

f. Explain the role of calcium ions (Ca^{2+}) in nerve transmission at chemical synapses.

g. The rationale for *one-way information transfer* at chemical synapses is that only synaptic end bulbs of *(pre? post?)*-synaptic neurons release neurotransmitters, and only *(pre? post?)*-synaptic neurons have receptors.

■ **D2.** Complete the following exercise about postsynaptic potentials.

a. If a neurotransmitter causes depolarization of a postsynaptic neuron, for example, from −70 to −65 mV, the postsynaptic potential (PSP) is called *(excitatory? inhibitory?)*; in other words, it is called an *(EPSP? IPSP?)*. EPSPs result primarily from inflow of *(K^+? Na^+? Ca^{2+}?)* through chemically gated chemicals.

b. Usually a single EPSP within a single neuron *(is? is not?)* sufficient to cause a threshold potential and initiate a nerve impulse. Instead, a single EPSP can cause *(partial? total?)* depolarization. Diagram a partial depolarization from −70 to −65 mV in green on Figure LG 12.3b.

c. If neurotransmitters are released from a number of presynaptic end bulbs at one time,

their combined effect may produce threshold EPSP of about _____ mV. This phenomenon is known as *(spatial? temporal?)* summation. Temporal summation is that due to accumulation of transmitters from *(one? many?)* presynaptic end bulb(s) over a period of time.

d. If a neurotransmitter causes inhibition of the postsynaptic neuron, the process is known as *(de? hyper?)*-polarization, and the PSP is *(excitatory? inhibitory?)*. IPSPs often result from influx of *(Cl^-? K^+? Na^+?)* and/or outflow of *(Cl^-? K^+? Na^+?)* through gated channels. An example of an IPSP would be change in the membrane PSP from −70 to *(−60? −80?)* mV. Diagram such an IPSP in red on Figure 12.3b.

e. In other words, a neuron with an IPSP is *(closer? farther?)* from threshold than a neuron at resting membrane potential (RMP) so a cell with an IPSP is less likely to have an action potential. An example of use of an IPSP is inhibition of your triceps

brachii muscle as the _____ is stimulated (with an EPSP) to contract, or innervation of the heart by the vagus nerve (cranial nerve X) which *(in? de?)*-creases heart rate.

■ **D3.** *A clinical challenge.* Once a neurotransmitter completes its job, it must be removed from the synaptic cleft. Describe two mechanisms for getting rid of these chemicals. Notice the consequences of alterations of these mechanisms.

a. Inactivation of a neurotransmitter may occur, such as breakdown of _____

(ACh) by the enzyme named _____. Certain drugs (such as physostigmine) destroy this enzyme, so postsynaptic neurons (or muscles) *(remain? are less?)* activated. Predict the effects of such a drug.

b. A neurotransmitter such as _____ may be recycled by the presynaptic neuron that had released it. Cocaine blocks reuptake of two neurotransmit-

ters, namely, _____ and _____. Describe the physiological consequences of cocaine use:

Short-term effects _____

Long-term effects _____

■ **D4.** In this Checkpoint, describe another mechanism besides IPSP that results in inhibition of nerve impulses.

a. An inhibitory neuron releases its inhibitory transmitter at a synapse with the synaptic

end bulb of a(n) _____ neuron, in other words, at an axoaxonal synapse. The result is *(in? de?)*-crease in release of excitatory transmitter.

b. This type of inhibition is known as _____ inhibition. It may last for *(milliseconds? minutes or hours?)*.

■ **D5.** Do this exercise about neurotransmitters.

a. We have already discussed acetylcholine (_____), a neurotransmitter released at synapses and neuromuscular junctions. ACh is *(excitatory? inhibitory?)* toward skeletal muscle, but ACh released from the vagus nerve is *(excitatory? inhibitory?)* toward cardiac muscle. So when the vagus nerve sends impulses to your heart, your pulse (heart rate) becomes *(faster? slower?)*.

b. GABA and glycine are both *(excitatory? inhibitory?)* neurotransmitters. They act by opening *(Cl⁻? Na⁺?)* channels, leading to IPSPS. The most common inhibitory transmit-

ter in the brain is *(GABA? glycine?)*, whereas _____ is more commonly released by neurons in the spinal cord.

c. Strychnine is a chemical that blocks *(GABA? glycine?)* receptors so that muscles are not properly inhibited (relaxed). Strychnine poisoning is likely to lead to death because the

muscles of the _____ cannot relax so that air that is high in carbon dioxide cannot be exhaled.

d. List three neurotransmitters that are classified as catecholamines.

e. Can a single neuron release more than one type of neurotransmitter from its synaptic

end bulbs? _____

■ **D6.** *A clinical challenge.* Alkalosis tends to *(stimulate? depress?)* the neurons of the central nervous system (CNS), leading to tingling, spasms, and possibly convulsions.

Acidosis tends to *(stimulate? depress?)* the CNS, leading to _____.
With this information in mind, write one sign/symptom of a person whose (arterial) blood pH is 7.28.

■ **D7.** Check your understanding of chemicals that affect transmission at synapses and neuromuscular or neuroglandular junctions by completing this activity. Write E if the effect is excitatory or I if the effect is inhibitory. The first one is done for you. (Lines following descriptions are for the *clinical challenge* activity below.)

_____I_____ a. A chemical that inhibits release of ACh ____3____

_____ b. A chemical that competes for the ACh receptor sites on muscle cells _____

_____ c. A chemical that inactivates acetylcholinesterase _____

_____ d. A chemical that increases threshold (for example, from −60 to −40 mV)

        of a neuron _____
_____ e. A chemical that decreases threshold _____

A clinical challenge. Match the following chemicals with related mechanisms of actions. Write the numbers of the chemicals on lines to the right of the above descriptions. One is done for you.

1. Hypnotics, tranquilizers, anesthetics
2. Caffeine, benzedrine, nicotine
3. Botulism toxin, inhibiting muscle contraction
4. Curare, a muscle relaxant
5. Nerve gas such as diisopropyl fluorophosphate or physostigmine

■ **D8.** Match the types of circuits in the box with related descriptions. Answers may be used more than once.

C. Converging   P. Parallel after-discharge
D. Diverging   R. Reverberating

_____ a. Impulse from a single presynaptic neuron causes stimulation of increasing numbers of cells along the circuit.

_____ b. An example is a single motor neuron in the brain that stimulates many motor neurons in the spinal cord, therefore activating many muscle fibers.

_____ c. Branches from a second and third neuron in a pathway may send impulses back to the first, so the signal may last for hours, as in coordinated muscle activities.

_____ d. One postsynaptic neuron receives impulses from several nerve fibers.

_____ e. A single presynaptic neuron stimulates intermediate neurons which synapse with a common postsynaptic neuron, allowing this neuron to send out a stream of impulses, as in precise mathematical calculations.

E. Regeneration and repair of nervous tissue; disorders (pages 355–356)

■ **E1.** Which of the following can regenerate if destroyed? Briefly explain why in each case.

a. Neuron cell body

b. CNS nerve fiber

c. PNS nerve fiber

E2. Discuss methods currently being investigated to promote regrowth of damaged neurons.

■ **E3.** Answer these questions about nerve regeneration.

a. In order for a damaged neuron to be repaired, it must have an intact cell body and also a

_____.

b. Can axons in the CNS regenerate? Explain.

c. When a nerve fiber (axon or dendrite) is injured, the changes that follow in the cell body are called *(chromatolysis? Wallerian degeneration?)*. Those that occur in the portion of

the fiber distal to the injury are known as _____.

E4. Explain how peripheral nerves regenerate by describing each of these events.

a. Chromatolysis

b. Wallerian degeneration

c. Retrograde degeneration

d. Accelerated protein synthesis

A2. (a) CNS. (b) Aff. (c) SNS or Eff (*Hint:* remember S A M E: *S*ensory = *A*fferent; *M*otor = *E*fferent). (d) ANS. (e) ANS. (f) PNS.

B1. (a) Neurons. (b) Neuroglia. (c) Neuroglia. (d) Neuroglia.

B2. (a) N, S. (b) N, O. (c–e) A. (f) M. (g) E. (h) S. (i) N.

B3. (a) PNS; up to 500. (b) A. (c) B C A. (d) B. (e) Regeneration or regrowth. (f) Myelin; saltatory; in. (g) No; injured CNS nerve fibers do not heal well; yes; oligodendrocytes. (h) A.

B4.

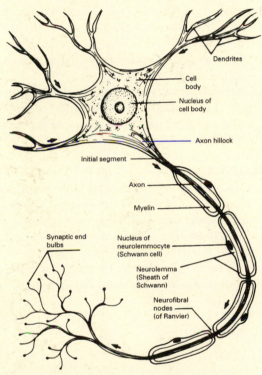

Figure LG 12.1A Structure of a typical neuron as exemplified by an efferent (motor) neuron.

B6. (a) CB. (b) L. (c) M. (d) NF. (e) CS. (f) D. (g) A. (h) T.

B7. (a) Neuron. (b) Nerve Fiber. (c) Nerve. (d) Tract. (e) Ganglion.

B10. (a) Multipolar. (b) Association.

B11. (a) SA. (b) GSA. (c) GVA. (d) GSA. (e) GVE. (f) GSE.

B12. (a) W. (b) G. (c) G. (d) W. (e) G. (f) G. (g) W. (h) W.

B13.

Structure	Color (Gray or White)	Composition (Cell Bodies or Nerve Fibers)	Location (CNS or PNS)
a. Nerve	White	Nerve fibers	PNS
b. Tract	White	Nerve fibers	CNS
c. Nucleus	Gray	Cell bodies	CNS
d. Ganglia	Gray	Cell bodies	PNS

C1. (a) Potential, current; ions. (b) Graded, action; ion; little.

C2. (a) Leakage, gated; K⁺, K⁺. (b) Light; chemically; mechanically.

C3. (a) Directly. (b) G protein and a second messenger system using molecules in cytosol.

C4. (a) Ions; see Figure LG 12.2A. (b) 50 to 100. (c) Negative, negative. (d) Anions of proteins, Cl^-; three, two; see Figure LG 12.2A (e) -70, -40 to -90; part of a neuron right next to the membrane, negative.

C5. (a) Size, stimulus; directly. (b) Ions; de; is; hyper.

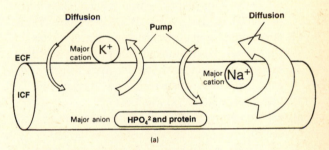

Figure 12.2A Diagram of a nerve cell.

(c) Synaptic; receptor, generator. (d) Dendrites or cell bodies; just a few micrometers, short or long distances.

C6. (a) More; channels; activation. (b) -70; positive; -55; depolarization; positive; action potential. (c) 0; reversed. (d) Inside, out; -70; re; after-hyper, -90. (e) Absolute; small, 4; small; relative, re. (f) Transmitted, Na⁺. (g) Such anesthetics prevent opening of voltage-gated Na⁺ channels in pain fibers.

C7. (a) B. (b) A. (c) F. (d) C. (e) D. (f) F. (g) E.

C10. (a) Myelinated; faster because the neurofibral nodes (of Ranvier) have a high density of voltage-gated Na^+ channels; less energy because only small regions (the nodes) have inflow of Na^+, which then must be pumped out by Na^+/K^+ pumps. (b) Largest, are; rapidly; large, sensory neurons, such as for touch, position of joints, and temperature, as well as nerves to skeletal muscles for quick reactions. (c) Smallest, are not; slowly; nerves that carry impulses to and from viscera. (d) Warm; ice or other cold applications can slow conduction of pain impulses.

C11.

Tissue	Typical Resting Membrane Potential	Duration of Nerve Impulse
a. Neuron	Higher (-70 mV)	Shorter (0.5–2 msec)
b. Muscle	Lower (-90mV)	Longer (1.0–5.0 msec: skeletal muscle; 10–300 msec: cardiac and smooth muscle)

D1. (a) Neuromuscular; synapse. (b) Electrical; gap, connexons; smooth and cardiac. (c) Do not; no; The nerve impulse travels along the presynaptic axon to the end of the axon, where synaptic vesicles are triggered to release neurotransmitters. These chemicals enter the synaptic cleft and then affect the postsynaptic neuron (either excite it or inhibit it). (d) B C A. (e) EB SV N SC NR. (f) The nerve impulse in the presynaptic neuron opens Ca^{2+} channels, creating an influx of Ca^{2+} into the presynaptic end bulbs. The Ca^{2+} here triggers exocytosis of synaptic vesicles with release of neurotransmitters. (g) Pre, post.

D2. (a) Excitatory, EPSP; Na^+. (b) Is not; partial; see Figure LG 12.3bA. (c) −55; spatial; one. (d) Hyper, inhibitory; Cl^-, K^+; −80; see Figure LG 12.3bA. (e) Farther; biceps, de.

D3. (a) Acetylcholine, acetylcholinesterase (ACHase); remain. Such drugs may help activate muscles of persons with myasthenia gravis because many of their neurotransmitter receptors are unavailable so the extra activation of ACh can enhance response of remaining receptors. On the other hand, use of some powerful anticholinesterase drugs (as "nerve gases") cause profound and lethal effects. See page 353 of the text. (b) Norepinephrine (NE); NE, dopamine (DA); short-term: increased heart rate, blood pressure, blood sugar, and body temperature, along with euphoria; long-term: depletion of dopamine (because it is not recycled) with harmful effects on heart and/or other organs.

D4. (a) Excitatory; de. (b) Presynaptic; minutes or hours.

D5. (a) ACh; excitatory, inhibitory; slower. (b) Inhibitory; Cl^-; GABA, glycine. (c) Glycine; diaphragm. (d) Norepinephrine (NE), epinephrine (epi), and dopamine (DA). (e) Yes.

D6. Stimulate; depress, decreased level of consciousness (LOC), such as lethargy, leading to possible coma and death; signs of acidosis are present at pH 7.28 (decreased LOC).

D7. (b) I (4). (c) E (5). (d) I (1). (e) E (2).

D8. (a) D. (b) D. (c) R. (d) C. (e) P.

E1. (a) No. About six months after birth, the mitotic apparatus is lost. (b) No. CNS nerve fibers lack the neurolemma necessary for regeneration, and scar tissue builds up by proliferation of astroglia cells. (c) Yes. PNS fibers do have a neurolemma.

E3. (a) Neurolemma. (b) No. They lack a neurolemma. (c) Chromatolysis, Wallerian degeneration.

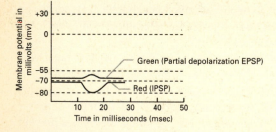

Figure LG 12.3bA Diagram of a facilitation EPSP and an IPSP.

WRITING ACROSS THE CURRICULUM: CHAPTER 12

1. Contrast the process of myelin formation in the CNS with that in the PNS.
2. Describe changes in myelin and impulse transmission, as well as signs or symptoms, that occur with multiple sclerosis (MS).
3. Contrast effects of the neurotransmitter acetylcholine on different types of muscle.
4. Contrast effects of the following neurotransmitters: glutamate, glycine, and catecholamines such as norepinephrine.
5. Describe the steps in repair of peripheral neurons.
6. Describe causes and symptoms of and methods for alleviating epileptic seizures.

MASTERY TEST: CHAPTER 12

Questions 1–2: Arrange the answers in correct sequence.

_____ _____ _____ 1. In order of transmission across synapse, from first structure to last:
A. Presynaptic end bulb
B. Postsynaptic neuron
C. Synaptic cleft

_____ _____ _____ 2. Membrane potential values, from most negative to zero:
A. Resting membrane potential
B. Depolarized membrane potential
C. Threshold potential

Questions 3–9: Circle the letter preceding the one best answer to each question.

3. Choose the one *false* statement.
   A. The membrane of a resting neuron has a membrane potential of −70 mV.
   B. In a resting membrane, permeability to K^+ ions is about 100 times less than permeability to Na^+ ions.
   C. C fibers are thin, unmyelinated fibers with a relatively slow rate of nerve transmission.
   D. C fibers are more likely to innervate the heart and bladder than structures (such as skeletal muscles) that must make instantaneous responses.

4. All of the following are listed with a correct function *except:*
   A. Neurolemmocyte—myelination of neurons in the PNS
   B. Oligodendrocytes—myelination of neurons in the CNS
   C. Astrocytes—form an epithelial lining of the ventricles of the brain
   D. Microglia—phagocytic

5. Synaptic end bulbs are located:
   A. At ends of axon terminals
   B. On axon hillocks
   C. On neuron cell bodies
   D. At ends of dendrites
   E. At ends of both axons and dendrites

6. Which of these is equivalent to a nerve fiber?
   A. A neuron
   B. A neurofibril
   C. An axon or dendrite
   D. A nerve, such as sciatic nerve

7. *ACh* is an abbreviation for:
   A. Acetylcholine
   B. Norepinephrine
   C. Acetylcholinesterase
   D. Serotonin
   E. Inhibitory postsynaptic potential

8. A term that means the same thing as *afferent* is:
   A. Autonomic
   B. Somatic
   C. Peripheral
   D. Motor
   E. Sensory

9. An excitatory transmitter substance that changes the membrane potential from −70 to −65 mV causes:
   A. Impulse conduction
   B. Partial depolarization
   C. Inhibition
   D. Hyperpolarization

Questions 10–20: Circle T (true) or F (false). If the statement is false, change the underlined word or phrase so that the statement is correct.

T F 10. The concentration of potassium ions (K+) is considerably <u>greater</u> inside a resting cell than outside of it.

T F 11. Because CNS fibers contain no neurolemma and the neurolemma produces myelin, CNS <u>fibers are all unmyelinated.</u>

T F 12. Neurotransmitter substances are released at <u>synapses and also at neuromuscular junctions.</u>

T F 13. Generally, release of excitatory transmitter by <u>a single presynaptic end bulb</u> is sufficient to develop an action potential in the postsynaptic neuron.

T F 14. <u>Epilepsy</u> is a condition that involves abnormal electrical discharges within neurons of the brain.

T F 15. In the <u>converging</u> circuit, a single presynaptic neuron influences several postsynaptic neurons (or muscle or gland cells) at the same time.

T F 16. Action potentials are measured in <u>milliseconds, which are thousandths of a second.</u>

T F 17. Nerve fibers with a short absolute refractory period can respond to <u>more rapid</u> stimuli than nerve fibers with a long absolute refractory period.

T F 18. Most neurons in the central nervous system (CNS) are classified as <u>unipolar.</u>

T F 19. A stimulus that is adequate will temporarily <u>increase permeability of the nerve membrane to Na±.</u>

T F 20. The brain and <u>spinal nerves</u> are parts of the peripheral nervous system (PNS).

Questions 21–25: Fill-ins. Complete each sentence with the word or phrase that best fits.

_____ 21. The _____ nervous system consists of the sympathetic and parasympathetic divisions.

_____ 22. Application of cold to a painful area can decrease pain in that area because _____

_____ 23. One-way nerve impulse transmission can be explained on the basis of release of transmitters only from the _____ of neurons.

_____ 24. The neurolemma is found only around nerve fibers of the _____ nervous system so that these fibers can regenerate.

_____ 25. In an axosomatic synapse, an axon's synaptic end bulb transmits nerve impulses to the _____ of a postsynaptic neuron.

ANSWERS TO MASTERY TEST: ★ CHAPTER 12

Arrange

1. A C B
2. A C B

Multiple Choice

3. B
4. C
5. A
6. C
7. A
8. E
9. B

True–False

10. T
11. F. May be myelinated because oligodendrocytes myelinate CNS fibers
12. T
13. F. A number of presynaptic end bulbs
14. T
15. F. Diverging
16. T
17. T
18. F. Multipolar
19. T
20. F. Spinal nerves (as well as some other structures; but not the brain)

Fill-ins

21. Autonomic
22. Cooling of neurons slows down the speed of nerve transmission, for example, of pain impulses
23. End bulbs of axons
24. Peripheral
25. Cell body (soma)

FRAMEWORK 13
The Spinal Cord & the Spinal Nerves

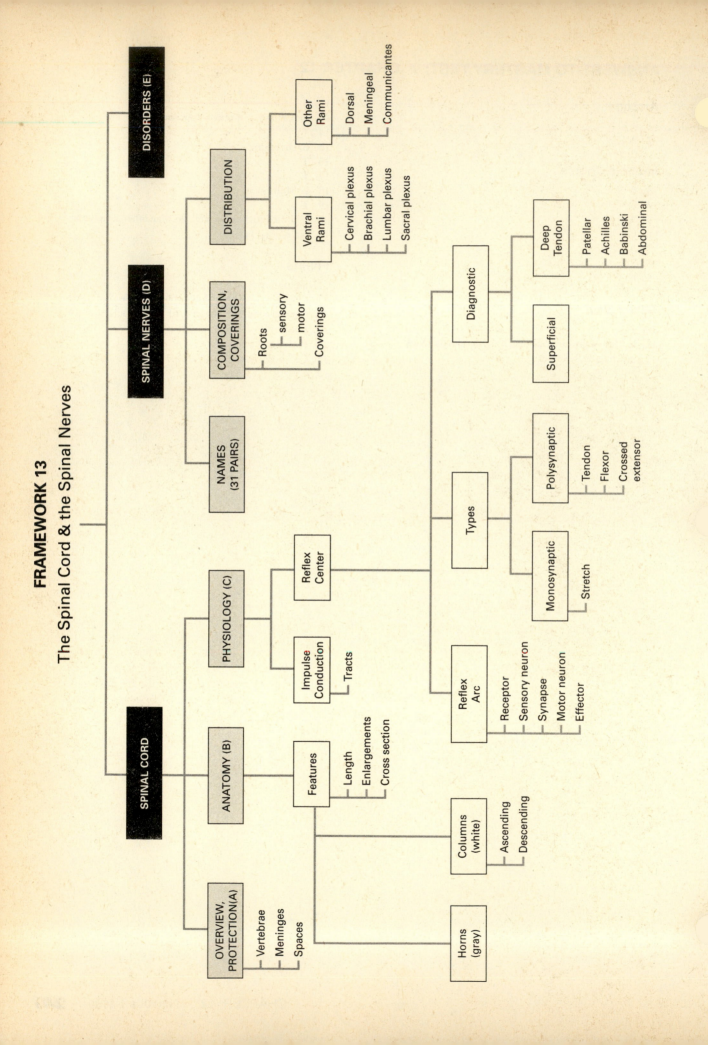

SPINAL CORD

DISORDERS (E)

SPINAL NERVES (D)

DISTRIBUTION

Other Rami
- Dorsal
- Meningeal
- Communicantes

Ventral Rami
- Cervical plexus
- Brachial plexus
- Lumbar plexus
- Sacral plexus

COMPOSITION, COVERINGS

Roots
- sensory
- motor
Coverings

NAMES (31 PAIRS)

PHYSIOLOGY (C)

Reflex Center

Impulse Conduction
- Tracts

Types

Monosynaptic
- Stretch

Polysynaptic
- Tendon
- Flexor
- Crossed extensor

Diagnostic

Superficial

Deep Tendon
- Patellar
- Achilles
- Babinski
- Abdominal

Reflex Arc
- Receptor
- Sensory neuron
- Synapse
- Motor neuron
- Effector

ANATOMY (B)

Features
- Length
- Enlargements
- Cross section

Columns (white)
- Ascending
- Descending

Horns (gray)

SPINAL CORD

OVERVIEW, PROTECTION (A)
- Vertebrae
- Meninges
- Spaces

The Spinal Cord and Spinal Nerves

The spinal cord and spinal nerves serve as the major links in the communication pathways between the brain and all other parts of the body. Nerve impulses are conveyed along routes (or tracts) in the spinal cord, laid out much as train tracks: some head north to regions of the brain, and others carry nerve messages south from the brain toward specific body parts. Spinal nerves branch off from the spinal cord, perhaps like a series of bus lines that pick up passengers (nerve messages) to or from train depots (points along the spinal cord) en route to their final destinations. Organization is critical in this nerve impulse transportation network. Any structural breakdowns—by trauma, disease, or other disorders—can lead to interruption in service with resultant chaos (such as spasticity) or standstill (such as sensory loss or paralysis).

As you begin your study of the spinal cord and spinal nerves, carefully examine the Chapter 13 Topic Outline and Objectives; check off each one as you complete it. To organize your study of the spinal cord and spinal nerves, glance over the Chapter 13 Framework now. Be sure to refer to the Framework frequently and note relationships among key terms in each section.

TOPIC OUTLINE AND OBJECTIVES

A. Spinal cord: overview; protection

B. Spinal cord: anatomy

☐ 1. Describe the protection and gross and anatomical features of the spinal cord.

C. Spinal cord: physiology

☐ 2. Describe the functions of the principal sensory and motor tracts of the spinal cord.

☐ 3. Describe the functional components of a reflex arc and the relationship of reflexes to homeostasis.

☐ 4. List and describe several clinically important reflexes.

D. Spinal nerves

☐ 5. Describe the composition and connective tissue coverings of a spinal nerve.

☐ 6. Define a plexus and identify the distribution of nerves of the cervical, brachial, lumbar, and sacral plexuses.

E. Disorders

☐ 7. Describe the clinical significance of dermatomes and myotomes.

☐ 8. Explain the causes and symptoms of neuritis, sciatica, shingles, and poliomyelitis.

WORDBYTES

Now become familiar with the language of this chapter by studying each wordbyte, its meaning, and an example of its use within a term. After you study the entire list, self-check your understanding by writing the meaning of each wordbyte on the line. As you continue through the *Learning Guide,* identify (and fill in) additional terms that contain the same wordbyte.

Wordbyte	Self-check	Meaning	Example(s)
arachn-	_____	spider	*arachn*oid
dura	_____	hard	*dura* mater
pia	_____	tender	*pia* mater
soma-	_____	body	*soma*tic

CHECKPOINTS

A. Spinal cord: overview; protection   (page 361)

■ **A1.** List three general functions of the spinal cord.

■ **A2.** The spinal cord is part of the *(central? peripheral?)* nervous system.

■ **A3.** Refer to Figure LG 13.1 and do the following exercise.

a. Color the meninges to match color code ovals.

b. Label these spaces: *epidural space, subarachnoid space,* and *subdural space.*

c. Write next to each label of a space the contents of that space. Choose from these answers: *cerebrospinal (CSF); lymphatic fluid; fat and connective tissue.*

d. Inflammation of meninges is a condition called _____.

e. The denticulate ligaments are extensions of *(dura mater? arachnoid? pia mater?)*

that attach laterally to _____ along the length of the cord. State functions of these ligaments.

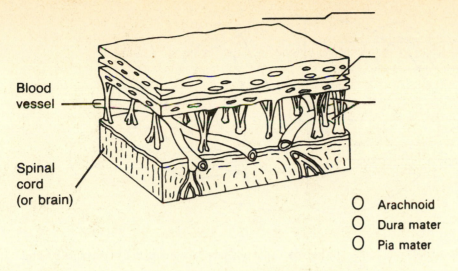

Blood vessel

Spinal cord (or brain)

○ Arachnoid
○ Dura mater
○ Pia mater

Figure LG 13.1   Meninges. Color and label as indicated in Checkpoint A3.

B. Spinal cord: anatomy   (pages 362–365)

■ **B1.** Do this activity about your own spinal cord.

a. Identity the location of your own spinal cord. It lies within the vertebral canal, extending from just inferior to the cranium to about the level of your *(waist? sacrum?)*. This level corresponds with about the level of *(L1–L2? L4–L5? S4–S5?)* vertebrae. In other words, the *(spinal cord? vertebral column?)* reaches a lower (more inferior) level in your body.

The spinal cord is about 42 to 45 cm (_____ inches) in length.

b. Circle the two regions of your spinal cord that have notable enlargements where nerves exit to your upper and lower limbs:

cervical   thoracic   lumbar   sacral   coccygeal

■ **B2.** Match the names of the structures listed in the box with descriptions given.

> Ca. Cauda equina   F. Filum terminale
> Co. Conus medullaris   S. Spinal segment

_____ a. Tapering inferior end of spinal cord

_____ b. Any region of spinal cord from which one pair of spinal nerves arises

_____ c. Nonnervous extension of pia mater; anchors cord in place

_____ d. "Horse's tail"; extension of spinal nerves in lumbar and sacral regions within subarachnoid space

■ **B3.** *A clinical challenge.* Answer these questions about meninges and related clinical procedures.

a. State two or more purposes of a spinal tap (or lumbar puncture).

Into which space does the needle enter? *(Epidural? Subdural? Subarachnoid?)*

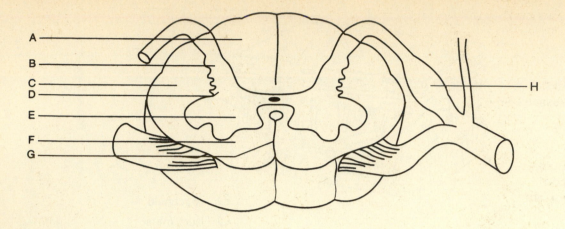

A ——————

B ——————

C ——————
D ——————

E ——————

F ——————
G ——————

————————— H

Figure LG 13.2 Outline of the spinal cord, roots, and nerves. Color and label according to Checkpoints B4 and D1.

b. At what level of the vertebral column (not spinal cord) is the needle for a spinal tap

(or lumbar puncture) inserted? Between _____ vertebrae. Explain why
this location is a relatively safe site for this procedure.

c. Identify level L4 of the vertebral column on yourself by listing two other landmarks

at this level. _____ _____

d. In some cases anesthetics are introduced into the epidural space, rather than into the
subarachnoid space (as in a spinal tap). Explain why an *epidural* is likely to exert its
effects on nerves with slower and more prolonged action than a spinal tap.

State one example of a procedure for which an epidural may be used. _____

■ **B4.** Identify the following structures on Figure LG 13.2.
a. Gray matter is found *(within? outside?)* the H-shaped outline, and white matter

is located _____ that outline. With a lead pencil, shade the gray
matter on the right side of the figure only.

b. Now label parts A–G on the left side of the figure.

c. On the right side of the figure, label H.

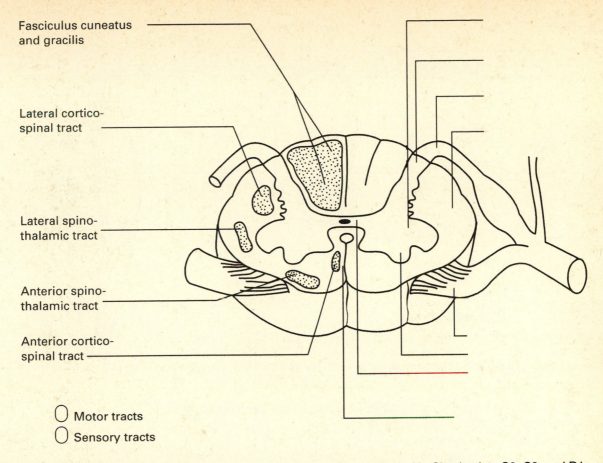

Fasciculus cuneatus
and gracilis

Lateral cortico-
spinal tract

Lateral spino-
thalamic tract

Anterior spino-
thalamic tract

Anterior cortico-
spinal tract

◯ Motor tracts
◯ Sensory tracts

Figure LG 13.3 Major tracts in the spinal cord. Color and complete as directed in Checkpoints C2, C3, and D1.

C. Spinal cord: physiology (pages 365–375)

■ **C1.** In Checkpoint A1, you listed main functions of the spinal cord. Expand that description in this Checkpoint.

a. One primary function of the spinal cord is to permit _____ between nerves in the periphery, such as arms and legs, and the brain by means of the *(tracts? nerves? ganglia?)* located in the white columns of the cord.

b. Another function of the cord is to serve as a _____ center by means of spinal

nerves. These are attached by _____ roots. The *(anterior? posterior?)* root contains sensory

nerve fibers, and the _____ root contains motor fibers.

■ **C2.** Refer to Figure LG 13.3 and do this activity on the conduction function of the spinal cord.

a. Tracts conduct nerve impulses in the *(central? peripheral?)* nervous system. Function-

ally they are comparable to _____ in the peripheral nervous system. Tracts are located in *(gray horns? white columns?)*. They appear white because they consist of bundles of *(myelinated? unmyelinated?)* nerve fibers.

b. Ascending tracts are all *(sensory? motor?)*, conveying impulses between the spinal cord

and the _____. All motor tracts in the cord are *(ascending? descending?)*.

c. Color the five tracts on the figure, using color code ovals to demonstrate sensory or motor function of tracts. (*Note:* Tracts are actually present on both sides of the cord but are shown on only one side here.)

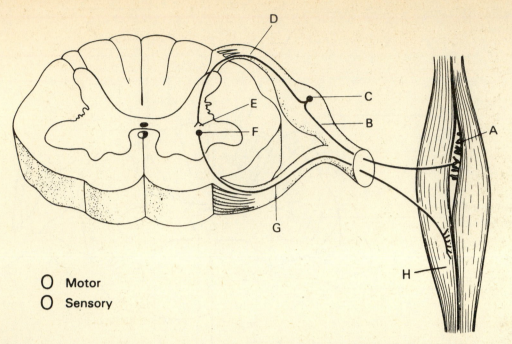

O Motor

O Sensory

Figure LG 13.4   Reflex arc: stretch reflex. Color and label as directed in Checkpoints C4 and C5.

d. Write next to the name of each tract its correct functions. Use these answers:
   Pain, temperature, crude touch, pressure
   Precise voluntary movements
   Proprioception, pressure, vibration, and two-point discriminative touch

e. The name *lateral corticospinal* indicates that the tract is located in the

   _____ white column, that it originates in the *(cerebral cortex?*

   thalamus? spinal cord?), and that it ends in the _____.

f. The lateral corticospinal tract is *(ascending, sensory? descending, motor?)* and it is a(n)
   (extrapyramidal? pyramidal?) tract.

g. The rubrospinal, tectospinal, and vestibulospinal tracts are *(extrapyramidal?*

   pyramidal?) tracts. These are involved with control of: _____.
   A. Precise, voluntary movements
   B. Automatic movements such as maintenance of muscle tone, posture, and equilibrium

h. Now label all structures with leader lines on the right side of the figure.

C3. *For extra review.* Draw and label the following tracts on Figure LG 13.3. Then color
them according to the color code ovals on the figure: *anterior spinocerebellar, posterior
spinocerebellar, rubrospinal, tectospinal,* and *vestibulospinal.*

■ **C4.** Checkpoint C1b describes the function of the spinal cord and spinal nerves in reflexes.
Color lightly the sensory and motor roots of a spinal nerve on Figure LG 13.4. Select
colors according to the color code ovals there.

■ **C5.** Complete this activity on the function of the spinal cord as a reflex center. Label
structures A–H on Figure LG 13.4, using the following terms: *effector, motor neuron axon,
motor neuron cell body, receptor, sensory neuron axon, sensory neuron cell body, sensory
neuron dendrite,* and *synapse* (integrating center). Note that these structures are lettered in
alphabetical order along the conduction pathway of a reflex arc. Add arrows showing the
direction of nerve transmission in the arc.

■ **C6.** Answer these questions about the reflex arc in Figure LG 13.4.

    a. How many neurons does this reflex contain? _____ The neuron that conveys the

        impulse toward the spinal cord is a _____ neuron; the one that carries

        the impulses toward the effector is a _____ neuron.

    b. This is a *(monosynaptic? polysynaptic?)* reflex arc. The synapse, like all somatic
       synapses, is located in the *(CNS? PNS?)*.

    c. Receptors, in this case located in skeletal muscle, are called _____.

       They are sensitive to changes in _____. This type of reflex might

       therefore be called a _____ reflex.

    d. Because sensory impulses enter the cord on the same side as motor impulses leave, the
       reflex is called *(ipsilateral? contralateral? intersegmental?)*.

    e. What structure is the effector? _____

    f. One example of a stretch reflex is the _____, in which stretching of
       the patellar tendon initiates the reflex.

■ **C7.** Explain how the brain may get the message that a stretch reflex (or other type of
reflex) has occurred.

■ **C8.** Do this exercise describing how tendon reflexes protect tendons.

    a. Receptors located in tendons are named *(muscle spindles? tendon organs?)*. They are
       sensitive to changes in muscle *(length? tension?)*, as when the hamstring muscles are
       contracted excessively, pulling on tendons.

    b. When this occurs, association neurons cause *(excitation? inhibition?)* of this same
       muscle, so that the hamstring fibers *(contract further? relax?)*.

    c. Simultaneously, other association neurons fire impulses that stimulate *(synergistic?
       antagonistic?)* muscles (such as the quadriceps in this case). These muscles then
       (contract? relax?).

    d. The net effect of such a *(mono? poly?)*-synaptic tendon reflex is that the tendons are
       (protected? injured?).

■ **C9.** Contrast stretch, flexor, and crossed extensor reflexes in this learning activity.

    a. A flexor reflex *(does? does not?)* involve association neurons, and so it is *(more? less?)*
       complex than a stretch reflex.

    b. A flexor reflex sends impulses to *(one? several?)* muscle(s), whereas a stretch reflex,
       such as the knee jerk, activates *(one? several?)* muscle(s), such as the quadriceps.

       A flexor reflex is also known as a _____ reflex, and it is

       _____-lateral.

c. If you simultaneously contract the flexor and extensor (biceps and triceps) muscles of

your forearm with equal effort, what action occurs? (Try it.) _____
In order for movement to occur, it is necessary for extensors (triceps) to be inhibited
while flexors (biceps) are stimulated. The nervous system exhibits such control by a

phenomenon known as _____ innervation.

d. Reciprocal innervation also occurs in the following instance. Suppose you step on a tack
under your right foot. You quickly withdraw that foot by *(flexing? extending?)* your
right leg using your hamstring muscles. (Stand up and try it.) What happens to your left
leg? You *(flex? extend?)* it. This is an example of reciprocal innervation involving a

_____ reflex.

e. A crossed extensor reflex is *(ipsilateral? contralateral?)*. It *(may? may not?)* be inter-
segmental. Therefore, many muscles may be contracted to extend your left thigh and leg
to shift weight to your left side and provide balance during the tack episode.

■ **C10.** Why are deep tendon reflexes (that involve stretching a tendon such as that of the
quadriceps femoris muscle in the patellar reflex) particularly helpful diagnostically?

■ **C11.** Complete the table about reflexes that are of clinical significance.

	Patellar Reflex	**Achilles Reflex**	**Babinski Sign**
a. Procedure used to demonstrate reflex	Tap patellar ligament		
b. Nature of positive response		Plantar flexion	
c. Spinal nerves and muscles evaluated by procedure in reflex			Determines if corticospinal tract is myelinated yet
d. Cause of negative response			Plantar flexion reflex: all toes curl under; normal after age 1-1/2
e. Cause of exaggerated response		Damage to motor tracts in S1 or S2 segments of cord	

D. Spinal nerves   (pages 375–386)

■ **D1.** Complete this exercise about spinal nerves.

   a. There are _____ pairs of spinal nerves. Write the number of pairs in each region.

      _____ Cervical _____ Thoracic _____ Lumbar _____ Sacral _____ Coccygeal

   b. Which of these spinal nerves form the cauda equina?

   c. Spinal nerves are attached by two roots. The posterior root is *(sensory? motor? mixed?)*,

      and the anterior root is _____, whereas the spinal nerve is _____.

   d. Individual nerve fibers are wrapped in a connective tissue covering known as *(endo-?
      epi-? peri-?)* neurium. Groups of nerve fibers are held in bundles (fascicles) by

      _____-neurium. The entire nerve is wrapped with _____-neurium.

   e. Spinal nerves branch when they leave the intervertebral foramen. These branches are

      called _____. Because they are extensions of spinal nerves, rami are
      (sensory? motor? mixed?).

   f. Which ramus is larger? *(Ventral? Dorsal?)* What areas does it supply?

   g. What area does the dorsal ramus innervate? Label it on Figure LG 13.3.

   h. Name two other branches (rami) and state their functions.

■ **D2.** Match the plexus names in the box with descriptions. Refer to Figure 13.2a (page 363
in your text) for help.

> B. Brachial   L. Lumbar
> C. Cervical   S. Sacral
> I. Intercostal

_____ a. Provides the entire nerve supply for the
            arm

_____ b. Contains origin of phrenic nerve
            (nerve that supplies diaphragm)

_____ c. Forms median, radial, and axillary nerves

_____ d. Not a plexus at all, but rather segmentally
            arranged nerves

_____ e. Supplies nerves to scalp, neck, and part
            of shoulder and chest

_____ f. Supplies fibers to the femoral nerve,
            which innervates the quadriceps, so injury
            to this plexus would interfere with actions
            such as touching the toes

_____ g. Forms the largest nerve in the body
            (the sciatic), which supplies posterior
            of thigh and the leg

D3. *For extra review.* Match names of spinal nerves in the box with their descriptions. On the line following the description, write the site of origin of the nerve. The first one is done for you.

> Axillary   Pudendal
> Inferior gluteal   Radial
> Musculocutaneous   Sciatic

_____**Sciatic**_____ a. Consists of two nerves, the tibial and common peroneal; supplies the hamstrings, abductors, and all muscles distal to the knee. **L4–S3**

_____ b. Supplies the deltoid muscle. _____

_____ c. Innervates the major flexors of the arm. _____

_____ d. Supplies most extensor muscles of the forearm, wrist, and fingers; may be damaged by extensive use

of crutches. _____

_____ e. Supplies the gluteus maximus muscle. _____

_____ f. May be anesthetized in childbirth because it innervates external genitalia and lower part of

the vagina. _____

D4. *For additional review* of nerve supply to muscles, complete the table below. Write the name of the plexus with which the nerve is associated, as well as the distribution of that nerve. The first one is done for you. (*Hint:* one answer involves a nerve that is not a spinal nerve, so it is not a part of a plexus.)

Name of Nerve	Plexus	Distribution
a. Lateral pectoral	Brachial	
b. Medial pectoral		Pectoralis major and minor muscles
c. Lesser occipital		
d. Ulnar		
e.		Latissimus dorsi muscle
f. Long thoracic		
g.		Trapezius and sternocleidomastoid muscles (*Hint:* See Exhibit 14.4, page 422 of text.)
h. Femoral		
i.	Lumbar	Cremaster muscle that pulls testes closer to pelvis
j. Obturator		
k.		Gastrocnemius, soleus, tibialis posterior
l.		Tibialis anterior

■ **D5.** *A clinical challenge.* If the cord were completely transected (severed) just below the C7 spinal nerves, how would the functions listed below be affected? (Remember that nerves that originate below this point would not communicate with the brain and so would lose much of their function.) Explain your reasons in each case. (*Hint:* Refer to Figure 13.2a, page 363 in the text.)

a. Breathing via diaphragm

b. Movement and sensation of thigh and leg

c. Movement and sensation of the arm

d. Use of muscles of facial expression and muscles that move jaw, tongue, eyeballs

D6. Describe the general pattern of dermatomes:

a. In the trunk

b. In the limbs

D7. Contrast *dermatome* with *myotome*.

E. Disorders (page 387)

■ **E1.** Shingles is an infection of the *(central? peripheral?)* nervous system. The causative virus is also the agent of *(chickenpox? measles? cold sores?)*. Following recovery from chickenpox, the virus remains in the body in the *(spinal cord? dorsal root ganglia?)*. At times it is activated and travels along *(sensory? motor?)* neurons, causing *(pain? paralysis?)*.

■ **E2.** Match names of disorders in the box with definitions below.

N.	Neuritis	Sc.	Sciatica
P.	Poliomyelitis	Sh.	Shingles

_____ a. Inflammation of a single nerve

_____ b. Also known as infantile paralysis; caused by a virus that may destroy motor cell bodies in the brainstem or in the anterior gray horn of the spinal cord

_____ c. Acute inflammation of the nervous system by *Herpes zoster* virus

_____ d. Neuritis of a nerve in the posterior of hip and thigh; often due to a slipped disc in the lower lumbar region

ANSWERS TO SELECTED CHECKPOINTS: CHAPTER 13

A1. Processing center for (reflexes), integration (summing) of afferent of efferent nerve impulses, and pathway for those impulses.

A2. Central (CNS).

A3.

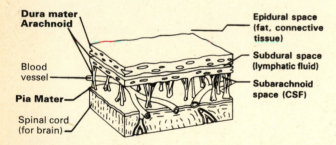

Figure LG 13.1A Meninges.

(d) Meningitis. (e) Pia mater, dura mater; anchor the cord and protect it from displacement.

B1. (a) Waist; L1–L2; vertebral column; 16–18.
(b) Cervical, lumbar.

B2. (a) Co. (b) S. (c) F. (d) Ca.

B3. (a) To insert anesthetics, antibiotics, chemotherapy or contrast media; to withdraw cerebrospinal fluid (CSF) for diagnostic purposes such as for analysis for blood or microorganisms; subarachnoid. (b) L3–L4 or L4–L5; the spinal cord ends at about L1–L2, so the cord is not likely to be injured at this lower level. (c) The iliac crest and umbilicus (navel) are both at about this level. (d) The anesthetic must penetrate epidural tissues and then all three layers of meninges before reaching nerve fibers, whereas an anesthetic in the subarachnoid space needs to penetrate only the pia mater to reach nerve tissue; childbirth (labor and delivery).

B4. (a) Within, outside; refer to Figure 13.3 of the text. (b) A, posterior columns; B, posterior horn; C, lateral columns; D, lateral horn; E, anterior horn; F, anterior columns; G, anterior median fissure. (c) Dorsal root ganglion.

C1. (a) Conduction (or a pathway), tracts. (b) Reflex; 2; posterior (dorsal), anterior (ventral). See Figure LG 13.3A.

C2. (a) Central; nerves; white columns; myelinated. (b) Sensory, brain; descending. (c–d) See Figure LG 13.3A. (e) Lateral, cerebral cortex, spinal cord (in anterior gray horn). (f) Descending, motor, pyramidal. (g) Extrapyramidal; B (h) See Figure LG 13.3A.

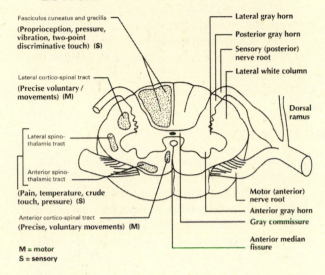

Fasciculus cuneatus and gracilis
(Proprioception, pressure, vibration, two-point discriminative touch) (S)

Lateral cortico-spinal tract
(Precise voluntary / movements) (M)

Lateral spino-thalamic tract

Anterior spino-thalamic tract

(Pain, temperature, crude touch, pressure) (S)

Anterior cortico-spinal tract
(Precise, voluntary movements) (M)

M = motor
S = sensory

Lateral gray horn
Posterior gray horn
Sensory (posterior) nerve root
Lateral white column
Dorsal ramus
Motor (anterior) nerve root
Anterior gray horn
Gray commissure
Anterior median fissure

Figure LG 13.3A Major tracts in the spinal cord.

C4. Refer to Figure LG 13.4A; A and D are sensory; F and G are motor.

C5.

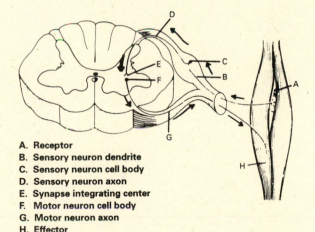

A. Receptor
B. Sensory neuron dendrite
C. Sensory neuron cell body
D. Sensory neuron axon
E. Synapse integrating center
F. Motor neuron cell body
G. Motor neuron axon
H. Effector

Figure LG 13.4A Reflex arc: stretch reflex.

C6. (a) 2; sensory, motor. (b) Monosynaptic; CNS. (c) Muscle spindles; length (or stretch); stretch. (d) Ipsilateral. (e) Skeletal muscle. (f) Knee jerk (patellar reflex).

C7. Branches from axons of sensory or association neurons travel through tracts to the brain.

C8. (a) Tendon organs; tension. (b) Inhibition, relax. (c) Antagonistic; contract. (d) Poly, protected.

C9. (a) Does, more. (b) Several, one; withdrawal, ipsi. (c) No action; reciprocal. (d) Flexing; extend; crossed extensor. (e) Contralateral; may.

C10. They can readily pinpoint a disorder of a specific spinal nerve or plexus, or portion of the cord, because they do not involve the brain.

C11.

	Patellar Reflex	Achilles Reflex	Babinski Sign
a. Procedure used to demonstrate reflex	Tap patellar ligament	Tap on calcaneal (Achilles tendon)	Light stimulation of outer margin of sole of foot
b. Nature of positive response	**Extension of leg**	Plantar flexion	Great toe extends, with or without fanning of other toes. Abnormal after age 1-1/2; shows incomplete myelination.
c. Spinal nerves and muscles evaluated by procedure in reflex	L2–L4 (quadriceps muscles)	L4–L5, S1–S3 (gastrocnemius and soleus muscles)	Determines if corticospinal tract is myelinated yet
d. Cause of negative response	Chronic diabetes mellitus, neuro-syphilis	Diabetes, neurosyphilis, alcoholism	Plantar flexion reflex: all toes curl under; normal after age 1-1/2
e. Cause of exaggerated response	Injury to corticospinal tracts	Damage to motor tracts in S1 or S2 segments of cord	Positive Babinski after age 1-1/2 indicates interruption of corticospinal tracts

D1. (a) 31; 8, 12, 5, 5, 1. (b) Lumbar, sacral, and coccygeal. (c) Sensory, motor, mixed. (d) Endo-; peri-; epi-. (e) Rami; mixed. (f) Ventral: all of the limbs and the ventral and lateral portions of the trunk; see Figure LG 13.3A. (g) Muscles and skin of the back; see Figure LG 13.3A. (h) Meningeal branch supplies primarily vertebrae and meninges; rami communicantes have autonomic functions.

D2. (a) B. (b) C. (c) B. (d) I. (e) C. (f) L. (g) S.

D3. (b) Axillary, C5–C6. (c) Musculocutaneous, C5–C7. (d) Radial, C5–C8, T1. (e) Inferior gluteal, L5–S2. (f) Pudendal, S2–S4.

D5. (a) Not affected because (phrenic) nerve to the diaphragm originates from the cervical plexus (at C3–C5), higher than the transection. So this nerve continues to receive nerve impulses from the brain. (b) Complete loss of sensation and paralysis because lumbar and sacral plexuses originate below the injury and therefore no longer communicate with the brain. (c) Most arm functions are not affected. As shown on Figure 13.2a, page 363 of the text, the brachial plexus originates from C5 through T1, so most nerves to the arm (those from C5 through C7) still communicate with the brain. (d) Not affected because all are supplied by cranial nerves which originate from the brain.

E1. Peripheral; chickenpox; dorsal root ganglia; sensory, pain.

E2. (a) N. (b) P. (c) Sh. (d) Sc.

WRITING ACROSS THE CURRICULUM: CHAPTER 13

1. Describe the anatomy and physiology of the meninges, as well as the spaces formed between and surrounding meninges.
2. State several reasons why a lumbar puncture might be done. Give a brief description of the procedure.
3. Describe a reflex arc, including the structure and function of its five components.
4. Contrast a stretch reflex with a flexor (withdrawal) reflex.
5. Describe how reflexes help you to maintain your balance when you quickly pick up your foot in response to stepping on a piece of glass.
6. Define the term plexus and describe what is meant by ventral rami that form a plexus. State examples of two plexuses and the major nerves they form.

MASTERY TEST: CHAPTER 13

Questions 1–4: Arrange the answers in correct sequence.

_____ _____ _____ 1. From superficial to deep:
   A. Subarachnoid space
   B. Epidural space
   C. Dura mater

_____ _____ _____ 2. From anterior to posterior in the spinal cord:
   A. Fasciculus gracilis and cuneatus
   B. Anterior spinothalamic tract
   C. Central canal of the spinal cord

_____ _____ _____ _____ 3. The plexuses, from superior to inferior:
   A. Lumbar
   B. Brachial
   C. Cervical
   D. Sacral

_____ _____ _____ _____ _____ 4. Order of structures in a conduction pathway, from origin to termination:
   A. Motor neuron
   B. Sensory neuron
   C. Integrative center
   D. Receptor
   E. Effector

Questions 5–10: Circle the letter preceding the one best answer to each question.

5. Herniation (or "slipping" of the disc between L4 and L5 vertebrae is most likely to result in damage to the _____ nerve.
   A. Femoral
   B. Sciatic
   C. Radial
   D. Musculocutaneous

6. All of these tracts are sensory *except:*
   A. Anterior spinothalamic
   B. Lateral spinothalamic
   C. Fasciculus cuneatus
   D. Lateral corticospinal
   E. Posterior spinocerebellar

7. Choose the *false* statement about the spinal cord.
   A. It has enlargements in the cervical and lumbar areas.
   B. It lies in the vertebral foramen.
   C. It extends from the medulla to the sacrum.
   D. It is surrounded by meninges.
   E. In cross section an H-shaped area of gray matter can be found.

8. All of these structures are composed of white matter *except:*
   A. Posterior root (spinal) ganglia
   B. Tracts
   C. Lumbar plexus
   D. Sciatic nerve
   E. Ventral ramus of a spinal nerve

9. Which is a *false* statement about the patellar reflex?
   A. It is also called the knee jerk.
   B. It involves a two-neuron, monosynaptic reflex arc.
   C. It results in extension of the leg by contraction of the quadriceps femoris.
   D. It is contralateral.

10. The cauda equina is:
    A. Another name for the cervical plexus
    B. The lumbar and sacral nerves extending below the end of the cord and resembling a horse's tail
    C. The inferior extension of the pia mater
    D. A denticulate ligament
    E. A canal running through the center of the spinal cord

Questions 11–20: Circle T (true) or F (false). If the statement is false, change the underlined word or phrase so that the statement is correct.

T F 11. The layer of the meninges that gets its name from its delicate structure, which is much like a spider's web, is the <u>arachnoid.</u>

T F 12. The two main functions of the spinal cord are that it serves as a <u>reflex center</u> and it is <u>the site where sensations are felt.</u>

T F 13. Dorsal roots of spinal nerves are <u>sensory, ventral roots are motor, and spinal nerves are mixed.</u>

T F 14. A tract is a bundle of nerve fibers <u>inside the central nervous system (CNS).</u>

T F 15. Synapses <u>are</u> present in posterior (dorsal) root ganglia.

T F 16. After a person reaches 18 months, the Babinski sign should be <u>negative, as indicated by plantar flexion (curling under of toes and foot).</u>

T F 17. Visceral reflexes are used diagnostically <u>more often than somatic ones because it is easy to stimulate most visceral receptors.</u>

T F 18. Transection (cutting) of the spinal cord at level C6 will result in <u>greater</u> loss of function than transection at level T6.

T F 19. The ventral root of a spinal nerve contains <u>axons and dendrites</u> of <u>both motor and sensory</u> neurons.

T F 20. A lumbar puncture (spinal tap) is usually performed at about the level of vertebrae <u>L1 to L2</u> because the cord ends between about <u>L3 and L4.</u>

Questions 21–25: Fill-ins. Complete each sentence with the word or short answer that best fits.

_____ 21. The phrenic nerve innervates the _____.

_____ 22. An inflammation of the dura mater, arachnoid, and/or pia mater is known as _____.

_____ 23. Tendon reflexes are _____-synaptic and _____-lateral.

_____ 24. Lateral gray horns are found only in _____ regions of the spinal cord.

_____ 25. The filum terminale and denticulate ligaments are both composed of _____ mater.

ANSWERS TO MASTERY TEST: ★ CHAPTER 13

Arrange

1. B C A
2. B C A
3. C B A D
4. D B C A E

Multiple Choice

5. B
6. D
7. C
8. A
9. D
10. B

True–False

11. T
12. F. Reflex center, conduction site
13. T
14. T
15. F. Are not
16. T
17. F. Less often than somatic ones because it is difficult to stimulate most visceral receptors
18. T
19. F. Axons of motor
20. F. L3–L4, L1–L2

Fill-ins

21. Diaphragm
22. Meningitis
23. Poly, ipsi
24. Thoracic, upper lumbar, sacral (further explanation in Chapter 17)
25. Pia

FRAMEWORK 14
Brain & Cranial Nerves

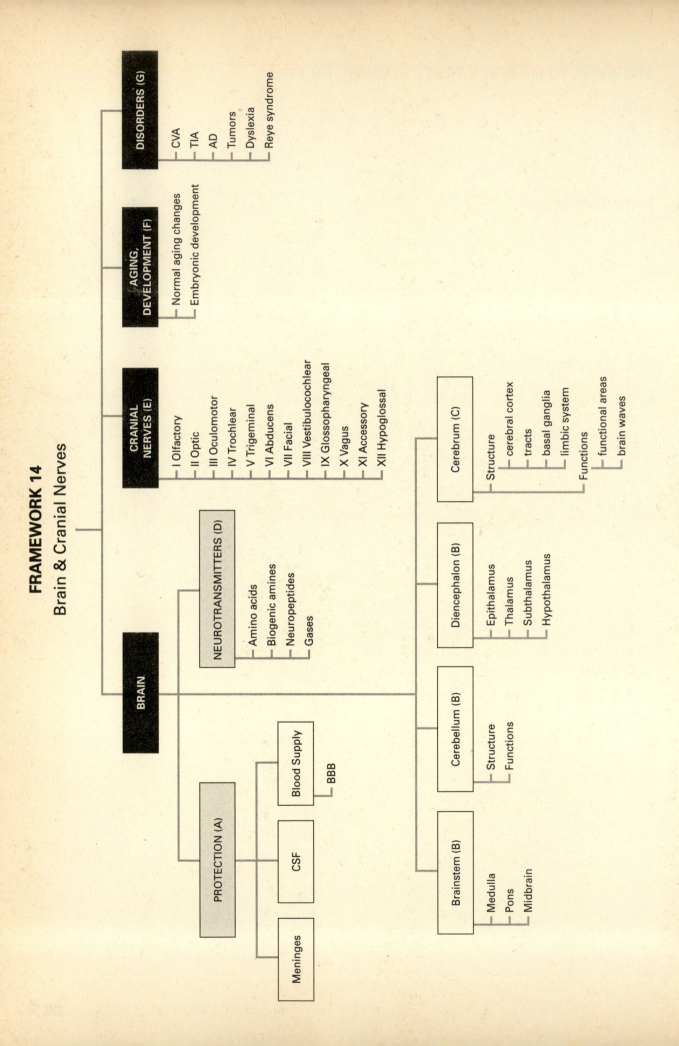

BRAIN

PROTECTION (A)
- Meninges
- CSF
- Blood Supply
  - BBB

NEUROTRANSMITTERS (D)
- Amino acids
- Biogenic amines
- Neuropeptides
- Gases

Brainstem (B)
- Medulla
- Pons
- Midbrain

Cerebellum (B)
- Structure
- Functions

Diencephalon (B)
- Epithalamus
- Thalamus
- Subthalamus
- Hypothalamus

Cerebrum (C)
- Structure
  - cerebral cortex
  - tracts
  - basal ganglia
  - limbic system
- Functions
  - functional areas
  - brain waves

CRANIAL NERVES (E)
- I Olfactory
- II Optic
- III Oculomotor
- IV Trochlear
- V Trigeminal
- VI Abducens
- VII Facial
- VIII Vestibulocochlear
- IX Glossopharyngeal
- X Vagus
- XI Accessory
- XII Hypoglossal

AGING, DEVELOPMENT (F)
- Normal aging changes
- Embryonic development

DISORDERS (G)
- CVA
- TIA
- AD
- Tumors
- Dyslexia
- Reye syndrome

The Brain and Cranial Nerves

The brain is the major control center for the global communication network of the body: the nervous system. The brain requires round-the-clock protection and maintenance afforded by bones, meninges, cerebrospinal fluid, a special blood–brain barrier, along with a fail-safe blood supply. This vital control center consists of four major substructures: the brain stem, diencephalon, cerebrum, and cerebellum. Each brain part carries out specific functions and each releases specific chemical neurotransmitters. Twelve pairs of cranial nerves convey information to and from the brain.

Structural defects may occur in construction (brain development) and also with the normal wear and tear that accompanies aging. Just as a giant computer network may experience minor disruptions in service or major shutdowns, disorders within the brain or cranial nerves may lead to minor, temporary changes in nerve functions or profound and fatal outcomes.

As you begin your study of the brain, carefully examine the Chapter 14 Topic Outline and Objectives; check off each one as you complete it. To organize your study of the brain, glance over the Chapter 14 Framework now. Be sure to refer to the Framework frequently and note relationships among key terms in each section.

TOPIC OUTLINE AND OBJECTIVES

A. Brain: introduction; protection and coverings, blood supply

☐ 1. Identify the principal parts of the brain and describe how the brain is protected.

☐ 2. Explain the formation and circulation of cerebrospinal fluid (CSF).

☐ 3. Describe the blood supply to the brain and the concept of the blood–brain barrier (BBB).

B. Brain: brain stem, cerebellum and diencephalon

C. Brain: cerebrum

☐ 4. Compare the structure and functions of the brain stem, cerebellum, diencephalon, and cerebrum,

D. Neurotransmitters in the brain

☐ 5. Discuss the various neurotransmitters found in the brain, as well as the different types of neuropeptides and their functions.

E. Cranial nerves

☐ 6. Define a cranial nerve and identify the 12 pairs of cranial nerves by name, number, type, location, and function.

F. Aging and developmental anatomy of the nervous system

☐ 7. Describe the effects of aging on the nervous system.

☐ 8. Describe the development of the nervous system.

G. Disorders, medical terminology

☐ **9.** List the clinical symptoms of these disorders of the nervous system: cerebrovascular accidents (CVA), transient ischemic attacks (TIA), Alzheimer's disease, brain tumors, dyslexia, and Reye syndrome (RS).

☐ **10.** Define medical terminology associated with the central nervous system.

WORDBYTES

Now become familiar with the language of this chapter by studying each wordbyte, its meaning, and an example of its use within a term. After you study the entire list, self-check your understanding by writing the meaning of each wordbyte on the line. As you continue through the *Learning Guide,* identify (and fill in) additional terms that contain the same wordbyte.

Wordbyte	Self-check	Meaning	Example(s)
-algia	_____	pain	neur*algia*, an*alge*sia
cauda-	_____	tail	*cauda*te
cephalo-	_____	head	hydro*cephalic*
cortico-	_____	bark	cerebral *cortex*
enceph-	_____	brain	di*enceph*alon
falx	_____	sickle	*falx* cerebri
glossi-	_____	tongue	hypo*glossa*l
hemi-	_____	half	*hemi*sphere
ophthalm-	_____	eye	*ophthalm*ic
pons	_____	bridge	*pons*

CHECKPOINTS

A. Brain: introduction; protection and coverings, blood supply   (pages 391–397)

■ **A1.** Identify numbered parts of the brain on Figure LG 14.1. Then complete this exercise.

a. Structures 1–3 are parts of the _____.

b. Structures 4 and 5 together form the _____.

c. Structure 6 is the largest part of the brain, the _____.

d. The second largest part is structure 7, the _____.

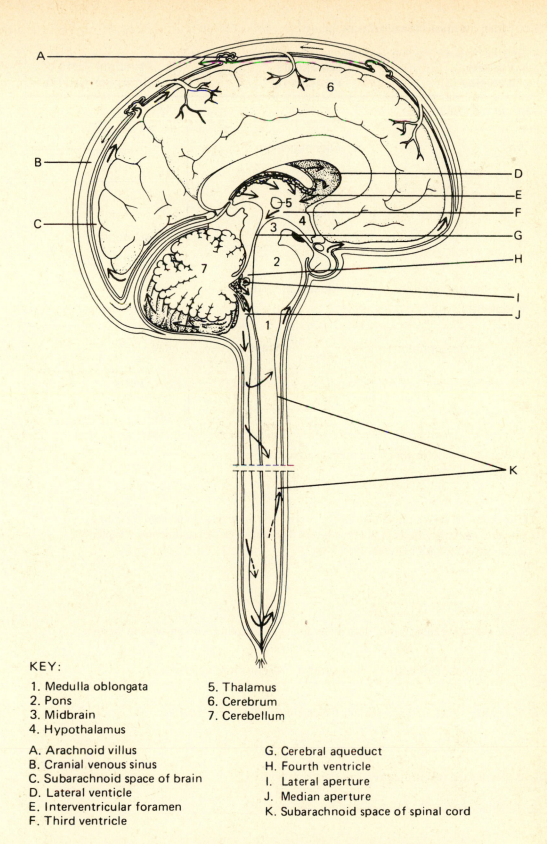

KEY:

1. Medulla oblongata
2. Pons
3. Midbrain
4. Hypothalamus
5. Thalamus
6. Cerebrum
7. Cerebellum

A. Arachnoid villus
B. Cranial venous sinus
C. Subarachnoid space of brain
D. Lateral venticle
E. Interventricular foramen
F. Third ventricle
G. Cerebral aqueduct
H. Fourth ventricle
I. Lateral aperture
J. Median aperture
K. Subarachnoid space of spinal cord

Figure LG 14.1 Brain and meninges seen in sagittal section. Parts of the brain are numbered; refer to Checkpoint A1. Letters indicate pathway of cerebrospinal fluid (CSF); refer to Checkpoint A7.

A2. Fill in the blanks in this table outlining brain development.

a. Primary Vesicles (3)	b. Secondary Vesicles (5)	c. Principal Parts of Brain Formed (7)
Prosencephalon (_____-brain) →	1. Diencephalon → 2. _____ →	1A. _____ 1B. _____ 1C. _____ 1D. _____ 2. Cerebrum
_____-encephalon → (Midbrain)	3. Mesencephalon →	3A. _____
_____-encephalon (_____) →	4. _____-encephalon → 5. Met-_____ →	4A. _____ 5A. Pons 5B. _____

A3. List three ways in which the brain is protected.

A4. Review the layers of the meninges covering the brain and spinal cord by listing them here. *For extra review.* Label the layers on Figure LG 14.1 and review Chapter 13, Checkpoint A3 (pages LG 250–251).

A5. Identify the extensions of dura mater in this Checkpoint. Use answers in the box.

Falx cerebelli   Falx cerebri   Tentorium cerebelli

a. A tentlike structure that separates the cerebrum from the cerebellum:

_____.

b. Separates the two hemispheres of the cerebellum: _____

c. Separates the two hemispheres of the cerebrum: _____

■ **A6.** Circle all correct answers about CSF.

    a. The entire nervous system contains 80–150 ml of CSF. This amount is equal to approximately:
        A. 1 to 2 tablespoons
        B. $\frac{1}{3}$ to $\frac{2}{3}$ cup
        C. 1 to 2 cups
        D. 1 quart

    b. The color of CSF is:
        A. Yellow
        B. Clear, colorless
        C. Red
        D. Green

    c. Choose the function(s) of CSF.
        A. Serves as a shock absorber for brain and cord
        B. Contains red blood cells
        C. Contains white blood cells called lymphocytes
        D. Contains nutrients

    d. Which statement(s) describe its formation?
        A. It is formed by diffusion of substances from blood.
        B. It is formed by filtration and secretion.
        C. It is formed from blood in capillaries called choroid plexuses.
        D. It is formed by ependymal cells that line all four ventricles.

    e. Which statement(s) describe its pathway?
        A. It circulates around the brain but not the cord.
        B. It flows inferior to the end of the spinal cord.
        C. It bathes the brain by flowing through the epidural space.
        D. It passes via projections (villi) of the arachnoid into blood vessels (venous sinuses) surrounding the brain.
        E. It is formed initially from blood and finally flows back to blood.

    f. An excessive accumulation of CSF within the ventricles is a condition known as:
        A. Hydrarthrosis
        B. Hydrophobia
        C. Hydrocephalus
        D. Hydrocholecystitis

■ **A7.** To check your understanding of the pathway of cerebrospinal fluid (CSF), list in order the structures through which it passes. Use key letters on Figure LG 14.1. Start at the site of formation of CSF.

A8. Briefly state results of oxygen starvation of the brain for about 4 minutes.

Explain the role of lysosomes in this process.

Now list effects of glucose deprivation of the brain.

■ **A9.** Describe the blood–brain barrier (BBB) in this exercise.

a. Blood capillaries supplying the brain are *(more? less?)* leaky than most capillaries of the body. This fact is related to a large number of *(gap? tight?)* junctions, as well as an

abundance of neuroglia named _____ pressed against capillaries.

b. What advantages are provided by this membrane?

c. List two or more substances needed by the brain that do normally cross the BBB.

d. What problems may result from the fact that some chemicals cannot cross this barrier?

e. List several substances that may harm the brain that *are* able to cross this barrier.

■ **A10.** Describe *circumventricular organs* (CVOs) in this exercise.

a. Name three organs that are CVOs.

b. CVOs are unique in the brain region in that they *(have? lack?)* the blood–brain barrier. State one advantage and one disadvantage of this structural feature.

B. Brain: brain stem, cerebellum, and diencephalon (pages 397–406)

■ **B1.** Describe the principal functions of the medulla in this exercise.

a. The medulla serves as a _____ pathway for all ascending and descending tracts. Its white matter therefore transmits *(sensory? motor? both sensory and motor?)* impulses.

b. Included among these tracts are the triangular _____ tracts which are the principal *(sensory? motor?)* pathways. The main fibers that pass through the pyramids are the *(spinothalamic? corticospinal?)* tracts.

c. Crossing (or _____) of fibers occurs in the medulla. This explains why movements of your right hand are initiated by motor neurons that originate in the *(right? left?)* side of your cerebrum. The *(axons? dendrites?)* of these motor neurons decussate in the medulla to proceed down the right lateral

_____-spinal tract.

d. The medulla contains gray areas as well as white matter. Several important nuclei lie in the medulla. Two of these are synapse points for sensory axons that have ascended in

the posterior columns of the cord. These two nuclei, named _____

and _____, transmit impulses for sensations of _____,

_____, and _____.

e. A hard blow to the base of the skull can be fatal because the medulla is the site of three

vital centers: the _____ center, regulating the heart; the _____

center, adjusting the rhythm of breathing; and the _____ center, regulating blood pressure by altering the diameter of blood vessels.

f. Some input to the medulla arrives by means of cranial nerves; these nerves may serve motor functions also. Which cranial nerves are attached to the medulla?

_____

(Functions of these nerves will be discussed later in this chapter.)

■ **B2.** Summarize important aspects of the pons in this learning activity.

a. The name *pons* means _____. It serves as a bridge in two ways. It contains longitudinally arranged fibers that connect the

_____ and _____ with the upper parts of the brain.

It has transverse fibers that connect the two sides of the _____.

b. Cell bodies associated with fibers in cranial nerves numbered

_____ lie in nuclei in the pons.

c. *(Respiration? Heartbeat? Blood pressure?)* is controlled by the pneumotaxic and apneustic areas of the pons.

■ **B3.** Relate the midbrain to the pons and medulla in this exercise.

a. Like the pons and medulla, the midbrain is about 1 inch (_____ cm) long.

b. The midbrain is more *(anterior and superior? posterior and inferior?)* compared to the pons and medulla.

■ **B4.** You maintain a conscious state, or you wake up from sleep, thanks to the regions of

the brain known as the RAS, or R_____ A_____

S_____. These areas are especially sensitive to sensations such as

_____ from ears and _____ or

_____ from skin. The RAS is part of the reticular formation.
Where is the reticular formation located?

B5. Describe the cerebellum
a. Where is it located?
b. Describe its structure. Include these terms in your description: *vermis, lobes, cortex,* and *arbor vitae.*

c. Describe its functions. Use these key terms: *coordinated movements, posture, equilibrium,* and *emotions.*

■ **B6.** Identify locations of each of the tracts known as *peduncles* that permit communication between the cerebellum and other brain parts. Fill in lines with the terms in the box.

| Inferior Middle Superior |

a. Cerebellum → _____ cerebullar peduncle → Midbrain

b. Cerebellum → _____ cerebullar peduncle → Pons

c. Cerebellum → _____ cerebullar peduncle → Medulla

■ **B7.** Describe the *diencephalon* in this Checkpoint.

a. If you look for the diencephalon in most external views of the brain, you will not find it. It is located *(deep within? on the surface of?)* the cerebrum. The diencephalon consists of *(two? four? eight?)* main structures: the thalamus, as well as the

_____thalamus, the _____thalamus, and the _____thalamus. Which part

comprises 80% of the diencephalon? _____

b. The epithalamus forms the *(roof? lateral walls? floor?)* of the third ventricle. The

epithalamus includes the _____ gland and the _____

nuclei, as well as the _____ plexus of the third ventricle. The pineal

gland secretes the hormone named _____ and serves as the body's biological

_____. The habenular nucleus is involved in *(hearing? smell? vision?)*.

c. The subthalamus works in controlling *(sensations? movements?)*.

d. The *(two? four?)* lobes of the thalamus form parts of the *(roof? lateral walls? floor?)* of the third ventricle. A "crossbar" between the two lobes passes through the center of the

_____ ventricle. (See Figure LG 14.1.)

e. The thalamus is the principal relay station for *(motor? sensory?)* impulses. For example, spinothalamic and lemniscal tracts convey general sensations such as pain,

_____, _____, _____, and temperature to the thalamus where they are relayed to the cerebral cortex. Special sense impulses (for vision and hearing) are relayed through the *(geniculate? reticular? ventral posterior?)* nuclei of the thalamus.

f. Although the thalamus plays a primary role in conduction of sensory impulses, it also has other functions. Name two or more.

g. The name hypothalamus indicates that this structure lies *(above? below?)* the thalamus,

forming the floor and part of the lateral walls of the _____ ventricle.

B8. Expanding on the key words listed below, write a sentence describing major hypothalamic functions.

a. Regulator of visceral activities via control of the ANS

b. Regulating factors to anterior pituitary

c. Feelings (rage)

d. Temperature

e. Thirst

f. Feeding and satiety center

g. Reticular formation: arousal and consciousness

■ **B9.** *For extra review.* Check your understanding of these parts of the brain stem and diencephalon by matching them with the descriptions given below.

H.	Hypothalamus	P.	Pons
Med.	Medulla	T.	Thalamus
Mid.	Midbrain		

_____ a. It is the principal regulator of visceral activities because it acts as a liaison between the cerebral cortex and autonomic nerves that control viscera.

_____ b. It is the site of origin of rubrospinal tracts concerned with muscle tone and posture.

_____ c. Cranial nerves III–IV attach to this brain part.

_____ d. Cranial nerves V–VIII attach to this brain part.

_____ e. Cranial nerves VIII–XII attach to this brain part.

_____ f. Feelings of hunger, fullness, and thirst stimulate centers here so that you can respond accordingly.

_____ g. All sensations are relayed through here.

_____ h. Regulation of heart, blood pressure, and respiration occurs by centers located here.

_____ i. It constitutes four-fifths of the diencephalon.

_____ j. It lies under the third ventricle, forming its floor.

_____ k. It forms most of side walls of the third ventricle.

_____ l. Tumor in this region could compress the cerebral aqueduct and cause internal hydrocephalus.

_____ m. It contains centers for coughing, sneezing, hiccuping, swallowing, and vomiting.

_____ n. The olivary and vestibular nuclei associated with equilibrium and posture control are located here.

■ **B10.** *For extra review* of specific locations and functions of nuclei and other masses of gray matter, complete this table. Choose answers from the lists of brain parts and functions. Provide two answers in parts of the table indicated by (2).

Brain part	Functions
Cer. Cerebellum	Emotions, memory
Epi. Epithalamus	Hearing
Hypo. Hypothalamus	Movements in general
Med. Medulla	Movements of head or neck in response
Mid. Midbrain	to sights or sounds
Sub. Subthalamus	Regulation of anterior pituitary
Thal. Thalamus	Regulation of posterior pituitary
	Smell
	Taste
	Touch, pressure, vibrations
	Vision

Name of Nucleus	Brain Part	Functions
a. Habenular nucleus	Epi	Smell
b. Subthalamic nucleus		
c. Red nucleus	(2)	
d. Colliculi		
e. Nuclei cuneatus and gracilis		
f. Anterior nucleus		
g. Lateral nucleus		
h. Medial geniculate nucleus		
i. Ventral posterior nucleus		(2)
j. Mammillary bodies		
k. Median eminence of tuber cinereum		
l. Supraoptic region		

C. Brain: cerebrum (pages 406–414)

■ **C1.** Complete this exercise about cerebral structure.

a. The outer layer of the cerebrum is called _____. It is composed of *(white? gray?)* matter. This means that it contains mainly *(cell bodies? tracts?)*.

b. In the margin, draw a line the same length as the thickness of the cerebral cortex. Use a metric ruler. Note how thin the cortex is.

c. The surface of the cerebrum looks much like a view of tightly packed mountains or

ridges, called _____. The parts where the cerebral cortex dips

down into valleys are called _____ (deep valleys) or

_____ (shallow valleys).

d. The cerebrum is divided into halves called _____. Connecting them

is a band of *(white? gray?)* matter called the _____. Notice this
structure in Figure LG 14.1 and in Figure 14.9 on page 404 of your text.

e. The falx cerebri is composed of *(nerve fibers? dura mater?)*. Where is it located?

_____ At its superior and inferior margins, the falx is dilated to
form channels for venous blood flowing from the brain; these enclosures are

called _____.

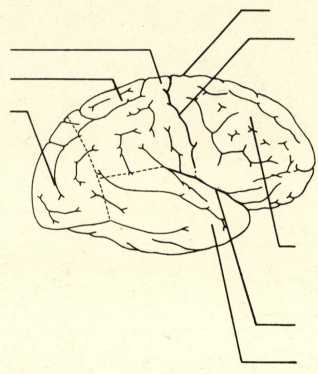

O   Motor speech (Broca's) area   O   Primary somatosensory (general sensory) area
O   Premotor area   O   Primary visual area
O   Primary auditory area   O   Auditory association area
O   Primary motor area

Figure LG 14.2   Right lateral view of lobes and fissures of the cerebrum. Label and color as directed in
Checkpoints C2 and C8.

■ **C2.** Label the following structures using leader lines on Figure LG 14.2: *frontal lobe, occipital lobe, parietal lobe, temporal lobe, central sulcus, lateral cerebral sulcus, precentral gyrus, postcentral gyrus.*

■ **C3.** Match the three types of white matter fibers with these descriptions.

A. Association	C. Commissural	P. Projection

_____ a. The corpus callosum contains these fibers and connects the two cerebral hemispheres.

_____ b. Sensory and motor fibers passing between cerebrum and other parts of the CNS are this type of fiber; the internal capsule is an example.

_____ c. These fibers transmit impulses among different areas of the same hemisphere.

C4. List names of structures that are considered parts of the basal ganglia.

Now list two functions of the basal ganglia.

C5. Describe the limbic system in this exercise.

a. This system consists of portions of the cerebrum, thalamus, and hypothalamus. List several component structures.

b. Explain why the limbic system is sometimes called the "visceral" or "emotional" brain.

c. One other function of the limbic system is _____. Forgetfulness, such as inability to recall recent events, results partly from impairment of this system.

■ **C6.** Which condition can be described as visible bruising of the brain with extended loss of consciousness? *Concussion? Contusion?*

■ **C7.** Draw two important generalizations about brain functions in this activity.

a. In general, the anterior of the cerebrum is more involved with *(motor? sensory?)* control, whereas the posterior of the cerebrum is more involved with *(motor? sensory?)* functions. (Take a moment to visualize those activities taking place in the front and back of your own brain.)

b. As a general rule, *(primary? association?)* sensory areas receive sensations and *(primary? association?)* sensory areas are involved with interpretation and memory of sensations. For example, your ability to see the outline of the cerebrum, as in Figure LG 14.2, depends on your *(primary? association?)* visual areas. The fact that you can distinguish this diagram as a cerebrum (not a hand or heart), along with your memory of its structure for your next test, is based on the health of the neurons in your *(primary? association?)* visual areas.

■ **C8.** Color functional areas of the cerebral cortex listed with color code ovals on Figure 14.2.

■ **C9.** Check your understanding of functional areas of the cerebral cortex by doing this matching exercise. Answers may be used more than once.

F. Frontal eye field	PM. Primary motor area
G. Gnostic area	PO. Primary olfactory area
M. Motor speech (Broca's) area	PS. Primary somatosensory area
PA. Primary auditory area	PV. Primary visual area

_____ a. In the occipital lobe

_____ b. In the postcentral gyrus

_____ c. Receives sensations of pain, touch, pressure, and temperature

_____ d. In the parietal lobe

_____ e. In the temporal lobe; permits hearing

_____ f. Integrates general and special sensations to form a common thought about them

_____ g. In precentral gyrus of the frontal lobe; controls specific muscles or groups of muscles

_____ h. In the frontal lobe; translates thoughts into speech

_____ i. Controls smell

_____ j. Controls scanning movements of eyes, such as searching for a name in a telephone book

■ **C10.** Do this exercise about control of speech.

a.. Language areas are usually located in the *(left? right?)* cerebral hemisphere. This is true for most *(persons who are right-handed? persons, regardless of handedness?)*.

b. The term *aphasia* refers to inability to _____. *Fluent aphasia* means inability to *(articulate or form? understand?)* words. *Word blindness* refers to inability to understand *(spoken? written?)* words, whereas *word deafness* is inability to understand _____ words.

■ **C11.** Identify the type of brain waves associated with each of the following situations.

A. Alpha	D. Delta
B. Beta	T. Theta

_____ a. Occur in persons experiencing stress and in certain brain disorders

_____ b. Lowest frequency brain waves, normally occurring when adult is in deep sleep; presence in awake adult indicates brain damage

_____ c. Highest frequency waves, noted during periods of mental activity

_____ d. Present when awake but resting, these waves are intermediate in frequency between beta and delta waves

■ **C12.** *A clinical challenge.* You are helping to care for Laura, a hospital patient with a contusion of the right side of the brain. Which of the following would you be most likely to observe in Laura? Circle letters of correct answers.

A. Her right leg is paralyzed.
B. Her left arm is paralyzed.
C. She cannot speak out loud to you nor write a note to you.
D. She used to sing well but cannot seem to stay on tune now.
E. She has difficulty with concepts involving numbers.

D. Neurotransmitters in the brain   (pages 414–418)

■ **D1.** Answer these questions about neurotransmitters in the brain.

a The number of known (or strongly suspected) neurotransmitters in the brain is about
(5? 30? 60?). A *(large? tiny?)* amount of neurotransmitter is released at a single synapse.

b. Which type of neurotransmitter acts more quickly at a synapse? One that *(opens or closes ion channels in the neuron membrane? acts by a second messenger system that influences reactions inside the postsynaptic cell?)*

c. *(All? Not all?)* neurotransmitters are made by neurons. Name one type of cell other than

a neuron that can synthesize neurotransmitters. _____ Name one

type of cell within the brain that can make hormones. _____
Identify a category of chemicals that can affect distant neurons by intensifying their

responses to neurotransmitters. _____

D2. Match the neurotransmitter or neuromodulator with the description that best fits.
The first one is done for you. Lines at the end of each description are for Checkpoint D3.

ACh. Acetylcholine	GABA. Gamma aminobutyric acid
DA. Dopamine	Ser. Serotonin
EED. Enkephalins, endorphins, and dynorphins	SP. Substance P

__ACh__ a. Released at neuromuscular junctions and by many neurons, including those of
           pyramidal tracts **ACh**

_____ b. Produced by neurons that degenerate in Parkinson's disease _____

_____ c. The most common inhibitory neurotransmitter in the brain _____

_____ d. Made by neurons involved with emotional responses and automatic movements of

           skeletal muscles _____

_____ e. Transmits pain-related impulses _____

_____ f. Morphinelike chemicals that are the body's natural painkillers; suppress release of

           substance P _____
_____ g. Concentrated in brainstem neurons, this chemical helps regulate sleep,

           temperature, and mood _____
_____ h. Reduces depression by its extended activation of postsynaptic neurons due to
           action of the antidepressant Prozac (which inhibits its reuptake into the

           presynaptic neuron) _____

_____ i Previously produced by neurons that are destroyed in Alzheimer's disease _____

■ **D3.** Select answers in the box below to identify the correct category of each of the neurotransmitters and neuromodulators listed in Checkpoint D2. Write answers on lines after each of those descriptions. The first one is done for you.

ACh.	Acetylcholine	B.	Biogenic amines
AA.	Amino acids	N.	Neuropeptides

■ **D4.** Identify the neurotransmitters destroyed by the following enzymes.

a. Acetylcholinesterase (AChE) destroys the neurotransmitter named

_____.

b. Catechol-*o*-methyltransferase (COMT) and monoamine oxidase (MAO) destroy neurotransmitters in the *catecholamine* category. List three such neurotransmitters.

_____ _____ _____

D5. Answer these questions about opioid peptides such as enkephalins.

a. Briefly explain the relationship between these chemicals and acupuncture.

b. Besides serving as painkillers, enkephalins have several other functions. List four or more of these functions.

■ **D6.** *For extra review* of neurotransmitters more recently identified or producing diverse and widespread effects on the body, complete this Checkpoint. Use answers listed in the box.

ADH.	Antidiuretic hormone	NO.	Nitric oxide
Ang. II.	Angiotensin II	OT.	Oxytocin
CCK.	Cholecystokinin		

_____ a. Regulates pancreatic secretion, as well as messages to "stop eating" in response to "feeling full."

_____ b. Also known as vasopressin because it has some effect on blood pressure.

_____ c. Stimulates uterine contractions during labor and delivery, as well as milk "letdown" in breast-feeding mothers.

_____ d. Stimulates release of the adrenal cortex hormone aldosterone.

_____ e. Produced by endothelial cells lining blood vessels, it was originally identified as EDRF (endothelial-derived relaxing factor).

_____ f. Synthesized on demand, it acts immediately and is inactivated within seconds.

_____ g. Appear to affect memory (4 answers).

E. Cranial nerves   (pages 418–422)

■ **E1.** Answer these questions about cranial nerves.

a. There are _____ pairs of cranial nerves. They are all attached to

the _____; they leave the _____ via foramina.

b. They are numbered by Roman numerals in the order that they leave the cranium. Which is most anterior? *(I? XII?)* Which is most posterior? *(I? XII?)*

c. All spinal nerves are *(purely sensory? purely motor? mixed?)*. Are all cranial nerves mixed? *(Yes? No?)*

■ **E2.** Complete the table about the 12 pairs of cranial nerves. (*For extra review* on locations of these nerves, identify their entrance or exit points through the cranial bones in Chapter 7, Checkpoint B7, page 130.)

Number	Name	Functions
a.	Olfactory	
b.		Vision (not pain or temperature of the eye)
c. III		
d.	Trochlear	
e. V		
f.		Stimulates lateral rectus muscle to abduct eye; proprioception of the lateral rectus
g.	Facial	
h.		Hearing; equilibrium
i. IX		
j.	Vagus	
k. XI		
l.		Supplies muscles of the tongue with motor and sensory fibers

■ **E3.** *The Big Picture: Looking Ahead and Looking Back.* Refer to figures and exhibits in your text that emphasize structure and function of cranial nerves.

a. Refer to Figure 16.1, page 454 and identify the olfactory nerves, which form part of the pathway for *(smell? taste?)*. These are located in mucosa of *(inferior? superior?)* portions of the nose. After passing through the cribriform plate of the *(ethmoid? frontal? sphenoid?)* bone, axons of olfactory nerves synapse with neuron cell bodies located in the olfactory *(bulb? tract?)*. Axons of these neurons then pass through the olfactory *(bulb? tract?)*, which is located at the base of the *(frontal? parietal?)* lobes of the cerebrum.

b. Figure 16.5, page 460, shows the optic nerves, which are cranial nerves number *(I? II? VIII?)*. Optic nerves form part of the pathway for *(hearing? taste? vision?)*; each eye has *(one? many?)* optic nerve(s). Notice on Figure 16.7, page 462, that optic nerves consist of axons of *(rod or cone? bipolar? ganglion?)* cells within the retina of the eye. These axons may cross in the optic *(chiasma? tract?)* (Figure 16.15, page 471) or remain uncrossed before they end in the *(thalamus? hypothalamus?)*. There they synapse with neurons that terminate in the *(occipital? temporal?)* lobe of the cerebral cortex.

c. Cranial nerve VIII is associated with the *(ear? tongue?)* (Figure 16.16, page 472). The *(cochlear? vestibular?)* portion of this nerve conveys impulses associated with hearing from the snail-shaped *(cochlea? vestibule?)*. The three semicircular canals contain receptors that sense body position and movement; these neurons send axons to the thalamus via the *(cochlear? vestibular?)* branch of cranial nerve VIII.

d Cranial nerve VII is the *(facial? trigeminal?)*; it can be seen on Figure 14.5, page 398, lying right next to cranial nerve *(II? VIII?)*. In fact, these two nerves both pass through the cranium via the same foramen, the *(internal? external?)* auditory meatus (Figure 16.16, page 472). The facial nerve is primarily involved with *(sensation? movement?)* of the face (Exhibit 11.3, pages 277–278). It also carries sensations of taste from the *(anterior? posterior?)* two-thirds of the tongue (page 457 of the text).

e. Four cranial nerves transmit impulses as part of the autonomic nervous system (Figure 17.1, page 490). These are cranial nerves numbers _____, _____, _____,

and _____. Which of these nerves stimulates production of tears from lacrimal

glands? _____ Which two pairs of nerves stimulate salivary glands?

_____ _____ Which pair of nerves innervates organs such as airways, heart,

and intestine? _____

f. Which three pairs of cranial nerves control eye movements (Exhibit 11.5, page 282)?

_____, _____, and _____ Which of these also stimulates the sphincter muscle of

the iris to control the size of the pupil (Figure 17.1, page 490)? _____

■ **E4.** Check your understanding of cranial nerves by completing this exercise.
Write the name of the correct cranial nerve following the related description.

a. Differs from all other cranial nerves in that it originates from the brainstem and from the

spinal cord: _____

b. Eighth cranial nerve (VIII): _____

c. Is widely distributed into neck, thorax, and abdomen: _____

d. Senses toothache, pain under a contact lens, wind on the face: _____

e. The largest cranial nerve; has three parts (ophthalmic, maxillary, and mandibular):

_____

f. Two nerves that contain taste fibers and autonomic fibers to salivary glands:

_____, _____

g. Three purely sensory cranial nerves:

_____, _____, _____

■ **E5.** Write the number of the cranial nerve related to each of the following disorders.

_____ a. Bell's palsy

_____ b. Inability to shrug shoulders or turn head

_____ c. Anosmia

_____ d. Strabismus and diplopia

_____ e. Blindness

_____ f. Trigeminal neuralgia (tic douloureux)

_____ g. Paralysis of vocal cords; loss of sensation of many organs

_____ h. Vertigo and nystagmus

F. Aging and developmental anatomy of the nervous system (pages 423–424)

■ **F1.** Do this exercise describing effects of aging on the nervous system.

a. The total number of nerve cells _____-creases with age. Because conduction velocity *(increases? slows down?)*, the time required for a typical reflex is

_____-creased.

b. Parkinson's disease is the most common *(motor? sensory?)* disorder in the elderly.

c. The sense of touch is likely to be *(impaired? heightened?)* in old age, placing the older person at *(higher? lower?)* risk for burns or other skin injury.

■ **F2.** Check your understanding of the early development of the nervous system by completing this exercise.

a. The nervous system begins to develop during the third week of gestation when the

_____-derm forms a thickening called the _____

plate. Soon a longitudinal depression is found in the plate; this is the neural

_____. As neural folds on the sides of the groove grow and meet,

they form the neural _____.

b. Three types of cells form the walls of this tube. Two of these are the marginal layer which forms *(gray? white?)* matter and the mantle layer which becomes *(gray? white?)* matter.

c. The neural crest forms *(peripheral? central?)* nervous system structures such as ganglia

as well as spinal and cranial _____.

d. The anterior portion of the neural plate and tube develops into three enlarged

_____ by the fourth week. These fluid-filled cavities eventually

become the _____ which are filled with _____ fluid.

e. By week five the primary vesicles have flexed (bent) to form a total of *(five? ten?)* secondary vesicles, described in Checkpoint A2, page LG 270.

G. Disorders, medical terminology (pages 424–426)

G1. CVA refers to_____.

CVAs are *(rare? common?)* brain disorders. Describe three causes of CVAs.

G2. Contrast CVA with TIA.

G3. Describe the following aspects of Alzheimer's disease (AD):

a. Progressive changes in behavior

b. Pathological findings of brain tissue

c. The possible roles of amyloid and A68 in the causation of AD and Down syndrome.

G4. Brain tumors are more likely to result from *(neurons? neuroglia?)*.
Write three signs or symptoms of brain tumors.

■ **G5.** *For extra review.* Match the name of the disorder with the related description.

Agnosia	Neuralgia
Alzheimer's disease (AD)	Reye Syndrome
Cerebrovascular accident (CVA)	Stupor
Dyslexia	Transient ischemic attack (TIA)
Lethargy	

a. Characterized by loss of neurons, amyloid plaques, and neurofibrillary tangles:

_____ ____

b. Inability to recognize significance of sensory stimuli, such as tactile or gustatory:

_____

c. Temporary cerebral dysfunction caused by interference of blood supply to the brain.

_____

d. Most common brain disorder; also called

stroke: _____

e. Characterized by difficulty in handling words and symbols, for example, reversal of letters (*b* for *d*); cause unknown:

_____

f. Attack of pain along an entire nerve, for example, tic douloureux: _____

g. A disease of children causing swelling of brain cells and fatty infiltration of liver; usually follows viral infection, especially if aspirin is

taken: _____

h. Functional sluggishness:

_____

i. Unresponsiveness; can be aroused only briefly and by repeated, vigorous stimulation:

_____

ANSWERS TO SELECTED CHECKPOINTS: CHAPTER 14

A1. (a) Brain stem. (b) Diencephalon. (c) Cerebrum. (d) Cerebellum.

A2.

a. Primary Vesicles (3)	b. Secondary Vesicles (5)	c. Principal Parts of Brain Formed (7)	
Prosencephalon			
(Fore -brain) →	1. Diencephalon →	1A. Epithalamus	
		1B. Thalamus	
		1C. Subthalamus	
		1D. Hypothalamus	
	2. Telencephalon →	2A. Cerebrum	
Mes -encephalon → (Midbrain)	3. Mesencephalon →	3A. Midbrain	
Rhomb -encephalon (Hindbrain) →	4. Myel -encephalon →	4A. Medulla oblongata	
	5. Met- encephalon →	5A. Pons	
		5B. Cerebellum	

A3. Skull bones, meninges, cerebrospinal fluid (CSF).

A5. (a) Tentorium cerebelli. (b) Falx cerebelli. (c) Falx cerebri.

A6. (a) B. (b) B. (c) A, C, D. (d) B, C, D. (e) B, D, E. (f) C.

A7. D E F G H I J K C A B.

A9. (a) Less; tight, astrocytes. (b) The brain is protected from many substances that could harm the brain but are kept from the brain by this barrier. (c) Glucose, oxygen, water, ions such as Na^+ and K^+, and anesthetics. (d) Certain helpful medications, such as most antibiotics, and chemotherapeutic chemicals cannot cross this barrier. (Fortunately, at the time they are most needed—

as in a brain infection—the BBB is more permeable and so may permit passage of these drugs.) (e) Alcohol, caffeine, nicotine, and heroin.

A10. (a) Hypothalamus, pineal gland, and pituitary. (b) Lack; advantage: because their capillaries are permeable to most substances, they can monitor homeostatic activities such as fluid balance and hunger; disadvantage: they are permeable to harmful substances, for example, possibly the AIDS virus.

B1. (a) Conduction; both sensory and motor. (b) Pyramidal, motor; corticospinal. (c) Decussation; left; axons, cortico. (d) Cuneatus, gracilis, two-point discriminatory touch, proprioception, vibrations. (e) Cardiac, respiratory, vasomotor (vasoconstrictor). (f) Part of VIII, as well as IX–XII.

B2. (a) Bridge; spinal cord, medulla; cerebellum. (b) V–VII and part of VIII. (c) Respiration.

B3. (a) 2.5. (b) Anterior and superior.

B4. Reticular activating system; sound, temperature, pain; medulla, pons, and midbrain, as well as portions of the diencephalon (thalamus and hypothalamus) and spinal cord.

B6. (a) Superior. (b) Middle. (c) Inferior. (Note that the names of these tracts are logical, based on locations of the midbrain, pons, and medulla.)

B7. (a) Deep within; four; epi-, sub-, hypo-; thalamus. (b) Roof; pineal, habenular, choroid; melatonin, clock; smell. (c) Movements. (d) Two, lateral walls; third. (e) Sensory; touch, pressure, proprioception; geniculate. (f) Motor control, emotions, and memory. (g) Below, third.

B9.   (a) H. (b) Mid. (c) Mid. (d) P. (e) Med. (f) H.
(g) T. (h) Med. (i) T. (j) H. (k) T. (l) Mid. (m)
Med. (n) Mid.

B10.

Name of Nucleus	Brain Part	Functions
a. Habenular nucleus	Epi	Smell
b. Subthalamic nucleus	Sub	Movements in general
c. Red nucleus	(2) Mid, sub	Movements in general
d. Colliculi	Mid	Movements of head or neck in response to sight or sounds
e. Nuclei cuneatus and gracilis	Med	Touch, pressure, vibrations
f. Anterior nucleus	Thal	Emotions, memory
g. Lateral nucleus	Thal	Vision
h. Medial geniculate nucleus	Thal	Hearing
i. Ventral posterior nucleus	Thal	(2) Taste; touch, pressure, vibrations
j. Mammillary bodies	Hypo	Smell
k. Median eminence of tuber cinereum	Hypo	Regulation of anterior pituitary
l. Supraoptic region	Hypo	Regulation of posterior pituitary

C1. (a) Cerebral cortex; gray; cell bodies. (b) 2 to
4 mm. (c) Gyri; fissures, sulci. (d) Hemispheres;
white, corpus callosum. (e) Dura mater; between
cerebral hemispheres; superior and inferior sagittal sinuses (see Figure 14.2, page 393 in the text).

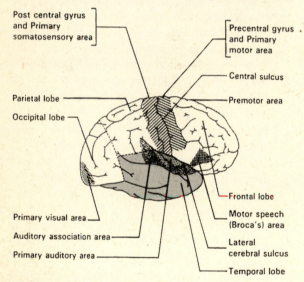

Figure LG 14.2A Right lateral view of lobes and
fissures of the cerebrum.

C2. See Figure LG 14.2A.
C3. (a) C. (b) P. (c) A.
C6. Contusion.
C7. (a) Motor, sensory. (b) Primary, association;
primary; association.
C8. See Figure LG 14.2A.
C9. (a) PV. (b) PS. (c) PS. (d) PS. (e) PA. (f) G.
(g) PM. (h) M. (i) PO. (j) F.
C10. (a) Left; persons, regardless of handedness.
(b) Speak; understand; written, spoken.
C11. (a) T. (b) D. (c) B. (d) A.
C12. B D
D1. (a) 60; tiny. (b) Opens or closes ion channels in
the neuron membrane. (c) Not all; endocrine
cells; neurosecretory; neuromodulators.
D2, D3. (b) DA (B). (c) GABA (AA). (d) DA (B).
(e) SP (N). (f) EED (N). (g–h) Ser (B).
(i) ACh (ACh).
D4. (a) Acetylcholine (ACh). (b) Norepinephrine
(NE), epinephrine, and dopamine (DA).
D6. (a) CCK. (b) ADH. (c) OT. (d) Ang. II. (e–f) NO.
(g) ADH, Ang. II, NO, OT.
E1. (a) 12; brain, cranium. (b) I; XII. (c) Mixed; no
(three are purely sensory).
E2. Refer to answers to LG Chapter 7, Activity B7,
page 138.
E3. (a) Smell; superior; ethmoid, bulb; tract, frontal.
(b) II; vision, one; ganglion; chiasma, thalamus;
occipital. (c) Ear; cochlear, cochlea; vestibular.
(d) Facial, VIII; internal; movement; anterior. (e)
III, VII, IX, X; VII; VII, IX; X. (f) III, IV, VI; III.
E4. (a) Accessory. (b) Vestibulocochlear. (c) Vagus.
(d) Trigeminal. (e) Trigeminal. (f) Facial,
glossopharyngeal. (g) Olfactory, optic, vestibulocochlear.
E5. (a) VII. (b) XI. (c) I. (d) III. (e) II. (f) V. (g) X.
(h) VIII (vestibular branch).
F1. (a) De; slows down, in. (b) Motor. (c) Impaired,
higher.
F2. (a) Ecto, neural; groove; tube. (b) White, gray.
(c) Peripheral, nerves. (d) Primary vesicles;
ventricles of brain, cerebrospinal. (e) Five.
G5. (a) AD. (b) Agnosia. (c) TIA. (d) CVA.
(e) Dyslexia. (f) Neuralgia. (g) Reye syndrome.
(h) Lethargy (i) Stupor.

WRITING ACROSS THE CURRICULUM: CHAPTER 14

1. Describe the embryonic development of the brain.
2. Describe the structure and functions of the following: (a) limbic system, (b) diencephalon, (c) reticular formation.
3. Contrast motor and sensory functions of cranial nerves V, VII, and IX.
4. Identify structural changes of the brain that are associated with Alzheimer's disease. Describe roles of the following proteins in these changes: beta amyloid, APP, and A68.
5. Contrast a CVA with a TIA.

MASTERY TEST: CHAPTER 14

Questions 1–9: Circle the letter preceding the one best answer to each question.

1. All of these are located in the medulla *except:*
   A. Nucleus cuneatus and nucleus gracilis
   B. Cardiac center, which regulates heart
   C. Site of decussation (crossing) of pyramidal tracts
   D. The olive
   E. Origin of cranial nerves V to VIII

2. Which of these is a function of the postcentral gyrus?
   A. Controls specific groups of muscles, causing their contraction
   B. Receives general sensations from skin, muscles, and viscera
   C. Receives olfactory impulses
   D. Primary visual reception area
   E. Somatosensory association area

3. All of these structures contain cerebrospinal fluid *except:*
   A. Subdural space
   B. Ventricles of the brain
   C. Central canal of the spinal cord
   D. Subarachnoid space

4. Damage to the accessory nerve would result in:
   A. Inability to turn the head or shrug shoulders
   B. Loss of normal speech function
   C. Hearing loss
   D. Changes in heart rate
   E. Anosmia

5. Which statement about the trigeminal nerve is *false?*
   A. It sends motor fibers to the muscles used for chewing.
   B. Pain in this nerve is called trigeminal neuralgia (tic douloureux).
   C. It carries sensory fibers for pain, temperature, and touch from the face, including eyes, lips, and teeth area.
   D. It is the smallest cranial nerve.

6. All of these are functions of the hypothalamus *except:*
   A. Control of body temperature
   B. Release of chemicals (regulating factors) that affect release or inhibition of hormones
   C. Principal relay station for sensory impulses
   D. Center for mind-over-body (psychosomatic) phenomena
   E. Involved in maintaining sleeping or waking state

Questions 10–13: Arrange the answers in correct sequence.

____ ____ ____ 10. From superior to inferior:
  A. Thalamus
  B. Hypothalamus
  C. Corpus callosum

____ ____ ____ 11. Order in which impulses are relayed in conduction pathway for vision:
  A. Optic nerve
  B. Optic tract
  C. Optic chiasma

____ ____ ____ ____ 12. Pathway of cerebrospinal fluid, from formation to final destination:
  A. Choroid plexus in ventricle
  B. Subarachnoid space
  C. Cranial venous sinus
  D. Arachnoid villi

____ ____ ____ 13. From anterior to posterior:
  A. Fourth ventricle
  B. Pons and medulla
  C. Cerebellum

Questions 14–20: Circle T (true) or F (false). If the statement is false, change the underlined word or phrase so that the statement is correct.

T F 14. The thalamus, hypothalamus, and cerebrum are all developed from the <u>forebrain (prosencephalon)</u>.

T F 15. <u>Endorphins, enkephalins, and dopamine</u> are all chemicals that are considered the "body's own pain killers."

T F 16. The language areas are located in the <u>cerebellar cortex</u>.

T F 17. Cranial nerves, I, III, and VIII are all <u>purely sensory</u> nerves.

T F 18. The limbic system functions in control of <u>emotional aspects of behavior</u>.

T F 19. <u>GABA, serotonin, and acetylcholine may</u> all function as excitatory transmitter substances.

T F 20. A cerebrovascular accident (CVA) can be expected to produce <u>more lasting</u> effects than a transient ischemic attack (TIA).

Questions 21–25: Fill-ins. Complete each sentence with the word or phrase that best fits.

_____ 21. The _____ is the extension of dura mater that separates the cerebral hemispheres, whereas the _____ separates the cerebrum from the cerebellum.

_____ 22. The superior and inferior colliculi, associated with movements of eyeballs and head in response to visual stimuli, are located in the _____

_____ 23. If the brain is deprived of oxygen for more than 4 minutes, _____ of brain cells release enzymes that cause these cells to self-destruct.

_____ 24. The _____ controls wakefulness.

_____ 25. The corpus striatum, including the caudate and lentiform nuclei, is part of the _____ .

ANSWERS TO MASTERY TEST: ⭐ CHAPTER 14

Multiple Choice

1. E
2. B
3. A
4. A
5. D
6. C
7. B
8. D
9. A

Arrange

10. C A B
11. A C B
12. A B D C
13. B A C

True-False

14. T
15. F. Endorphins, enkephalins, and dynorphin
16. F. Cerebral cortex
17. F. I, II, and VIII
18. T
19. F. Serotonin and acetylcholine
20. T

Fill-ins

21. Falx cerebri; tentorium cerebelli
22. Midbrain
23. Lysosomes
24. Reticular formation or reticular activating system (RAS)
25. Basal ganglia

FRAMEWORK 15
Sensory, Motor & Integrative Systems

SENSORY SYSTEMS

SENSATIONS (A)
- Levels of sensation
- Modality
- Components
- Receptors
  - selectivity
  - classification
  - types of potentials
  - adaptations

SOMATIC SENSES (B)
- Cutaneous
  - tactile: touch, pressure, vibration
  - thermal
  - pain
- Proprioceptive

SOMATIC SENSORY PATHWAYS (C)

Examples of Major Sensory Pathways
- Posterior Column
  - Discriminative touch
  - Stereognosis
  - Proprioception
  - Weight discrimination
  - Vibratory
- Spinothalamic
  - Pain
  - Somatosensory cortex
- Spinocerebellar
  - Subconscious muscle sense

Integration of Sensory Pathways With Motor Pathways

SOMATIC MOTOR PATHWAYS (D)

DIRECT (PYRAMIDAL)
- Corticospinal
  - lateral (90%) crosses in medulla
  - anterior (10%) crosses in cord
- Corticobulbar
- Upper motor neuron
- Lower motor neuron
- Precise movements

INDIRECT (EXTRAPYRAMIDAL)
- Begin in basal ganglia or reticular formation
- Lead to lower motor neurons
- Muscle tone, posture equilibrium, visual responses

INTEGRATIVE FUNCTIONS (E)
- Memory
  - short term
  - long term
- Sleep
  - NREM
  - REM

DISORDERS (F)
- Spinal cord injury
- Cerebral palsy
- Parkinson's disease

Sensory, Motor, and Integrative Systems

The last three chapters have introduced the components of the nervous system and demonstrated the arrangement of neurons in the spinal cord, spinal nerves, brain, and cranial nerves. Chapter 15 integrates this information in the study of major nerve pathways including those for (a) general senses (such as touch, pain, and temperature), (b) motor control routes for precise, conscious movements and for unconscious movements, and (c) complex integrative functions such as memory and sleep.

As you begin your study of sensory, motor, and integrative systems, carefully examine the Chapter 15 Topic Outline and Objectives; check off each one as you complete it. To organize your study of this chapter glance over the Chapter 15 Framework now. Be sure to refer to the Framework frequently and note relationships among key terms in each section.

TOPIC OUTLINE AND OBJECTIVES

A. Sensations

☐ 1. Define a sensation and discuss the levels and components of sensation.
☐ 2. Describe the classification of receptors.

B. Somatic senses

☐ 3. List the location and function of the receptors for tactile sensations (touch, pressure, vibration), thermal sensations (warmth and cold), and pain.
☐ 4. Distinguish somatic, visceral, referred, and phantom pain.
☐ 5. Identify the proprioceptive receptors and indicate their functions.

C. Somatic sensory pathways

☐ 6. Discuss the neuronal components and functions of the posterior column–medial lemniscus, anterolateral, and spinocerebellar pathways.
☐ 7. Describe the integration of sensory input and motor output.

D. Somatic motor pathways

☐ 8. Compare the locations and functions of the direct and indirect motor pathways.
☐ 9. Explain how the basal ganglia and cerebellum are related to motor responses.

E. Integrative functions

☐ 10. Compare integrative functions such as learning, memory, wakefulness, and sleep.

F. Disorders

☐ 11. Compare the following types of paralysis: monoplegia, diplegia, paraplegia, hemiplegia, and quadriplegia; define complete transection, hemisection, and spinal shock.
☐ 12. List the clinical symptoms of spinal cord injury, cerebral palsy, and Parkinson's disease.

WORDBYTES

Now become familiar with the language of this chapter by studying each wordbyte, its meaning, and an example of its use within a term. After you study the entire list, self-check your understanding by writing the meaning of each wordbyte on the line. As you continue through the *Learning Guide,* identify (and fill in) additional terms that contain the same wordbyte.

Wordbyte	Self-check	Meaning	Example(s)
a-, an-	_____	without	*an*esthesia
-ceptor	_____	receiver	proprio*ceptor*
-esthesia	_____	sensation	an*esthesia,* par*esthesia*
hemi-	_____	half	*hemi*plegia, *hemi*sphere
noci-	_____	hurt, pain	*noci*ceptor
para-	_____	abnormal, beyond	*par*esthesia
-plegia	_____	paralysis or stroke	hemi*plegia*
proprio-	_____	one's own	*proprio*ception
quad-	_____	four	*quad*riplegia
rhiz-	_____	root	*rhizo*tomy

CHECKPOINTS

A. Sensations   (pages 430–432)

■ **A1.** Review the three basic functions of the nervous system by listing them in sequence here:

a. Receiving _____ input

b. _____, _____, _____ information

c. Transmitting _____ impulses to _____

   or _____

A2. For a moment, visualize what your life would be like if you were unable to experience any sensations. List the three types of sensations that you believe you would miss most.

_____ _____ _____ Now write a sentence describing how your health and safety might be endangered by lack of ability to perceive sensations.

■ **A3.** Do this activity about levels of sensation.

a. *(Perception? Sensation?)* is the awareness (either conscious or unconscious) of

stimuli, such as the act of seeing a bird, whereas _____ is the conscious awareness and interpretation of sensation, such as the recognition of that bird as abluebird and not a robin.

b. Identify levels of sensation by matching parts of the nervous system listed in the box with descriptions below.

```
B. Brain stem          S. Spinal cord
C. Cerebral cortex     T. Thalamus
```

_____ 1. Immediate reflex action is possible without involvement of the brain.

_____ 2. Involves subconscious motor reactions, such as those that facilitate balance and equilibrium.

_____ 3. Offers a general (or "crude") awareness of locale and type of sensations, such as

awareness of pressure within the leg or that something is being heard; offers no specifics.

_____ 4. Gives precise information about sensations, such as distinguishing Bach from Baez from Buffett, as well as memories of hearing a particular musical work.

■ **A4.** Arrange in correct order the components in the pathway of sensation.

```
IC. Impulse generation and conduction    S. Stimulation
Int. Integration                         T. Transduction
```

1._____ → 2._____ → 3._____ → 4._____

Now match each of these components of sensation with the correct description below.

_____ a. Activation of a sensory neuron

_____ b. Stimulus → generator potential

_____ c. Generator potential → nerve impulse; impulse is conveyed along a first-order neuron into the spinal cord or brainstem up to a higher level of the brain.

_____ d. Nerve impulse → sensation; usually occurs in the cerebral cortex

■ **A5.** Do this exercise on the characteristics of sensations.

a. The quality that makes the sensation of pain different from the sense of touch or which

makes hearing different from vision is called the _____ of sensation. Any single sensory neuron carries sensations of *(one modality? many modalities?)*. Different modalities of sensations are distinguished precisely at the level of the *(spinal cord? thalamus? cerebral cortex?)*.

b. Recall getting dressed this morning. The fact that you felt the shirt touching your back just after you put on that shirt but that the awareness of that touch has dissipated over

time is known as the characteristic of _____. Circle the type of receptors that adapt most slowly. *(Pain? Pressure? Smell? Touch?)* Explain how such slow adaptation is advantageous to you.

c. Describe what is meant by the *selectivity* of receptors.

d. Arrange these receptors from simplest to most complex structurally: _____ _____ _____
   A. Receptors for special senses such as vision or hearing
   B. Free nerve endings, such as receptors for pain
   C. Receptors for touch or pressure

■ **A6.** Identify the class of receptor that fits each description.

E. Exteroceptor	I. Interoceptor	P. Proprioceptor

_____ a. These receptors inform you that you are hungry or thirsty; these also inform your brain of need for adjustment of blood pressure.

_____ b. Your ears, eyes, and receptors in your skin for pain, touch, hot, and cold are of this type.

_____ c. With your eyes closed, you can tell your exact body position, thanks to these receptors.

■ **A7.** Complete these sentences describing types of receptors based on nature of stimulus detected.

a. You can maintain balance by means of inner ear _____-receptors sensitive to change in position.

b. Different tastes and smells are distinguished with the help of

_____-receptors in nose and mouth.

c. _____-receptors inform you of the temperature around you.

■ **A8.** Contrast generator potential (GP) with receptor potential (RP) by completing this table.

	GP	RP
a. Triggers action potential. (*Yes? No?*)		
b. Most receptors for (*general? special?*) senses are of this type.		
c. (*Always excitatory? Always inhibitory? Excitatory or inhibitory?*)		

B. Somatic senses   (pages 432–437)

■ **B1.** List six examples of cutaneous sensations.

■ **B2.** Do the following exercise about tactile sensations.

a. Name the tactile sensations.

These sensations are all sensed by _____-receptors. Receptors for these sensations are (*evenly? unevenly?*) distributed throughout the skin of the body.

b. (*Discriminative? Crude?*) touch refers to ability to determine that something has touched the skin, but its precise location, shape, size, or texture cannot be distinguished.

c. In general, pressure receptors are more (*superficial? deep?*) in location than touch receptors.

d. Sensations of _____ result from rapid repetition of sensory impulses from tactile receptors.

■ **B3.** Match names of receptors with their descriptions. Answers may be used more than once.

C. Corpuscles of touch (Meissner's corpuscles) H. Hair root plexuses L. Lamellated (Pacinian) corpuscles N. Nociceptors	Type I. Type I cutaneous mechanoreceptors (tactile or Merkel discs) Type II. Type II cutaneous mechanoreceptors (end organs of Ruffini)

_____ a. Egg-shaped masses located in dermal papillae, especially in fingertips, palms of hands, and soles of feet

_____ b. Onion-shaped structures sensitive to pressure and high-frequency vibration

_____ c. Free nerve endings that sense pain

_____ d. Slowly adapting touch receptors (two answers)

_____ e. May respond to any type of stimulus if stimulus is strong enough to cause tissue damage

_____ f. Surround hair follicles; can detect breeze blowing against hairs on skin.

B4. *For extra review.* Refer to Figure LG 5.1 (page 91). Identify and label types of receptors on that figure. Add your own drawings of three other types of receptors in appropriate layers of skin. Label those receptors also.

■ **B5.** For additional review of cutaneous sensations, do this Checkpoint.

a. Explain how the sensation of itch develops.

b. Besides pain, what other sensations involve free nerve endings? Name four.

c. Name two chemicals that may be released from injured tissues that may stimulate pain

receptors. _____, _____

d. Acute pain is felt in *(deep? superficial?)* tissues of the body. Circle the qualities that are associated with acute (rather than chronic) pain.

   Aching   Burning   Fast   Pricking   Sharp   Slow   Throbbing

e. Which type of pain impulses travel along myelinated A nerve fibers? *(Acute? Chronic?)*
f. Circle the type of pain that arises from receptors in skin.

   Deep somatic pain   Superficial somatic pain   Visceral pain

■ **B6.** *The Big Picture: Looking Ahead and a clinical challenge.* Refer to figures in upcoming chapters of your text to help you determine why patients experiencing visceral pain (as during a heart attack or gallbladder attack) may feel pain in locations quite distant from these two organs.

a. Pain impulses that originate from the heart, for example, during a "heart attack" involving the left ventricle (Figure 20.7, page 588), enter the spinal cord at the same level as do sensory fibers

   from skin covering the _____. This level of the cord is about _____ to _____.
b. Refer to Figure 15.2 (page 434 in your text). Notice that some liver and gallbladder pain is felt

   in a region quite distant from these organs, that is, in the _____ region. These

   organs lie just inferior to the dome-shaped _____, which is supplied by the phrenic nerve from the *(cervical? brachial? lumbar?)* plexus. A painful gallbladder can send impulses to the cord via the phrenic nerve, which enters the cord at the same level

   (about C _____ or C _____) as nerves from the neck and shoulder. So gallbladder

   pain is said to be _____ to the neck and shoulder area.

■ **B7.** *A clinical challenge.* Mr. Hamilton was in a motorcycle accident that resulted in his left leg being severed above the knee. Months later he is referred to the university pain clinic for an assessment. He states, "I feel this pain in my left foot, and sometimes my toes itch like crazy, but they're not there to scratch!" How might you interpret his experience?

■ **B8.** *A clinical challenge.* Match methods of pain relief listed in the box with the following descriptions.

A. Anesthetic (Novocaine)   O. Opioids such as morphine,
 AA1. Aspirin, acetaminophen (Tylenol),   codeine, or meperidine (Demerol)
      ibuprofen (Motrin, Advil)   R. Rhizotomy
      C. Cordotomy

_____ a. Surgical technique that severs spinal cord pathways for pain, such as spinothalamic tracts

_____ b. Surgical technique involving cutting of posterior (sensory) root of spinal nerve(s)

_____ c. Medication that alters the quality of pain perception so that pain feels less noxious

_____ d. Medication that inhibits formation of chemicals (prostaglandins) that stimulate pain receptors

_____ e. Medication that blocks conduction of nerve impulses in first-order neurons

B9. Close your eyes for a moment. Reflect on the position of your arms and legs as well as the degree of tension in your muscles and tendons. This awareness of body position and

movements is known as _____. Now match names of proprioceptors listed in the box with their descriptions below. Answers may be used more than once.

> H. Hair cells of the inner ear   M. Muscle spindles
> J. Joint kinesthetic receptors   T. Tendon organs

_____ a. Type Ia and type II fibers that are sensitive to stretch

_____ b. Consists of a thin capsule penetrated by type Ib fibers

_____ c. Sensitive to tension; monitor the force of contraction

_____ d. Located within articular capsules of synovial joints

B10. State advantages and disadvantages of each type of anesthesia:

a. General anesthesia

b. Spinal anesthesia

C. Somatic sensory pathways   (pages 437–440)

C1. Write terms listed in the box after correct definitions below to identify types of sensations you experience.

> Crude touch   Stereognosis
> Kinesthesia   Two-point discriminative touch
> Proprioception

a. Ability to sense touch on the back of your neck but inability to specify precise location

of the touch: _____

b. Ability to recognize exact location of a sharp pencil touched to your fingertip and ability to distinguish whether one or two sharp pencils are touching nearby points on

that fingertip: _____

c. Ability to tell—with your eyes closed—the exact position of your arms and legs:

_____

d. Awareness of precise movements of your fingers as you type a paper:

_____

e. Ability to distinguish a dime from a button based on size, shape, and texture:

_____

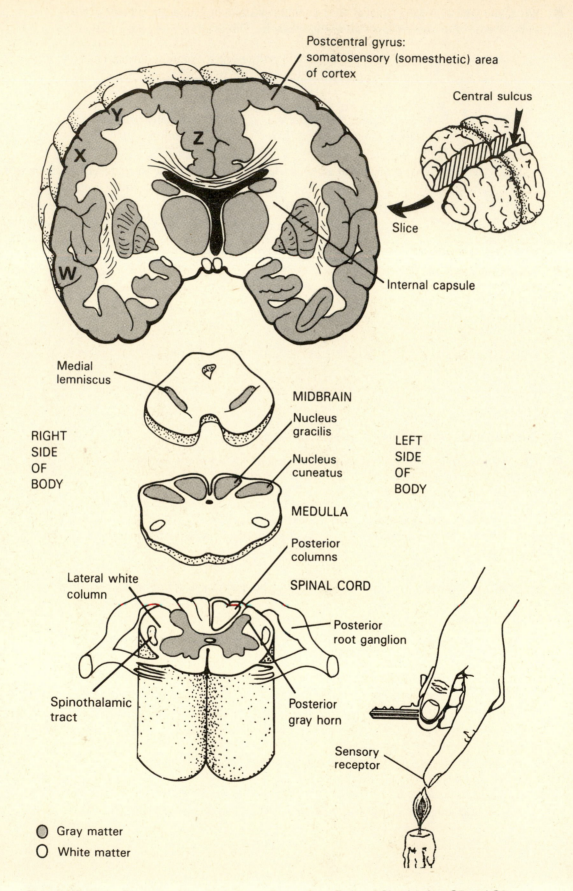

Postcentral gyrus: somatosensory (somesthetic) area of cortex

Central sulcus

Slice

Internal capsule

Medial lemniscus

MIDBRAIN

Nucleus gracilis

Nucleus cuneatus

MEDULLA

RIGHT SIDE OF BODY

LEFT SIDE OF BODY

Posterior columns

SPINAL CORD

Lateral white column

Posterior root ganglion

Spinothalamic tract

Posterior gray horn

Sensory receptor

○ Gray matter
○ White matter

Figure LG 15.1 Diagram of the central nervous system. Refer to Checkpoints C5 and C6.

■ **C2.** Contrast functions of major sensory pathways by selecting answers from the box that match the types of sensations they transmit.

> APSC. Anterior and posterior spinocerebellar tract
> AST. Anterior spinothalamic tract
> LST. Lateral spinothalamic tract
> PCML. Posterior column and medical lemniscus tract

_____ a. Pain and temperature

_____ b. Crude touch or pressure

_____ c. Discriminative, two-point discrimination

_____ d. Proprioception, kinesthesia, stereognosis

_____ e. Weight discrimination and vibration

_____ f. Tickle and itch

_____ g. Subconscious proprioceptive impulses to the cerebellum needed for posture, balance, and coordination

■ **C3.** Sensory pathways involve *(one? two? three?)*-neuron sets that allow nerve impulses from receptors, for example, in skin or muscles, to reach the primary somatosensory area of the cerebral cortex. Identify first-order (I), second-order (II), and third-order (III) neurons by writing I, II, or III next to the related neuron.

_____ a. Crosses to the opposite side of the spinal cord or brainstem and ascends to the thalamus

_____ b. Reaches the spinal cord or brainstem

_____ c. Passes to the postcentral gyrus of the cerebral cortex

■ **C4.** Write in locations of cell bodies of the three neurons in these two pathways.

Pathway	First-Order	Second-Order	Third-Order
a. Posterior column–medial lemniscus			
b. Anterolateral (spinothalamic)			

■ **C5.** Check your understanding of pathways by drawing them according to the directions below. Try to draw them from memory; do not refer to the text figures or the table until you finish. Label the *first-, second-,* and *third-order neurons* as I, II, and III.

a. Draw in *blue* on Figure LG 15.1 the pathway that allows you to accomplish stereognosis so that you can recognize the shape of a key in your left hand. (*Hint:* The correct recognition of the key involves discriminative touch, proprioception, and weight discrimination.)

b. Now draw in *red* the pathway for sensing pain and temperature as the index finger of the left hand comes too close to a flame. For simplification, draw only the lateral (not anterior) component.

■ **C6.** Do this learning activity about sensory reception in the cerebral cortex.

a. Picture the brain as roughly similar in shape to a small, rounded loaf of bread. The crust

would be comparable to the cerebral _____.

b. Suppose you cut a slice of bread in midloaf that represents the postcentral gyrus. It looks somewhat like the "slice" or section of brain shown at the top of Figure LG 15.1.

Locate the Z on that figure. This portion (next to the _____ fissure) receives sensations from the *(face? hand? hip? foot?)*. (Refer to Figure 15.5, page 439 of your text, for help.)

c. Suppose a person suffered a loss of blood supply (CVA or "stroke") to the area of the brain marked X in Figure LG 15.1. Loss of sensation in the *(face? hand? leg?)* area would result. Damage to the area marked W would be likely to lead to loss of the sense of *(vision? hearing? taste?)* because the W is located on the temporal lobe.

d. The letters W, X, Y, and Z are marked on the *(right? left?)* side of the brain. (Note that this is an anterior view of this brain section. If a friend stands and faces you, the *right* side of your friend's brain will be in your *left* field of view.) What effects would brain damage to area Y of the postcentral gyrus have? Loss of *(sensation? movement?)* to the *(face? hand? foot?)* on the *(right? left?)* side of the body.

e. Certain regions of the body are represented by large areas of the _____

_____ cortex, indicating that these areas *(are? are not?)* very sensitive. Name two of these areas.

C7. Review locations of tracts in the spinal cord by referring to Chapter 13, Checkpoint C2, page LG 253.

BG. Basal ganglia	CC. Cerebral cortex
C. Cerebellum	S. Spinal cord

_____ a. Quick withdrawal of a hand when it touches a hot object

_____ b. Unconscious responses to proprioceptive impulses to permit smooth, coordinated movements (two answers)

_____ c. Playing piano or writing a letter

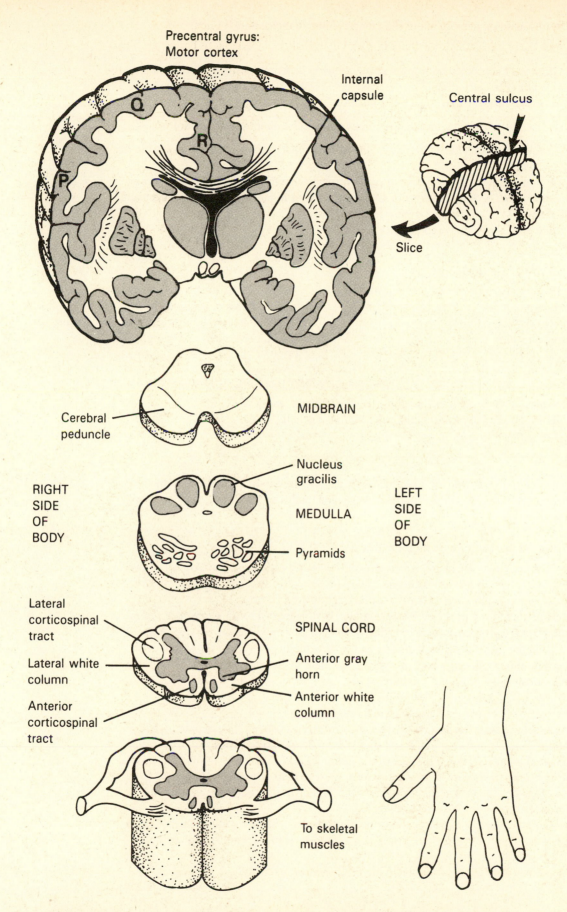

Precentral gyrus:
Motor cortex

Internal
capsule

Central sulcus

Slice

Cerebral
peduncle

MIDBRAIN

Nucleus
gracilis

RIGHT
SIDE
OF
BODY

MEDULLA

LEFT
SIDE
OF
BODY

Pyramids

Lateral
corticospinal
tract

SPINAL CORD

Lateral white
column

Anterior gray
horn

Anterior
corticospinal
tract

Anterior white
column

To skeletal
muscles

Figure LG 15.2 Diagram of the central nervous system. Refer to Checkpoint D1 and D3.

■ **C8.** Does every bit of sensory input that reaches the CNS elicit a motor response? *(Yes? No?)* Discuss integration of sensory input and motor output.

■ **C9.** The linking of sensory to motor response occurs at different levels in the central nervous system. Indicate which is the highest level required for each of these responses.

D. Somatic motor pathways (pages 440–446)

■ **D1.** Refer to Figure LG 15.2 and do the following exercise about motor control.

a. This "slice" or section of the brain containing the primary motor cortex is located in the *(frontal? parietal?)* lobe.
b. Area P in this figure controls *(sensation? movement?)* to the *(left? right?)* *(arm? leg? side of the face?)*. Motor neurons in area P would send impulses to the pons to activate cranial nerve _____ to facial muscles.
c. Describe effects of damage to area R.

A. Anterior gray horn (lower motor neuron)   E. Lateral corticospinal tract
B. Midbrain and pons   F. Medulla, decussation site
C. Effector (skeletal muscle)   G. Precentral gyrus (upper motor neuron)
D. Internal capsule   H. Ventral root of spinal nerve

■ **D2.** Show the route of impulses along the principal pyramidal pathway by listing in correct sequence the structures that comprise the pathway. Refer to Figure 15.6 (page 441 in the text) if you have difficulty with this activity.

_____ _____ _____ _____ _____ _____ _____ _____

■ **D3.** Now draw this pathway on Figure LG 15.2. Show the control of muscles in the left hand by means of neurons in the lateral corticospinal pathway. Label the *upper* and *lower motor neurons.*

■ **D4.** Complete this exercise on motor pathways.

a. Corticospinal tracts are classified as *(direct? indirect?)* pathways; they control *(precise, voluntary? involuntary?)* movements. Based on their structure, corticospinal tracts are also known as *(pyramidal? extrapyramidal?)* pathways.
b. Axons of about 90% of upper motor neurons (UMNs) pass through *(anterior? lateral?)* corticospinal tracts; these are axons that *(decussated or crossed? did not decussate or cross?)* in the medulla. Lateral corticospinal tracts on the left side of the spinal cord consist of axons that originated in upper motor neurons on the *(right? left?)* side of the motor cortex.

c. About _____% of UMN axons pass through _____ corticospinal tracts. Because axons of these UMNs *(do? do not?)* decussate in the medulla, anterior corticospinal tracts on the left side of the cord consist of axons that originated on the *(right? left?)* side of the motor cortex.
d. Corticobulbar tracts consist of UMNs that terminate in the *(brainstem? anterior gray horn of the cord?)*. Lower motor neurons (LMNs) there pass out through *(cranial? spinal?)* nerves.
e. Destruction of *(upper? lower?)* motor neurons results in spastic paralysis because the brain is not controlling movements. Flaccid paralysis results from damage of *(upper? lower?)* motor neurons.
f. Why are lower motor neurons called the *final common pathway?*

■ **D5.** *The Big Picture: Looking Ahead.* Refer to Figure 21.20c, page 641, as you determine potential effects of cerebrovascular accidents (CVAs or "strokes") involving specific arteries to the brain.

    a. Locate the anterior cerebral arteries on the figure. These proceed anteriorly and pass between the two cerebral hemispheres, supplying area R of Figure LG 15.2. Paralysis and/or sensory loss of *(arm and hand? leg and foot? face?)* is most likely to result from interruption of blood flow through this vessel.

    b. Locate the middle cerebral arteries. These proceed laterally and then pass medial to the temporal lobe to supply pre- and postcentral gyri. A stroke involving this artery is likely to cause effects on *(sensation only? movement only? sensation and movement?)*

    c. Identify the vertebral and basilar arteries and their branches. Strokes involving these vessels *(are? are not?)* likely to cause effects on extrapyramidal pathways. State your rationale.

■ **D6.** Describe locations of indirect (extrapyramidal) tracts in this exercise:

    a. All of these tracts lie *(within? outside of?)* corticospinal tracts. All of them originate in

    the *(brainstem? cerebral cortex?)* and end in the _____.

    b. Fill in names of extrapyramidal tracts.

        1. Midbrain → _____-spinal → spinal cord

        2. Midbrain → _____-spinal → spinal cord

        3. Pons → _____-spinal → spinal cord

        4. Medulla → _____-spinal → spinal cord

        5. Medulla → _____-spinal → spinal cord

D7. Summarize how the basal ganglia and cerebellum, together with associated neurons, help you to carry out coordinated, precise movements, such as those involved in a game of tennis.

■ **D8.** *A clinical challenge.* Describe two conditions that involve damage to basal ganglia in this exercise.

    a. *(Huntington's? Parkinson's?)* disease involves changes in basal ganglia associated with decrease in the neurotransmitter *(acetylcholine? dopamine? GABA?)*. Signs of this condition are *(jerky movements and facial twitches? muscle rigidity and tremors?)*.

    b. A predominant sign of Huntington's disease (HD) is jerky movements (or "dance")

    known as _____ related to loss of neurons that make the

    inhibitory transmitter named _____. Signs of this progressive condition are most likely to be first noted at about age *(4 years? 40 years? 70 years?)*. A child of a parent with HD has *(almost no? a 50–50?)* chance of having this condition.

D9. Assess a study partner for cerebellar damage. Then describe your assessment here. *Note:* Signs of such damage can be expected to be *(contra-? ipsi?)*-lateral.

E. Integrative functions   (pages 446–449)

■ **E1.** List three types of activities that require complex integration processes by the brain.

E2. Define these terms: *learning, memory, plasticity.*

E3. List several parts of the brain that are considered to assist with memory.

■ **E4.** Do this exercise on memory.

a. Looking up a phone number, dialing it, and then quickly forgetting it is an example of *(short-term? long-term?)* memory. One theory of such memory is that the phone

number is remembered only as long as a _____ neuronal circuit is active.

b. Repeated use of a telephone number can commit it to long-term memory. Such rein-

forcement is known as memory _____. *(Most? Very little?)* of the information that comes to conscious attention goes into long-term memory because the brain is selective about what it retains.

c. Some evidence indicates that *(short-term? long-term?)* memory involves electrical and chemical events rather than anatomical changes. Supporting this idea is the fact that chemicals used for anesthesia and shock treatments interfere with *(recent? long-term?)* memory.

E5. Discuss changes in the following anatomical structures or chemicals that may be associated with enhanced long-term memory:

a. Presynaptic terminals, synaptic end bulbs, dendritic branches

b. Glutamate, NMDA glutamate receptors, nitric oxide (NO)

c. Acetylcholine (ACh)

d. RNA

■ **E6.** Complete this exercise about sleep and wakefulness.

a. _____ rhythm is a term given to the usual daily pattern of sleep and wakefulness. This rhythm appears to be established by a nucleus within the *(basal ganglia? hypothalamus? thalamus?)*.

b. The RAS (or R_____ A_____

   S_____) is a portion of the reticular formation. List three types of sensory input that can stimulate the RAS.

c. Stimulation of the reticular formation leads to *(increased? decreased?)* activity of the cerebral cortex, as indicated by _____ recordings of brain waves.

d. Continued feedback between the RAS, the _____,

   and the _____ maintains the state called *consciousness*. Inactivation of the RAS produces a state known as _____.

e. Name one chemical that may activate the RAS: _____. Damage to the RAS produces the state of _____. List several factors that may cause coma.

■ **E7.** Do the following learning activity about sleep.

a. Name two kinds of sleep.

b. In NREM sleep, a person gradually progresses from stage *(1 to 4? 4 to 1?)* into deep sleep. Alpha waves are present in stage(s) *(1? 3 and 4?)*, whereas slow delta waves characterize stage(s) *(1? 3 and 4?)* sleep. Sleep spindles characterize stages _____ and _____ of NREM sleep.

c. After moving from stage 1 to stage 4 sleep, a person *(moves directly into? ascends to stages 3 and 2 and then moves into?)* REM sleep.

d. REM is much *(more active? slower?)* sleep than NREM sleep. Respirations and pulse

are _____-creased compared to their levels in stage 4. Alpha waves are present as in

stage *(1? 4?)* sleep. The name REM indicates that _____ movements can be observed during REM sleep. Dreaming occurs during *(stages 3–4 of NREM? REM?)* sleep.

e. Periods of REM and NREM sleep alternate throughout the night in about

_____-minute cycles. During the early part of an 8-hour sleep period, REM periods last about *(5–10? 30? 50?)* minutes; they gradually increase in length until the final

REM period lasts about _____ minutes.

f. With aging, the percentage of REM sleep normally _____-creases.

F. Disorders (pages 449–450)

F1. List several causes of spinal cord injury.

■ **F2.** Match the terms in the box with descriptions below.

```
H.  Hemiplegia          P.  Paraplegia
M.  Monoplegia          Q.  Quadriplegia
```

_____ a. Paralysis of one extremity only

_____ b. Paralysis of both legs

_____ c. Paralysis of both arms and both legs

_____ d. Paralysis of the arm, leg, and trunk on one side of the body

■ **F3.** Describe changes associated with transection of the spinal cord in this activity.

a. Hemisection of the cord refers to transection of *(half? all?)* of the spinal cord.

b. If the posterior columns (cuneatus and gracilis) and lateral corticospinal tracts on the right side of the cord are severed at level T11–T12, symptoms of loss of awareness of muscle sensations (proprioception), loss of touch sensations, and paralysis are likely to occur in the *(right arm? left arm? right leg? left leg?)*. (Circle all that apply.)

c. The period of spinal shock is likely to last for several *(days to weeks? years?)*. During this time, the person is likely to experience *(areflexia? hyperreflexia?)*, meaning *(exaggerated? no?)* reflexes.

d. When reflexes gradually return to the injured person, they are likely to return in a specific sequence. Arrange these reflexes in order from first to last, according to their

return. _____ _____ _____
A. Stretch reflex
B. Flexion reflex
C. Crossed extensor reflex

■ **F4.** Cerebral palsy (CP) is a group of disorders affecting *(movements? sensations?)*. This condition results from brain damage *(in fetal life, at birth, or in infancy? in early adulthood? in later years?)*. CP *(is? is not?)* a progressive disease; it *(is? is not?)* reversible.

■ **F5.** Complete this Checkpoint about Parkinson's disease (PD).

a. Parkinson's typically affects persons *(in fetal life, at birth, or in infancy? in early adulthood? in later years?)*. PD *(is? is not?)* a progressive disease; it *(is? is not?)* usually hereditary.

b This condition involves a decrease in the release of the neurotransmitter

_____. This chemical is normally produced by cell bodies located in

the substantia nigra, which is part of the _____. Axons lead from here

to the _____, where dopamine (DA) is released. DA is an *(excitatory? inhibitory?)* transmitter.

c. The basal ganglia neurons produce a neurotransmitter _____ (ACh), which is *(excitatory? inhibitory?)* to skeletal muscles. The result is a relative increase in the ratio of ACh to DA in Parkinson's patients. Explain how this imbalance leads to muscle tremor and rigidity.

d. Voluntary movements may be slower than normal; this is the condition of *(brady? tachy?)*-kinesia.

e. Because persons with parkinsonism have a deficit in dopamine, it seems logical to treat the condition by administration of this neurotransmitter. Does this work? *(Yes? No?)* Explain.

f. Why do levodopa and/or anticholinergic medications provide some relief to these patients?

g. How do MAO-inhibitor drugs help PD patients?

A1. (a) Sensory. (b) Integrating, associating, and storing. (c) Motor (efferent), muscles or glands.

A3. (a) Sensation, perception. (b) 1: S. 2: B. 3: T. 4: C.

A4. 1. S → 2. T → 3. IC → 4. Int. (a) S. (b) T. (c) IC. (d) Int.

A5. (a) Modality; one modality; cerebral cortex.
(b) Adaptation; pain; protects you by continuing to provide warning signals of painful stimuli.
(c) Vigorous response of each type of receptor to a specific type of stimulus, such as touch receptor response to mechanical energy. (d) B C A.

A6. (a) I. (b) E. (c) P.

A7. (a) Mechano-. (b) Chemo-. (c) Thermo-.

A8.

	GP	RP
a. Triggers action potential. (*Yes? No?*)	Yes	No
b. Most receptors for (*general? special?*) senses are of this type.	General (such as pain) *Hint: remember G as in "generator for general senses."*)	Special (such as vision)
c. (*Always excitatory? Always inhibitory? Excitatory or inhibitory?*)	Always excitatory	Excitatory or inhibitory

B1. Touch, pressure, vibration, cold, heat, and pain.

B2. (a) Touch, pressure, vibration, as well as itch and tickle; mechano; unevenly. (See Mastery Test Question 6.) (b) Crude. (c) Deep. (d) Vibration.

B3. (a) C. (b) L. (c) N. (d) Type I and Type II. (e) N. (f) H.

B5. (a) A local inflammation, for example, from a mosquito bite, causes release of chemicals such as bradykinin, which stimulates free nerve endings. (b) Itch, tickle, and temperature (warmth and coolness). (c) Prostaglandins and kinins. (d) Superficial; sharp, fast, pricking. (e) Acute. (f) Superficial somatic pain.

B6. (a) Medial aspects of left arm; T1, T4. (b) Shoulder and neck (right side); diaphragm, cervical; 3, 4; referred.

B7. Phantom pain (phantom limb sensation). This results from irritation of nerves in the stump. Pathways from the stump to the cord and brain are still intact and may be stimulated even though the original sensory receptors in his foot are missing. Remember that he feels pain or itching in his *brain,* not in his leg or foot.

B8. (a) C. (b) R. (c) O. (d) AA1. (e) A.

B9. Proprioception. (a) M. (b) T. (c) T. (d) J.

C1. (a) Crude touch. (b) Two-point discriminative touch. (c) Proprioception. (d) Kinesthesia. (e) Stereognosis.

C2. (a) LST. (b) AST. (c–e) PCML. (f) AST. (g) APSC.

C3. Three. (a) II. (b) I. (c) III.

C4.

Pathway	First-Order	Second-Order	Third-Order
a. Posterior column–medial lemniscus	Posterior root ganglion	Medulla: nucleus cuneatus and nucleus gracilis	Thalamus
b. Anterolateral (spinothalamic)	Posterior root ganglion	Posterior gray horn of spinal cord	Thalamus

C5. See Figure LG 15.1A.

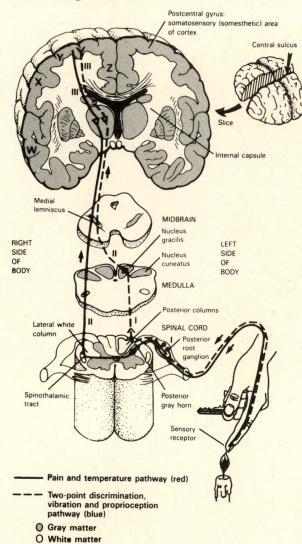

Figure LG 15.1A Diagram of the central nervous system.

C6. (a) Cortex. (b) Longitudinal, foot. (c) Face; hearing. (d) Right; sensation, hand, left. (e) General sensory or somatosensory, are; lips, face, thumb and other fingers (see Figure 15.5, page 439 in the text).

C9. (a) S. (b) C and BG. (c) CC.

D1. (a) Frontal. (b) Movement, left side of the face; VII. (c) Movement of left foot affected (spasticity).

D2. G D B F E A H C.

D3. See Figure 15.2A.

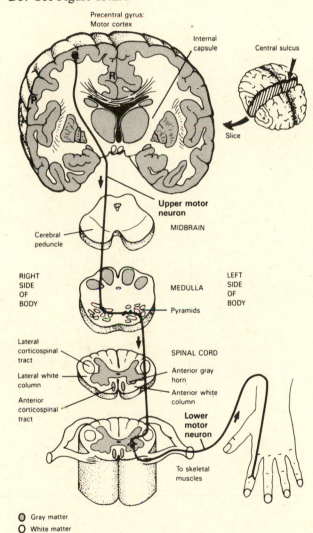

Precentral gyrus: Motor cortex

Internal capsule

Central sulcus

Slice

Upper motor neuron

MIDBRAIN

Cerebral peduncle

RIGHT SIDE OF BODY

MEDULLA

LEFT SIDE OF BODY

Pyramids

Lateral corticospinal tract

SPINAL CORD

Lateral white column

Anterior gray horn

Anterior corticospinal tract

Anterior white column

Lower motor neuron

To skeletal muscles

○ Gray matter
○ White matter

Figure LG 15.2A   Diagram of the central nervous system.

D4. (a) Direct; precise, voluntary; pyramidal. (b) Lateral, decussated or crossed; right. (c) 10, anterior; do not, left. (d) Brainstem; cranial. (e) Upper; lower. (f) Because these neurons, which are activated by input from a variety of sources (brain or spinal nerves), must function for *any* movement to occur.

D5. (a) Leg and foot. (b) Sensation and movement. (c) Are, because the brainstem (including reticular formation) and cerebellum are sites of origin of these pathways.

D6. (a) Outside of; brainstem, spinal cord. (b) 1, 2. Rubro and tecto; 3. Anterior reticulo; 4–5. Vestibulo and lateral reticulo.

D8. (a) Parkinson's, dopamine; muscle rigidity and tremors. (b) Chorea, GABA; 40 years; 50–50.

E1. Memory, sleep/wakefulness, emotions.

E4. (a) Short-term; reverberating. (b) Consolidation; very little. (c) Short-term, recent.

E6. (a) Circadian; hypothalamus. (b) Reticular activating system; pain, proprioception, visual or auditory stimuli. (c) Increased, EEG. (d) Cerebral cortex, spinal cord; sleep (or coma). (e) Amphetamines; coma; poisoning, hypoxia, ischemia, hypoglycemia, infection, disorders of acid–base balance or electrolytes.

E7. (a) Nonrapid eye movement (NREM) and rapid eye movement (REM). (b) 1 to 4; 1; 3 and 4; 3, 4. (c) Ascends to stages 3 and 2 and then moves into. (d) More active; in; 1; rapid eye; REM. (e) 90; 5–10; 50. (f) De.

F2. (a) M. (b) P. (c) Q. (d) H.

F3. (a) Half. (b) Right leg (same side as cord is severed, and in areas served by nerves inferior to the T11–T12 transection such as the right leg. (c) Days to weeks; areflexia, no. (d) A B C.

F4. (a) Movements; in fetal life, at birth, or in infancy; is not; is not.

F5. (a) In later years; is; is not. (b) Dopamine (DA); midbrain; basal ganglia; inhibitory. (c) Acetylcholine, excitatory; lack of inhibitory DA causes a double-negative effect so that extra, unnecessary movements occur (tremor) that may then interfere with normal movements (such as extension at the same time as flexion), resulting in rigidity and masklike facial appearance. (d) Brady. (e) No, DA cannot cross the blood–brain barrier. (f) Levodopa (L-dopa) is a precursor (leads to formation) of DA and anticholinergic medications counteract effects of ACh. (g) MAO (monoamine oxidase) is an enzyme made in the body that normally destroys catecholamines such as norepinephrine or dopamine. Use of MAO inhibitors permits the dopamine to continue acting for longer periods.

WRITING ACROSS THE CURRICULUM: CHAPTER 15

1. Describe structures and mechanisms that help prevent overstretching of your muscles, tendons, and joints. Be sure to include roles of receptors and describe nerve pathways.
2. Contrast the sensations of proprioception, kinesthesia, and stereognosis. Tell what information these senses provide to you.
3. Contrast these two pathways: (a) posterior columns–medial lemniscus and (b) anterolateral (spinothalamic). Include the types of sensations transmitted by each pathway as well as locations of cell bodies and axons of the three-neuron set in each pathway.
4. Describe the pathological changes and treatment associated with Parkinson's disease.
5. Contrast pyramidal and extrapyramidal pathways according to their specific functions and also locations of the upper and lower motor neurons in each pathway.
6. Discuss whether a CVA affecting a 0.5 cm^2 area of the cerebral cortex or a 0.5 cm^2 area of the internal capsule would be likely to cause greater loss of motor function.

MASTERY TEST: CHAPTER 15

Questions 1–5: Circle the letter preceding the one best answer to each question.

1. All of these sensations are conveyed by the posterior column–medial lemniscus pathway *except:*
   A. Pain and temperature
   B. Proprioception
   C. Fine touch, two-point discrimination
   D. Vibration
   E. Stereognosis
2. Choose the *false* statement about REM sleep.
   A. Infant sleep consists of a higher percentage of REM sleep than does adult sleep.
   B. Dreaming occurs during this type of sleep.
   C. The eyes move rapidly behind closed lids during REM sleep.
   D. EEG readings are similar to those of stage 4 of NREM sleep.
   E. REM sleep occurs periodically throughout a typical 8-hour sleep period.
3. The hallmark of cerebellar disease is known as:
   A. Ataxia
   B. Parkinson's disease
   C. Huntington's disease
   D. Paresthesia (tingling sensations)
4. Choose the *false* statement about pain receptors.
   A. They may be stimulated by any type of stimulus, such as heat, cold, or pressure.
   B. They have a simple structure with no capsule.
   C. They are characterized by a high level of adaptation.
   D. They are found in almost every tissue of the body.
   E. They are important in helping to maintain homeostasis.
5. "Change in sensitivity to a long-lasting stimulus" is a description of the characteristic of sensation called:
   A. Adaptation
   B. Modality
   C. Selectivity
   D. Proprioception

Questions 6–9: Arrange the answers in correct sequence.

_____ _____ _____ 6. Density of sense receptors, from most dense to least:
   A. In tip of tongue
   B. In tip of finger
   C. In back of neck

_____ _____ _____ _____ 7. In order of occurrence in a sensation:
   A. Transduction of a stimulus to a generator potential
   B. Integration of the impulse into a sensation
   C. Impulse generation and conduction along a nervous pathway
   D. A stimulus

_____ _____ _____ _____ 8. Levels of sensation, from those causing simplest, least precise reflexes to those causing most complex and precise responses:
   A. Thalamus
   B. Brain stem
   C. Cerebral cortex
   D. Spinal cord

_____ _____ _____ _____ _____ 9. Pathway for conduction of most of the impulses for voluntary movement of muscles:
   A. Anterior gray horn of the spinal cord
   B. Precentral gyrus
   C. Internal capsule
   D. Location where decussation occurs
   E. Lateral corticospinal tract

Questions 10–20: Circle T (true) or F (false). If the statement is false, change the underlined word or phrase so that the statement is correct.

T F 10. In general, the <u>left</u> side of the brain controls the right side of the body.

T F 11. The final common pathway consists of <u>upper motor</u> neurons.

T F 12. Damage to the final common pathway is likely to result in <u>flaccid paralysis.</u>

T F 13. The fasciculus cuneatus and the nucleus cuneatus form parts of the pathway for <u>proprioception and most tactile sensations from trunk and legs.</u>

T F 14. The fasciculus cuneatus and the nucleus cuneatus within posterior columns contain <u>second-order neurons</u> in a <u>sensory</u> pathway.

T F 15. The medial lemniscus contains <u>second-order neurons in a sensory</u> pathway.

T F 16. The neuron that crosses to the opposite side in sensory pathways is usually the <u>second-order</u> neuron.

T F 17. Damage to the left lateral spinothalamic tract would be most likely to result in loss of awareness of <u>pain and vibration sensations in the left</u> side of the body.

T F 18. Conscious sensations, such as those of sight and touch, can occur only in the <u>cerebral cortex.</u>

T F 19. Pain experienced by an amputee as if the amputated limb were still there is an example of <u>phantom pain.</u>

T F 20. Circadian rhythm pertains to the <u>complex feedback circuits involved in producing coordinated movements.</u>

Questions 21–25: Fill-ins. Complete each sentence with the word or phrase that best fits.

_____ 21. _____ refers to the awareness of muscles, tendons, joints, balance, and equilibrium.

_____ 22. Inactivation of the _____ results in the state of sleep.

_____ 23. Rubrospinal, tectospinal, and vestibulospinal tracts are all classified as _____ tracts.

_____ 24. A CVA affecting the medial portion of the postcentral gyrus of the right hemisphere (area Z in Figure LG 15.1, page 300) is most likely to result in symptoms such as _____.

_____ 25. Voluntary motor impulses are conveyed from motor cortex to neurons in the spinal cord by_____ pathways.

ANSWERS TO MASTERY TEST: ★ CHAPTER 15

Multiple Choice

1. A
2. D
3. A
4. C
5. A

Arrange

6. A B C
7. D A C B
8. D B A C
9. B C D E A

True-False

10. T
11. F. Lower motor
12. T
13. F. Proprioception and most tactile sensations from neck, arms, and upper chest.
14. F. First-order neurons in a sensory.

15. T
16. T
17. F. Pain and temperature sensations in the right
18. T
19. T
20. F. Events that occur at approximately 24-hour intervals, such as sleeping and waking

Fill-ins

21. Proprioception
22. Reticular activating system (RAS)
23. Extrapyramidal
24. Loss of sensation of left foot
25. Direct (pyramidal or corticospinal) pathways

FRAMEWORK 16
The Special Senses

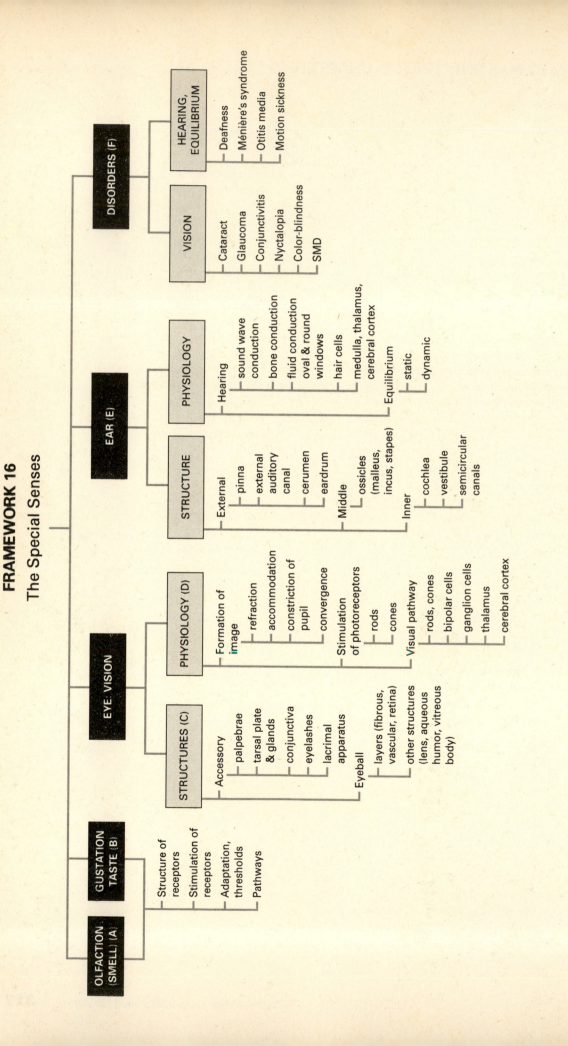

The Special Senses

CHAPTER
16

All sensory organs contain receptors for increasing sensitivity to the environment. In Chapter 15 you considered the relatively simple receptors and pathways for the general senses of touch, pressure, temperature, pain, and vibration. The special senses include smell, taste, vision, hearing, and equilibrium. Special afferent pathways in many ways resemble general afferent pathways. However, a major point of differentiation is the arrangement of special sense receptors in complex sensory organs, specifically the nose, tongue, eyes, and ears.

As you begin your study of the special senses, carefully examine the Chapter 16 Topic Outline and Objectives; check off each one as you complete it. To organize your study of special senses, glance over the Chapter 16 Framework now. Be sure to refer to the Framework frequently and note relationships among key terms in each section.

TOPIC OUTLINE AND OBJECTIVES

A. Olfactory sensations

☐ 1. Locate the receptors for olfaction and describe the neural pathway for smell.

B. Gustatory sensations

☐ 2. Identify the gustatory receptors and describe the neural pathway for taste.

C. Visual sensations: anatomy

☐ 3. List and describe the accessory structures of the eye and the structural divisions of the eyeball.
☐ 4. Discuss image formation by describing refraction, accommodation, and constriction of the pupil.

D. Visual sensations: physiology

☐ 5. Describe how photoreceptors and photo-pigments function in vision.
☐ 6. Describe the retinal processing of visual input and the neural pathway of light impulses to the brain.

E. Auditory sensations and equilibrium

☐ 7. Describe the anatomical subdivisions of the ear.
☐ 8. List the principal events in the physiology of hearing.
☐ 9. Identify the receptor organs for equilibrium and how they function.

F. Disorders, medical terminology

☐ 10. Contrast the causes and symptoms of glaucoma, senile macular degeneration, deafness, Ménière's syndrome, otitis media, and motion sickness.
☐ 11. Define medical terminology associated with the sense organs.

WORDBYTES

Now become familiar with the language of this chapter by studying each wordbyte, its meaning, and an example of its use within a term. After you study the entire list, self-check your understanding by writing the meaning of each wordbyte on the line. As you continue through the *Learning Guide,* identify (and fill in) additional terms that contain the same wordbyte.

Wordbyte	Self-check	Meaning	Example(s)
aqua-	_____	water	*aque*ous humor
blephar-	_____	eyelid	*blephar*oplasty
kerat-	_____	cornea	radial *kerat*otomy
macula	_____	spot	*macula* lutea
med-	_____	middle	otitis *med*ia
ocul-	_____	eye	*ocul*omotor nerve
olfact-	_____	smell	*olfact*ory
ophthalm-	_____	eye	*ophthalm*ologist
opt-	_____	eye	*opt*ic nerve, *opt*ician
orbit-	_____	eye socket	supra*orbit*al
ossi-	_____	bone	*ossi*cle
ot(o)	_____	ear	*oto*lith
presby-	_____	old	*presby*opia
sclera	_____	hard	*sclera*, oto*scler*osis
tympano-	_____	drum	*tympan*ic membrane
vitr-	_____	glassy	*vitr*eous humor

CHECKPOINTS

A. Olfactory sensations (pages 454–455)

■ **A1.** Describe olfactory receptors in this exercise.

a. Receptors for smell are located in the *(superior? inferior?)* portion of the nasal cavity. Receptors consist of

_____-polar neurons which are located between _____ cells. The distal end of each

olfactory receptor cell consists of dendrites that have cilia known as olfactory _____.

b. What is unique about the function of basal cells?

c. What is the function of olfactory (Bowman's) glands in the nose?

d. Because smell is a _____ sense, receptors in olfactory hair membranes respond to

different chemical molecules, leading to a _____ and _____.

e. Adaptation to smell occurs *(slowly? rapidly?)* at first and then happens at a much slower rate.

■ **A2.** Complete this exercise about the nerves associated with the nose.

a. Upon stimulation of olfactory hairs, impulses pass to cell bodies and axons; the axons of these olfactory cells form cranial nerves *(I? II? III?)*, the olfactory nerves. These pass from the nasal cavity to the cranium

to terminate in the olfactory _____ located just inferior to the _____

lobes of the cerebrum.

b. Cell bodies in the olfactory bulb then send axons through the olfactory *(nerves? tracts?)* along two pathways. One route extends to the limbic system. What is the effect of projection of sensations of smell to this area?

c. The other olfactory pathway passes to the thalamus and then to the frontal lobes of the

_____ cortex.

B. Gustatory sensations   (pages 455–457)

■ **B1.** Describe receptors for taste in this exercise.

a. Receptors for taste, or _____ sensation, are located in taste buds. These receptors

are protected by a _____ formed of supporting cells, as well as the

_____ cells that produce the supporting cells. The latter cells ultimately become receptor cells, which have a life span of *(10 days? 10 years?)*.

b. Within each taste bud are about *(5? 50?)* gustatory receptor cells, each with a gustatory hair

(or _____) that projects from the taste bud pore. Stimulation of this hair appears to

initiate a _____ potential, which results in release of _____
from receptor cells. This chemical then activates sensory neurons in cranial nerves (see Checkpoint B2d).

c. Taste buds are located on elevated projections of the tongue called _____.
The largest of these are located in a V-formation at the back of the tongue. These are

_____ papillae. _____ papillae are mushroom shaped and are

located on the sides and tip of the tongue. Pointed _____ papillae cover the

_____ of the tongue, and these *(do? do not?)* contain many taste buds.

■ **B2.** Do the following exercise about taste.

a. There are only four primary taste sensations: _____, _____,

_____, and _____. The taste buds at the tip of the tongue

are most sensitive to _____ and _____ tastes, whereas the

posterior portion is most sensitive to _____ taste.

b. The four primary tastes are modified by _____ sensations to produce the wide variety of
"tastes" experienced. If you have a cold (with a "stuffy nose"), you may not be able to discriminate the

usual variety of tastes, largely because of loss of the sense of _____.

c. The threshold for smell is quite *(high? low?)*, meaning that *(a large? only a small?)* amount of odor must
be present in order for smell to occur. The threshold varies for the four primary tastes; the threshold for

(sour? sweet and salt? bitter?) is lowest, while those for _____ are highest.

d. Taste impulses are conveyed to the brainstem by cranial nerves _____, _____, and _____;

there, nerve messages travel to lower parts of the brain, including the _____,

_____, and _____, and finally stimulate the cerebral cortex.

C. Visual sensations: anatomy (pages 457–467)

C1. Examine your own eye structure in a mirror. With the help of Figure 16-3 (page 457 in your text),
identify each of the accessory structures of the eye listed below.

Conjunctiva	Palpebrae
Eyebrow	Tarsal glands
Eyelashes	Tarsal plate
Lacrimal glands: identify site	

■ **C2.** *For extra review.* Match the structures listed above with the following descriptions. Write the name of the structure on the line provided.

   a. Eyelid: _____

   b. Thick fold of connective tissue that forms much of the eyelid: _____

   c. Covers the orbicularis oculis muscle; provides protection: _____

   d. Short hairs; infection of glands at the base of these hairs is *sty:* _____

   e. Located in superolateral region of orbit; secrete tears: _____

   f. Secrete oil; infection is *chalazion:* _____
   g. Mucous membrane lining the eyelids and covering anterior surface of the eye:

   _____

■ **C3.** Do this exercise on accessory eye structures.

   a. Lacrimal glands produce lacrimal fluid, also known as _____,

   at the rate of about _____ ml per day. Identify the location of these glands on yourself.

   Arrange in correct order these structures in the pathway of tears: _____ _____ _____ _____
   A. Lacrimal canals and nasolacrimal duct
   B. Lacrimal glands
   C. Excretory lacrimal ducts
   D. Lacrimal punctae

   State two or more functions of tears.

   b. Each eye is moved by *(3? 4? 6?)* extrinsic eye muscles; these are innervated by cranial

   nerves _____, _____, and _____. *(For extra review,* see Chapter 11, Checkpoint C4, page LG 206.)

   c. The eye sits in the bony socket known as the _____ with only the anterior *(one-sixth? one-half?)* of the eyeball exposed.

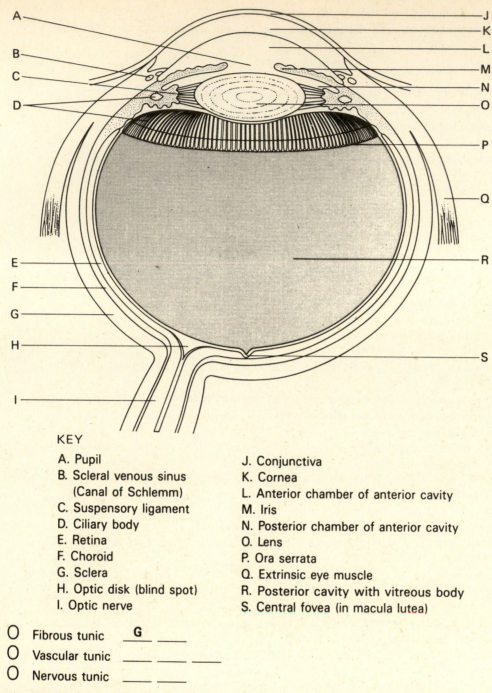

KEY

A. Pupil	J. Conjunctiva
B. Scleral venous sinus (Canal of Schlemm)	K. Cornea
	L. Anterior chamber of anterior cavity
C. Suspensory ligament	M. Iris
D. Ciliary body	N. Posterior chamber of anterior cavity
E. Retina	O. Lens
F. Choroid	P. Ora serrata
G. Sclera	Q. Extrinsic eye muscle
H. Optic disk (blind spot)	R. Posterior cavity with vitreous body
I. Optic nerve	S. Central fovea (in macula lutea)

O Fibrous tunic   __G__ ____

O Vascular tunic   ____ ____ ____

O Nervous tunic   ____ ____

Figure LG 16.1 Structure of the eyeball in horizontal section. Color as directed in Checkpoint C4.

■ **C4.** On Figure LG 16.1, color the three layers (or tunics) of the eye using color code ovals. Next to the ovals write the letters of the labeled structures that form each layer. One is done for you. Then label all structures in the eye and check your answers against the key.

■ **C5.** *A clinical challenge.* A cornea that is surgically transplanted *(is? is not?)* likely to be rejected. State your rationale.

■ **C6.** Complete this exercise about blood vessels of the eye.

    a. Retinal blood vessels *(can? cannot?)* be viewed through an ophthalmoscope. Of what advantage is this?

    b. The central retinal artery enters the eye at the *(optic disc? central fovea?)*.

■ **C7.** Contrast the two types of photoreceptor cells by writing R for rods or C for cones before the related descriptions.

    _____ a. About 6 million in each eye; most concentrated in the central fovea of the macula lutea.

    _____ b. Over 120 million in each eye; located mainly in peripheral regions of the eye.

    _____ c. Sense color and acute (sharp) vision.

    _____ d. Used for night vision.

■ **C8.** Do this exercise that describes the retina.

    a. After light strikes the back of the retina, nerve impulses pass anteriorly through three zones of neurons.

    First is the region of rods and cones, or _____ zone;

    next is the _____ layer; most anterior is the _____ layer.

    b. Axons of the ganglion layer converge to form the optic _____.

    This nerve exits from the eye at the point known as the _____ disc,

    or _____. The name indicates that no image formation can occur here because no rods or cones are present.

    c. What function(s) do *amacrine cells* and *horizontal cells* serve?

    d. The point of sharpest vision is known as the area of highest visual _____. This site is the *(central fovea? optic disc?)*. It contains only the photoreceptor cells named *(rods? cones?)* and *(no? many?)* bipolar and ganglion cells.

■ **C9.** Describe the lens and the cavities of the eye in this exercise.

    a. The lens is composed of *(lipid? protein?)* arranged in layers, much like an onion. Normally the lens is *(clear? cloudy?)*. A loss of transparency of the lens occurs in

    the condition called _____.

    b. The lens divides the eye into anterior and posterior *(chambers? cavities?)*. The posterior

    cavity is filled with _____ body, which has a *(watery? jellylike?)* consistency.

    This body helps to hold the _____ in place. Vitreous body is formed during embryonic life and *(is? is not?)* replaced in later life.

c. The anterior cavity is subdivided by the _____ into two chambers. Which is larger? *(Anterior chamber? Posterior chamber?)* A watery fluid called

_____ humor is present in the anterior cavity.

This fluid is formed by the _____ processes, and it is normally replaced about every 90 *(minutes? days?)*.

d. How is the formation and final destination of aqueous fluid similar to that of cerebrospinal fluid (CSF)?

e. State two functions of aqueous humor.

f. Aqueous fluid creates pressure within the eye; in glaucoma this pressure *(increases? decreases?)*. What effects may occur?

■ **C10.** *For extra review.* Check your understanding by matching eye structures in the box with descriptions below. Use each answer once.

CF. Central fovea   OD. Optic disc
Cho. Choroid   P. Pupil
CM. Ciliary muscle   R. Retina
Cor. Cornea   S. Sclera
I. Iris   SVS. Scleral venous sinus
L. Lens   (canal of Schlemm)

_____ a. "White of the eye"

_____ b. Normally clear; if cloudy, it is a cataract

_____ c. Blind spot; area in which there are no cones or rods

_____ d. Area of sharpest vision; area of densest concentration of cones

_____ e. Anteriorly located, avascular coat responsible for about 75% of all focusing done by the eye

_____ f. Consists of a pigment epithelium and a neural portion; if these two portions separate (detach), blindness may result

_____ g. Brown-black layers prevent reflection of light rays; also nourishes eyeball because it is vascular

_____ h. A hole; appears black, like a circular doorway leading into a dark room

_____ i. Regulates the amount of light entering the eye; colored part of the eye

_____ j. Attaches to the lens by means of radially arranged fibers called the suspensory ligaments

_____ k. Located at the junction of iris and cornea; drains aqueous humor

C11. Define each of these processes involved in *image formation* on the retina.

a. *Refraction* of light

b. *Accommodation* of the lens

c. *Constriction* of the pupil

d. *Convergence* of the eyes

■ **C12.** Answer these questions about formation of an image on the retina.

a. Bending of light rays so that images focus exactly upon the retina is a process known as

_____. The *(cornea? lens?)* accomplishes most (75%) of the refraction within the eye. Explain why you think this is so.

b. Draw how a letter "e" would look as it is focused on the retina. _____

c. Explain why you see things right side up even though refraction produces inverted images on your retina.

■ **C13.** Explain how *accommodation* enables your eyes to focus clearly.

a. A *(concave? convex?)* lens will bend light rays so that they converge and finally intersect. The lens in each of your eyes is *(biconcave? biconvex?)*.

b. When you bring your finger toward your eye, light rays from your finger must converge *(more? less?)* so that they will focus on your retina. Thus for near vision, the anterior surface of your lens must become *(more? less?)* convex. This occurs by a thickening and bulging forward of the lens caused by *(contraction? relaxation?)* of

the _____ muscle. In fact, after long periods of close work (such as reading), you may experience eye strain.

c. What is the *near point of vision* for a young adult? _____ cm (_____ inches)

For a 60-year-old? _____ cm (_____ inches) What factor explains this difference?

■ **C14.** *A clinical challenge.* Describe errors in refraction and accommodation in this exercise.

a. The normal (or _____ eye) can refract rays from a

distance of _____ meters (_____ feet) and form a clear image on the retina.

b. In the nearsighted (or _____) eye, images focus *(in front of? behind?)* the retina. This may be due to an eyeball that is too *(long? short?)*. Corrective lenses should be slightly *(concave? convex?)* so that they refract rays less and allow them to focus farther back on the retina.

c. The farsighted person can see *(near? far?)* well but has difficulty seeing *(near? far?)*

without the aid of corrective lenses. The farsighted (or _____) person requires *(concave? convex?)* lenses. After about age 40, most persons lose the

ability to see near objects clearly. This condition is known as _____.

■ **C15.** Choose the correct answers to complete each statement.

a. Both accommodation of the lens and constriction of the pupil involve *(extrinsic? intrinsic?)* muscles, whereas convergence of the eyes involves *(extrinsic? intrinsic?)* muscles.

b. The pupil *(dilates? constricts?)* in the presence of bright light; the pupil *(dilates? constricts?)* in stressful situations when sympathetic nerves stimulate the iris.

D. Visual sensations: physiology   (pages 467–471)

■ **D1.** List three principal processes that are necessary for vision to occur.

a.

b.

c.

■ **D2.** Do this exercise on photoreceptors and photopigments.

a. The basis for the names *rods* and *cones* rests in the shape of the *(inner? outer?)* segments of these cells, that is, the segment closest to the *(choroid? vitreous body?)*. In which segment are photopigments such as rhodopsin located? *(Inner? Outer?)* In which segment can the nucleus, Golgi complex, and mitochondria be found? *(Inner? Outer?)*

b. How many types of photopigment are found in the human retina? _____ Rhodopsin is the photopigment found in *(rods? cones?)*. How many types of photopigments are found in cones?

c. Photopigments are also known as _____ pigments and consist of colored *(proteins? lipids?)* in outer segment plasma membranes. In rods, these chemicals pinch off to form stacks of *(bubbles? discs?)*. A new disc is formed every 20–60 *(minutes? days? months?)*; old discs are destroyed by the process of

_____ .

d. Any photopigment consists of two parts: *retinal* and *opsin*. What is the function of retinal?

e. Retinal is a vitamin _____ derivative and is formed from carotenoids. List several carotenoid-rich foods.

Tell how such foods may help to prevent *nyctalopia* (night blindness)?

■ **D3.** Refer to Figure 16.12, page 468 in your text, and complete this Checkpoint about response of photopigments to light and darkness.

a. *(Retinal? Opsin?)* appears as a colored, dumbell-shaped protein embedded in a disc in a rod or cone. Retinal fits snugly into opsin when retinal is in a *(bent? straightened?)* shape. This form of retinal is known as the *(cis? trans?)* form, and it is found when eyes are in *(darkness? light?)*.

b. Conversion of *cis-* to *trans*-retinal is known as _____ . This occurs when retinal *(absorbs light? sits in darkness?)*. Within a minute of exposure to light, the *(cis? trans?)*-retinal slips out of place in opsin. The separate products (opsin and *trans*-retinal) are *(colored? colorless or bleached?)*. This is the first step in the process

of conversion of a _____ (light) into a generator potential; this process is known as *(transduction? integration?)*. As a result, bipolar and then ganglion cells will be activated to begin nerve transmission along the visual pathway (see Checkpoints D6–D9).

c. To make further visual transductions possible, *trans*-retinal must be converted

back to *cis*-retinal, a process known as _____ . This process

requires *(light? darkness?)* and action of an enzyme known as retinal _____ . Which photopigment regenerates more quickly? *(Rhodopsin? Cone photopigments?)*

■ **D4.** Explain what happens in your retinas after you walk from a bright lobby into a darkened movie theater.

    a. In the lobby the rhodopsin in your rods had been bleached (broken into _____-
       retinal and opsin) as the initial step in vision involving rods.

    b. Only a period of darkness will permit adequate regeneration of retinal in the *cis* form

       that can bind to opsin and form _____. After complete bleaching of
       rhodopsin, it takes *(5? 1.5?)* minutes to regenerate 50% of rhodopsin and about

       _____ minutes for full regeneration of this photopigment. This process is
       called *(light? dark?)* adaptation.

    c. Also necessary for rhodopsin formation is adequate vitamin A. Explain why a
       detached retina can result in reduced rhodopsin formation and nyctalopia.

■ **D5.** Explain how cones help you to see in "living color" by completing this paragraph.

    Cones contain pigments which *(do? do not?)* require bright light for breakdown (and
    therefore for a generator potential). Three different pigments are present, sensitive to

    three colors: _____, _____, and

    _____. A person who is red-green color-blind lacks some of the

    cones receptive to two colors (_____ and _____)
    and so cannot distinguish between these colors. This condition occurs more often in
    (males? females?).

■ **D6.** In Checkpoint D3 you reviewed the transduction in photoreceptor cells in response to
a stimulus (dim light). Refer to Figure LG 16.2 and do this activity describing what
happens next in the pathway to vision.

    a. First color arrows indicating pathways of light and of nerve impulses (visual data
       processing) and all cells on the figure according to the color code ovals.

    b. At point ∗ in the figure, a *(rod? cone?)* synapses with a *(bipolar? ganglion?)* cell.
       In darkness, an inhibitory neurotransmitter named *(cyclic GMP? glutamate?)* is
       continually released at this synapse. The effect is to *(de? hyper?)*-polarize bipolar
       cells so that they *(become activated? stay inactive?)* and no visual information is
       transmitted to the brain.

    c. What causes glutamate to be continually released at this synapse in darkness? Inflow

       of *(Ca^{2+}? Na$^+$? K$^+$?)*, known as the "_____ current," causes release
       of glutamate from synaptic terminals of rods. But what keeps these Na$^+$ channels open
       during darkness is a chemical called cyclic *(AMP? GMP?)*, which is formed in the

       presence of an enzyme known as _____ cyclase. This enzyme must, in turn,

       be activated by an enzyme called guanylate cyclase _____ _____ .
       Review these steps by referring to the left side of Figure 16.13, page 469 of your text.

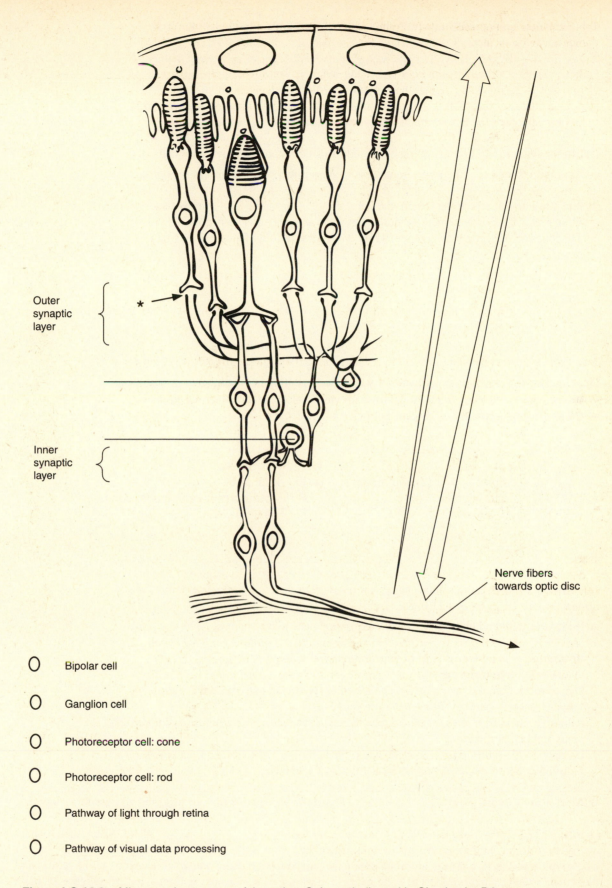

Outer
synaptic
layer

*

Inner
synaptic
layer

Nerve fibers
towards optic disc

O Bipolar cell

O Ganglion cell

O Photoreceptor cell: cone

O Photoreceptor cell: rod

O Pathway of light through retina

O Pathway of visual data processing

Figure LG 16.2 Microscopic structure of the retina. Color as indicated in Checkpoint D6.

d. Now review events (right side of Figure 16.13) that follow a stimulus of dim light exposure to rods. Effects will be roughly *(the same as? opposite to?)* those just described in (b) and (c). An enzyme named *transducin* activates an enzyme named

_____, which breaks down GMP. Na⁺ channels are therefore *(opened? closed?)*, and the inhibitory glutamate is *(still? no longer?)* released so that bipolar cells (and the rest of the visual pathway) can become activated.

■ **D7.** Refer again to Figure LG 16.2 and complete this Checkpoint about retinal cells.

a. The figure shows that the retina contains a larger number of *(photoreceptor? bipolar*

and ganglion?) cells. In fact, there are between _____ and _____ rods for every one bipolar cell. This arrangement demonstrates *(convergence? divergence?)*.

b. Do cones also exhibit convergence? *(Yes? No?)* State the significance of this fact.

c. What functions do *horizontal cells* and *amacrine cells* perform?

■ **D8.** Describe the conduction pathway for vision by arranging these structures in sequence. Write the letters in correct order on the lines provided. Also, indicate the four points where synapsing occurs by placing an asterisk (*) between the letters.

B. Bipolar cells	ON. Optic nerve
C. Cerebral cortex (visual areas)	OT. Optic tract
G. Ganglion cells	P. Photoreceptor cells (rods, cones)
OC. Optic chiasma	T. Thalamus

_____ _____ _____ _____ _____ _____ _____ _____

■ **D9.** Answer these questions about the visual pathway to the brain. It may be helpful to refer to Figure 16.15 (page 471 in the text).

a. Hold your left hand up high and to the left so you can still see it. Your hand is in the

(temporal? nasal?) visual field of your left eye and in the _____ visual field of your right eye.

b. Due to refraction, the image of your hand will be projected onto the *(left? right?)* *(upper? lower?)* portion of the retinas of your eyes.

c. All nerve fibers from these areas of your retinas reach the *(left? right?)* side of your thalamus and cerebral cortex.

d. Damage to the right optic tract, right side of thalamus, or right visual cortex would result in loss of sight of the *(left? right?)* visual fields of each eye.

■ **D10.** *The Big Picture: Looking Back.* Refer to Figure 14.5, page 398 of your text, and address this *clinical*

challenge. The optic chiasm is located just anterior to the _____ gland. Visual disturbances are likely to signal a tumor in this gland. Pressure of the tumor is likely to be exerted against the more medial fibers within the chiasm. These are fibers that have passed from *(nasal? temporal?)* regions of the retinas and that enter "crossed fiber" regions of optic tracts to reach opposite sides of the brain. (Refer to Figure 16.14, page 470 in your text.) As a result of loss of function of nasal retinas, the patient would experience blindness in *(nasal? temporal?)* visual fields, a condition known as "tunnel vision."

E. Auditory sensations and equilibrium (pages 472–483)

■ **E1.** Refer to Figure LG 16.3 and do the following exercise.

   a. Color the three parts of the ear using color code ovals. Then write on lines next to the ovals the letters of structures that are located in each part of the ear. One is done for you.
   b. Label each lettered structure using leader lines on the figure.

■ **E2.** *For extra review.* Select the ear structures in the box that fit the descriptions below. Not all answers will be used.

A. Auricle (pinna)	OW. Oval window
AT. Auditory (Eustachian) tube	RW. Round window
I. Incus	S. Stapes
M. Malleus	TM. Tympanic membrane

_____ a. Tube used to equalize pressure on either side of tympanic membrane

_____ b. Eardrum

_____ c. Structure on which stapes exerts pistonlike action

_____ d. Ossicle adjacent to eardrum

_____ e. Anvil-shaped ear bone

_____ f. Portion of the external ear shaped like the flared end of a trumpet

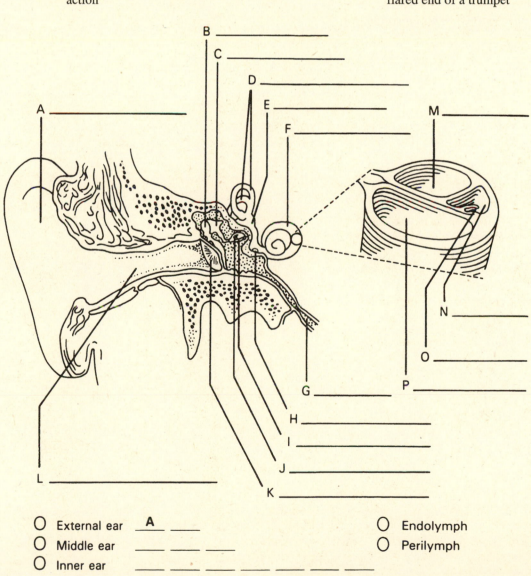

○ External ear ___A___ ____

○ Middle ear ____ ____ ____

○ Inner ear ____ ____ ____ ____ ____ ____ ____

○ Endolymph

○ Perilymph

Figure LG 16.3 Diagram of the ear in frontal section. Color and label as directed in Checkpoints E1 and E5.

E3. Contrast locations, anatomy, and physiology of the following structures:

a. *External auditory canal—auditory (Eustachian) tube*

b. *Tensor tympani muscle—stapedius muscle*

c. *Oval window—round window*

d. *Bony labyrinth—membranous labyrinth*

e. *Utricle—saccule*

f. *Modiolus—helicotrema*

g. *Vestibular membrane—tectorial membrane*

h. *Ossicles—otoliths*

E4. Describe the structure of the spiral organ (organ of Corti).

Explain how hair cells of the spiral organ may be damaged by loud noises.

■ **E5.** Color *endolymph* and *perilymph* and related color code ovals in Figure LG 16.3.

■ **E6.** Fill in the blanks and circle correct answers about sound waves.

a. Sound waves heard by humans range from frequencies of 20 to 20,000 cycles/sec (Hz).

Humans can best hear sounds in the range of _____ Hz.

b. A musical high note has a *(higher? lower?)* frequency than a low note. So frequency is *(directly? indirectly?)* related to pitch.

c. Sound intensity (loudness) is measured in units called _____.

Normal conversation is at a level of about _____ dB, whereas sounds at _____ dB can cause pain.

■ **E7.** Summarize events in the process of hearing in this activity. It may help to refer to Figure 16.20 (page 479 in the text).

a. Sound waves travel through the _____ and strike the

_____ membrane. Sound waves are magnified by the action of

the three _____ in the middle ear.

b. The ear bone named *(malleus? incus? stapes?)* strikes the *(round? oval?)* window, setting up waves in *(endo-? peri-?)* lymph. This pushes on the floor of the upper *scala (vestibuli? tympani?)*. As a result the cochlear duct is moved, and so is the perilymph in the lower canal, the *scala (vestibuli? tympani?)*. The pressure of the perilymph is finally expended by bulging out the *(round? oval?)* window.

c. As the cochlear duct moves, tiny hair cells embedded in the floor of the duct are

stimulated. These hair cells are part of the _____ organ; its name is

based on its spiral arrangement on the _____ membrane all the way around the 2¾ coils of the cochlear duct. *(High? Low?)*-pitched sounds are sensed best toward the apex (helicotrema region) of the cochlea.

d. As spiral organ hair cells are moved by waves in endolymph, hairs move against the gelatinous

_____ membrane. This movement generates receptor potentials

in hair cells that excite nearby sensory neurons of the _____ branch

of cranial nerve _____. The pathway continues to the brainstem, _____

(relay center), and finally to the _____ lobe of the cerebral cortex.

■ **E8.** Check your understanding of the pathway of fluid conduction by placing the following structures in correct sequence. Write the letters on the lines provided.

B. Basilar membrane	ST. Scala tympani (perilymph)
C. Cochlear duct (endolymph)	SV. Scala vestibuli (perilymph)
O. Oval window	VM. Vestibular membrane
RW. Round window	

_____   _____   _____   _____   _____   _____   _____

E9. Describe the roles of K^+ and Ca^{2+} in stimulating hair cells and generating nerve impulses related to hearing.

■ **E10.** Contrast receptors for hearing and for equilibrium in this summary of the inner ear.
a. Receptors for hearing and equilibrium are all located in the *(middle? inner?)* ear. All

consist of supporting cells and _____ cells that are covered

by a _____ membrane.

b. In the spiral organ, which senses _____, the gelatinous membrane is

called the _____ membrane. Hair cells move against this membrane as a result of *(sound waves? change in body position?)*.

c. In the macula, located in the *(semicircular canals? vestibule?)*, the gelatinous membrane

is embedded with calcium carbonate crystals called _____. These respond to gravity in such a way that the macula is the main receptor for *(static? dynamic?)* equilibrium. An example of such equilibrium occurs as you are aware of your *(position while lying down? change in position on a careening roller coaster?)*.

d. In the semicircular canals the gelatinous membrane is called the _____. Its shape is *(flat? like an inverted cup?)*. The cupula is part of the *(crista? saccule?)*

located in the ampulla. Change in direction (as in a roller coaster) causes _____ to bend hairs in the cupula. Cristae in semicircular canals are therefore receptors primarily for *(static? dynamic?)* equilibrium.

E11. Describe vestibular pathways, including the role of the cerebellum in maintaining equilibrium.

F. Disorders, medical terminology   (pages 483–484)

■ **F1.** Match the name of the disorder with the description.

B. Blepharitis	Myo. Myopia
Cat. Cataract	Myr. Myringitis
Con. Conjunctivitis	Pr. Presbyopia
G. Glaucoma	Pt. Ptosis
H. Hyperacusia	Str. Strabismus
K. Keratitis	T. Tinnitus
Men. Ménière's syndrome	V. Vertigo

_____ a. Condition requiring corrective lenses to focus distant objects

_____ b. Excessive intraocular pressure resulting in blindness; second most common cause of blindness

_____ c. Ringing in the ears

_____ d. Pinkeye

_____ e. Abnormally sensitive hearing

_____ f. Inflammation of the eardrum

_____ g. Inflammation of the eyelid

_____ h. Disturbance of the inner ear with excessive endolymph

_____ i. Loss of transparency of the lens

_____ j. Farsightedness due to loss of elasticity of lens, especially after age 40

_____ k. Inflammation of the cornea

_____ l. Sense of spinning or whirling

_____ m. Drooping of an eyelid (or other organ, such as kidney)

F2. Discuss causes and treatments of:

a. Otitis media

b. Motion sickness

F3. Contrast sensorineural deafness with conduction deafness.

ANSWERS TO SELECTED CHECKPOINTS: CHAPTER 16

A1. Superior; bi, supporting; hairs. (b) They continually produce new olfactory receptors that live for only about a month. Because receptor cells are neurons, this "breaks the rule" that mature neurons are never replaced. (c) Produce mucus that acts as a solvent for odoriferous substances. (d) Chemical, generator potential, nerve impulse. (e) Rapidly.

A2. (a) I; bulb, frontal. (b) Tracts; awareness of certain smells, such as putrid odors or fragrance of roses, may lead to emotional responses. (c) Cerebral.

B1. (a) Gustatory; capsule, basal; 10 days. (b) 50, microvillus; receptor, neurotransmitter. (c) Papillae; circumvallate; fungiform; filiform, anterior, do not.

B2. (a) Sweet, sour, salt, bitter; sweet and salty, bitter. (b) Olfactory; smell. (c) Low, only a small; bitter; sweet and salt. (d) VII, IX, X, thalamus, hypothalamus, limbic system.

C2. (a) Palpebra. (b) Tarsal plate. (c) Eyebrow. (d) Eyelashes. (e) Lacrimal glands. (f) Tarsal glands. (g) Conjunctiva.

C3. (a) Tears, 1; superiolateral to each eyeball: see Figure 16.4b, page 458 of your text; B C D A; tears protect against infection via the bactericidal enzyme lysozyme and by flushing away irritating substances; tears are signs of emotions. (b) 6; III, IV, VI. (c) Orbit, one-sixth.

C4. Fibrous tunic: G, K; vascular tunic: D, F, M; nervous tunic: E, S.

C5. Is not. Because the cornea is avascular, antibodies that cause rejection do not circulate there.

C6. (a) Can; health of blood vessels can be readily assessed, for example, in persons with diabetes. (b) Optic disc.

C7. (a) C. (b) R. (c) C. (d) R.

C8. (a) Photoreceptor, bipolar, ganglion. (b) Nerve; optic, blind spot. (c) They modify signals transmitted along optic pathways. (d) Acuity; central fovea; cones, no.

C9. (a) Protein; clear; cataract. (b) Cavities; vitreous, jellylike; retina; is not. (c) Iris, anterior chamber; aqueous; ciliary, minutes. (d) Both are formed from blood vessels (choroid plexuses) and the fluid finally returns to venous blood. (e) Provides nutrients and oxygen and removes wastes from the cornea and lens; also provides pressure to separate the cornea from lens. (f) Increases; damage to the retina and optic nerve with possible blindness.

C10. (a) S. (b) L. (c) OD. (d) CF. (e) Cor. (f) R. (g) Cho. (h) P. (i) I. (j) CM. (k) SVS.

C12. (a) Refraction; cornea; the density of the cornea differs considerably from the air anterior to it. (b) "ɘ" ("e" inverted 180° and much smaller). (c) Your brain learned early in your life how to "turn" images so that what you see corresponds with locations of objects you touch.

C13. (a) Convex; biconvex. (b) More; more; contraction, ciliary. (c) 10, 4, 80, 31. The lens loses elasticity (presbyopia) with aging.

C14. (a) Emmetropic, 6, 20. (b) Myopic, in front of; long; concave. (c) Far, near; hypermetropic, convex; presbyopia.

C15. (a) Intrinsic, extrinsic. (b) Constricts; dilates.

D1. (a) Formation of an image on the retina. (b) Stimulation of photoreceptors so that a light stimulus is converted into an electrical stimulus (receptor potential and nerve impulse). (c) Transmission of the impulse along neural pathways to the thalamus and visual cortex.

D2. (a) Outer, choroid; outer; inner. (b) 4; rods; one in each of three types of cones that are sensitive to different colors. (c) Visual, proteins; discs; minutes (one to three per hour), phagocytosis. (d) Retinal is the light-absorbing portion of all photopigments (rhodopsin and the three opsins in cones), so retinal begins the process of transduction of a stimulus (light) into a receptor potential. (e) A; carrots, yellow squash, broccoli, spinach, and liver; these foods lead to production of retinal in rods or cones.

D3. (a) Opsin; bent, *cis,* darkness. (b) Isomerization; absorbs light; *trans;* colorless or bleached; stimulus, transduction. (c) Regeneration; darkness, isomerase; cone photopigments.

D4. (a) *Trans.* (b) Rhodopsin; 5; 30–40; dark. (c) Because the retina has detached from the adjacent pigment epithelium, which is the storage area for vitamin A needed for synthesis of retinal for rhodopsin; difficulty seeing in dim light (nyctalopia) may result.

D5. Do; yellow to red, green, blue; red, green; males.

D6.　(a) See Figure LG 16.2A.

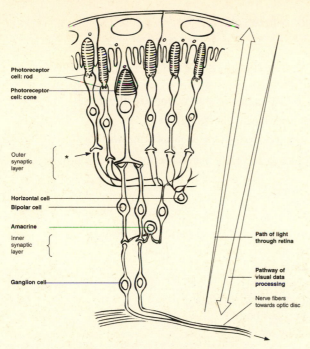

Figure LG 16.2A　Microscopic structure of the retina.

(b) Rod, bipolar; glutamate; hyper, stay inactive.
(c) Na⁺, dark; GMP, guanylate; stimulating
factor. (d) Opposite to; PDE; closed, no longer.

D7.　(a) Photoreceptor; 6, 600; convergence.
(b) No; one-on-one (cone to bipolar cell)
synapsing allows for greater acuity (sharpness)
of vision. (c) Enhancement of contrast and
differentiation of colors.

D8.　P * B * G ON OC OT * T * C. (Note that some
crossing but no synapsing occurs in the optic
chiasma.)

D9.　(a) Temporal, nasal. (b) Right, lower. (c) Right.
(d) Left.

D10.　Pituitary; nasal; temporal.

E1.

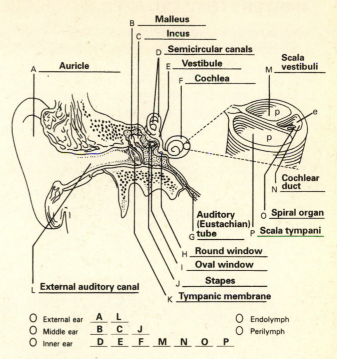

Figure LG 16.3A　Diagram of the ear in frontal section.

E2.　(a) AT. (b) TM. (c) OW. (d) M. (e) I. (f) A.

E5.　See Figure LG 16.3A

E6.　(a) 1,000–4,000. (b) Higher; directly. (c) Decibels
(dB); 45, 115–120.

E7.　(a) External auditory canal, tympanic; ossicles.
(b) Stapes, oval, peri-; vestibuli; tympani; round.
(c) Spiral; basilar; low. (d) Tectorial; cochlear,
VIII; thalamus, temporal.

E8.　O SV VM C B ST RW.

E10.　(a) Inner; hair, gelatinous. (b) Hearing, tectorial;
sound waves. (c) Vestibule, otoliths; static;
position while lying down. (d) Cupula; like an
inverted cup; crista; endolymph; dynamic.

F1.　(a) Myo. (b) G. (c) T. (d) Con. (e) H. (f) Myr. (g)
B. (h) Men. (i) Cat. (j) Pr. (k) K. (l) V. (m) Pt.

WRITING ACROSS THE CURRICULUM: CHAPTER 16

1. Contrast smell and taste according to the following
criteria: (a) rate of adaptation, (b) threshold, (c)
cranial nerve pathways for each. Also describe how
you could demonstrate that much of what most
people think of as taste is actually smell.

2. Contrast roles of the following professionals:
(a) ophthalmologist, (b) optometrist, (c) optician.

3. Discuss causes and potential damage to the eye
associated with the following conditions:

(a) detachment of the retina, (b) glaucoma,
(c) cataracts, (d) trachoma.

4. Contrast roles of ions (Na⁺, K⁺, and Ca²⁺) in activation of photoreceptor cells of the retina versus hair
cells of the cochlea.

5. Explain how cochlear implants function to take the
place of cells of the spiral organ (organ of Corti).

6. Contrast the anatomy of receptors for static equilibrium with those for dynamic equilibrium.

MASTERY TEST: CHAPTER 16

Questions 1–5: Circle T (true) or F (false). If the statement is false, change the underlined word or phrase so that the statement is correct.

T F 1. <u>Parasympathetic</u> nerves stimulate the circular iris muscle (constrictor pupillae) to constrict the pupil, whereas <u>sympathetic</u> nerves cause dilation of the pupil.

T F 2. <u>Convergence and accommodation are both results</u> of contraction of smooth muscle of the eye.

T F 3. <u>Crista, macula, otolith, and spiral organ</u> are all structures located in the inner ear.

T F 4. <u>Both the aqueous body and vitreous humor are</u> replaced constantly throughout your life.

T F 5. The receptor organs for special senses are <u>less</u> complex structurally than those for general sense.

Questions 6–12: Arrange the answers in correct sequence.

_____ _____ _____ 6. Layers of the eye, from superficial to deep:
  A. Sclera
  B. Retina
  C. Choroid

_____ _____ _____ _____ 7. From anterior to posterior:
  A. Vitreous body
  B. Optic nerve
  C. Cornea
  D. Lens

_____ _____ _____ _____ 8. Pathway of aqueous humor, from site of formation to destination:
  A. Anterior chamber
  B. Scleral venous sinus
  C. Ciliary body
  D. Posterior chamber

_____ _____ _____ _____ _____ 9. Pathway of sound waves and resulting mechanical action:
  A. External auditory canal
  B. Stapes
  C. Malleus and incus
  D. Oval window
  E. Tympanic membrane

_____ _____ _____ _____ _____ 10. Pathway of tears, from site of formation to entrance to nose:
  A. Lacrimal gland
  B. Excretory lacrimal duct
  C. Nasolacrimal duct
  D. Surface of conjunctiva
  E. Lacrimal punctae and lacrimal canals

_____ _____ _____ _____ _____ 11. Order of impulses along conduction pathway for smell:
  A. Olfactory bulb
  B. Olfactory hairs
  C. Olfactory nerves
  D. Olfactory tract
  E. Primary olfactory area of cortex

_____ _____ _____ _____ _____ 12. From anterior to posterior:
  A. Anterior chamber
  B. Iris
  C. Lens
  D. Posterior cavity
  E. Posterior chamber

13. In darkness, Na⁺ channels are held open by a molecule called:
    A. Glutamate
    B. Cyclic GMP
    C. PDE
    D. Transducin

14. Infections in the throat (pharynx) are most likely to lead to ear infections in the following manner. Bacteria spread through the:
    A. External auditory meatus to the external ear
    B. Auditory (eustachian) tube to the middle ear
    C. Oval window to the inner ear
    D. Round window to the inner ear

15. Choose the *false* statement about rods.
    A. There are more rods than cones in the eye.
    B. Rods are concentrated in the fovea and are less dense around the periphery.
    C. Rods enable you to see in dim (not bright) light.
    D. Rods contain rhodopsin.
    E. No rods are present at the optic disc.

16. Choose the *false* statement about the lens of the eye.
    A. It is biconvex.
    B. It is avascular.
    C. It becomes more rounded (convex) as you look at distant objects.
    D. Its shape is changed by contraction of the ciliary muscle.
    E. Change in the curvature of the lens is called accommodation.

17. Choose the *false* statement about the middle ear.
    A. It contains three ear bones called ossicles.
    B. Infection in the middle ear is called otitis media.
    C. It functions in conduction of sound from the external ear to the inner ear.
    D. The cochlea is located here.

18. Choose the *false* statement about the semicircular canals.
    A. They are located in the inner ear.
    B. They sense acceleration or changes in position.
    C. Nerve impulses begun here are conveyed to the brain by the vestibular branch of cranial nerve VIII.
    D. There are four semicircular canals in each ear.
    E. Each canal has an enlarged portion called an ampulla.

19. Destruction of the left optic tract would result in:
    A. Blindness in the left eye
    B. Loss of left visual field of each eye
    C. Loss of right visual field of each eye
    D. Loss of lateral field of view of each eye ("tunnel vision")
    E. No effects on the eye

20. Mr. Frederick has a detached retina of the lower right portion of one eye. As a result he is unable to see objects in which area in the visual field of that eye?
    A. High in the left
    B. High in the right
    C. Low in the left
    D. Low in the right

Questions 21–25: Fill-ins. Complete each sentence with the word or phrase that best fits.

_____ 21. _____ is the chemical within rods and cones that absorbs light.

_____ 22. Name the four processes necessary for formation of an image on

the retina: _____ of light rays, _____ of the lens, _____

of the pupil, and _____ of the eyes.

_____ 23. _____ is the study of the structure, functions, and diseases of the eye.

_____ 24. The names of three ossicles are _____, _____, and _____.

_____ 25. Two functions of the ciliary body are _____ and _____.

ANSWERS TO MASTERY TEST: ☆ CHAPTER 16

True–False

1. T
2. F. Accommodation, but not convergence, is a result.
3. T
4. F. Aqueous humor, but not vitreous humor, is
5. F. More

Arrange

6. A C B
7. C D A B
8. C D A B
9. A E C B D
10. A B D E C
11. B C A D E
12. A B E C D

Multiple Choice

13. B
14. B
15. B
16. C
17. D
18. D
19. C
20. A

Fill-ins

21. Retinal
22. Refraction, accommodation, constriction, convergence
23. Ophthalmology
24. Malleus, incus, and stapes
25. Production of aqueous humor and alteration of lens shape for accommodation

FRAMEWORK 17

The Autonomic Nervous System (ANS)

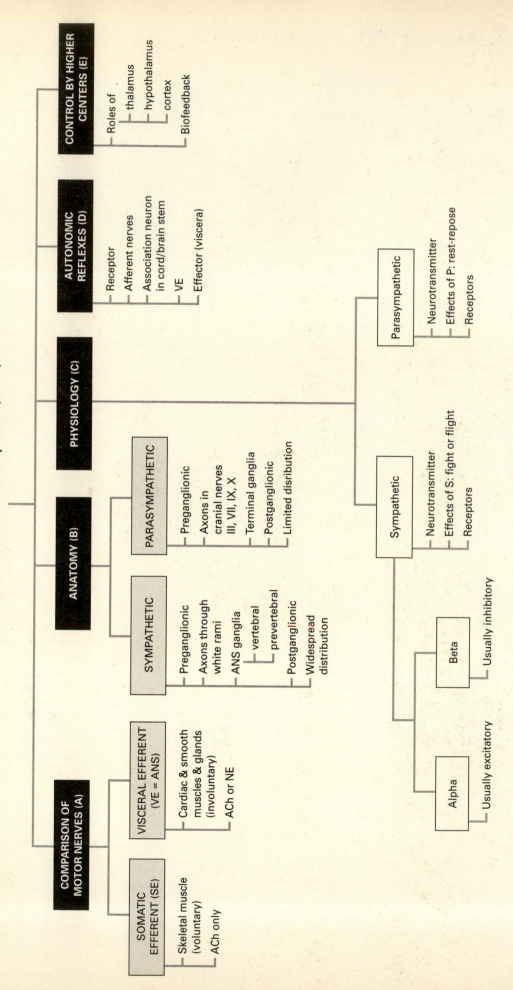

The Autonomic Nervous System (ANS)

The autonomic nervous system (ANS) consists of the special branch of the nervous system that exerts unconscious control over viscera. The ANS regulates activities such as heart rate and blood pressure, glandular secretion, and digestion. The two divisions of the ANS—the sympathetic and parasympathetic—carry out a continual balancing act, readying the body for response to stress or facilitating rest and relaxation, according to momentary demands. ANS neutrotransmitters hold particular significance clinically because they are often mimicked or inhibited by medications.

As you begin your study of the autonomic nervous system, carefully examine the Chapter 17 Topic Outline and Objectives; check off each one as you complete it. To organize your study of the autonomic nervous system, glance over the Chapter 17 Framework now. Be sure to refer to the Framework frequently and note relationships among key terms in each section.

TOPIC OUTLINE AND OBJECTIVES

A. Comparison of somatic and autonomic nervous systems

☐ 1. Compare the structural and functional differences of the somatic and autonomic nervous systems.

B. Anatomy

☐ 2. Identify the principal structural features of the autonomic nervous system.

☐ 3. Compare the sympathetic and parasympathetic divisions of the autonomic nervous system in terms of anatomy, physiology, and neurotransmitters released.

C. Physiological effects of the ANS

☐ 4. Describe the various neurotransmitters and receptors involved in autonomic responses.

D. Autonomic reflexes

☐ 5. Describe the components of an autonomic reflex.

E. Control by higher centers

☐ 6. Explain the relationship of the hypothalamus to the autonomic nervous system.

WORDBYTES

Now become familiar with the language of this chapter by studying each wordbyte, its meaning, and an example of its use within a term. After you study the entire list, self-check your understanding by writing the meaning of each wordbyte on the line. As you continue through the *Learning Guide,* identify (and fill in) additional terms that contain the same wordbyte.

Wordbyte	Self-check	Meaning	Example(s)
auto-	_____	self	*auto*nomic
nomos	_____	law, governing	auto*nomic*
para-	_____	near, beside	*para*vertebral, *para*thyroid
post-	_____	after	*post*ganglionic
pre-	_____	before	*pre*ganglionic
ram(i)-	_____	branch	white *ram*us
splanch-	_____	viscera	*splanch*nic

CHECKPOINTS

A. Comparison of somatic and autonomic nervous systems   (page 488)

■ **A1.** Why is the autonomic nervous system (ANS) so named? Is the ANS entirely independent of higher control centers? Explain.

■ **A2.** Contrast the somatic and autonomic nervous systems in this table.

	Somatic	Autonomic
a. Sensations and movements are mostly (*conscious? unconscious/automatic?*)		
b. Types of tissue innervated by efferent nerves	Skeletal muscle	
c. Efferent neurons are excitatory (E), inhibitory (I), or both (E + I)d.		
d. Examples of sensations (input)		Mostly visceral sensory (such as changes in CO_2 level of blood or stretching of visceral wall)

	Somatic	Autonomic
e. Number of neurons in efferent pathway		
f. Neurotransmitter(s) released by motor neurons	Acetylcholine (ACh)	

■ **A3.** Name the two divisions of the ANS.

_____ _____

■ **A4.** Many visceral organs have dual innervation by the autonomic system.

a. What does dual innervation mean?

b. How do the sympathetic and parasympathetic divisions work in harmony to control viscera?

B. Anatomy (pages 488–494)

■ **B1.** Complete this exercise describing structural differences between visceral efferent and somatic efferent pathways.

a. Between the spinal cord and effector, somatic pathways include _____ neuron(s),

whereas visceral pathways require _____ neuron(s), known as the pre-_____

and _____ neurons.

b. Somatic pathways begin at *(all? only certain?)* levels of the cord, whereas visceral routes

begin at _____ levels of the cord.

■ **B2.** Contrast preganglionic with postganglionic fibers in this exercise.

a. Consider the sympathetic division first. Its preganglionic cell bodies lie in the lateral

gray of segments _____ to _____ of the cord. For this reason, the sympathetic

division is also known as the _____ outflow.

b. Preganglionic axons *(are? are not?)* myelinated and therefore they appear white. These

axons may branch to reach a number of different _____
where they synapse. These ganglia contain *(pre-? post-?)* ganglionic neuron cell bodies
whose axons proceed to the appropriate organ. Postganglionic axons are *(gray? white?)*
because they *(do? do not?)* have a myelin covering.

c. Parasympathetic preganglionic cell bodies are located in two areas. One is the

_____, from which axons pass out in cranial nerves _____,

_____, _____, and _____. A second location is the *(anterior? lateral?)*

gray horns of segments _____ through _____ of the cord.

d. Based on location of *(pre? post?)*-ganglionic neuron cell bodies, the parasympathetic

division is also known as the _____ outflow. The axons of these neurons extend to *(the same? different?)* autonomic ganglia from those in sympathetic pathways (see Checkpoint B4). There, postganglionic cell bodies send out short *(gray? white?)* axons to innervate viscera.

e. Most viscera *(do? do not?)* receive fibers from both the sympathetic and the parasympathetic divisions. However, the origin of these divisions in the CNS and the pathways taken to reach viscera *(are the same? differ?)*.

■ **B3.** Contrast these two types of ganglia.

	Posterior Root Ganglia	**Autonomic Ganglia**
a. Contains neurons that are *(afferent? efferent?)*.		
b. Contains neurons in *(somatic? visceral? both?)* pathways.		
c. Synapsing *(does? does not?)* occur here.		

■ **B4.** Complete the table contrasting types of autonomic ganglia.

	Sympathetic Trunk	**Prevertebral**	**Terminal**
a. Sympathetic or para-sympathetic		Sympathetic	
b. Alternate name	Paravertebral ganglia or vertebral chain		
c. General location			Close to or in walls of effectors

■ **B5.** Refer to Figure 17.2 (page 491 in your text). Trace with your finger the route of a preganglionic neuron. Now describe the pathway common to *all* sympathetic preganglionic neurons by listing the structures in correct sequence. _____ _____ _____ _____

| A. Ventral root of spinal nerve |
| B. Sympathetic trunk ganglion |
| C. Lateral gray of spinal cord (between T1 and L2) |
| D. White ramus communicans |

■ **B6.** Suppose you could walk along the route of nerve impulses in sympathetic pathways. Start at the sympathetic trunk ganglion. It may be helpful to refer to the "map" provided by Figure 17.2, page 491 in your text.

a. What is the shortest possible path you could take to reach a synapse to a postganglionic neuron cell body?

b. You could then ascend and/or descend the sympathetic chain to reach trunk ganglia at other levels. This is important because sympathetic preganglionic cell bodies are limited in location to _____ to _____ levels of the cord, and yet the sympathetic trunk extends from _____ to _____ levels of the vertebral column. Sympathetic fibers must be able to reach these distant areas to provide the entire body with sympathetic innervation.

c. Now "walk" the sympathetic nerve pathway to sweat glands, blood vessels, or hair muscles in skin or extremities. Again trace this route on Figure 17.2, page 491 in the text. Preganglionic neurons synapse with postganglionic cell bodies in sympathetic trunk ganglia (at the level of entry or after ascending or descending). Postganglionic fibers then pass through _____ , which connect to _____ and then convey impulses to skin or extremities.

d. Some preganglionic fibers do not synapse as described in (b) or (c) but pass on through trunk ganglia without synapsing there. They course through _____ nerves to _____ ganglia. Follow their route as they synapse with postganglionic neurons whose axons form _____ en route to viscera.

e. Prevertebral ganglia are located only in the _____. In the neck, thorax, and pelvis the only sympathetic ganglia are those of the trunk (see Figure 17.1, page 490 in the text). As a result, cardiac nerves, for example, contain only *(preganglionic? postganglionic?)* fibers, as indicated by broken lines in the figure.

f. A given sympathetic preganglionic neuron is likely to have *(few? many?)* branches; these may take any of the paths you have just "walked." Once a branch synapses, it *(can? cannot?)* synapse again, because any autonomic pathway consists of just _____ neurons (preganglionic and postganglionic).

■ **B7.** Test your understanding of these routes by completing the sympathetic pathway shown in Figure LG 17.1.1. A preganglionic neuron located at level T5 of the cord has its axon drawn as far as the white ramus. Finish this pathway by showing how the axon may branch and synapse to eventually innervate these three organs: the heart, a sweat gland in skin of the thoracic region, and the stomach. Draw the preganglionic fibers in solid lines and the postganglionic fibers in broken lines.

■ **B8.** In Checkpoint B4 you contrasted two types of sympathetic trunk ganglia, including sympathetic trunk and prevertebral ganglia. For more details on these ganglia, complete this Checkpoint and also B9.

a. The sympathetic trunk ganglia lie in two chains close to vertebrae, ribs, and sacrum. Each chain has *(12? 22? 31?)* ganglia. *(Three? Seven? Eight?)* ganglia lie in the cervical region. The *(superior? middle? inferior?)* cervical ganglion lies next to the second cervical vertebra; its postganglionic fibers innervate *(muscles and glands of the head? the heart?)*.

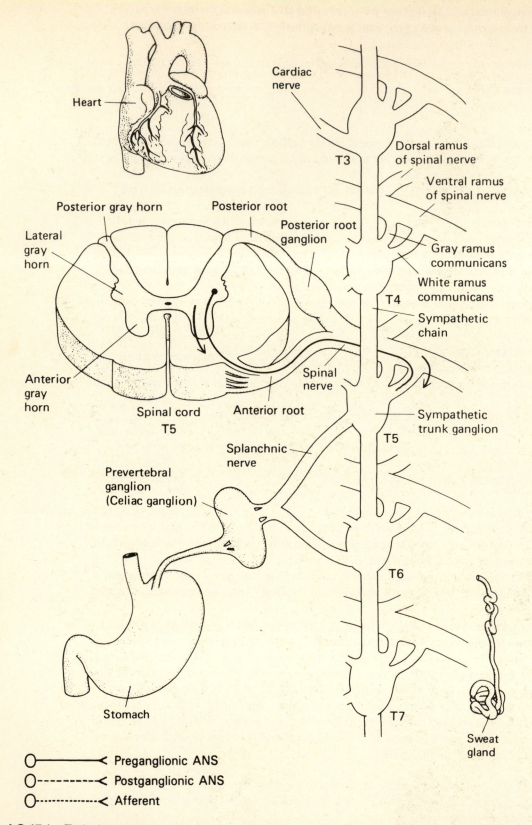

Heart

Cardiac nerve

T3

Dorsal ramus of spinal nerve

Ventral ramus of spinal nerve

Posterior gray horn

Posterior root

Posterior root ganglion

Lateral gray horn

Gray ramus communicans

White ramus communicans

Sympathetic chain

T4

Anterior gray horn

Spinal nerve

Spinal cord T5

Anterior root

Sympathetic trunk ganglion

T5

Splanchnic nerve

Prevertebral ganglion (Celiac ganglion)

T6

Stomach

T7

Sweat gland

O————< Preganglionic ANS
O----------< Postganglionic ANS
O·············< Afferent

Figure LG 17.1 Typical sympathetic pathway beginning at level T5 of the cord. Complete pathways according to directions in Checkpoints B7 and D2.

b. The middle and inferior cervical ganglia lie near the *(third and fourth?*
sixth and seventh?) vertebrae. Postganglionic fibers from these ganglia innervate

the _____.

c. There are *(10? 11? 14?)* trunk ganglia in the thoracic area; they lie against the *(anterior?*
posterior?) of ribs. List six types of viscera innervated by postganglionic fibers that arise
from these 11 pairs of ganglia.

d. There are *(4? 5? 6?)* sympathetic trunk ganglia in the lumbar region, and _____ in the sacral region.

■ **B9.** Do this Checkpoint on autonomic pathways involving splanchnic nerves. These nerves
contain *(sympathetic? parasympathetic?)* fibers. Splanchnic nerves are composed of
(pre? post?)-ganglion fibers that pass through trunk ganglia and *(do? do not?)* synapse
there; instead they extend to *(prevertebral? terminal?)* ganglia. Postganglionic cells there

pass axons on to viscera located in the _____. Complete this table about
these nerves.

Splanchnic Nerve	Prevertebral Ganglion: Site of Postganglionic Cell Bodies	Viscera Supplied
a. Greater splanchnic n.		
b. Lesser splanchnic n.		
c. Lumbar splanchnic n.		

■ **B10.** In the past several Checkpoints, you have been focusing on the *(sympathetic?*
parasympathetic?) division of the ANS. Now that you are somewhat familiar with the rather
complex sympathetic pathways, you will notice that parasympathetic pathways are much
simpler. This is due to the fact that there is no parasympathetic chain of ganglia; the only

parasympathetic ganglia are _____ _____, which are in
or close to the organs innervated. Also, parasympathetic preganglionic fibers synapse with
(more? fewer?) postganglionic neurons. What is the significance of the structural simplicity of
the parasympathetic system?

■ **B11.** Complete the table about the cranial portion of the parasympathetic system.

Cranial Nerve		Name of Terminal Ganglion	Structures Innervated
Number	**Name**		
a.			Iris and ciliary muscle of eye
b.		1. Pterygopalatine 2. Submandibular	1. 2.
c.	Glossopharyngeal		
d.		Ganglia in cardiac and pulmonary plexuses and in abdominal plexuses	

■ **B12.** *For extra review.* Write S if the description applies to the sympathetic division of the ANS, P if it applies to the parasympathetic division, and P, S if it applies to both.

_____ a. Also called thoracolumbar outflow

_____ b. Has long preganglionic fibers leading to terminal ganglia and very short post-ganglionic fibers

_____ c. Celiac and superior mesenteric ganglia are sites of postganglionic neuron cell bodies

_____ d. Sends some preganglionic fibers through cranial nerves

_____ e. Has some preganglionic fibers synapsing in vertebral chain (trunk)

_____ f. Has more widespread effect in the body, affecting more organs

_____ g. Has some fibers running in gray rami to supply sweat glands, hair muscles, and blood vessels

_____ h. Has fibers in white rami (connecting spinal nerve with vertebral chain)

_____ i. Contains fibers that supply viscera with motor impulses

_____ j. The only division to supply kidneys and adrenal glands

■ **B13.** In Activity B1, we emphasized that autonomic pathways require two neurons.

One exception to this rule is the pathway to the _____ glands. Because this pair of glands function like modified *(sympathetic? parasympathetic?)* ganglia, no "postganglionic neurons" lead out to any other organ.

C. Physiological effects of the ANS (pages 494–498)

■ **C1.** *(All? Most? No?)* viscera receive both sympathetic and parasympathetic innervation,

known as _____ innervation of the ANS. List exceptions here:

a. Sympathetic only: _____

_____

b. Parasympathetic only: _____

■ **C2.** Fill in the blanks to indicate which neurotransmitters are released by sympathetic (S) or parasympathetic (P) neurons or by somatic neurons. (It may help to refer to Figure LG 17.2.) Use these answers:

> ACh. Acetylcholine (cholinergic)
> NE. Norepinephrine = noradrenalin, or possibly epinephrine = adrenaline (adrenergic)

_____ a. All preganglionic neurons (both S and P) release this neurotransmitter

_____ b. Most S postganglionic neurons

_____ c. A few S postganglionic neurons, namely those to sweat glands and some blood vessels

_____ d. All P postganglionic neurons

_____ e. All somatic neurons

For extra review, color the neurotransmitters on Figure LG 17.2, matching colors you use in the key with those you use in the figure.

■ **C3.** Do this activity summarizing the effects of autonomic neurotransmitters.

a. Axons that release the transmitter acetylcholine (ACh) are known as *(adrenergic? cholinergic?).* Those that release norepinephrine (NE, also called noradrenalin) are

called _____. Most sympathetic postganglionic nerves

are said to be *(cholinergic? adrenergic?),* whereas all parasympathetic nerves are

_____ .

b. During stress, the *(sympathetic? parasympathetic?)* division of the ANS prevails, so stress responses are primarily *(adrenergic? cholinergic?)* responses.

c. Another source of NE as well as epinephrine is the gland known as the

_____ _____. Therefore, chemicals released by this gland (as in stress) will mimic the action of the *(sympathetic? parasympathetic?)* division of the ANS.

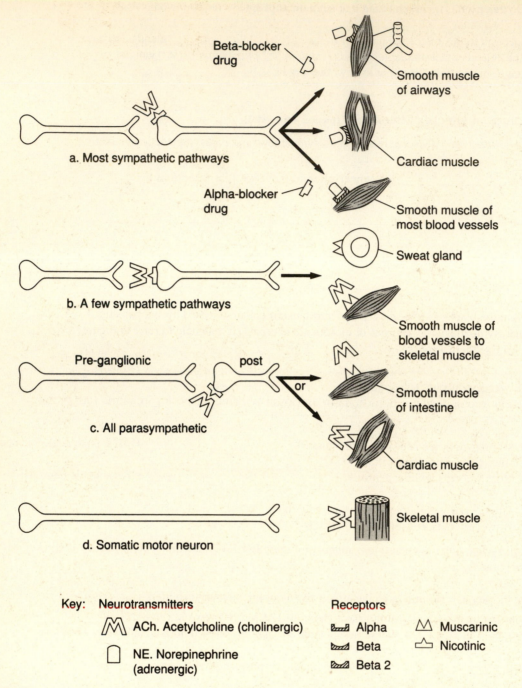

Beta-blocker drug

Smooth muscle of airways

a. Most sympathetic pathways

Cardiac muscle

Alpha-blocker drug

Smooth muscle of most blood vessels

Sweat gland

b. A few sympathetic pathways

Smooth muscle of blood vessels to skeletal muscle

Pre-ganglionic   post

or

Smooth muscle of intestine

c. All parasympathetic

Cardiac muscle

Skeletal muscle

d. Somatic motor neuron

Key: Neurotransmitters

ACh. Acetylcholine (cholinergic)

NE. Norepinephrine (adrenergic)

Receptors

Alpha

Beta

Beta 2

Muscarinic

Nicotinic

Figure LG 17.2   Diagram of autonomic and somatic neurons, neurotransmitters, and receptors. Note that neurotransmitters and receptors are relatively enlarged. Refer to Checkpoints C2 and C5 to C7.

■ **C4.** Do this exercise relating the fate of ANS neurotransmitters and overall effects of the two divisions of the ANS.

a. On the arrows below, write the initials of the enzymes that destroy each transmitter:

ACh ⟶ ~~ACh~~   NE ⟶ ~~NE~~

b. The ANS neurotransmitter that "hangs around" longer (has more lasting effect) is *(ACh? NE?)*. This fact provides one explanation for the longer-lasting effects of *(sympathetic/adrenergic? parasympathetic/cholinergic?)* neurons.

c. State two other rationales for the more lasting and widespread effects of the sympathetic system:

1.

2.

■ **C5.** Again refer to Figure LG 17.2 and do this exercise on neurotransmitters and related receptors. Be sure to match the color you use for receptor symbols in the key to the color you use for receptors in the figure. (The symbol S denotes *sympathetic* and P indicates *parasympathetic* in the following statements.)

a. First identify all nicotinic receptors in the figure. Notice that these receptors are found on cell bodies of *(P postganglionic neurons? S postganglionic neurons? autonomic effector cells? somatic effector cells [skeletal muscle]?)*. (Circle *all* that apply.) In other words, nicotinic receptors are all sites for action of the neurotransmitter *(ACh? NE? either ACh or NE?)*. Activation of nicotinic receptors is *(always? sometimes? never?)* excitatory.

b. On what other types of receptors can ACh act? _____. Identify these receptors on the figure. As you do, notice that these receptors are found on *all effectors* stimulated by *(P? S?)* postganglionic neurons. In addition, muscarinic receptors are found

on a few S effectors, such as _____. Activation of muscarinic receptors is *(always? sometimes? never?)* excitatory. For example, as P nerves (vagus) release ACh to a muscarinic receptor in cardiac muscle, heart activity *(increases? decreases?)*, but when P nerves (vagus) release ACh to a muscarinic receptor in the wall of the stomach or intestine, gastrointestinal activity *(increases? decreases?)*.

c. Explain what accounts for the names *nicotinic* and *muscarinic* for ACh receptors. The

names are based on the fact that the action of _____ on these recep-

tors is mimicked by action of _____ on nicotinic receptors and by a

poison named _____ from mushrooms on muscarinic receptors.

d. *Alpha* and *beta* receptors are found only on effectors innervated by *(P? S?)* nerves; these effectors are stimulated by *(ACh? NE and epinephrine?)*. NE and epinephrine are

categorized as _____ neurotransmitters.

e. *A clinical challenge.* Alex receives word that his brother has been seriously injured in an automobile accident. Among other immediate physical responses, his skin "goes pale." Explain this response. (Refer to Exhibit 17.3, page 497 in your text, for help.) Which type of receptor is found in smooth muscle of blood vessels of skin? *(Alpha? Beta₁? Beta₂?)* Alpha receptors are usually *(excitatory? inhibitory?).* As sympathetic (the "stress system") nerves release NE, smooth muscle of blood vessels are excited, causing them to *(constrict/narrow? dilate/widen?).* As a result, blood flows *(into? out of?)* skin, and the skin pales. How may this response help Alex?

■ **C6.** *A clinical challenge.* Certain medications can mimic the effects of the body's own transmitters. Again refer to Figure LG 17.2 and do this exercise.

a. A beta₁-stimulator (or beta-exciter) mimics *(ACh? NE?),* causing the heart rate and

strength of contraction to _____-crease. Name one such drug.

(For help, refer to a pharamacology text.) _____ Such a drug has a structure that is close enough to the shape of NE that it can "sit on" the beta₁ receptor as NE would and activate it.

b. A beta-blocker (or beta-inhibitor) also has a shape similar to that of NE and so can sit on

_____ receptors. Name one such drug. _____. However, this drug cannot activate the receptor but simply prevents NE from doing so.

So a beta-blocker _____-creases heart rate and force of contraction.

c. Other drugs, known as anticholinergics (or ACh-blockers), can take up residence on ACh receptors so that ACh being released from vagal nerve stimulation cannot exert its normal effects. A person taking such a medication may exhibit a(n) *(increased? decreased?)* heart rate.

■ **C7.** *For extra review.* Do this additional exercise on more effects of medications on ANS nerves.

a. What types of receptors are located on smooth muscle of bronchi and bronchioles (airways)? *(Alpha? Beta₁? Beta₂?)* (For help, refer to Exhibit 17.3, page 497 in your text.) The effect of the sympathetic stimulation of lungs is to cause *(dilation? constriction?)* of airways, making breathing *(easier? more difficult?)* during stressful times.

b. The effect of NE on the smooth muscle of the stomach, intestines, bladder, and uterus is

also *(contraction? relaxation?)* so that during stress, the body _____-creases activity of these organs and can focus on more vital activities such as heart contractions.

c. From Activity C5e, you may recognize that the sympathetic response during stress causes many blood vessels, such as those in skin, to *(constrict? dilate?).* Prolonged stress may therefore lead to *(high? low?)* blood pressure. One type of medication used to lower blood pressure is an alpha-*(blocker? stimulator?)* because it tends to dilate these vessels and "pool" blood in nonessential areas like skin, thereby avoiding overloading the heart with blood flow.

C8. Use arrows to show whether parasympathetic (P) or sympathetic (S) fibers stimulate (↑) or inhibit (↓) each of the following activities. Use a dash (—) to indicate that there is no parasympathetic innervation. The first one is done for you.

a. P ___↓___   S ___↑___   Dilation of pupil

b. P _____   S _____   Heart rate and blood flow to coronary (heart muscle) blood vessels

c. P _____   S _____   Constriction of skin blood vessels

d. P _____   S _____   Salivation and digestive organ contractions

e. P _____   S _____   Erection of genitalia

f. P _____   S _____   Dilation of bronchioles for easier breathing

g. P _____   S _____   Contraction of bladder and relaxation of internal urethral sphincter causing urination

h. P _____   S _____   Contraction of pili of hair follicles causing "goose bumps"

i. P _____   S _____   Contraction of spleen which transfers some of its blood to general circulation, causing increase in blood pressure

j. P _____   S _____   Release of epinephrine and norepinephrine from adrenal medulla

k. P _____   S _____   Coping with stress, fight-or-flight response

C9. Write a paragraph outlining the changes effected by the ANS during a fight-or-flight response, such as running in fear.

D. Autonomic reflexes   (page 498)

D1. Autonomic neurons can be stimulated by *(somatic afferents? visceral afferents? either somatic or visceral afferents?)*.

D2. On Figure LG 17.1 draw an afferent neuron (in contrasting color or with a dotted line) to show how the sympathetic neuron could be stimulated by stomach pain.

■ **D3.** Arrange in correct sequence structures in the pathway for a painful stimulus at your fingertip to cause an autonomic (visceral) reflex, such as sweating. Write the letters of the structures in order on the lines provided.

____ ____ ____ ____ ____ ____ ____ ____ ____ ____

A. Association neuron in spinal cord	F. Nerve fiber in anterior root of spinal nerve
B. Nerve fiber in gray ramus	
C. Pain receptor in skin	G. Nerve fiber in white ramus
D. Cell body of postganglionic neuron in trunk ganglion	H. Fiber in spinal nerve in brachial plexus
	I. Sweat gland
E. Cell body of preganglionic neuron in lateral gray of cord	J. Sensory neuron

■ **D4.** Most visceral sensations *(do? do not?)* reach the cerebral cortex, so most visceral sensations are at *(conscious? subconscious?)* levels. Hunger and nausea are exceptions.

E. Control by higher centers (pages 498–499)

E1. Explain the role of each of these structures in control of the autonomic system.

a. Hypothalamus (Which part controls sympathetic nerves? Which controls parasympathetic?)

b. Cerebral cortex (When is it most involved in ANS control? In stress or nonstress situations?)

E2. Explain how biofeedback is used to help in each of the following cases.

a. Tension headache

b. Childbirth

ANSWERS TO SELECTED CHECKPOINTS: CHAPTER 17

A1. It was thought to be autonomous (self-governing). However, it is regulated by brain centers such as the hypothalamus and medulla, with input from the limbic system and other parts of the cerebrum.

A2.

	Somatic	Autonomic
a. Sensations and movements are mostly (*conscious? unconscious/automatic?*)	**Conscious**	**Unconscious/ automatic**
b. Types of tissue innervated by efferent nerves	Skeletal muscle	**Cardiac muscle, smooth muscle, and most glands**
c. Efferent neurons are excitatory (E), inhibitory (I), or both (E + I)	E	E + I
d. Examples of sensations (input)	**Special senses, general somatic senses, or proprioceptors, and possibly visceral sensory, as when pain of appendicitis or menstrual cramping causes a person to curl up (flexion of thighs, legs)**	Mostly visceral sensory (such as changes in CO_2 level of blood or stretching of visceral wall)
e. Number of neurons in efferent pathway	**One (from spinal cord to effector)**	**Two (pre- and postganglionic)**
f. Neurotransmitter(s) released by motor neurons	Acetylcholine (ACh)	**ACh or norepinephrine (NE)**

A3. Sympathetic and parasympathetic.

A4. (a) Most viscera are innervated by both sympathetic and parasympathetic divisions. (b) One division excites and the other division inhibits the organ's activity.

B1. (a) 1, 2, ganglionic, postganglionic. (b) All, only certain (see Checkpoint B2).

B2. (a) T1, L2; thoracolumbar. (b) Are; ganglia; post-; gray, do not. (c) Brainstem, III, VII, IX, X; lateral, S2, S4. (d) Pre-, craniosacral; different; gray. (e) Do; differ.

B3. (a) Afferent, efferent. (b) Both, visceral. (c) Does not, does.

B4.

	Sympathetic Trunk	Prevertebral	Terminal
a. Sympathetic or para-sympathetic	**Sympathetic**	Sympathetic	**Parasympathetic**
b. Alternate name	Paravertebral ganglia or vertebral chain	**Collateral or prevertebral (celiac, superior and inferior mesenteric)**	**Intramural (in walls of viscera)**
c. General location	**In vertical chain along both sides of vertebral bodies from base of skull to coccyx**	In three sites anterior to spinal cord and close to major abdominal arteries	Close to or in walls of effectors

B5. C A D B.

B6. (a) Immediately synapse in trunk ganglion. (b) T1, L2 or L3, C3, sacral. (c) Gray rami, spinal nerves. (d) Splanchnic, prevertebral; plexuses. (e) Abdomen; postganglionic. (f) Many; cannot, 2.

B7.

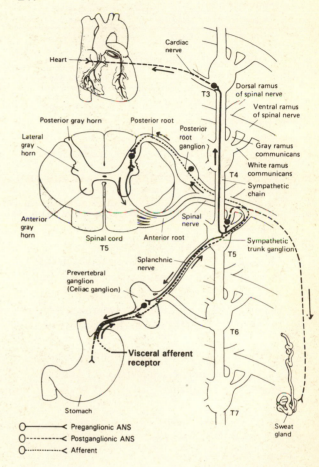

Figure LG 17.1A   Typical sympathetic pathway beginning at level T5 of the cord.

B8. (a) 22; three; superior, muscles and glands of the head. (b) Sixth and seventh; heart. (c) 11, posterior; heart, lungs, bronchi, sweat glands, smooth muscle of blood vessels, and hair (arrector pili) muscles. (d) 4, 4.

B9. Sympathetic; pre, do not; prevertebral; abdomen or pelvis.

Splanchnic Nerve	Prevertebral Ganglion: Site of Postganglionic Cell Bodies	Viscera Supplies
a. Greater splanchnic n.	Celiac	**Stomach, spleen, liver, kidneys, small intestine**
b. Lesser splanchnic n.	**Superior mesenteric**	**Small intestine, colon**
c. Lumbar splanchnic n.	**Inferior mesenteric**	**Distal colon, rectum; urinary bladder, genitalia**

B10. Sympathetic; terminal ganglia; fewer. Parasympathetic effects are much less widespread than sympathetic, distributed to fewer areas of the body, and characterized by a response of perhaps just one organ, rather than an integrated response in which many organs respond *in sympathy* with one another.

B11.

Cranial Nerve		Name of Terminal Ganglion	Structures Innervated
Number	**Name**		
a. III	Oculomotor	Ciliary	Iris and ciliary muscle of eye
b. VII	Facial	1. Pterygopalatine 2. Submandibular	1. **Nasal mucosa, palate, pharynx, lacrimal glands** 2. **Submandibular and submaxillary salivary glands**
c. IX	Glossopharyngeal	Otic	**Parotid salivary gland**
d. X	Vagus	Ganglia in cardiac and pulmonary plexuses and in abdominal plexuses	**Thoracic and abdominal viscera**

B12. (a) S. (b) P. (c) S. (d) P. (e) S. (f) S. (g) S. (h) S. (i) P, S. (j) S.

B13. Adrenal (medulla); sympathetic.

C1. Most, dual. (a) Sweat glands, arrector pili (hair) muscles, fat cells, kidneys, adrenal gland, and most blood vessels. (b) Lacrimal (tear) glands. (Refer to Exhibit 17.3 in your text, page 495.)

C2. (a) ACh. (b) NE. (c–e) ACh. (Refer to Figure LG 17.2.)

C3. (a) Cholinergic; adrenergic; adrenergic, cholinergic. (b) Sympathetic, adrenergic. (c) Adrenal medulla; sympathetic.

C4. (a)

$$\text{ACh} \xrightarrow{\text{AChE}} \text{ACh}$$
$$\text{NE} \xrightarrow{\text{COMT or MAO}} \text{NE}$$

(b) NE; sympathetic/adrenergic. (c) 1. Greater divergence of sympathetic neurons, for example, through the extensive sympathetic trunk chain and through gray rami to all spinal nerves. 2. Mimicking of sympathetic responses via release of catecholamines (epinephrine and NE) from the adrenal medulla.

C5. (a) Refer to key for Figure LG 17.2; both P and S postganglionic neurons and somatic effector cells (skeletal muscle); ACh; always.
(b) Muscarinic (*Hint:* Take time to notice that in this figure and in Figure 17.3, page 495 of your text, the *m*uscarinic receptors are ones shaped so that ACh "sits down" in an "M" [for muscarinic] position, whereas nicotinic receptors have ACh attach in the "W" position); P (notice this at top of Exhibit 17.3, page 497 in text); blood vessels that supply skeletal muscles and many sweat glands; sometimes; decreases, increases.
(c) ACh, nicotine, muscarine. (d) S, NE and epinephrine; catecholamine. (e) Alpha; excitatory; constrict/narrow; out of; blood flows into major vessels to increase blood pressure, which may assist in a stress response.

C6. (a) NE, in; isoproterinol (Isoprel) or epinephrine. (b) Beta; propranolol (Inderal); de. (c) Increased (a double negative: the inhibiting vagus is itself inhibited).

C7. (a) Beta$_2$; dilation, easier. (b) Relaxation, de. (c) Constrict; high; blocker.

C8. (b) P ↓, S ↑. (c) P—, S ↑. (d) P ↑, S ↓. (e) P ↑, S ↓. (f) P ↓, S ↑. (g) P ↑, S ↓. (h) P—, S ↑. (i) P—, S ↑. (j) P—, S ↑. (k) P ↓, S ↑.

D1. Either somatic or visceral afferents.

D2. See Figure LG 17.1A above.

D3. C J A E F G D B H I.

D4. Do not, subconscious.

WRITING ACROSS THE CURRICULUM: CHAPTER 17

1. Identify eight specific organs in your own body that are supplied by autonomic neurons.
2. Explain why the names craniosacral and thoracolumbar are given to the two divisions of the ANS.
3. Contrast locations of autonomic ganglia in sympathetic and parasympathetic divisions of the ANS.

4. Contrast white rami with gray rami with regard to locations, numbers, and functions.
5. Describe a "sympathetic response" that you might exhibit when faced with a threatening situation. Include specific changes likely to occur in your viscera.

MASTERY TEST: CHAPTER 17

Questions 1–6: Circle letters preceding all correct answers to each question.

1. Choose all *true* statements about the vagus nerve.
   A. Its autonomic fibers are sympathetic.
   B. It supplies ANS fibers to viscera in the thorax and abdomen but not in the pelvis.
   C. It causes digestive glands to increase secretions.
   D. Its ANS fibers are mainly preganglionic.
   E. It is a cranial nerve originating from the medulla.
2. Which activities are characteristic of the stress response, or fight-or-flight reaction?
   A. The liver breaks down glycogen to glucose.
   B. The heart rate decreases.
   C. Kidneys increase urine production because blood is shunted to kidneys.
   D. There is increased blood flow to genitalia, causing erect state.
   E. Hairs stand on end ("goose pimples") due to contraction of arrector pili muscles.
   F. In general, the sympathetic system is active.
3. Which fibers are classified as autonomic?
   A. Any visceral efferent nerve fiber
   B. Any visceral afferent nerve fiber
   C. Nerves to salivary glands and sweat glands
   D. Pain fibers from ulcer in stomach wall
   E. Sympathetic fibers carrying impulses to blood vessels
   F. All nerve fibers within cranial nerves
   G. All parasympathetic nerve fibers within cranial nerves

4. Choose all *true* statements about gray rami.
   A. They contain only sympathetic nerve fibers.
   B. They contain only postganglionic nerve fibers.
   C. They carry impulses from trunk ganglia to spinal nerves.
   D. They are located at all levels of the vertebral column (from Cl to coccyx).
   E. They carry impulses between lateral ganglia and collateral ganglia.
   F. They carry preganglionic neurons from anterior ramus of spinal nerve to trunk ganglion.
5. Which of the following structures contain some sympathetic preganglionic nerve fibers?
   A. Splanchnic nerves
   B. White rami
   C. Sciatic nerve
   D. Cardiac nerves
   E. Ventral roots of spinal nerves
   F. The sympathetic chains
6. Which are structural features of the parasympathetic system?
   A. Ganglia close to the CNS and distant from the effector
   B. Forms the craniosacral outflow
   C. Distributed throughout the body, including extremities
   D. Supplies nerves to blood vessels, sweat glands, and adrenal gland
   E. Has some of its nerve fibers passing through lateral (paravertebral) ganglia

Questions 7–10: Circle the letter preceding the one best answer to each question.

7. All of the following axons are cholinergic *except:*
   A. Parasympathetic preganglionic
   B. Parasympathetic postganglionic
   C. Sympathetic preganglionic
   D. Sympathetic postganglionic to sweat glands
   E. Sympathetic postganglionic to heart muscle
8. All of the following are collateral ganglia *except:*
   A. Superior cervical
   B. Prevertebral ganglia
   C. Celiac ganglion
   D. Superior mesenteric ganglion
   E. Inferior mesenteric ganglion

9. Which region of the cord contains no preganglionic cell bodies at all?
   A. Sacral
   B. Lumbar
   C. Cervical
   D. Thoracic

10. Which statement about postganglionic neurons is *false?*
    A. They all lie entirely outside of the CNS.
    B. Their axons are unmyelinated.
    C. They terminate in visceral effectors.
    D. Their cell bodies lie in the lateral gray matter of the cord.
    E. They are very short in the parasympathetic system.

Questions 11–20: Circle T (true) or F (false). If the statement is false, change the underlined word or phrase so that the statement is correct.

T F 11. Control of the ANS by the cerebral cortex occurs primarily during times when a person is <u>relaxed (nonstressed)</u>.

T F 12. Synapsing occurs in both <u>sympathetic and parasympathetic</u> ganglia.

T F 13. The sciatic, brachial, and femoral nerves all contain <u>sympathetic postganglionic</u> nerve fibers.

T F 14. Biofeedback and meditation confirm that the ANS is <u>independent of</u> higher control centers.

T F 15. Under stress conditions the <u>sympathetic system</u> dominates over the parasympathetic.

T F 16. When one side of the body is deprived of its sympathetic nerve supply, as in Horner's syndrome, the following symptoms can be expected: <u>constricted pupil (miosis) and lack of sweating (anhidrosis)</u>.

T F 17. Sympathetic cardiac nerves <u>stimulate</u> heart rate, and the vagus <u>slows down</u> heart rate.

T F 18. Viscera <u>do have sensory nerve fibers, and they are included in the autonomic nervous system</u>.

T F 19. <u>All</u> viscera have dual innervation by sympathetic and parasympathetic divisions of the ANS.

T F 20. The sympathetic system has a <u>more</u> widespread effect in the body than the parasympathetic does.

Questions 21–25: Fill-ins. Complete each sentence with the word or phrase that best fits.

_____ 21. The three types of tissue (effectors) innervated by the ANS nerves are _____.

_____ 22. _____ nerves convey impulses from the sympathetic trunk ganglia to collateral ganglia.

_____ 23. About 80% of the craniosacral outflow (parasympathetic nerves) is located in the _____ nerves.

_____ 24. Alpha (α) and beta (β) receptors are stimulated by the transmitter _____.

_____ 25. Drugs that block (inhibit) beta receptors in the heart will cause _____-crease in heart rate and blood pressure.

ANSWERS TO MASTERY TEST: ★ CHAPTER 17

Multiple Answers

1. B, C, D, E
2. A, E, F
3. A, B, C, D, E, G
4. A, B, C, D
5. A, B, E, F
6. B

Multiple Choice

7. E
8. A
9. C
10. D

True–False

11. F. Stressed
12. T
13. T
14. F. Dependent on
15. T
16. T
17. T
18. T
19. F. Most (or some)
20. T

Fill-ins

21. Cardiac muscle, smooth muscle, and glandular epithelium
22. Splanchnic
23. Vagus
24. NE (and also epinephrine)
25. De

FRAMEWORK 18

The Endocrine System

OVERVIEW (A)
- Endocrine vs. exocrine
- Endocrine vs. nervous
- Receptors
- Circulating vs. local
- Chemical classes (4)
- Hormone transport

HORMONAL ACTION (B)
- Hormonal action
  - 1st & 2nd messengers
  - DNA activation
- Control of hormones
  - nerves
  - chemical feedback
  - hormonal feedback
  - hypothalamic hormones

ENDOCRINE GLANDS

AGING, DEVELOPMENT (G)

OTHER ENDOCRINE TISSUES (G)
- GI tract
- Placenta
- Kidneys
- Skin
- Heart

STRESS (H)
- Alarm: adrenal medulla
- Resistance: adrenal cortex
- Exhaustion

SUMMARY, DISORDERS (I)

HYPOTHALAMUS (C)
- Hypophysiologic hormones via pituitary portal vessels
  - GHRH, TRH, CRH, GnRH, PRH
  - HGIH, PIH
- Neurons

PITUITARY (C)
- Neurohypophysis (posterior pituitary)
  - OT
  - ADH
- Adenohypophysis (anterior pituitary)
  - hGH
  - TSH
  - ACTH ⎫
  - FSH ⎬ Tropic
  - LH   ⎭
  - PRL
  - MSH

THYROID (D)
- Thyroid hormones
  - T₃ (T_3)
  - T₄ (T_4)
- Calcitonin

PARATHYROID (D)
- PTH

ADRENAL (E)
- Cortex
  - Mineralocorticoids — aldosterone
  - Glucocorticoids — cortisol
  - Gonadocorticoids — androgens
- Medulla
  - Epinephrine
  - NE

OTHER (F)
- Pancreas
  - insulin
  - glucagon
- Ovaries
- Testes
- Pineal
- Thymus

The Endocrine System

CHAPTER

18

Hormones produced by at least 18 organs exert widespread effects on just about every body tissue. Hormones are released into the bloodstream, which serves as the vehicle for distribution throughout the body. In fact, analysis of blood levels of hormones can provide information about the function of specific endocrine glands. How do hormones "know" which cells to affect? How does the body "know" when to release more or less of a hormone? Mechanisms of action and regulation of hormone levels are discussed in this chapter. The roles of all major hormones and effects of excesses or deficiencies in those hormones are also included. The human body is continually exposed to stressors, such as rapid environmental temperature change, a piercing sound, or serious viral infection. Adaptations to stressors are vital to survival; however, certain adaptations lead to negative consequences. A discussion of stress and adaptation concludes Chapter 18.

As you begin your study of the endocrine system, carefully examine the Chapter 18 Topic Outline and Objectives; check off each one as you complete it. To organize your study of the endocrine system, glance over the Chapter 18 Framework now. Be sure to refer to the Framework frequently and note relationships among key terms in each section.

TOPIC OUTLINE AND OBJECTIVES

A. Overview: functions, receptors, chemistry

☐ **1.** Define the components of the endocrine system and discuss the functions of the endocrine and nervous systems in maintaining homeostasis.

☐ **2.** Describe how hormones interact with target cell receptors.

☐ **3.** Compare the four chemical classes of hormones.

B. Mechanisms of hormonal action and regulation

☐ **4.** Explain the two general mechanisms of hormonal action.

☐ **5.** Describe the control of hormonal secretions via feedback cycles and give several examples.

C. Hypothalamus and pituitary (hypophysis)

☐ **6.** Explain why the hypothalmus is considered to be an endocrine gland.

D. Thyroid and parathyroids

E. Adrenals

F. Other endocrine glands: pancreas, gonads, pineal, and thymus; endocrine tissues

☐ **7.** Describe the location, histology, hormones, and functions of the following endocrine glands: pituitary, thyroid, parathyroids, adrenals, pancreas, ovaries, testes, pineal, and thymus.

G. Aging and developmental anatomy of the endocrine system; other endocrine tissues

☐ **8.** Describe the effects of aging on the endocrine system.

☐ **9.** Describe the development of the endocrine system.

H. Stress and homeostasis

☐ **10.** Define the general adaptation syndrome (GAS) and compare homeostatic responses and stress responses.

I. Summary of hormones, disorders

☐ **11.** Discuss the symptoms of pituitary dwarfism, giantism, acromegaly, diabetes insipidus, cretinism, myxedema, Grave's disease, and goiter; hypoparathyroidism; aldosteronism, Addison's disease, Cushing's syndrome, congenital adrenal hyperplasia and adrenal tumors; and pheochromocytoma, diabetes mellitus, and hyperinsulinism.

WORDBYTES

Now become familiar with the language of this chapter by studying each wordbyte, its meaning, and an example of its use within a term. After you study the entire list, self-check your understanding by writing the meaning of each wordbyte on the line. As you continue through the *Learning Guide,* identify (and fill in) additional terms that contain the same wordbyte.

Wordbyte	Self-check	Meaning	Example(s)
adeno-	_____	gland	*adeno*hypophysis
crin-	_____	to secrete	endo*crine*
endo-	_____	within	*endo*crine
exo-	_____	outside	*exo*crine
gen-	_____	to create	diabeto*genic*
horm-	_____	excite, urge on	*horm*one
insipid-	_____	without taste	diabetes *insipid*us
mellit-	_____	sweet	diabetes *mellit*us
oxy(s)-	_____	swift	*oxy*tocin
para-	_____	around	*para*thyroid, *para*crine
-tocin	_____	childbirth	pi*tocin*
trop-	_____	turn	thyro*tropin*

CHECKPOINTS

A. Overview: functions, receptors, chemistry (pages 503–506)

■ **A1.** Complete the table contrasting exocrine and endocrine glands.

	Secretions Transported In	Examples
a. Exocrine		
b. Endocrine		

■ **A2.** Refer to Figure LG 18.1 and do this exercise.

    a. Color and label each endocrine gland next to letters A–G.

    b. Next to numbers 1–12, label each organ that contains endocrine tissue but is not an endocrine gland exclusively.

■ **A3.** Compare the ways in which the nervous and endocrine systems exert control over the body. Write N (nervous) or E (endocrine) next to descriptions that fit each system.

_____ a. Sends messages to muscles, glands, and neurons only.

_____ b. Sends messages to virtually any part of the body.

_____ c. Effects are generally faster and shorter lived.

A4. Briefly list seven functions of hormones.

1. _____   5. _____

2. _____   6. _____

3. _____   7. _____

4. _____

■ **A5.** Write a sentence explaining how a specific hormone "knows" to affect a specific target organ, for example, how thyroid-stimulating hormone (TSH) "knows" to affect the thyroid gland rather than cells of pancreas, ovaries, or muscles.

How does this same principle explain the effectiveness of RU486 (mifepristone)?

■ **A6.** Describe regulation of hormones in this exercise.

    a. When excessive amounts of hormones are present, the number of related receptors is likely to *(decrease? increase?)* so that the overabundant hormone is less effective. This effect is called *(down? up?)*-regulation.

    b. Up-regulation makes a target organ *(less? more?)* sensitive to hormones by increase of receptors, for example, when hormone level is *(deficient? excessive?)*.

■ **A7.** Contrast types of hormones by filling in blanks. Use these answers: *auto, endo, para.*

    a. Circulating hormones that exert effects on distant target cells are called _____-crines.

    b. Local hormones include _____-crines such as interleukin-2 that act upon the same cell that secreted the hormone, as well as _____-crines that act on nearby cells.

    c. Which hormones act fastest and have longest lasting effects? _____

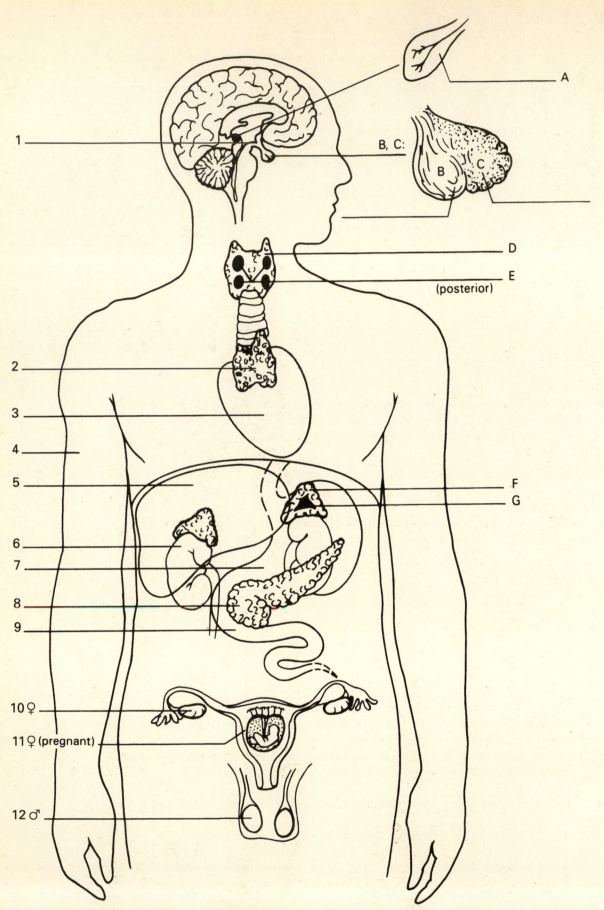

Labels visible in figure:

A

B, C:

B C

D

E
(posterior)

F

G

1

2

3

4

5

6

7

8

9

10 ♀

11 ♀ (pregnant)

12 ♂

Figure LG 18.1 Diagram of endocrine glands in black (A–G) and organs containing endocrine tissue (1–12) in gray. Label and color as directed in Checkpoints A2 and C1.

■ **A8.** Carefully examine Exhibit 18.2, page 505 in your text. Match the following chemical classes with descriptions of hormones listed below.

> BA. Biogenic amines   PP. Peptide, protein
> E. Eicosanoid   S. Steroid

_____ a. Chains of 3 to about 200 amino acids that may form glycoproteins; most hormones are in this category.

_____ b. Chemically, the simplest hormones; examples include thyroid hormones (T_3 and T_4) and catecholamines (epinephrine and norepinephrine).

_____ c. Derived from 20-carbon fatty acid; examples are prostaglandins and leukotrienes.

_____ d. Derived from cholesterol; includes hormones from adrenal cortex, ovary, and testis.

■ **A9.** *A clinical challenge.* Give a rationale for *parenteral* (by injection) rather than *p.o.* (by mouth) administration of insulin.

■ **A10.** Contrast transport mechanisms of hormones. Use these answers.

> FF. Water-soluble hormone that circulates in *free form* in blood
> TP. Lipid-soluble hormone that circulates attached to *transport protein*

The lines following each hormone group are for Activity B1.

_____ a. Thyroid hormone _____

_____ b. Steroid, such as hydrocortisone or progesterone _____

_____ c. Catecholamine _____

_____ d. Protein or peptide, including all hypothalamic and pituitary hormones _____

B. Mechanisms of hormonal action and regulation   (pages 506–509)

■ **B1.** Look back at Activity A10. On the lines following descriptions of chemical groups of hormones (a–d), write letters indicating the expected mechanism of hormonal action. (*Hint:* Remember that each action depends on the chemistry of the hormone.) Use these answers:

I. Because it is a lipid-soluble hormone, it can enter cell to use *intracellular receptors* that activate DNA to direct synthesis of new proteins.

P. Because it is not lipid-soluble, the hormone uses *plasma membrane receptors* to activate second messengers such as cyclic AMP.

■ **B2.** Study the cyclic AMP mechanism in Figure 18.4 (page 507 in your text). Then complete this exercise.

a. A hormone such as antidiuretic hormone (ADH) acts as the _____ messenger

as it carries a message from the _____ where it is secreted (in this

case the hypothalamus) to the outside of the _____ (in this case kidney cells).

b. The hormone then binds to a receptor on the *(inner? outer?)* surface of the plasma

membrane, causing activation of _____-proteins.

c. Such activation increases activity of the enzyme _____ , located on the
(inner? outer?) surface of the plasma membrane. This enzyme catalyzes conversion of

ATP to _____ , which is known as the _____ messenger.

d. Cyclic AMP then activates one or more enzymes known as _____ to
help transfer *(calcium? phosphate?)* from ATP to a protein, usually an enzyme. The resulting
chemical can then set off the target cell's response. In the case of ADH, this is an increase in

permeability of kidney cells, with a resulting decrease in _____ production.

e. Effects of cyclic AMP are *(short? long?)* lived. This chemical is rapidly degraded by

an enzyme named _____ .

f. A number of hormones are known to act by means of the cyclic AMP mechanism.

Included are most of the _____ -soluble hormones. Name several.

g. Name two or more chemicals besides cyclic AMP that may act as second messengers.

B3. Explain what is meant by amplification of hormone effects.

■ **B4.** Describe hormonal interactions in this activity.

a. The interaction by which effects of progesterone on the uterus in preparation for
pregnancy are enhanced by earlier exposure to estrogen is known as a(n) *(antagonistic?
permissive? synergistic?)* effect.

b. Insulin and glucagon exert _____ effects on blood glucose levels.

■ **B5.** Precise regulation of hormone levels is critical to homeostasis. Briefly describe control mechanisms for each hormone listed below. Use these answers.

> C. Blood level of a chemical that is controlled by the hormone
> H. Blood level of another hormone
> N. Nerve impulses

_____ a. Epinephrine   _____ c. Cortisol

_____ b. Parathyroid hormone

■ **B6.** *(Negative? Positive?)* feedback mechanisms are utilized for most hormonal regulation. For example, a low level of Ca^{2+} circulating through the parathyroid gland causes a(n) *(increase? decrease?)* in release of parathyroid hormone (PTH), which will increase blood level of Ca^{2+}. State one or more example(s) of hormone control by *positive* feedback.

C. Hypothalamus and pituitary (hypophysis)   (pages 509–520)

■ **C1.** Complete this exercise about the pituitary gland.

a. The pituitary is also known as the _____.
b. Defend or dispute this statement: "The pituitary is the master gland."

c. Where is it located?

d. Seventy-five percent of the gland consists of the *(anterior? posterior?)* lobe, called the

_____-hypophysis. The posterior lobe

(or _____-hypophysis) is somewhat smaller.
e. The anterior pituitary is known to secrete *(2? 5? 7?)* different hormones. Write abbreviations for each of these hormones next to the diagram of the anterior pituitary on Figure LG 18.1.

f. Arrange in correct sequence from first to last the vessels that form the blood pathway connecting the hypothalamus to the anterior pituitary. (*Hint:* it may help to refer to Figure 18.5, page 510, in your text.)

> AHV. Anterior hypophyseal veins
> HPV. Hypophyseal portal veins
> PPC. Primary plexus of capillaries
> SPC. Secondary plexus of capillaries
> SHA. Superior hypophyseal arteries

_____ _____ _____ _____ _____

Hypothalamic releasing and inhibiting factors are known as _____ hormones. Now place an asterisk (∗) next to the one answer listed above that consists of vessels located within the anterior pituitary (adenohypophysis) that serve as an exchange site for both hypophysiotropic hormones and anterior pituitary hormones between blood and the anterior pituitary.

g. What advantage is afforded by this special arrangement of blood vessels that link the hypothalamus with the anterior pituitary?

h. Figure 18.5, page 510 in your text, shows that the inferior hypophyseal artery, plexus of the infundibular process, and posterior hypophyseal veins form the blood supply to the *(anterior? posterior?)* lobe of the pituitary.

i. List functions of these five types of anterior pituitary cells.

   1. Corticotrophs: _____

   2. Gonadotrophs: _____

   3. Lactotrophs: _____

   4. Somatotrophs: _____

   5. Thyrotrophs: _____

j. Define *tropic* hormones.

Which four of the anterior pituitary hormones are tropic hormones?

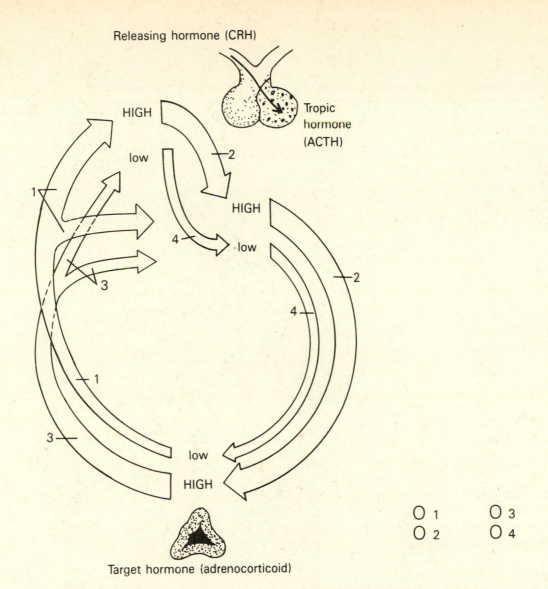

Releasing hormone (CRH)

HIGH

low

Tropic hormone (ACTH)

HIGH

low

Target hormone (adrenocorticoid)

○ 1   ○ 3
○ 2   ○ 4

Figure LG 18.2 Control of hormone secretion: example of releasing, tropic, and target hormones. Thickness of arrows indicates amounts of hormones. Color arrows as directed in Checkpoint C2. The branching of arrows 1 and 3 indicates that target hormones affect both the hypothalamus (releasing hormones) and the anterior pituitary (tropic hormones).

■ **C2.** Tropic hormones are involved in feedback mechanisms summarized in steps 1 to 4 in Figure LG 18.2. Trace the pathway of those four steps as you do this exercise.

  a. Color the arrow for step 1, in which a low blood level of target hormone, in this example,

     adrenocorticoid such as cortisone produced by the target gland _____, triggers the hypothalamus to secrete a *(high? low?)* level of releasing hormone (CRH). Simultaneously, the low level of target hormones directly stimulates the anterior pituitary to produce a *(high? low?)* level of ACTH.

  b. Color the two arrows for step 2. First the high level of releasing hormone (CRH) passes

     through blood vessels to the _____ and stimulates more *release* of ACTH. A high level of this *(tropic? target?)* hormone in blood passing through the adrenal cortex then stimulates secretion of a *(high? low?)* level of target hormone.

  c. Color steps 3 and 4, and note thickness of arrows. Both steps 1 and 3 are *(positive? negative?)* feedback mechanisms, whereas steps 2 and 4 are *(positive? negative?)* feedback mechanisms.

■ **C3.** Complete this table listing releasing-tropic-target hormone relationships. Fill in the name of a hormone next to each letter and number.

Hypothalamic Hormone →	Anterior Pituitary Hormone →	Target Hormone
a. GHRH (growth hormone releasing hormone or somatocrinin)		None
b. CRH (corticotropin releasing hormone)	1. 2.	1. 2. None
c.	1. TSH (thyrotropin) 2.	1. 2. None
d. GnRH (gonadotropic releasing hormone)	1. FSH (follicle stimulating hormone) 2. LH (luteinizing hormone) or ICSH (interstitial cells stimulating hormone)	1. 2. or Testosterone
e.	Prolactin	None

■ **C4.** Fill in this table naming two hypothalamic inhibiting hormones and their related anterior pituitary hormones. Hypothalamic inhibiting hormones provide additional regulation for the anterior pituitary hormones that *(are? are not?)* tropic hormones. (Notice this in the right column in the table below.)

Hypothalamic Hormone →	Anterior Pituitary Hormone →	Target Hormone
a. GHIH (growth hormone inhibiting hormone or somatostatin)		None
b.	Prolactin	None

■ **C5.** Refer to Figure LG 18.3 as you do this Checkpoint about human growth hormone (hGH).

a. Effects of hGH are likely to be *(direct? indirect?)* because hGH may work via

_____-like growth factors (IGFs). Previously called

_____, IGFs are much like _____ in structure, with many functions similar but *(more? less?)* potent than insulin. List four sites in the body where these chemicals are made in response to hGH.

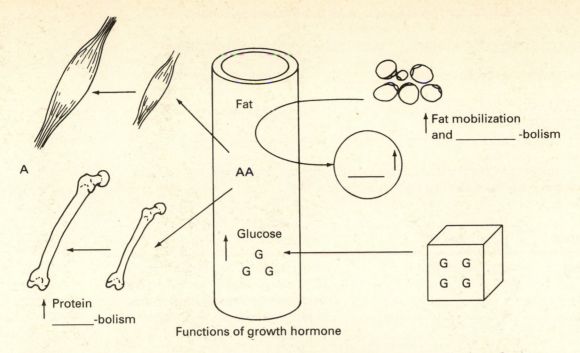

Fat

AA

A

Glucose
G
G G

↑ Protein
_____-bolism

↑ Fat mobilization
and _____ -bolism

G G
G G

Functions of growth hormone

Figure LG 18.3 Functions of growth hormone in regulating metabolism. AA, amino acid; G, glucose. Complete as directed in Checkpoint C5.

b. A main function of hGH and IGFs is to stimulate growth and maintain size of

_____ and _____. The alternative name

for hGH, _____, indicates its function: to "turn" *(trop-)* the body *(soma-)* to growth. Note that somatotropic hormone *(is? is not?)* classified as a true tropic hormone because it *(does? does not?)* stimulate another endocrine gland to produce a hormone.

c. hGH stimulates growth by *(accelerating? inhibiting?)* entrance of amino acids into cells

where they can be used for _____ synthesis. Therefore, growth hormone stimulates protein *(anabolism? catabolism?)*. Indicate this by filling in the blank on the left side of Figure LG 18.3.

d. hGH promotes release of stored fats and *(ana? cata?)*-bolism of fats, leading to

production of _____. Show this by filling in the blank on the right side of the figure.

e. Because cells now turn to fats for energy, the level of glucose in blood _____-creases and may lead to a condition known as *(hyper? hypo?)*-glycemia. In this capacity, action of hGH is *(similar? opposite?)* to action of insulin. In fact, because hGH-induced hyperglycemia may overstimulate and "burn out" *(alpha? beta?)* cells of the pancreas, hGH

may be considered _____-genic.

f. hGH secretion is controlled by two regulating factors: _____ and

_____. One stimulus that promotes release of hGH is low blood sugar (for example, during growth when cells require much energy). This condition, *(hyper-? hypo-?)*glycemia, stimulates secretion of *(GHRH? GHIH?)*, which then causes release of *(high? low?)* levels of hGH.

g. Hypersecretion of GH in early years leads to *(dwarfism? giantism? acromegaly?)*,

whereas deficiency causes pituitary _____. Excess hGH after

closure of epiphyseal plates causes _____, characterized by bones

in three areas: _____, _____, and

_____.

C6. *For extra review.* Complete the table describing factors that affect production of each
of the following hormones. The first one is done for you.

Hormone	How Affected	Factors
a. hGH	Decreased by	HGIH, hGH (by negative feedback), REM sleep, obesity, increased fatty acids and decreased amino acids in blood; emotional deprivation; low thyroid
b. hGH	Increased by	
c. ACTH	Increased by	
d. ADH	Increased by	

■ C7. Do this exercise about hormonal control of the mammary glands.

a. Milk is produced in mammary glands following stimulation by the hormone

_____ (_____) which is secreted by the *(anterior? posterior?)*

pituitary. A different hormone causes ejection of milk from the glands at the time of the

baby's suckling. This hormone, named _____, is released by the
(anterior? posterior?) pituitary.

b. During pregnancy, prolactin levels *(increase? decrease?)* due to increased levels of
(PIH? PRH?). During most of a woman's lifetime prolactin levels are low as a result of
high levels of *(PIH? PRH?)*. An exception is the time just prior to each menstrual flow
when PIH levels *(increase? decrease?)*. Due to this lack of inhibition, prolactin secretion increases, activating breast tissues (and causing tenderness).

■ **C8.** Answer these questions about posterior pituitary hormones.

a. The time period from the start of a baby's suckling at the breast to the delivery of milk

to the baby is about _____-seconds. Although afferent impulses (from breast to hypothalamus) require only *(a fraction of a second? 45 seconds?),* passage of the

oxytocin through the _____ to the breasts requires about _____ seconds.

b. Explain how oxytocin is involved in positive feedback cycles involving the uterus and the breasts.

c. On a day when your body becomes dehydrated, for example, by loss of much sweat,

your ADH production is likely to _____-crease. ADH causes kidneys to *(retain?*

eliminate?) more water, _____-creasing urine output; ADH also _____-creases sweat production. ADH *(dilates? constricts?)* small blood vessels, which allows more blood to be retained in large arteries. As a result of all of these mechanisms, ADH

_____-creases blood pressure.

d. Effects of diuretic medications are *(similar to? opposite of?)* ADH. In other words,

diuretics _____-crease urine production and _____-crease blood volume and blood pressure.

e. Inadequate ADH production, as occurs in diabetes *(mellitus? insipidus?)* results in *(enormous? minute?)* daily volumes of urine. Unlike the urine produced in diabetes mellitus, this urine *(does? does not?)* contain sugar. It is bland or *insipid.*

f. Alcohol *(stimulates? inhibits?)* ADH secretion, which contributes to

_____-creased urine production after alcohol intake. *For extra review* of other factors affecting ADH production, refer to Checkpoint C6.

D. Thyroid and parathyroids   (pages 520–527)

■ **D1.** Describe the histology and hormone secretion of the thyroid gland in this exercise.

a. The thyroid is composed of two types of glandular cells. *(Follicular? Parafollicular?)*

cells produce the two hormones _____ and _____.

Parafollicular cells manufacture the hormone _____.

b. The ion *(Cl⁻? I⁻? Br⁻?)* is highly concentrated in the thyroid because it is an essential component of T_3 and T_4. Each molecule of thyroxine (T_4) contains four of the I⁻ ions,

whereas T_3 contains _____ ions of I⁻. *(T_3? T_4?)* is found in greater abundance in thyroid hormone, but *(T_3? T_4?)* is more potent.

c. Once formed, thyroid hormones *(are released immediately? may be stored for months?).* T_3 and T_4 travel in the bloodstream bound to plasma *(lipids? proteins?).*

D2. Briefly describe steps in the formation of T_3 and T_4.

■ **D3.** Fill in blanks and circle answers about functions of thyroid hormone contrasted to those of growth hormone.

a. Like growth hormone, thyroid hormone increases protein _____-bolism.

b. Unlike hGH, thyroid hormone increases most aspects of carbohydrate

_____-bolism. Therefore, thyroid hormone *(increases? decreases?)* basal metabolic rate (BMR), releasing heat and *(increasing? decreasing?)* body temperature. This is

known as the _____ effect of thyroid hormones.

c. Thyroid hormone *(increases? decreases?)* heart rate and blood pressure and *(increases? decreases?)* nervousness.

d. This hormone is important for development of normal bones and muscles, as well as brain tissue. Therefore, deficiency in thyroid hormone during childhood (a condition

called _____) *(does? does not?)* lead to retardation as well as small stature. Note that in childhood hGH deficiency, retardation of mental development *(occurs also? does not occur?)*.

D4. Release of thyroid hormone is under influence of releasing factor

_____ and tropic hormone _____.
Review the table in Checkpoint C3. List factors that stimulate TRF and TSH production.

■ **D5.** Do this exercise about thyroid disorders.

a. A simple goiter is *(an enlargement? a decrease in size?)* of the thyroid gland. It results from the futile attempts of the gland to produce thyroid hormone when iodine is present in *(excessive? deficient?)* quantities.

b. Myxedema is *(hypo? hyper?)*-thyroidism during adult years. Symptoms are *(increased? decreased?)* heart rate, body temperature, and energy level.

c. Graves disease is a form of *(hypo? hyper?)*-thyroidism. One symptom is exophthalmos,

which is protrusion of _____. This condition is an autoimmune

disease in which the person produces _____ that act like TSH in overstimulating the thyroid.

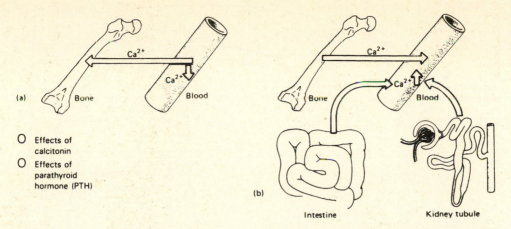

(a) Bone Ca²⁺ Ca²⁺ Blood

○ Effects of calcitonin
○ Effects of parathyroid hormone (PTH)

(b) Bone Ca²⁺ Ca²⁺ Blood

Intestine   Kidney tubule

Figure LG 18.4 Hormonal control of calcium ion (Ca2+). (a) Increased calcium storage in bones lowers blood calcium (hypocalcemic effect). (b) Calcium from three sources increases blood calcium level (hypercalcemic effect). Arrows show direction of calcium flow. Color arrows as directed in Checkpoint D6.

■ **D6.** Color arrows on Figure LG 18.4 to show hormonal control of calcium. This figure emphasizes that CT and PTH are *(antagonists? synergists?)* with regard to effects on calcium. One way that PTH increases blood calcium level is by stimulating kidney

production of the active form of vitamin _____ (also known as _____),

which facilitates _____
absorption in the intestine. Both CT and PTH act synergistically to *(lower? raise?)* blood phosphate levels.

■ **D7.** Hypocalcemia can result from *(hypo? hyper?)*-parathyroid conditions. Decreased

blood calcium results in abnormal _____-crease in nerve impulses to muscles, a

condition known as _____. In other words, there is a(n) *(direct? inverse?)* relationship between blood level of calcium and nerve and muscle activity.

E. Adrenals (pages 527–533)

E1. In the space below, draw an outine of the adrenal cortex. Label the *adrenal medulla* and the *adrenal cortex*. Then name the three zones *(zona)* of the adrenal cortex from outer-most to innermost. Finally, label your figure with the primary hormones(s) secreted by each zone.

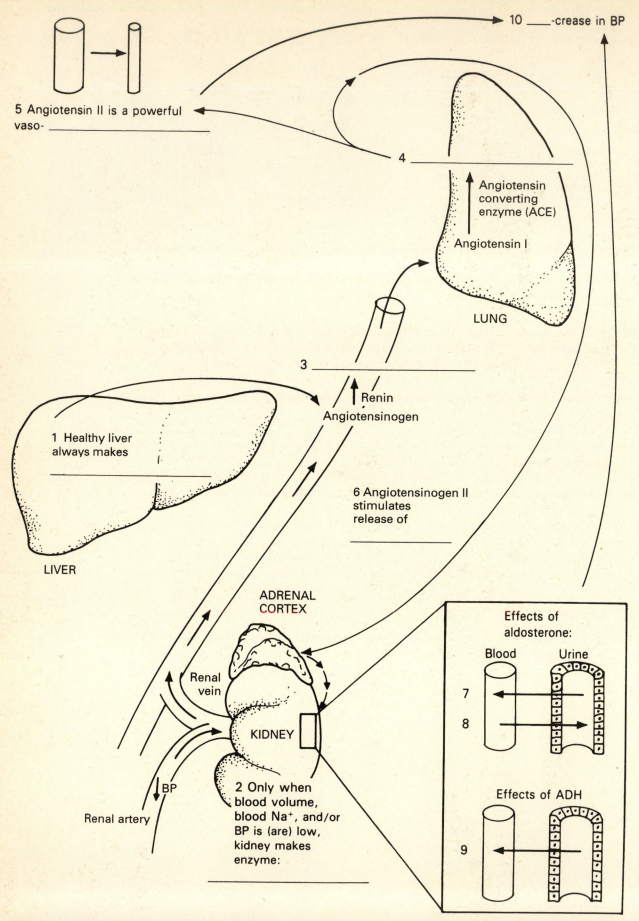

10 ____ -crease in BP

5 Angiotensin II is a powerful
vaso- _____

4 _____

Angiotensin
converting
enzyme (ACE)

Angiotensin I

LUNG

3 _____

↑ Renin
Angiotensinogen

1 Healthy liver
always makes

_____

LIVER

6 Angiotensinogen II
stimulates
release of

_____

ADRENAL
CORTEX

Renal
vein

KIDNEY

↓ BP

Renal artery

2 Only when
blood volume,
blood Na⁺, and/or
BP is (are) low,
kidney makes
enzyme:

_____

Effects of
aldosterone:

Blood   Urine

7

8

Effects of ADH

9

Figure LG 18.5   Hormonal control of fluid and electrolyte balance and blood pressure (BP) by renin, angiotensin, aldosterone, and ADH. Fill in blanks 1–10 as directed in Checkpoint E2.

■ **E2.** Fill in blank lines on Figure LG 18.5 to show mechanisms that control fluid balance and blood pressure.

■ **E3.** *A clinical challenge.* Determine whether each of the following conditions is likely to *increase* or *decrease* blood pressure.

    a. Excessive renin production: _____

    b. Use of an angiotensin converting enzyme (A.C.E.)-inhibitor medication: _____

    c. Aldosteronism (for example, due to a tumor of the zona glomerulosa): _____

■ **E4.** State two factors that stimulate aldosterone production.

E5. Name three glucocorticoids. Circle the one that is most abundant in the body.

■ **E6.** In general, glucocorticoids are involved with metabolism and resistance to stress. Describe their effects in this exercise.

    a. These hormones promote _____-bolism of proteins. In this respect their effects are *(similar? opposite?)* to those of both growth hormone and thyroid hormone.
    b. However, glucocorticoids may stimulate conversion of amino acids and fats to glucose,

        a process known as _____. In this way glucocorticoids (like hGH) *(increase? decrease?)* blood glucose.
    c. Refer to Figure 18.26 (page 544 in your text). Note that glucocorticoid (and also hGH) production *(increases? decreases?)* during times of stress. By raising blood glucose levels, these hormones make energy available to cells most vital for a response to stress. During stress glucocorticoids also act to *(increase? decrease?)* blood pressure and to *(enhance? limit?)* the inflammatory process. (They are *anti-inflammatory.*)
    d. *For extra review* of regulation of glucocorticoids, refer to Activities C2, C3, and C6.

■ **E7.** Write A (for Addison's disease) or C (for Cushing's syndrome) next to the manifestations of those adrenal cortex disorders listed below.

_____ a. Abdominal striae   _____ d. Decreased resistance to stress and infection

_____ b. Hyperglycemia   _____ e. Low blood pressure

_____ c. "Moon face" and "buffalo hump"   _____ f. High level of ACTH

E8. Explain why congenital adrenal hyperplasia in which cortisol production is inadequate leads to the following manifestations.

a. Increased levels of ACTH

b. Enlarged adrenal glands

c. Virilism

■ **E9.** Do this exercise on the adrenal medulla.

a. Name the two principal hormones secreted by the adrenal medulla.

Circle the one that accounts for 80% of adrenal medulla secretions.

These hormones are classified chemically as _____.

b. In general, effects of these hormones mimic the *(sympathetic? parasympathetic?)* nervous system. List three or more effects.

c. Which of the following hormones are essential for life? Hormones produced by the adrenal *(cortex? medulla?)*.

F. Other endocrine glands: pancreas, gonads, pineal, and thymus; endocrine tissues (pages 533–541)

■ **F1.** The islets of Langerhans are located in the _____. Do this exercise describing hormones produced there.

    a. The name *islets of Langerhans* suggests that these clusters of _____-crine cells lie amidst a "sea" of exocrine cells within the pancreas. Name the four types of islet cells.

    b. Refer to Figure LG 18.6a. Glucagon, produced by *(alpha? beta? delta?)* cells, *(increases? decreases?)* blood sugar in two ways. First, it stimulates the breakdown of

    _____ to glucose, a process known as *(glycogenesis? glycogenolysis? gluconeogenesis?)*. Second, it stimulates the conversion of amino acids and other

    compounds to glucose. This process is called _____.

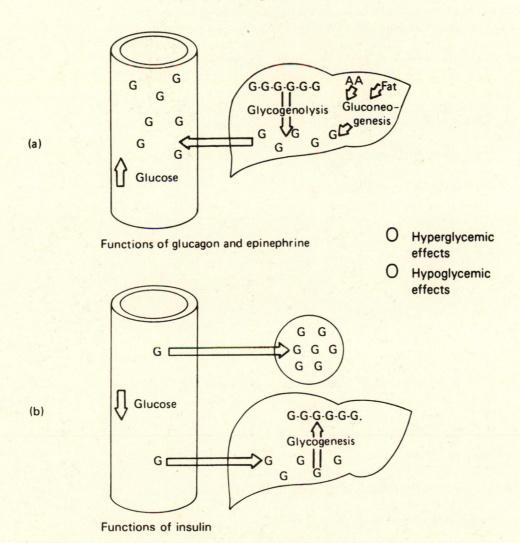

Figure LG 18.6 Hormones that regulate blood glucose level. AA, amino acid; G, glucose; G-G-G-G-G-G, glycogen. Color according to directions in Checkpoint F1.

c. Is glucagon controlled directly by an anterior pituitary *tropic* hormone? *(Yes? No?)*

  In fact, control is by effect of blood _____ level directly upon the pancreas. When blood glucose is low, then a *(high? low?)* level of glucagon will be produced. This will raise blood sugar.

d. *For extra review.* Define the terms *glycogen* and *glucagon.*

e. Now look at Figure LG 18.6b. The action of insulin is *(the same as? opposite?)* that of glucagon. In other words, *insulin decreases blood sugar.* (Repeat that statement three times; it is important!) Insulin acts by several mechanisms. Two are shown in the figure. First, insulin *(helps? hinders?)* transport of glucose from blood into cells. Second, it

  accelerates conversion of glucose to _____, the process

  called _____.

f. Hormones that raise blood glucose levels are called *(hypo? hyper?)*-glycemic hormones. Insulin, the one main hormone that lowers blood glucose level is said to be

  _____-glycemic. To show this, color arrows in Figure 18.6.

■ **F2.** List three or more factors that increase insulin secretion.

■ **F3.** Contrast type I and type II diabetes mellitus (DM) by writing I or II next to the related descriptions.

_____ a. Also known as maturity-onset DM

_____ b. The more common type of DM

_____ c. Related to insensitivity of body cells to insulin, rather than to absolute insulin deficiency

_____ d. Also known as insulin-dependent diabetes mellitus (IDDM)

_____ e. More likely to lead to serious complications such as ketoacidosis

■ **F4.** *A clinical challenge.* Mrs. Jefferson has diabetes mellitus. In the absence of sufficient insulin, her cells are deprived of glucose. Explain why she experiences the following symptoms.

    a. Hyperglycemia and glycosuria

    b. Increased urine production (poly-_____) and increased water intake

       (poly-_____)

    c. Ketoacidosis, a form of _____-osis

    d. Atherosclerosis

■ **F5.** Sex hormones are produced in several locations in the body. List four. (More about these in Chapter 28.)

F6. Where is the pineal gland located?

Secretion of one pineal hormone, melatonin, increases in *(light? darkness?)*. How may this hormone be related to the form of depression known as seasonal affect disorder (SAD)?

■ **F7.** What is the general function of the thymus? _____ Name four hormones made by this gland.

G. Aging and developmental anatomy of the endocrine system; other endocrine tissues (pages 541–542; 545–546)

■ **G1.** Name endocrine organs that often cause problems among the elderly.

_____   _____

■ **G2.** Match the endocrine gland with the correct description of its embryological origin.

AC. Adrenal cortex	P. Pancreas
AM. Adrenal medulla	PT. Parathyroid
AP. Anterior pituitary	T. Thymus

_____ a. Develops from the foregut area that later becomes part of small intestine

_____ b. Derived from the roof of the mouth; called the hypophyseal (Rathke's) pouch

_____ c. Originates from the neural crest which also produces sympathetic ganglia

_____ d. Derived from tissue from the same region that forms gonads

_____ e. Arise from pharyngeal pouches (two answers)

■ **G3.** *The Big Picture: Looking Ahead.* Refer to upcoming chapters and answer questions about hormones produced by tissues that not only contain endocrine cells, but that also have other functions. Select answers from the box.

Atrial natriuretic peptide	Erythropoietin
CCK, gastrin, GIP, and secretin	Human chorionic gonadotropin (HCG)
Eicosanoids: thromboxane A_2 and PG F_{2a}	Vitamin D

a. Chemical that promotes development of red blood cells (erythropoiesis)

(Figure 19.4, page 560): _____.

b. Chemical that decreases blood pressure by promoting loss of Na^+ and water, and by

dilating blood vessels (Exhibit 21.1, page 628): _____.

c. Chemicals that constrict blood vessels and therefore increase blood pressure

(page 628): _____.

d. Hormones that alter secretion of gastric or intestinal secretions and movements

(Exhibit 24.4, page 783): _____.

e. Hormone made in skin and kidney; aids in absorption of calcium

(Exhibit 25.6, page 840): _____.

f. Hormone made by the chorion layers of the placenta that stimulates production of estrogens and progesterone to maintain pregnancy (Figure 29.9, page 970):

_____.

G4. Write a brief description of functions of each of the following categories of chemicals produced by the body.

a. Leukotrienes (LTs)

b. Prostaglandins (PGs)

c. Thromboxane (TX)

■ **G5.** *A clinical challenge.* Aspirin and nonsteroidal anti-inflammatories (NSAIDs) inhibit synthesis of one of the categories of chemicals listed in Checkpoint G4. Which one?

_____ Write three therapeutic effects of NSAIDs.

■ **G6.** List seven hormones that stimulate cell growth and cell division.

■ **G7.** Match growth factors listed in the box with descriptions below.

> EGF. Epidermal growth factor
> PDGF. Platelet-derived growth factor
> TAF. Tumor angiogenesis factor

_____ a. Produced by normal and tumor cells; stimulates growth of new blood vessels

_____ b. Produced in salivary glands; stimulates growth of epithelial cells, fibroblasts, and neurons

_____ c. Produced by platelets; appears to play a role in wound healing

H. Stress and homeostasis (pages 542–545)

H1. Define and give two examples of *stressors*.

■ **H2.** Summarize the body's response to stressors in this exercise.

a. The part of the brain that senses the stress and initiates response is the

_____. It responds by two main mechanisms.

b. First is the *(alarm? resistance?)* reaction involving the adrenal *(medulla? cortex?)*

and the _____ division of the autonomic nervous system.

c. Second is the _____ reaction. Playing key roles in this

response are three anterior pituitary hormones: _____,

_____, and _____ and their target hormones.
Therefore, the adrenal *(medulla? cortex?)* hormones are activated.

■ **H3.** Now describe events that characterize the first (or alarm) stage of response to stress. Complete this exercise.

a. The alarm reaction is sometimes called the _____ response.

b. During this stage blood glucose *(increases? decreases?)* by a number of hormonal

mechanisms, including those of adrenal medulla hormones _____

and _____. Glucose must be available for cells to have energy for the stress response.

c. Oxygen must also be available to tissues; the respiratory system *(increases? decreases?)* its activity. Heart rate and blood pressure *(increase? decrease?)*. Blood is shunted to

vital tissues such as the _____, _____,

and _____ and is directed away from reservoir organs such as the

_____, _____, and _____.

d. Nonessential activities such as digestion *(increase? decrease?)*. Sweating *(increases? decreases?)* in order to control body temperature and eliminate wastes.

H4. Describe the effects of each of these hormones in the resistance stage.

a. Mineralocorticoids

b. Glucocorticoids

c. Thyroxin

d. hGH

■ **H5.** Identify activities of the three stages of response stress in this Checkpoint. Use answers in the box; they may be used once, more than once, or not at all.

> A. Alarm stage
> E. Exhaustion stage
> R. Resistance stage

_____ a. Makes large amounts of glucose and oxygen available for a fight-or-flight response.

_____ b. This stage is short lived.

_____ c. Initiated by sympathetic nerve impulses and adrenal medulla hormones (epinephrine and norepinephrine).

_____ d. Mediated by hypophysiotropic hormones CRH, GHRH, and TRH, which then stimulate anterior pituitary hormones cortisol, hGH, and T_3-T_4.

_____ e. Sodium and water retention is promoted by effects of aldosterone in this stage.

_____ f. During this stage, blood chemistry returns to nearly normal.

_____ g. Depletion of body reserves and failure of resistance mechanisms occur.

I. Summary of hormones, disorders

■ **I1.** *For extra review.* Test your understanding of these hormones by writing the name (or abbreviation where possible) of the correct hormone after each of the following descriptions.

a. Stimulates release of growth hormone: _____

b. Stimulates testes to produce testosterone: _____

c. Promotes protein anabolism, especially of bones and muscles: _____

d. Tropic hormone for thyroxin: _____

e. Stimulates development of ova in females and sperm in males: _____

f. Present in bloodstream of nonpregnant women to prevent lactation:

_____

g. Stimulates ovary to release egg and change follicle cells into corpus luteum:

_____

h. Stimulates uterine contractions and also breast milk let-down:

_____

i. Mimics many effects of sympathetic nerves: _____
j. Stimulates kidney tubules to produce small volume of concentrated urine:

_____

k. Target hormone of ACTH: _____

l. Increases blood calcium and decreases blood phosphate: _____

m. Decreases blood calcium and phosphate: _____

n. Contains iodine as an important component: _____
o. Raises blood glucose (three or more answers):

_____, _____, _____

p. Lowers blood glucose: _____

q. Serves anti-inflammatory functions: _____

■ **I2.** *For extra review.* Match the disorder with the hormonal imbalance.

Ac.	Acromegaly	M.	Myxedema
Ad.	Addison's disease	PC.	Pheochromocytoma
C.	Cretinism	PD.	Pituitary dwarfism
DI.	Diabetes insipidus	S.	Simple goiter
DM.	Diabetes mellitus	T.	Tetany
G.	Graves' disease		

_____ a. Deficiency of hGH in child; slow bone growth

_____ b. Excess of hGH in adult; enlargement of hands, feet, and jawbones

_____ c. Deficiency of ADH; production of enormous quantities of "insipid" (nonsugary) urine

_____ d. Deficiency of effective insulin; hyperglycemia and glycosuria (sugary urine)

_____ e. Deficiency of thyroxin in child; short stature and mental retardation

_____ f. Deficiency of thyroxin in adult; edematous facial tissues, lethargy

_____ g. Excess of thyroxin; protruding eyes, "nervousness," weight loss

_____ h. Deficiency of thyroxin due to lack of iodine; most common reason for enlarged thyroid gland

_____ i. Result of deficiency of PTH; decreased calcium in blood and fluids around muscles resulting in abnormal muscle contraction

_____ j. Deficiency of adrenocorticoids; increased K^+ and decreased Na^+ resulting in low blood pressure and dehydration

_____ k. Tumor of adrenal medulla causing sympathetic-type responses, such as increased pulse and blood pressure, hyperglycemia, and sweating

■ **I3.** Summarize effects of hormones on electrolytes by completing this table. Write *In* for increase and *De* for decrease to indicate the effect of a hormone on blood levels of each ion. Omit portions of the table with—.

Hormone	Ca^2	Mg^{2+}	HOP_4^{2-}, PO_4^{3-}	K^+	Na^+
a. PTH	In			—	—
b. Calcitonin		—		—	—
c. Aldosterone	—	—	—		

■ **I4.** Summarize effects of different hormones on metabolism by completing this table.

		Stimulates Protein	Stimulates Lipolysis?	Hyperglycemic Effect?
a.	hGH	_____-bolism	Yes? No?	Yes? No?
b.	Thyroid hormone	_____-bolism	Yes? No?	Yes? No?
c.	Cortisol	_____-bolism	Yes? No?	Yes? No?
d.	Insulin	_____-bolism	Yes? No?	Yes? No?

A1.

	Secretions Transported In	Examples
a. Exocrine	Ducts	Sweat, oil, mucus, digestive juices
b. Endocrine	Bloodstream	Hormones

A2. (a) A, pineal; B, posterior pituitary; C, anterior pituitary; D, thyroid; E, parathyroid; F, adrenal cortex; G, adrenal medulla. (b) 1, hypothalamus; 2, thymus; 3, heart; 4, skin; 5, liver; 6, kidney; 7, stomach; 8, pancreas; 9, intestine; 10, ovary; 11, placenta; 12, testis.

A3. (a) N. (b) E. (c) N.

A5. Hormones affect only cells with specific receptors that "fit" the hormone. RU486 blocks receptors for the hormone progesterone and therefore prevents pregnancy because this hormone is needed to establish and maintain a luxuriant uterine lining required for embryonic development.

A6. (a) Decrease; down. (b) More deficient.

A7. (a) Endo. (b) Auto, para. (c) Endo.

A8. (a) PP. (b) BA. (c) E. (d) S.

A9. Insulin is a protein that would be destroyed by digestive enzymes.

A10. (a–b) TP. (c–d) FF.

B1. (a–b) I. (c–d) P.

B2. (a) First, endocrine gland, target cell. (b) Outer, G. (c) Adenylate cyclase, inner; cyclic AMP, second. (d) Protein kinases, phosphate; urine. (e) Short; phosphodiesterase. (f) Water; ADH, OT, FSH, LH, TSH, ACTH, CT, PTH, glucagon, hypothalamic regulating hormones, epinephrine, and NE. (g) Ca^{2+}, cGMP, IP_3, or DAG.

B4. (a) Permissive. (b) Antagonistic.

B5. (a) N. (b) C. (c) H.

B6. Negative; increase. Examples of positive feedback: uterine contractions stimulate more oxytocin, or high estrogen level stimulates high LH level and ovulation.

C1. (a) Hypophysis. (b) The pituitary regulates many body activities; yet this gland is "ruled" by releasing and inhibiting hormones from the hypothalamus and also by negative feedback mechanisms. (c) In the sella turcica just inferior to the hypothalamus. (d) Anterior, adeno; neuro. (e) 7; hGH, PRL, MSH, ACTH, TSH, FSH, LH. (f) SHA PPC HPV SPC* AHV; hypophysiotropic. (g) Hypothalamic hormones reach the pituitary quickly and without dilution because they do not have to traverse systemic circulation. (h) Posterior. (i) 1, Corticotrophs synthesize ACTH, and some make MSH; 2, gonadotrophs make FSH and LH; 3, lactotrophs produce PRL; 4, somatotrophs make hGH; 5, thyrotrophs produce TSH. (j) Hormones that stimulate release of other hormones; ACTH, FSH, LH, and TSH.

C2. (a) Adrenal cortex, high; high. (b) Anterior pituitary; tropic, high. (c) Negative, positive.

C3.

Hypothalamic Hormone →	Anterior Pituitary Hormone →	Target Hormone
a. GHRH (growth hormone releasing hormone or somatocrinin)	hGH (human growth hormone)	None
b. CRH (corticotropin releasing hormone)	1. ACTH (corticotropin or adrenocorticotropic hormone) 2. MSH (melanocyte stimulating hormone)	1. Cortisol 2. None
c. TRH (thyrotropin releasing hormone)	1. TSH (thyrotropin) or thyroid stimulating hormone) 2. hGH	1. Thyroxine 2. None
d. GnRH (gonadotropic releasing hormone)	1. FSH (follicle stimulating hormone) 2. LH (luteinizing hormone) or ICSH (interstitial cells stimulating hormone)	1. Estrogen 2. Estrogen, progesterone or Testosterone
e. PRH (prolactin releasing hormone)	Prolactin	None

C4. Are not. (a) hGH (human growth hormone). (b) PIH (prolactin inhibiting hormone), which is dopamine; none.

C5. (a) Indirect, insulin; somatomedins, insulin, more; liver, muscle, cartilage, and bone. (b) Bones, skeletal muscles; somatotropin; is not, does not. (c) Accelerating, protein; anabolism. (d) Cata, ATP; write *cata* on the line and *ATP* in the circle. (e) In, hyper; opposite; beta, diabeto. (f) GHRH (somatocrinin), GHIH (somatostatin); hypo, GHRH, high. (g) Giantism, dwarfism; acromegaly, hands, feet, face.

C7. (a) Prolactin (or lactogenic hormone) (PRL), anterior; oxytocin (OT), posterior. (b) Increase, PRH; PIH; decrease.

C8. (a) 30–60; a fraction of a second, bloodstream, 30–60. (b) Uterine contractions cause sensory impulses which stimulate secretion of more oxytocin (and therefore more contractions). Suckling of the infant at breast stimulates sensory nerves in the nipple which signal release of more oxytocin (and therefore more milk). (c) In; retain, de, de; constricts; in. (d) Opposite of; in, de. (e) Insipidus, enormous; does not. (f) Inhibits; in.

D1. (a) Follicular, thyroxine, triiodothyronine; calcitonin. (b) I^-; 3; T_4, T_3. (c) May be stored for months; proteins.

D3. (a) Ana. (b) Cata; increases, increasing; calorigenic. (c) Increases, increases. (d) Cretinism, does; does not occur.

D5. (a) An enlargement; deficient. (b) Hypo; decreased. (c) Hyper; eyes; antibodies.

D6. Labels on Figure LG 18.4: (a) Calcitonin (CT). (b) Parathyroid hormone (PT). Antagonists; D, calcitriol or 1, 25-dihydroxy vitamin D_3, calcium, phosphate, and magnesium; lower.

D7. Hypo; in, tetany; inverse.

E2. 1, The protein angiotensinogen; 2, renin; 3, angiotensin I; 4, angiotensin II; 5, constrictor; 6, aldosterone; 7, Na^+ (and H_2O by osmosis, and Cl^- and HCO_3^- to achieve electrochemical balance); 8, K^+ and H^+; 9, H_2O; 10, In.

E3. (a) Increase (known as *high renin hypertension*). (b) Decrease, such as the antihypertensive captopril (Capoten). (c) Increase.

E4. Angiotensin II (resulting from renin-angiotensin pathway) and increased blood level of K^+ (hyperkalemia).

E6. (a) Cata; opposite. (b) Gluconeogenesis; increase. (c) Increases; increase, limit.

E7. (a–d) C. (e–f) A.

E9. (a) Epinephrine or adrenaline (80%) and norepinephrine or noradrenaline; catecholamines. (b) Sympathetic; fight-or-flight response such as increase of heart rate, force of heart contraction, and blood pressure; constrict blood vessels and dilate airways; increase blood glucose level and metabolism; decrease digestion. (c) Cortex.

F1. Pancreas. (a) Endo; alpha, beta, delta, and F. (b) Alpha, increases; glycogen, glycogenolysis; gluconeogenesis. (c) No; glucose; high. (d) *Glycogen* is a storage form of glucose; *glucagon* is a hormone that stimulates breakdown of glycogen to glucose. (e) Opposite; helps; glycogen, glycogenesis. (f) Hyper; hypo; glucagon and epinephrine are hyperglycemic; insulin is hypoglycemic.

F2. Hyperglycemia, as well as hormones that lead to hyperglycemia (hGH, ACTH, cortisol, glucagon, epinephrine, or NE), increased blood level of amino acids, and parasympathetic (vagal) nerve impulses.

F3. (a–c) II. (d–e) I.

F4. (a) Lack of insulin prevents glucose from entering cells, so more is left in blood and spills into urine. (b) -Uria, -dipsia; glucose in urine draws water (osmotically), and as result of water loss, the patient is thirsty. (c) Acid; because cells have little glucose to use, they turn to excessive fat catabolism which leads to keto-acid formation (discussed in Chapter 25). (d) As fats are transported in extra amounts (for catabolism), some fats are deposited in walls of vessels; fatty (= *athero*) thickening (= *sclerosis*) results.

F5. Ovaries, testes, adrenal cortex, and placenta.

F7. Immunity; thymosin, thymic humoral factor, thymic factor, and thymopoietin.

G1. Pancreas and thyroid, as well as testes and ovaries.

G2. (a) P. (b) AP. (c) AM. (d) AC. (e) PT, T.

G3. (a) Erythropoietin. (b) Atrial natriuretic peptide. (c) Eicosanoids: thromboxane A_2 and PG F_2a. (d) CCK, gastrin, GIP, and secretin. (e) Vitamin D. (f) Human chorionic gonadotropin (HCG).

G5. Prostaglandins (PGs); reduce fever, pain, and inflammation.

G6. Erythopoietin, hGH, insulin, IGFs, prolactin, thymosin, thyroid hormone.

G7. (a) TAF. (b) EGF. (c) PDGF.

H2. (a) Hypothalamus. (b) Alarm, medulla, sympathetic. (c) Resistance; ACTH, hGH, TSH; cortex.

H3. (a) Fight-or-flight. (b) Increases, norepinephrine, epinephrine. (c) Increases; increase; skeletal muscles, heart, brain; spleen, GI tract, skin. (d) Decrease; increases.

H5. (a–c) A. (d–f) R. (g) E.

I1. (a) GHRH. (b) LH (or ICSH). (c) hGH, thyroid hormone (T_3 and T_4), and testosterone. (d) TSH. (e) FSH. (f) PIH. (g) LH. (h) OT. (i) Epinephrine and norepinephrine. (j) ADH. (k) Adrenocorticoid or glucocorticoid, such as cortisol. (l) PTH. (m) CT. (n) Thyroid hormone (T_3 and T_4). (o) Epinephrine, norepinephrine, glucagon, hGH (and indirectly CRH, ACTH, GHRH). (p) Insulin. (q) Adrenocorticoid or glucocorticoid, such as cortisol (and indirectly CRH and ACTH).

I2. (a) PD. (b) Ac. (c) DI. (d) DM. (e) C. (f) M. (g) G. (h) S. (i) T. (j) Ad. (k) PC.

I3.

Hormone	Ca^2	Mg^{2+}	HOP_4^{2-}, PO_4^{3-}	K^+	Na^+
a. PTH	In	In	De	—	—
b. Calcitonin	De	—	De	—	—
c. Aldosterone	—	—	—	De	In

I4.

	Stimulates Protein	Stimulates Lipolysis?	Hyperglycemic Effect?
a. hGH	Anabolism	Yes	Yes
b. Thyroid hormone	Anabolism	Yes	No (increases use of glucose for ATP production)
c. Cortisol	Catabolism	Yes	Yes
d. Insulin	Anabolism	No	No

1. Describe the roles of feedback systems in the following disorders: (a) a condition in which dietary iodine is decreased, leading to enlargement of the thyroid gland (simple goiter); (b) a condition in which cortisol is decreased (congenital adrenal hyperplasia), leading to enlarged adrenal glands.
2. Describe hormonal regulation of blood levels of the following ions: (a) calcium; (b) potassium.
3. Contrast blood supply and hormone production of the anterior and posterior lobes of the pituitary gland.
4. Describe the renin-angiotensin-aldosterone mechanism for controlling blood pressure.
5. Discuss advantages and disadvantages of anti-inflammatory effects of glucocorticoids.
6. Discuss roles of interleukin-1 (IL-1) in stress and resistance to disease.
7. Discuss the hormone imbalance (whether excessive or deficient) and symptoms of each of the following conditions: acromegaly, Addison's disease, Cushing's syndrome, diabetes insipidus, diabetes mellitus, and myxedema.

MASTERY TEST: CHAPTER 18

Questions 1–2: Arrange the answers in correct sequence.

_____ _____ _____ _____ _____

1. Steps in action of TSH upon a target cell:
   A. Adenylate cyclase breaks down ATP.
   B. TSH attaches to receptor site.
   C. Cyclic AMP is produced.
   D. Protein kinase is activated to cause the specific effect mediated by TSH, namely, stimulation of thyroid hormone production.
   E. G-protein is activated.

_____ _____ _____ _____

2. Steps in renin-angiotensin mechanism to increase blood pressure:
   A. Angiotensin I is converted to angiotensin II.
   B. Renin converts angiotensinogen, a plasma protein, to angiotensin I.
   C. Angiotensin II causes vasoconstriction and stimulates aldosterone production which causes water conservation and raises blood pressure.
   D. Low blood pressure or low blood Na+ level causes secretion of enzyme renin from cells in kidneys.

Questions 3–10: Circle the letter preceding the one best answer to each question.

3. All of these compounds are synthesized in the hypothalamus *except:*
   A. ADH
   B. GHRH
   C. CT
   D. PIH
   E. Oxytocin
4. Choose the *false* statement about endocrine glands.
   A. They secrete chemicals called hormones.
   B. Their secretions enter extracellular spaces and then pass into blood.
   C. Sweat and sebaceous glands are endocrine glands.
   D. Endocrine glands are ductless.
5. All of the following correctly match hormonal imbalance with related signs or symptoms *except:*
   A. Deficiency of ADH—excessive urinary output
   B. Excessive aldosterone—high blood pressure
   C. Deficiency of aldosterone—low blood level of potassium and muscle weakness
   D. Excess of parathyroid production—demineralization of bones; high blood level of calcium.
6. All of the following hormones are secreted by the anterior pituitary *except:*
   A. ACTH
   B. FSH
   C. Prolactin
   D. Oxytocin
   E. hGH
7. Which one of the following is a function of adrenocorticoid hormones?
   A. Lower blood pressure
   B. Raise blood level of calcium
   C. Convert glucose to amino acids
   D. Lower blood level of sodium
   E. Anti-inflammatory

8. All of these hormones lead to increased blood glucose *except*:
   A. ACTH
   B. Insulin
   C. Glucagon
   D. Growth hormone
   E. Epinephrine
9. Choose the *false* statement about the anterior pituitary.
   A. It secretes at least seven hormones.
   B. It develops from mesoderm.
   C. It secretes tropic hormones.
   D. It is stimulated by releasing factors from the hypothalamus.
   E. It is also known as the adenohypophysis.

10. Choose the *false* statement.
    A. Both ADH and aldosterone tend to lead to retention of water and increase in blood pressure.
    B. PTH activates vitamin D, which enhances calcium absorption.
    C. PTH activates osteoclasts, leading to bone destruction.
    D. Tetany is a sign of hypocalcemia.
    E. ADH and OT pass from hypothalamus to posterior pituitary via pituitary portal veins.

Questions 11–20: Circle T (true) or F (false). If the statement is false, change the underlined word or phrase so that the statement is correct.

T F 11. Polyphagia, polyuria, and hypoglycemia are all symptoms associated with insulin deficiency.

T F 12. Hypophysiotropic hormones are all secreted by the anterior pituitary and affect the hypothalamus.

T F 13. The hormones of the adrenal medulla mimic the action of parasympathetic nerves.

T F 14. Prostaglandins are classified as eicosanoids and have a broad range of effects.

T F 15. Epinephrine and cortisol are both adrenocorticoids.

T F 16. Secretion of growth hormone, insulin, and glucagon is controlled (directly or indirectly) by blood glucose level.

T F 17. Most of the feedback mechanisms that regulate hormones are positive (rather than negative) feedback mechanisms.

T F 18. The alarm reaction occurs before the resistance reaction to stress.

T F 19. Most hormones studied so far appear to act by a mechanism involving cyclic AMP rather than by the gene activation mechanism.

T F 20. A high level of thyroxine circulating in the blood will tend to lead to a low level of TRH and TSH.

Questions 21–25: Fill-ins. Complete each sentence with the word or phrase that best fits.

_____ 21. _____ is a hormone used to induce labor because it stimulates contractions of smooth muscle of the uterus.

_____ 22. _____ and _____ are two hormones that are classified chemically as bioamines.

_____ 23. _____ is a term that refers to the control mechanism by which the number of receptors for a hormone will be decreased when that hormone is present in excessive amounts.

_____ 24. Type _____ diabetes mellitus is a non-insulin-dependent condition in which insulin level is adequate but cells have decreased sensitivity to insulin.

_____ 25. Three factors that stimulate release of ADH are _____.

ANSWERS TO MASTERY TEST: ☆ CHAPTER 18

Arrange
1. B E A C D
2. D B A C

Multiple Choice

3. C	7. E
4. C	8. B
5. C	9. B
6. D	10. E

True–False
11. F. Polyphagia and polyuria are both or Polyphagia, poluria, and hyperglycemia are all
12. F. Hypothalamus and affect the anterior pituitary
13. F. Sympathetic
14. T
15. F. Cortisol, aldosterone, and androgens are all
16. T
17. F. Negative (rather than positive)
18. T
19. T
20. T

Fill-ins
21. Oxytocin (OT or pitocin)
22. Thyroid hormones (T_3 and T_4), epinephrine or norepinephrine
23. Down-regulation
24. II (maturity onset)
25. Decreased extracellular water, pain, stress, trauma, anxiety, nicotine, morphine, tranquilizers

UNIT IV

Maintenance
of the Human Body

FRAMEWORK 19
Blood

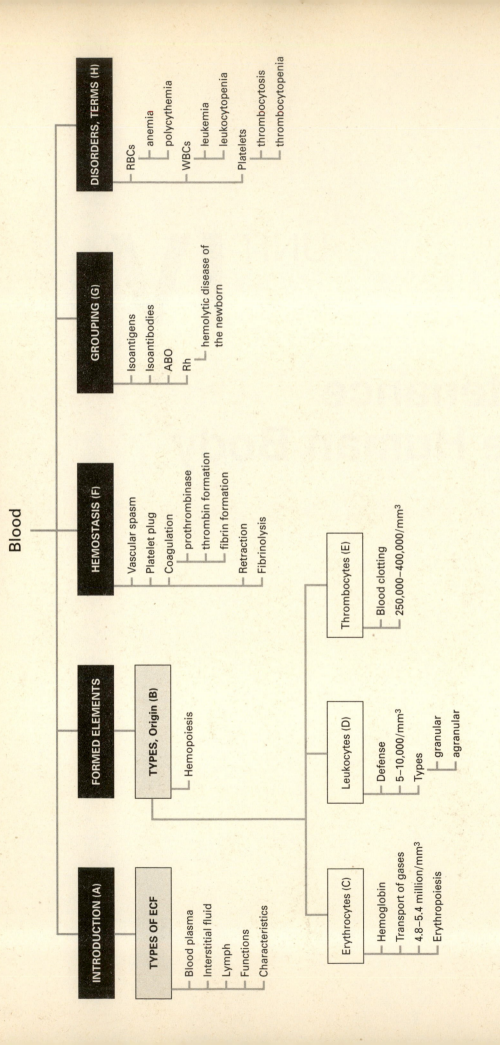

INTRODUCTION (A)

TYPES OF ECF
- Blood plasma
- Interstitial fluid
- Lymph
- Functions
- Characteristics

FORMED ELEMENTS

TYPES, Origin (B)
- Hemopoiesis

Erythrocytes (C)
- Hemoglobin
- Transport of gases
- 4.8–5.4 million/mm³
- Erythropoiesis

Leukocytes (D)
- Defense
- 5–10,000/mm³
- Types
  - granular
  - agranular

Thrombocytes (E)
- Blood clotting
- 250,000–400,000/mm³

HEMOSTASIS (F)
- Vascular spasm
- Platelet plug
- Coagulation
  - prothrombinase
  - thrombin formation
  - fibrin formation
- Retraction
- Fibrinolysis

GROUPING (G)
- Isoantigens
- Isoantibodies
- ABO
- Rh
  - hemolytic disease of the newborn

DISORDERS, TERMS (H)
- RBCs
  - anemia
  - polycythemia
- WBCs
  - leukemia
  - leukocytopenia
- Platelets
  - thrombocytosis
  - thrombocytopenia

The Cardiovascular System: The Blood

CHAPTER
19

The major role of the blood is transportation through an intricate system of channels—blood vessels—that reach virtually every part of the body. Blood constantly courses through the vessels, carrying gases, fluids, nutrients, electrolytes, hormones, and wastes to and from body cells. Most of these chemicals "ride" within the fluid portion of blood (plasma). Some, such as oxygen, piggyback on red blood cells. White blood cell "passengers" take short trips in the blood and depart from the vessels at sites calling for defense. Platelets and other clotting factors congregate to plug holes in the network of vessels.

As you begin your study of the blood, carefully examine the Chapter 19 Topic Outline and Objectives; check off each one as you complete it. To organize your study of blood, glance over the Chapter 19 Framework now. Be sure to refer to the Framework frequently and note relationships among key terms in each section.

TOPIC OUTLINE AND OBJECTIVES

A. Types of body fluids, functions and characteristics of blood

☐ 1. Contrast the general roles of blood, lymph, and interstitial fluid in maintaining homeostasis.
☐ 2. Define the functions and physical characteristics of the various components of blood.
☐ 3. List the components of plasma and explain their importance.

B. Formation of blood cells

☐ 4. Compare the origins, histology, and functions of the formed elements in blood.

C. Erythrocytes (red blood cells)

D. Leukocytes (white blood cells)

E. Platelets

F. Hemostasis

☐ 5. Identify the stages involved in blood clotting and explain the various factors that promote and inhibit blood clotting.

G. Blood groups and blood types

☐ 6. Explain ABO and Rh blood grouping.

H. Disorders, medical terminology

☐ 7. Identify the clinical symptoms of several types of anemia, polycythemia, infectious mononucleosis (IM), and leukemia.
☐ 8. Define medical terminology associated with blood.

WORDBYTES

Now become familiar with the language of this chapter by studying each wordbyte, its meaning, and an example of its use within a term. After you study the entire list, self-check your understanding by writing the meaning of each wordbyte on the line. As you continue through the *Learning Guide,* identify (and fill in) additional terms that contain the same wordbyte.

Wordbyte	Self-Check	Meaning	Example(s)
a-, an-	_____	not, without	*anemia*
-crit	_____	to separate	hemato*crit*
-cyte	_____	cell	leuko*cyte*
dia-	_____	through	*dia*phragm
-emia	_____	blood	leuk*emia*
erythro-	_____	red	*erythro*poiesis
heme-, hemo-	_____	blood	*hemo*stasis
leuko-	_____	white	*leuko*cytosis
mega-	_____	large	*mega*karyocyte
-osis	_____	condition	leukocyt*osis*
-pedesis	_____	moving	dia*pedesis*
-penia	_____	want, lack	leukocyto*penia*
-phil-	_____	love	hemo*phil*iac
-poiesis	_____	formation	hemo*poiesis*
-rrhag-	_____	burst forth	hemo*rrhage*
-stasis	_____	standing still	hemo*stasis*
thromb-	_____	clot	*thromb*us, pro*thromb*in

CHECKPOINTS

A. Overview: functions and characteristics of blood (pages 553–554)

A1. Review relationships among these three body fluids: blood, interstitial fluid, and lymph. (Refer to Figure LG 1.1, page 9 of the *Learning Guide.*)

■ **A2.** Contrast body fluids in the table.

	Interstitial Fluid and Lymph	Plasma
a. Amount of protein (more or less)		
b. Formed elements present		Red blood cells, white blood cells, platelets
c. Location	Between cells and in lymph vessels	

■ **A3.** Name the structures that comprise the:

a. Cardiovascular system

_____   _____   _____

b. Lymphatic system

_____   _____   _____

_____   _____

■ **A4.** Describe functions of blood in this Checkpoint.

a. Blood transports many substances. List six or more.

_____   _____   _____

_____   _____   _____

b. List four aspects of homeostasis regulated by blood. One is done for you.

_____**pH**_____   _____

_____   _____

c. Explain how your blood protects you.

■ **A5.** Circle the answers that correctly describe characteristics of blood.

a. Average temperature:
36°C (96.8°C)   37°C (98.6°C)   38°C (100.4°C)

b. pH:
6.8   7.0–7.1   7.35–7.45

c. Volume of blood in average adult:
1.5–3 liters   5–6 liters   8–10 liters

d. Three hormones that regulate volume and osmotic pressure of blood:
ADH   aldosterone   ANP   epinephrine   PTH

A6. Contrast the following procedures: *venipuncture/arterial stick/finger stick.*

■ **A7.** Refer to Figure 19.1a, page 555 in your text, and complete this checkpoint. Use these answers.

E. Erythrocytes	LP. Leukocytes and platelets	P. Plasma

a. In centrifuged blood, layers of blood from top to bottom are:_____ _____ _____

b. In centrifuged blood, volume of layers of blood from greatest to least are: _____ _____ _____

c. Which layer contains the red blood cells? _____

d. Which layer forms the buffy coat? _____

■ **A8.** Match the names of components of plasma with their descriptions.

A. Albumins	GAF. Glucose, amino acids, and fats
E. Electrolytes	HE. Hormones and enzymes
F. Fibrinogen	UC. Urea, creatinine
G. Globulins	W. Water

_____ a. Makes up about 92% of plasma

_____ b. Regulatory substances carried in blood

_____ c. Cations and anions carried in plasma

_____ d. Constitutes about 54% of plasma protein

_____ e. Made by liver; a protein used in clotting

_____ f. Antibody protein

_____ g. Wastes carried to kidneys or sweat glands

_____ h. Food substances carried in blood

_____ i. Of albumin, fibrinogen, and globulin, found in least abundance in plasma

■ **A9.** Contrast antigens (Ag) and antibodies (Ab) in this exercise by writing *Ag* or *Ab* next to related descriptions below.

_____ a. Viruses or bacteria that enter the body as "invaders"

_____ b. Chemicals made by the body to defend against invaders

_____ c. Immunoglobulins (Ig's) produced by plasma cells and found in plasma

■ **A10.** Check your understanding of formed elements of blood in this Checkpoint. Use the answers in the box.

> E. Erythrocytes   L. Leukocytes   P. Platelets

_____ a. Not truly cells, but fragments of cells   _____ d. Also known as white blood cells

_____ b. Also known as thrombocytes   _____ e. Makes up over 99% of all formed elements

_____ c. Include two subcategories: granular and agranular

B. Formation of blood cells (pages 554–558)

■ **B1.** Check your knowledge about blood formation in this activity.

a. Regulation of *(erythrocyte? leukocyte?)* number occurs by a negative feedback mechanism, whereas alteration of *(erythrocyte? leukocyte?)* number occurs in response to invading microorganisms and antigens.

b. Blood formation is a process known as _____.

All blood cells arise from _____ _____ cells that give rise to five types of blast cells. For example, myeloblasts will form granular *(erythrocytes? leukocytes? platelets?)*, whereas proerythroblasts will form mature *(erythrocytes? leukocytes? platelets?)*. (For more review of this content, see Checkpoint B3 below.)

c. Name several structures where blood formation takes place before birth.

d. After birth, most hemopoiesis takes place in the *(red bone marrow? liver? lymph nodes?)*. Name six or more bones in which red blood cells are formed after birth.

e. Erythropoietin is a hormone made mostly in the _____. This hormone stimulates production of *(RBCs? WBCs? platelets?)*.

f. Two types of cytokines that stimulate WBC formation are _____-

_____ factors (CSFs) and _____ (ILs).

■ **B2.** *A clinical challenge.* Explain how the following chemicals produced via recombinant DNA may be helpful clinically:

a. *EPO* for patients with kidney failure

b. *Neupogen* for patients with immune deficiency

■ **B3.** *For extra review* of hemopoiesis, carefully study Figure 19.2, page 557 in your text, and then identify the types of mature blood cells (or other cells) in the box that are destined to form from the cells listed below.

Basophils	Erythrocytes	Platelets
B-lymphocytes	Monocytes	T lymphocytes
Eosinophils	Neutrophils	Wandering macrophages

a. Reticulocytes   →   _____

b. Megakaryocytes   →   _____

c. Lymphoblasts (more than one answer)   →   _____

d. Basophilic myelocyte   →   _____

e. Monocyte   →   _____

f. Hemopoietic stem cell (more than one answer)   →   _____

C. Erythrocytes (red blood cells)   (pages 558–562)

■ **C1.** Refer to Figure LG 19.1. Which diagram represents a mature erythrocyte? _____ Note that it

(does? does not?) contain a nucleus. The chemical named _____ accounts for the color of red blood cells (RBCs). Label and color the RBC.

■ **C2.** Describe hemoglobin in this exercise.
   a. The hemoglobin molecule consists of a central portion, which is the protein *(heme? globin?)* with four polypeptide chains called *(hemes? globins?)*.

   b. Each heme contains one _____ atom on which a molecule of *(oxygen? carbon dioxide?)* can be transported so that each hemoglobin molecule can carry *(1? 4?)* oxygen molecule(s). Almost all oxygen is transported in this manner.
   c. Carbon dioxide has one combining site on hemoglobin; it is the amino acid portion of the *(heme? globin?)*; therefore, it contributes to formation of

   _____-hemoglobin. About *(23? 70? 97?)* % of carbon dioxide is carried in this state. Most carbon dioxide is transported

   (in blood cells? in plasma?) within the _____ (HCO_3^-) ion.

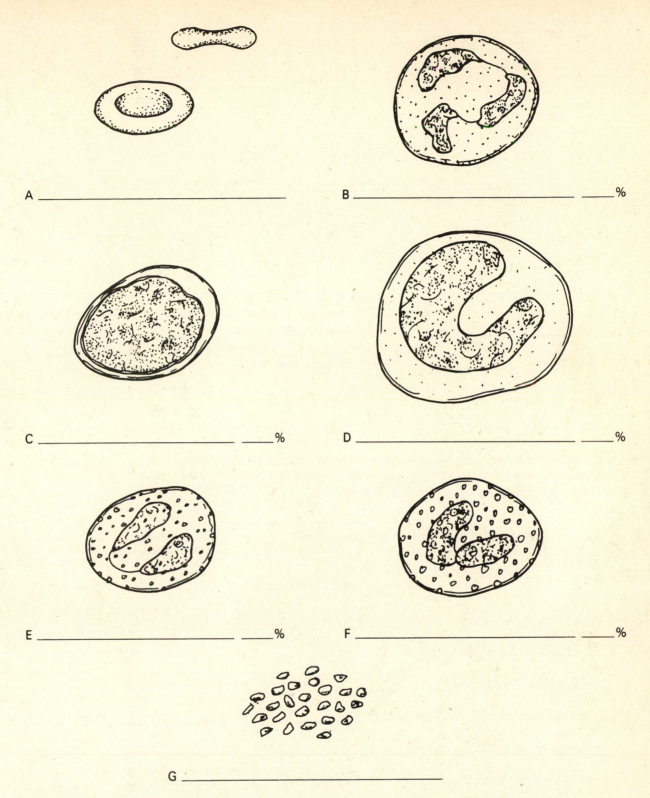

A _____

B _____ ____%

C _____ ____%

D _____ ____%

E _____ ____%

F _____ ____%

G _____

Figure LG 19.1 Diagrams of blood cells. Label, color, and complete as directed in Checkpoints C1, D1, and E2.

■ **C3.** Explain the significance of the structure and shape of red blood cells (RBCs) in this Checkpoint.

a. RBCs *(do? do not?)* have a nucleus and mitochondria. State one advantage of this fact.

State one disadvantage.

b. RBCs are normally shaped like _____ disks.
What advantage does this shape offer?

c. What shape may RBCs assume in the disorder sickle-cell anemia (SCA)?

_____ Under what conditions are the RBCs likely

to do this? _____ What are the consequences?

■ **C4.** Fill in blanks in the following description of the RBC life cycle. Select answers from the box. Each answer will be used once except one which will be used three times. It may help to refer to Figure 19.4, page 560 in your text.

Amino acids	Ferritin and hemosiderin	Stercobilin
Bile	Iron	Transferrin
Bilirubin	Red bone marrow	Urobilin
Biliverdin	Spleen, liver, or red bone marrow	Urobilinogen

a. Old RBCs are phagocytosed by macrophages in the _____.

b. Hemoglobin is then recycled: globin is broken into _____ and hemes

are degraded into the metal _____ and a noniron portion; see (e) below.

c. Iron is carried through blood attached to the protein _____; iron may

be stored in muscles attached to proteins _____. Once released from

muscles, iron is again combined with _____. Iron absorbed in the

intestine is also attached to _____ for transport through blood.

d. Ultimately iron reaches _____, where it is used for formation

of new _____ molecules.

e. What happens to the noniron portion of heme? It is converted to the green pigment named

_____ and the orange pigment _____.

f. Bilirubin has two possible fates. It may pass in blood from bone marrow to liver

and be used to form _____, or bilirubin may be transformed by

intestinal bacteria into _____. Urobilinogen may be used to

form _____ that gives feces its brown color or _____
that contributes to the yellow color of urine.

■ **C5.** Identify the stages of cells in erythrocyte formation in this exercise. It may help to refer to Figure 19.2, page 557 in your text. Use each answer in the box once.

Polychromatophilic erythroblasts	Proerythroblasts
Acidophilic erythroblasts	Reticulocytes

a. _____ are direct descendants of hematopoietic stem cells.

b. Hemoglobin is first synthesized in _____.

c. Hemoglobin synthesis is maximal in _____.

d. _____ have lost their nuclei and are just about to leave
the red bone marrow and enter the blood stream.

■ **C6.** After you study Figure 19.5, page 561 in your text, explain the roles of the terms in the box in red blood cell production.

B_{12}	H. Hypoxia
E. Erythropoietin	IF. Intrinsic factor

_____ a. Decrease of oxygen in cells; serves as a signal that erythropoiesis is needed (to help provide more oxygen to tissues)

_____ b. Hormone produced by kidneys when they are hypoxic; stimulates erythropoiesis in red bone marrow

_____ c. Vitamin necessary for normal erythropoiesis

_____ d. Substance produced by the stomach lining and necessary for normal vitamin B_{12} absorption

■ **C7.** Circle the most normal blood values. (Note that values vary slightly according to age and gender.)

a. Average life of a red blood cell: 4 hours   4 days   4 months   4 years

b. RBC count in cube this size: ■ (mm³)   500   5,000   250,000   5 million   250 million

c. Hemoglobin in adults (g/100 ml blood):   1   8   15   27   41

■ **C8.** Write the correct terms that fit each description on lines provided; then write the value that best matches each term. Select answers from the box; not all answers will be used.

> Terms: anemia, hematocrit, hemoglobin, polycythemia, reticulocyte count
> Values: 1 8 15 27 41 65

a. Percentage of RBCs that are not quite mature in the circulating blood; this value

increases during rapid erythropoiesis: _____ _____ %

b. Percentage of blood that consists of RBCs; value is usually about 3 times the person's

hemoglobin value: _____ _____ %

c. Condition in which a person has a higher than normal red blood cell count:

_____ _____ %

■ **C9.** *A clinical application.* Explain the rationale behind each of the following practices of some athletes in attempts to increase oxygen-carrying capacity of blood and, ultimately, athletic performance.

a. Training in a high-altitude city such as Mexico City or Denver for a competition to be held in one of those cities.

b. *Induced erythrocytemia (blood doping):* removal of RBCs a certain period before competition with reinjection of the cells shortly before the competition.

D. Leukocytes (white blood cells)   (pages 562–564)

■ **D1.** Refer to Figure LG 19.1 and do this exercise about leukocytes.

a. Leukocytes *(have? lack?)* hemoglobin, and so these cells are known as *(red blood cells or RBCs? white blood cells or WBCs?)*.

b. Label each leukocyte (WBC) on Figure 19.1, and indicate what percentage of the total WBC count is accounted for by each type of WBC. Such a breakdown of white blood cells is

known as a _____ WBC count.

c. Each white blood cell (WBC) *(has? lacks?)* a nucleus. Which type of WBC has a large

kidney-shaped nucleus? _____ Which WBC has a nucleus that occupies

most of the cell? _____. Which is polymorphonuclear? _____

d. Which three WBCs are granulocytes? _____ _____ _____

e. *For extra review.* Color nucleus, cytoplasm, and granules of all WBCs.

D2. Contrast terms in each pair.

a. *Granular leukocytes/agranular leukocytes*

b. *Polymorphonuclear leukocytes (PMNs)/bands*

c. *Fixed macrophages/wandering macrophages*

■ **D3.** What are MHC antigens and how do they affect the success rate of transplants?

■ **D4.** Contrast WBCs with RBCs in this activity.

a. A normal RBC count is about *(700? 5,000–10,000? 250,000? 5 million?)* cells/mm³,

whereas a typical WBC count is _____ cells/mm³. In other words, the

ratio of RBCs to WBCs is about _____:1.

b. Which cells normally live longer? *(RBCs? WBCs?)* The typical lifespan of a WBC is a few *(days? months? years?).*

■ **D5.** Match the answers in the box with descriptions of processes involved in inflammation.

| A. Adhesion molecules | E. Emigration |
| C. Chemotaxis | P. Phagocytosis |

_____ a. The process by which WBCs stick to capillary wall cells, slow down, stop, and then squeeze out through the vessel wall

_____ b. Chemicals known as selectins made by injured or inflamed capillary cells stick to chemicals on neutrophils, slowing them down

_____ c. Attraction of phagocytes to chemicals such as kinins or CSFs in microbes or in inflamed tissue

_____ d. Ingestion and disposal of microbes or damaged tissue by neutrophils or macrophages

■ **D6.** Check your understanding of types of WBCs by matching names of WBCs with descriptions. Answers may be used more than once.

B. Basophils	M. Monocytes
E. Eosinophils	N. Neutrophils
L. Lymphocytes	

_____ a. Constitute the largest percentage of WBCs

_____ b. Produce defensins, lysozymes, and oxidants with antibiotic activity

_____ c. Types of these cells include B cells, T cells, and natural killer (NK) cells

_____ d. They leave the blood, develop into mast cells, and release heparin, histamine, and serotonin, which are involved in allergic and inflammatory responses

_____ e. Involved in allergic reactions, combat histamines, and provide protection against parasitic worms

_____ f. Form fixed and wandering macrophages that clean up sites of infection

_____ g. Important in phagocytosis (two answers)

_____ h. Classified as agranular leukocytes (two answers)

_____ i. Known as *bands* in immature state

_____ j. In stained cells, lilac-colored granules are visible

_____ k. In stained cells, red-orange granules are visible

_____ l. The only type of WBC that leaves the bloodstream to fight infection but then returns to the bloodstream again

■ **D7.** Contrast types of lymphocytes in this activity.

a. In response to invasion of foreign antigens, *(B? T? NK?)* lymphocytes become plasma

cells that secrete _____. B cells are especially effective against *(bacteria and their toxins? cancer or transplanted cells, viruses, fungi, and some bacteria?)*.

b. T cells are particularly effective against *(bacteria and their toxins? cancer or transplanted cells, viruses, fungi, and some bacteria?)*. Natural killer cells combat a *(few selected? wide variety of?)* microbes plus some cancer cells.

■ **D8.** *A clinical challenge.* Mrs. Doud arrives at a health clinic with a suspected acute infection. Complete this Checkpoint about her.

a. During infection, it is likely that Mrs. Doud's leukocyte count will *(in? de?)*-crease. A count of *(4000? 8000? 12,000?)* leukocytes/mm³ blood is most likely. This condition is known as *(leukocytosis? leukopenia?)*.

b. A differential increase in the number of WBCs named _____ is most indicative of enhanced phagocytic activity during infection. Neutrophils are most likely to account for *(48? 62? 76?)%* of the total white count in Mrs. Doud's blood. A sign of a chronic infection is increase in

the percentage of cells that can become macrophages; these are _____

■ **D9.** *A clinical challenge.* Do this exercise on diagnostic clues offered in a differential WBC count. Select the conditions that might be indicated for each specific WBC change listed below. Select answers from the box.

A. Allergic reactions
B. Prolonged, severe illness and/or immunosuppression
C. Lupus, vitamin B_{12} deficiency, radiation
D. Pregnancy, ovulation, stress, and hyperthyroidism

_____ a. Decrease in lymphocyte count   _____ c. Decrease in neutrophil count

_____ b. Decrease in basophil count   _____ d. Increase in basophil or eosinophil count

D10. Briefly describe the procedure of *bone marrow transplant.* Describe the process as well as conditions for which this procedure is used.

E. Platelets   (pages 564–565)

■ **E1.** Circle correct answers related to platelets.

a. Platelets are also known as:

antibodies   thrombocytes   red blood cells   white blood cells

b. Platelets are formed in:

bone marrow   tonsils and lymph nodes   spleen

c. Platelets are:

entire cells   chips off the old metamegakaryocytes

d. A normal range for platelet count is _____/mm³

5000–1000   4.5–5.5 million   250,000–400,000

e. The primary function of platelets is related to:

O_2 and CO_2 transport   blood clotting   defense   blood typing

■ **E2.** Label platelets on Figure LG 19.1.

■ **E3.** List the components of a complete blood count (CBC).

F. Hemostasis   (pages 565–570)

■ **F1.** Hemostasis literally means _____ *(hemo)*-

_____ *(stasis).* List the three basic mechanisms of hemostasis.

a. Vascular _____

b. _____ plug formation

c. _____ (clotting)

F2. List chemicals present within granules of platelets.

a. In alpha granules:

b. In dense granules:

■ **F3.** After reviewing Figure 19.8, page 566 in your text, check your understanding of the three steps of platelet plug formation in this Checkpoint.

a. When platelets snag on the inner lining of a damaged blood vessel, their shape is altered. They become *(small, disc-shaped? large, irregular-shaped?).* Additional platelets aggregate on the uneven projections. This phase is known as platelet *(adhesion? release reaction? plug?).*

b. Platelets then release chemicals. Some of these chemicals *(dilate? constrict?)* vessels. How does this help hemostasis?

Name two of these chemicals. _____, _____

c. State two roles of the ADP that is released by platelets.

d. The resulting platelet plug is useful in preventing blood loss in *(large? small?)* vessels.

■ **F4.** The third step in hemostasis, coagulation, requires clotting factors which are derived from three sources.

    a. Some of these factors are made in the liver and constantly circulate in blood
       (*plasma? serum?*).

    b. Others are released by *(red blood cells? platelets?).*

    c. One important factor, called _____ factor (TF) or thromboplastin, is
       released from damaged cells. If these cells are within blood or in the lining
       of blood vessels, then the *(extrinsic? intrinsic?)* pathway to clotting is initiated. If other
       tissues are traumatized and release tissue factor, then the *(extrinsic? intrinsic?)* pathway
       is initiated.

■ **F5.** Summarize the three major stages in the cascade of events in coagulation by filling in
    blanks A–E in Figure LG 19.2. Choose from the list of terms on the figure.

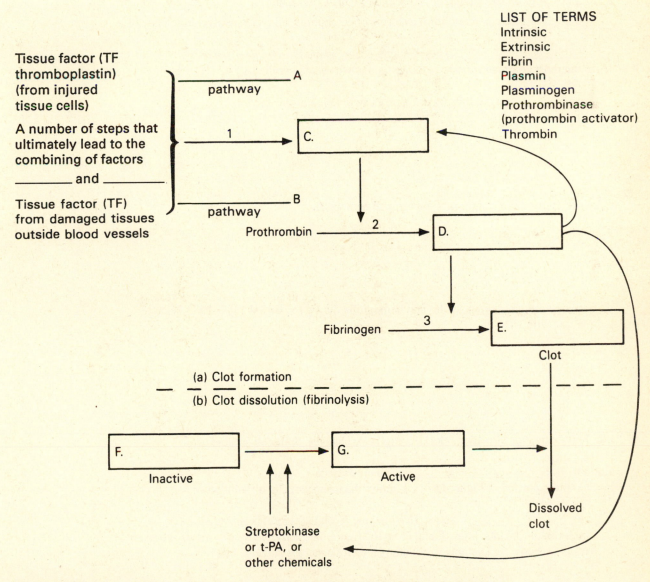

Figure LG 19.2 Summary of steps in clot formation and dissolution. Complete as directed in
Checkpoints F5, F6, and F8.

■ **F6.** *For extra review* refer to Figure LG 19.2 as well as Exhibit 19.3, page 567 in your text, and do this exercise.

    a. The figure indicates that thrombin exerts a *(positive? negative?)* feedback effect upon prothrombinase, causing the clot to *(enlarge? shrink?)*. Which chemical finally absorbs

       90% of the thrombin to prevent snowballing of the clot? _____

    b. The ion *(Ca²⁺? K⁺? Na⁺?)*, also known as factor _____, plays a critical role in many aspects of coagulation. Write this ion on the figure next to each of the three major steps.

    c. Fill in the two short lines on Figure LG 19.2 with Roman numerals to indicate which two factors combine to form prothrombinase.

    d. Write in factor numbers for other clotting chemicals shown on Figure LG 19.2.

■ **F7.** Check your understanding of hemophilia in this Checkpoint.

    a. Although there are several different forms of hemophilia, all types involve deficiency of

       _____. Types A and B occur primarily in *(males? females?)*.

    b. In the most common form of hemophilia, known as hemophilia *(A? B? C?)*,

       factor _____ is missing. State the primary function of this factor.

    c. Write three or more signs or symptoms of hemophilia.

    d. Explain why most persons who had hemophilia during the period 1982–1985 became infected with HIV.

■ **F8.** Following clot formation, clots undergo two changes. Summarize these.

    a. First, clot _____ or syneresis occurs.

    b. The lower portion of Figure LG 19.2 shows the next step: clot dissolution or

       _____. An inactive enzyme (F on the figure) is activated to (G), which dissolves clots. Label F and G on the figure. F is activated by chemicals produced within the body, for example, from damaged blood vessels and also by

       (D) _____.

■ **F9.** *A clinical challenge.* Match each chemical with its role in clot formation or dissolution.

> H. Heparin   S, t-PA. Streptokinase and t-PA
> K. Vitamin K   W. Warfarin (Coumadin)
> PGI₂. Prostacyclin

_____ a. Required for synthesis of factors II, VII, IX, and X in clot formation. Commercial preparations may be required for newborns who lack the intestinal bacteria that synthesize this chemical and also for persons with malabsorption of fat because this is a fat soluble chemical.

_____ b. Slow-acting anticoagulant that is antagonistic to vitamin K, so decreases synthesis of prothrombin in liver; used to prevent clots in persons with history of clot formation; also the active ingredient in rat poison.

_____ c. Fast-acting anticoagulant that blocks conversion of prothrombin to thrombin; produced in the body by mast cells and basophils, but also available in pharmacologic preparations.

_____ d. Commercial preparations introduced into coronary arteries can limit clot size and prevent severe heart attacks; classified as fibrinolytic agents because they activate plasminogen (shown on Figure LG 19.2) to dissolve clot.

_____ e. A chemical made by white blood cells and cells lining blood vessels; opposes actions of thromboxane A2.

■ **F10.** *A clinical challenge.* Match the correct term with the description.

> E. Embolus   T. Thrombus

_____ a. A blood clot

_____ b. A "clot-on-the-run," dislodged from the site at which it formed (usually a deep vein of the leg); also fat from broken bone, bubble of air, or amniotic fluid traveling through blood, possibly to lung (pulmonary) vessels

■ **F11.** *A clinical challenge.* Address factors related to clotting in this Checkpoint.

a. Mr. B, a chronic alcoholic, needs to have a surgical procedure. Explain why he might be at risk for excessive bleeding.

b. Aspirin is used to *(promote? prevent?)* hemostasis, so it can lower risk of thrombi in cerebral, coronary, and peripheral arteries. Circle the step(s) of hemostasis that it affects. *(Vasoconstriction? Platelet plug formation? Coagulation?)*

G. Blood groups and blood types   (pages 570–573)

■ **G1.** Contrast *agglutination* and *coagulation*.

a. Coagulation (or _____) involves coagulation factors found in platelets, plasma, or other tissue fluids. The process *(requires? can occur in absence of?)* red blood cells.

b. Agglutination (or "clumping") of erythrocytes is an antigen-_____ process that *(does? does not?)* require red blood cells because these are sites of

_____ used in the agglutination reaction.

■ **G2.** Do this exercise about factors responsible for blood groups.

a. The surface of RBCs contains chemicals that act as *(antigens? antibodies?)*, also known

as iso-_____. Based on these chemicals, blood can be categorized into at least *(2? 4? 24? 100?)* different blood groups. One of these groups, the ABO group, consists of four possible blood types. Name them: type

_____, type _____, type _____, and type _____.

b. Determining the ABO blood group are genes known as the *I* gene. A baby with blood type AB must have received an I^A gene from one parent and an

_____ gene from the other parent.

c. Type A blood has *(A? B? both A and B? neither A nor B?)* antigen on RBCs.

Genotypes _____ and _____ produce persons with type A blood.

d. Type A blood most likely has the *(anti-A? anti-B? both anti-A and anti-B? neither anti-A nor anti-B?)* antibodies in blood plasma. If a person with type A blood receives type B blood, the recipient's *(anti-A? anti-B?)* antibodies will attack the *(A? B?)* antigens on the donor's cells. Are the donor's anti-A antibodies likely to attack the recipient's type A cells and cause a major incompatibility reaction?

G3. Complete the table contrasting ABO blood types.

Blood Group	Percentage of White Population	Percentage of Black Population	Sketch of Blood Showing Correct Antigens and Antibodies	Can Donate Safely to	Can Receive Blood Safely from
a. Type A					A, O
b.	10				
c. Type AB		4	A B B ◯ A neither A B anti-A B A nor anti-B		
d.				A, B, AB, O	

■ **G4.** Type O is known as the universal *(donor? recipient?)* with regard to the ABO group,

because type O blood lacks _____ of the ABO group. Type _____ is
known as the universal recipient. Explain why.

G5. Discuss risks associated with blood transfusions and precautions taken to protect
recipients.

■ **G6.** Complete this exercise about the *Rh system.*
   a. The Rh *(+? -?)* group is more common. Rh *(+? -?)* blood has Rh antigens on the surfaces
      of RBCs.
   b. Under normal circumstances plasma of *(Rh⁺ blood? Rh⁻ blood? both Rh groups?*
      neither Rh group?) contains anti-Rh antibodies.
   c. Rh *(+? -?)* persons can develop these antibodies when they are exposed to Rh *(+? -?)*
      blood.
   d. An example of this occurs in fetal-maternal incompatibility when a mother who is Rh
      (+? -?) has a baby who is Rh *(+? -?)* and some of the baby's blood enters the mother's
      bloodstream. The mother develops anti-Rh antibodies that may cross the placenta in
      future pregnancies and hemolyze the RBCs of Rh *(+? -?)* babies. Such a condition is

      known as _____.
   e. *A clinical challenge.* Discuss precautions taken with Rh⁻ mothers soon after
      delivery, miscarriage, or abortion of an Rh⁺ baby to prevent future problems with
      Rh incompatibility.

H. Disorders, medical terminology   (pages 574–576)

■ **H1.** Match names of types of anemia with descriptions below.

A. Aplastic	N. Nutritional
Hl. Hemolytic	P. Pernicious
Hr. Hemorrhagic	S. Sickle cell

_____ a. Condition resulting from inadequate
diet, such as deficiency of iron,
vitamin B_{12}, or amino acids

_____ b. Condition in which intrinsic factor is
not produced, so absorption of
vitamin B_{12} is inadequate

_____ c. Inherited condition in which hemoglobin
forms stiff rodlike structures causing
erythrocytes to assume sickle shape and
rupture, reducing oxygen supply to tissues

_____ d. Rupture of red blood cell membranes due to
variety of causes, such as parasites, toxins,
or antibodies

_____ e. Condition due to excessive bleeding, as
from wounds, gastric ulcers, heavy
menstrual flow

_____ f. Inadequate erythropoiesis as a result of
destruction or inhibition of red bone
marrow.

H2. Persons with one gene for sickle cell anemia are said to have *(sickle cell trait? sickle cell anemia?)*. These individuals have *(greater? less?)* resistance to malaria. Explain why.

■ **H3.** *A clinical challenge.* Amy (27 years old) has a red blood count of 7 million/mm³. She has the condition known as *(anemia? polycythemia?)*. Amy has a hematocrit done. It is more likely to be *(under 32? over 45?)*. Her blood is *(more? less?)* viscous than normal, which is likely to cause *(high? low?)* blood pressure.

■ **H4.** Infectious mononucleosis is caused by the Epstein-Barr *(bacteria? virus?)*. It affects mostly *(teenagers? the elderly?)*. The name of this infection is based upon the fact that

B lymphocytes become abnormal in appearance, resembling _____.

A differential count shows a great increase in _____-cytes. Permanent damage usually *(does? does not?)* result.

■ **H5.** Leukemia is a form of cancer involving abnormally high production of *(erythrocytes? leukocytes?)*. Briefly explain why the following symptoms are likely to occur.
   a. Anemia

   b. Hemorrhage

   c. Infection

H6. Define and explain the clinical significance of each of these terms.
   a. Gamma globulin

   b. Thrombocytopenia

   c. Septicemia

A2.

		Interstitial Fluid and Lymph	Plasma
a.	Amount of protein (more or less)	Less	More
b.	Formed elements present	Contain some leukocytes but no red blood cells or platelets	Red blood cells, white blood cells, platelets
c.	Location	Between cells and in lymph vessels	Within blood vessels

A3. (a) Blood, heart, and blood vessels. (b) Lymph, lymph vessels, lymph nodes, tonsils, spleen.

A4. (a) Oxygen, carbon dioxide and other wastes, nutrients, hormones, enzymes, heat. (b) pH, temperature, water (fluids) and dissolved chemicals (such as electrolytes). (c) Contains phagocytic cells, blood clotting proteins, and defense proteins such as antibodies, complement, and interferon.

A5. (a) 38°C (100.4°C). (b) 7.35–7.45. (c) 5–6 liters. (d) ADH, aldosterone, and ANP.

A7. (a) P LP E. (b) P E LP. (c) E. (d) LP.

A8. (a) W. (b) HE. (c) E. (d) A. (e) F. (f) G. (g) UC. (h) GAF. (i) F.

A9. (a) Ag. (b) Ab. (c) Ab.

A10. (a–b) P. (c–d) L. (e) E.

B1. (a) Erythrocyte, leukocyte. (b) Hemopoiesis (or hematopoiesis); hematopoietic stem; leukocytes, erythrocytes. (c) Yolk sac, liver, spleen, thymus, lymph glands, bone marrow. (d) Red bone marrow; proximal epiphyses of femurs and humeri, flat bones of skull, sternum, ribs, vertebrae, hipbones. (e) Kidneys; RBCs. (f) Colony-stimulating factors, interleukins.

B2. EPO stimulates RBC production in persons whose failing kidneys no longer produce erythropoietin. (b) Neupogen (G-CSF) stimulates neutrophil production in persons whose bone marrow function is inadequate, for example, transplant or AIDS patients.

B3. (a) Erythrocytes. (b) Platelets. (c) B and T lymphocytes. (d) Basophils. (e) Wandering macrophages. (f) All answers in the box.

C1. A; does not; hemoglobin; see Figure 19.2, page 557 in your text.

C2. (a) Globin, hemes. (b) Iron (Fe), oxygen, 4. (c) Globin, carbamino-; 23; in plasma, bicarbonate.

C3. (a) Do not; absence of these organelles leaves more room for O_2 transport; RBCs have a limited life span, partly because the anuclear cell cannot synthesize new components. (b) Biconcave; greater plasma membrane surface area for better diffusion of gas molecules and flexibility for

squeezing through small capillaries. (c) Sickle; upon exposure to low oxygen; sickled cells are more rigid so they may lodge in and block small vessels.

C4. (a) Spleen, liver, or red bone marrow. (b) Amino acids, iron. (c) Transferrin, ferritin and hemosiderin; transferrin; transferrin. (d) Red bone marrow, heme. (e) Biliverdin, bilirubin. (f) Bile; urobilinogen; stercobilin, urobilin.

C5. (a) Proerythroblasts. (b) Polychromatophilic erythroblasts. (c) Acidophilic erythroblasts. (d) Reticulocytes.

C6. (a) H. (b) E. (c) B_{12}. (d) IF.

C7. (a) 4 months. (b) 5 million. (c) 15.

C8. (a) Reticulocyte count, 1. (b) Hematocrit, 41. (c) Polycythemia, 65.

C9. Mild hypoxia stimulates RBC formation. Hypoxia is induced in (a) by a move to high altitudes where the air is "thinner" (lower pressure, so lower in oxygen). Residents of high-altitude cities develop higher hematocrits. Removal of RBCs [in (b)] induces hypoxia and erythropoiesis, giving this athlete an unethical boost of hematocrit once original RBCs are reintroduced.

D1. (a) Lack, white blood cells or WBCs. (b) B, neutrophil (60–70); C, lymphocyte (20–25); D, monocyte (3–8); E, eosinophil (2–4); F, basophil (0.5–1.0); differential. (c) Has; monocyte (D); lymphocyte (C); neutrophil (B). (d) B E F. (e) See Figure 19.2, page 557 of your text.

D3. MHC antigens are proteins on cell surfaces; they are unique for each person. The greater the similarity between major histocompatibility (MHC) antigens of donor and recipient, the less likely is rejection of the transplant.

D4. (a) 5 million, 5,000–10,000; 700. (b) RBCs; days.

D5. (a) E. (b) A. (c) C. (d) P.

D6. (a) N. (b) N. (c) L. (d) B. (e) E. (f) M. (g) M, N. (h) L, M. (i) N. (j) N. (k) E. (l) L.

D7. (a) B, antibodies; bacteria and their toxins. (b) Cancer or transplanted cells, viruses, fungi, and some bacteria; wide variety of.

D8. (a) In; 12,000; leukocytosis. (b) Neutrophils (PMNs or polys); 76; monocytes.

D9. (a) B. (b) D. (c) C. (d) A.

E1. (a) Thrombocytes. (b) Bone marrow. (c) Chips off the old metamegakaryocytes. (d) 250,000–400,000. (e) Blood clotting.

E2. G.

E3. RBC, WBC, and platelet counts; differential WBC count; hemoglobin and hematocrit (H & H).

F1. Blood standing still (or blood stoppage). (a) Spasm. (b) Platelet. (c) Coagulation.

F3. (a) Large, irregular-shaped; adhesion. (b) Constrict; reduces blood flow into injured area; serotonin, thromboxane A2. (c) Activates platelets and makes them stickier, thereby inviting more platelets to the scene. This phenomenon is known as platelet aggregation. (d) Small.

F4. (a) Plasma. (b) Platelets. (c) Tissue; intrinsic; extrinsic.

F5. A, extrinsic. B, intrinsic. C, Prothombinase (prothrombin activator). D, Thrombin. E, Fibrin.

F6. (a) Positive, enlarge; fibrin. (b) Ca^{2+}, IV. (c) X and V. (d) I, Fibrinogen; II, Prothrombin; III, Tissue factor (thromboplastin).

F7. (a) Some clotting factor; males. (b) A, VIII; it is known as fibrin stabilizing factor and it does just that, forming fibrin threads into a sturdy clot. (c) Hemorrhaging, including nosebleeds; blood in urine (hematuria); and joint damage and pain. (d) Because the HIV virus became increasingly widespread after 1982, yet blood testing for the virus was not available until 1985, persons who received blood products during those years were at high risk for infection.

F8. (a) Retraction. (b) Fibrinolysis; F, plasminogen; G, plasmin; D, thrombin.

F9. (a) K. (b) W. (c) H. (d) S, t-PA. (e) PGI_2.

F10. (a) T. (b) E.

F11. (a) Most coagulation factors are normally synthesized in the liver, which is severely damaged by cirrhosis, a liver disease that can be caused by excessive consumption of alcohol. (b) Prevent; all three: vasoconstriction, platelet plug formation, and coagulation.

G1. (a) Clotting; can occur in absence of. (b) Antibody, does, antigens.

G2. (a) Antigens, antigens; 24; A, B, AB, and O. (b) I^B. (c) A; $I^A I^A$, $I^A i$. (d) Anti-B; anti-B, B; not likely to have major reaction because donor's antibodies are rapidly diluted in the recipient's plasma, but a minor (mild) reaction may occur.

G4. Donor, both anti-A and anti-B antibodies; AB; persons with type AB blood lack both anti-A and anti-B antibodies.

G6. (a) +; +. (b) Neither Rh group. (c) −, +. (d) −, +; +, hemolytic disease of the newborn (HDN) or erythroblastosis fetalis. (e) The chemical RhoGAM (anti-Rh antibodies), given soon after delivery, binds to any fetal Rh antigens within the mother, destroying them. Therefore the Rh⁻ mother will not produce anti-Rh antibodies that would attack Rh antigens of future Rh⁺ fetuses.

H1. (a) N. (b) P. (c) S. (d) H1. (e) Hr. (f) A.

H3. Polycythemia; over 45; more, high.

H4. Virus; teenagers; monocytes; lympho; does not.

H5. Focus on production of only immature leukocytes in abnormal bone marrow limits production of RBCs which leads to (a) anemia; platelet production also decreases (b: hemorrhage) and WBC production declines (c: infection).

WRITING ACROSS THE CURRICULUM: CHAPTER 19

1. Describe mechanisms for assuring homeostasis of red blood cell count.

2. Describe the recycling of red blood cells. Be sure to use all of these terms in your essay: hemoglobin, iron, transferrin, ferritin, and hemosiderin; bilirubin, bile, urobilinogen, urobilin, and stercobilin.

3. Describe roles of the following types of cells in inflammation (including phagocytosis) and immune respnses: neutrophils, monocytes, and lymphocytes.

4. Contrast advantages and disadvantages of blood clotting in the human body. Then describe control mechanisms that provide checks and balances so that clotting does not get out of hand.

MASTERY TEST: CHAPTER 19

Questions 1–2: Arrange the answers in correct sequence.

_____ _____ _____ 1. Events in the coagulation process:
    A. Retraction or tightening of fibrin clot
    B. Fibrinolysis or clot dissolution by plasma
    C. Clot formation

_____ _____ _____ 2. Stages in the clotting process:
    A. Formation of prothrombin activator (prothrombinase)
    B. Conversion of prothrombin to thrombin
    C. Conversion of fibrinogen to fibrin

Questions 3–17: Circle the letter preceding the one best answer to each question.

3. The normal RBC count in healthy females is
   _____ RBCs/mm³
   A. 4.8 million
   B. 2 million
   C. 0.5 million
   D. 250,000
   E. 8000

4. All of the following types of formed elements are produced in bone marrow *except:*
   A. Neutrophils
   B. Basophils
   C. Platelets
   D. Erythrocytes
   E. Lymphocytes

5. Which chemicals are forms in which bilirubin is excreted in urine or feces?
   A. Ferritin and hemosiderin
   B. Biliverdin and transferrin
   C. Urobilinogen and stercobilin
   D. Interleukin 5 and 7

6. Metamegakaryocytes are involved in formation of:
   A. Red blood cells
   B. Basophils
   C. Lymphocytes
   D. Platelets
   E. Neutrophils

7. Choose the *false* statement about Norine, who has type O blood.
   A. Norine has neither A nor B antigens on her red blood cells.
   B. She has both anti-A and anti-B antibodies in her plasma.
   C. She is called the universal donor.
   D. She can receive blood safely from both type O and type AB persons.

8. Choose the *false* statement about blood.
   A. Blood is thicker than water.
   B. It normally has a pH of 7.0.
   C. The human body normally contains about 5 to 6 liters of it.
   D. It normally consists of more plasma than cells.

9. Choose the *false* statement about factors related to erythropoiesis.
   A. Erythropoietin stimulates RBC formation.
   B. Oxygen deficiency in tissues serves as a stimulus for erythropoiesis.
   C. Intrinsic factor, which is necessary for RBC formation, is produced in the kidneys.
   D. A reticulocyte count of over 1.5 % of circulating RBCs indicates that erythropoiesis is occurring rapidly.

10. Choose the *false* statement about plasma.
    A. It is red in color.
    B. It is composed mainly of water.
    C. Its concentration of protein is greater than that of interstitial fluid.
    D. It contains plasma proteins, primarily albumins.

11. Choose the *false* statement about neutrophils.
    A. They are actively phagocytic.
    B. They are the most abundant type of leukocyte.
    C. Neutrophil count decreases during most infections.
    D. An increase in their number would be a form of leukocytosis.

12. All of the following correctly match parts of blood with principal functions *except:*
    A. RBC: carry oxygen and CO_2
    B. Plasma: carries nutrients, wastes, hormones, enzymes
    C. WBCs: defense
    D. Platelets: determine blood type

13. Which of the following chemicals is an enzyme that converts fibrinogen to fibrin?
    A. Heparin
    B. Thrombin
    C. Prothrombin
    D. Coagulation factor VI
    E. Tissue factor

14. A type of anemia that is due to a deficiency of intrinsic factor production, caused, for example, by alterations in the lining of the stomach, is:
    A. Aplastic
    B. Sickle cell
    C. Hemolytic
    D. Pernicious

15. A person with a hematocrit of 66 and hemoglobin of 22 is most likely to have the condition named:
    A. Anemia
    B. Polycythemia
    C. Infectious mononucleosis
    D. Leukemia

16. Which condition is best described as a malignant disorder of plasma cells of the bone marrow?
    A. Aplastic anemia
    B. Multiple myeloma
    C. Porphyria
    D. Leukemia

17. A hematocrit value is normally about _____ the value of the hemoglobin.
    A. The same
    B. One third
    C. Three times
    D. 700 times

Questions 18–20: Circle T (true) or F (false). If the statement is false, change the underlined word or phrase so that the statement is correct.

T F 18. Plasmin is an enzyme that facilitates clot <u>formation.</u>

T F 19. Erythroblastosis fetalis is most likely to occur with an Rh <u>positive mother and her Rh negative babies.</u>

T F 20. In both black and white populations in the U.S., type O is <u>most</u> common and type AB is <u>least</u> common of the ABO blood groups.

Questions 21–25: Fill-ins. Answer questions or complete sentences with the word or phrase that best fits.

_____ 21. Name five types of substances transported by blood.

_____ 22. Name three types of plasma proteins.

_____ 23. Name the type of leukocyte that is transformed into plasma cells which then produce antibodies.

_____ 24. Write a value for a normal leukocyte count: _____/mm^3.

_____ 25. All blood cells and platelets are derived from ancestor cells called _____.

ANSWERS TO MASTERY TEST: ✮ CHAPTER 19

Arrange

1. C A B
2. A B C

Multiple Choice

3. A
4. E
5. C
6. D
7. D
8. B
9. C
10. A
11. C
12. D
13. B
14. D
15. B
16. B
17. C

True–False

18. F. Dissolution or fibrinolysis
19. F. Negative mother and her Rh positive babies
20. T

Fill-ins

21. Oxygen, carbon dioxide, nutrients, wastes, sweat, hormones, enzymes
22. Albumins, globulins, and fibrinogens
23. Lymphocyte
24. 5,000–10,000
25. Hematopoietic stem cells

FRAMEWORK 20

The Heart

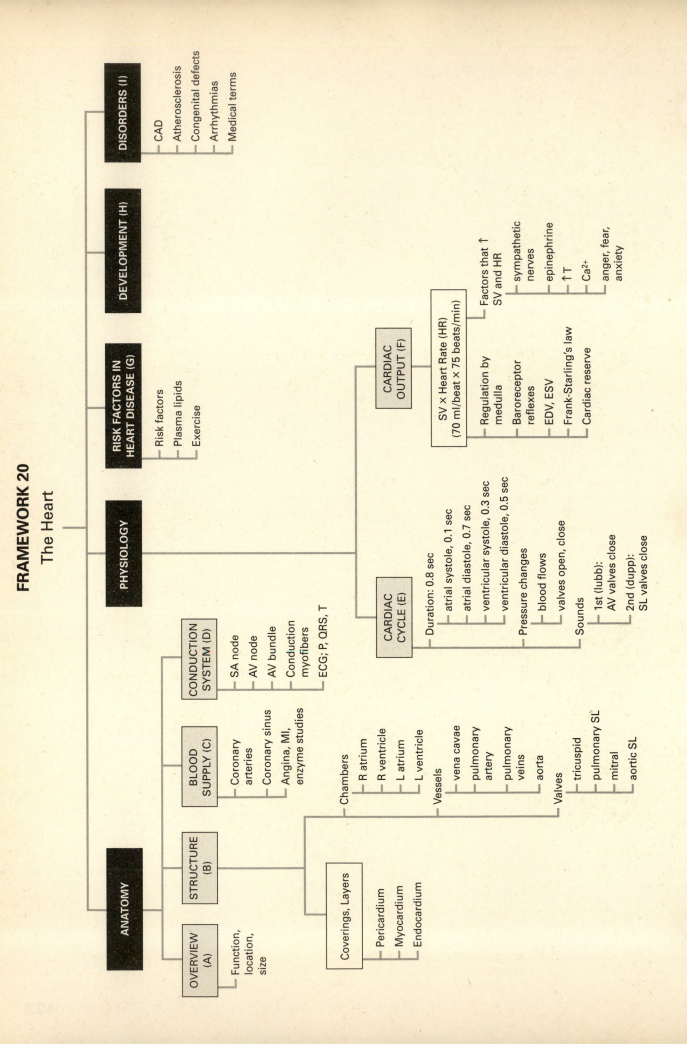

ANATOMY

OVERVIEW (A)
- Function, location, size

STRUCTURE (B)
- Coverings, Layers
  - Pericardium
  - Myocardium
  - Endocardium
- Chambers
  - R atrium
  - R ventricle
  - L atrium
  - L ventricle
- Vessels
  - vena cavae
  - pulmonary artery
  - pulmonary veins
  - aorta
- Valves
  - tricuspid
  - pulmonary SL
  - mitral
  - aortic SL

BLOOD SUPPLY (C)
- Coronary arteries
- Coronary sinus
- Angina, MI, enzyme studies

CONDUCTION SYSTEM (D)
- SA node
- AV node
- AV bundle
- Conduction myofibers
- ECG; P, QRS, T

PHYSIOLOGY

CARDIAC CYCLE (E)
- Duration: 0.8 sec
  - atrial systole, 0.1 sec
  - atrial diastole, 0.7 sec
  - ventricular systole, 0.3 sec
  - ventricular diastole, 0.5 sec
- Pressure changes
  - blood flows
  - valves open, close
- Sounds
  - 1st (lubb): AV valves close
  - 2nd (dupp): SL valves close

CARDIAC OUTPUT (F)
- SV × Heart Rate (HR) (70 ml/beat × 75 beats/min)
- Regulation by
  - medulla
  - Baroreceptor reflexes
  - EDV, ESV
  - Frank-Starling's law
  - Cardiac reserve
- Factors that ↑ SV and HR
  - sympathetic nerves
  - epinephrine
  - ↑ T
  - Ca^{2+}
  - anger, fear, anxiety

RISK FACTORS IN HEART DISEASE (G)
- Risk factors
- Plasma lipids
- Exercise

DEVELOPMENT (H)

DISORDERS (I)
- CAD
- Atherosclerosis
- Congenital defects
- Arrhythmias
- Medical terms

The Cardiovascular System: The Heart

CHAPTER
20

Make a fist, then open and squeeze tightly again. Repeat this about once a second as you read the rest of this Overview. Envision your heart as a muscular pump about the size of your fist. Unfailingly, your heart exerts pressure on your blood, moving it onward through the vessels to reach all body parts. The heart has one job; it is simply a pump. But this function is critical. Without the force of the heart, blood would come to a standstill, and tissues would be deprived of fluids, nutrients, and other vital chemicals. To serve as an effective pump, the heart requires a rich blood supply to maintain healthy muscular walls, a specialized nerve conduction system to synchronize actions of the heart, and intact valves to direct blood flow correctly. Heart sounds and ECG recordings as well as a variety of more complex diagnostic tools provide clues to the status of the heart.

Is your fist tired yet? The heart ordinarily pumps 24 hours a day without complaint and seldom reminds us of its presence. As you complete this chapter on the heart, keep in mind the value and indispensability of this organ. Start by studying the Chapter 20 Framework and refer to it frequently; note relationships among key terms in each section. As you begin your study of the heart, also carefully examine the Chapter 20 Topic Outline and Objectives; check off each one as you complete it.

TOPIC OUTLINE AND OBJECTIVES

A. Overview of circulation, location, and size of the heart

B. Heart structure

☐ 1. Describe the location of the heart and the structure and functions of the wall, chambers, great vessels, and valves of the heart.

C. Heart blood supply

☐ 2. Describe the blood supply of the heart.

D. Conduction system, cardiac muscle physiology; electrocardiogram

☐ 3. Explain the structural and functional features of the conduction system of the heart.
☐ 4. Describe the physiology of cardiac muscle contraction.
☐ 5. Explain the meaning of an electrocardiogram (ECG) and its diagnostic importance.

E. Cardiac cycle

☐ 6. Describe the phases, timing, and sounds associated with a cardiac cycle.

F. Cardiac output; regulation of the heart

☐ 7. Define cardiac output (CO) and describe the factors that affect it.
☐ 8. Explain how heart rate is regulated.

G. Risk factors in heart disease

☐ 9. List and explain the risk factors involved in heart disease.
☐ 10. Explain the relationship between plasma lipids and heart disease.
☐ 11. Explain the benefits of regular exercise on the heart.

423

H. Developmental anatomy

☐ 12. Describe the developmental anatomy of the heart.

I. Disorders, medical terminology

☐ 13. Define the following disorders: coronary artery disease (CAD), congenital defects, and arrhythmias.

☐ 14. Define medical terminology associated with the heart.

WORDBYTES

Now become familiar with the language of this chapter by studying each wordbyte, its meaning, and an example of its use within a term. After you study the entire list, self-check your understanding by writing the meaning of each wordbyte on the line. As you continue through the *Learning Guide,* identify (and fill in) additional terms that contain the same wordbyte.

Wordbyte	Self-check	Meaning	Example(s)
cardi-	_____	heart	*cardi*ologist
coron-	_____	crown	*coron*ary arteries
endo-	_____	within	*endo*carditis
myo-	_____	muscle	*myo*cardial infarction
peri-	_____	around	*peri*cardium
vascul-	_____	small vessel	cardio*vascul*ar

CHECKPOINTS

A. Overview of circulation, location, and size of the heart (pages 579–581)

■ **A1.** Do this exercise about heart functions.

a. What is the primary function of the heart? _____

b. Visualize a coffee cup; now imagine that sitting next to it is a gallon-size milk container.

   Keep in mind that a gallon holds about _____ quarts or approximately 4 liters.
   Now take a moment to imagine a number of containers lined up and holding
   the amount of blood pumped out by the heart.
   1. With each heartbeat, the heart pumps a stroke volume equal to about $\frac{1}{3}$ cup (70 ml)

   because 1 cup   =   about _____ ml.

   2. Each minute, the heart pumps enough blood to fill up 1.3 gallons (_____ liters).
   3. Within the 1440 minutes in a day, the heart would fill up about 1800 gallons

   (over _____ liters).

c. Reflecting on the work your heart is doing for you, what message would you give to your heart?

A2. Refer to Figure LG 20.1 and complete this Checkpoint on blood vessels.

a. Label vessels 1–7. Note that numbers are arranged in the sequence of the pathway of blood flow. Use these labels:

Aorta and other large arteries	Small artery
Arteriole	Small vein
Capillary	Venule
Inferior vena cava	

b. Color vessels according to color code ovals.
c. The human circulatory system is called a(n) *(open? closed?)* system. State the significance of this fact.

A3. On Figure LG 20.1, label the *pulmonary artery, pulmonary capillaries,* and a *pulmonary vein.* All blood vessels in the body other than pulmonary vessels are known as

_____ vessels because they supply a variety of body systems.

A4. Closely examine Figure 20.2, page 580 in your text. Consider the location, size, and shape of your heart as you do this exercise. Trace its outline on your body.

a. Your heart lies in the _____ portion of your thorax, between your

two _____. About *(one-third? one-half? two-thirds?)* of the mass of your heart lies to the left of the midline of your body.

b. Your heart is about the size and shape of your _____.

c. The pointed part of your heart, called the _____, lies in the

_____ intercostal space, about _____ cm (_____ inches) from the midline of your body.

d. The region of the heart that lies directly superior to the diaphragm is the right *(atrium? ventricle?)*.

B. Heart structure   (pages 581–588)

B1. Arrange in order from most superficial to deepest. _____ _____ _____ _____ _____ _____

E. Endocardium	PC. Pericardial cavity (site of pericardial fluid)
FP. Fibrous pericardium	PP. Parietal pericardium
M. Myocardium	VP. Visceral pericardium (epicardium)

For extra review of the pericardium, look again at Checkpoint D2 in Chapter 4, page 82 in the *Learning Guide.*

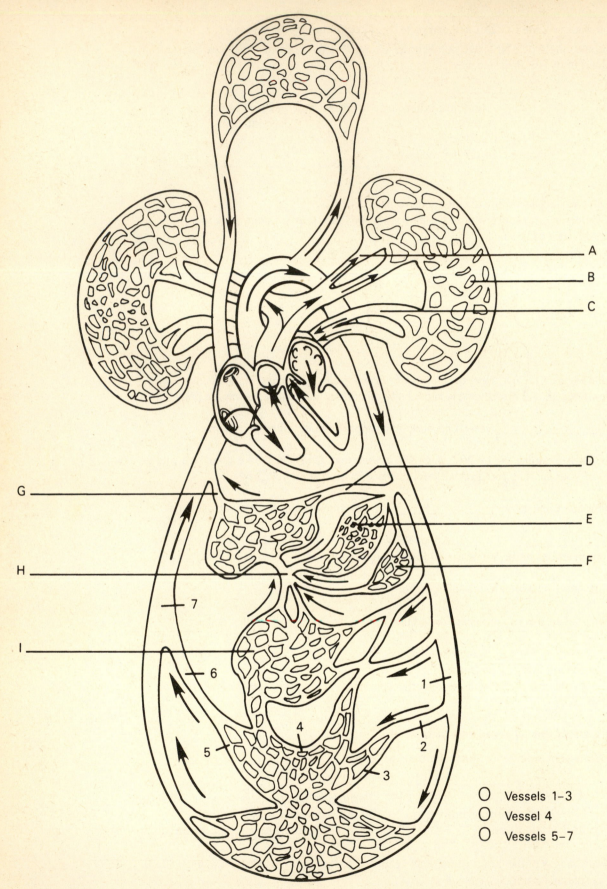

A

B

C

D

E

F

G

H

I

7

6

5

4

3

2

1

○ Vessels 1–3
○ Vessel 4
○ Vessels 5–7

Figure LG 20.1 Circulatory routes Numbers 1–7 follow the pathway of blood through the systemic blood vessels. Color and label numbered vessels according to directions in Checkpoint A2. Letters A–I refer to Checkpoints A2, A3, and Chapter 21 Checkpoint F1.

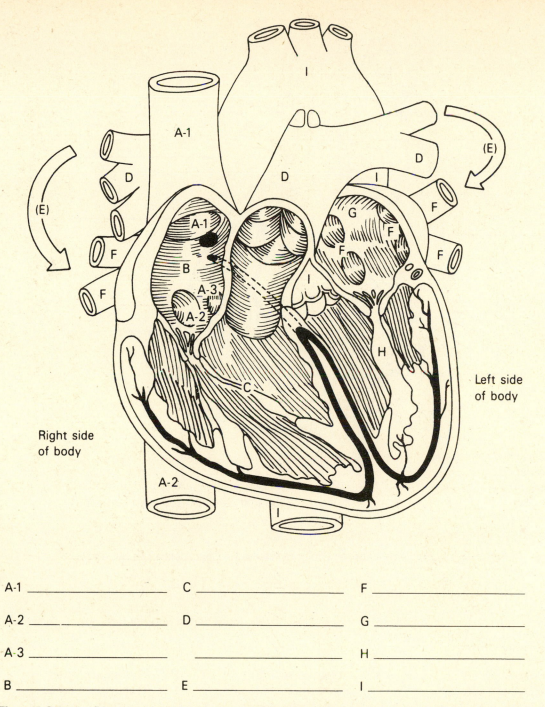

A-1

(E)

D

D

D

(E)

I

G

F

F

F

F

F

A-1

B

A-3

A-2

C

H

F

I

Left side
of body

Right side
of body

A-2

I

A-1 _____   C _____   F _____

A-2 _____   D _____   G _____

A-3 _____   _____   H _____

B _____   E _____   I _____

Figure LG 20.2 Diagram of a frontal section of the heart. Letters follow the path of blood through the heart. Label, color, and draw arrows as directed in Checkpoints B3, B7, and D1.

■ **B2.** Inflammation of the pericardium is known as _____. This condition

may lead to cardiac tamponade, which means _____

_____ .

■ **B3.** Refer to Figures LG 20.2 and 20.3. Identify all structures with letters A–I by writing
labels on lines A–I on Figure LG 20.2.

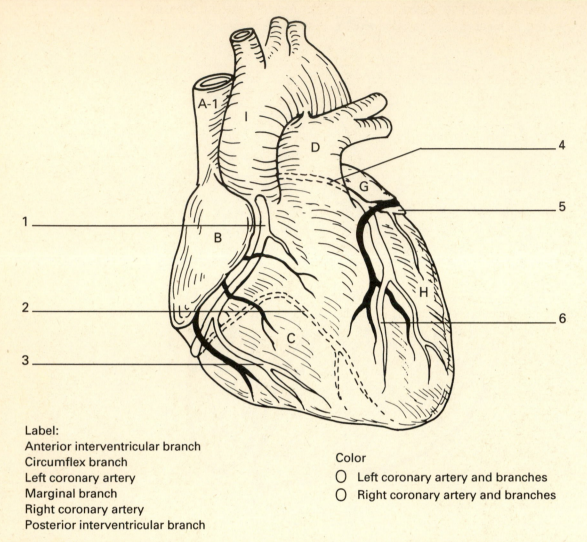

Label:
Anterior interventricular branch
Circumflex branch
Left coronary artery
Marginal branch
Right coronary artery
Posterior interventricular branch

Color
○ Left coronary artery and branches
○ Right coronary artery and branches

Figure LG 20.3 Anterior view of the heart. Label and color as directed in Checkpoints B3, B7, and C2. Cardiac veins are solid black.

■ **B4.** Check your understanding of heart structure by selecting the terms that fit descriptions below. Not all answers will be used.

A. Atria	IAS. Interatrial septum	PV. Pulmonary vein
CS. Coronary sulcus	IVS. Interventricular septum	TC. Trabeculae carnae
CT. Chordae tendineae	PA. Pulmonary artery	V. Ventricle

_____ a. Thin-walled chambers that receive blood from veins

_____ b. Blood vessel that carries blood that is rich in oxygen from lungs to left atrium

_____ c. Site of the fossa ovalis, the location of the foramen ovale in fetal life

_____ d. Strong wall separating the two ventricles

_____ e. Sites of most myocardium

_____ f. A groove separating left from right side of the heart; contains fat and coronary blood vessels

_____ g. Irregular ridges and folds of myocardium covered with smooth endocardium

_____ h. Strong tendons that anchor atrioventricular (AV) valves to ventricular muscle

■ **B5.** Check your understanding of the valves by doing this matching exercise. More than one answer may be required for each description.

> A. Aortic semilunar   P. Pulmonary semilunar
> B. Bicuspid   T. Tricuspid

_____ a. Also called the mitral valve

_____ b. Prevents backflow of blood from right ventricle to right atrium

_____ c. Prevents backflow from pulmonary trunk to right ventricle

_____ d. Prevents backflow of blood into left atrium

_____ e. Have half-moon-shaped leaflets or cusps (two answers)

_____ f. Also called atrioventricular (AV) valves (two answers)

■ **B6.** *A clinical challenge.* Name the microorganism that causes rheumatic fever (RF).

_____ What condition is likely to signal the presence of this microbe and warn of possible effects (sequelae) upon the heart?

Which parts of the heart are most likely to be affected?

A thought to ponder: as you look at heart physiology, speculate about why these two valves might be more susceptible.

■ **B7.** Refer to Figures LG 20.2 and 20.3 and check your understanding of the pathway of blood through the heart in this Checkpoint.

a. Draw arrows on Figure LG 20.2 to indicate direction of blood flow.
b. Color red the chambers of the heart and vessels that contain highly oxygenated blood; color blue the regions in which blood is low in oxygen and high in carbon dioxide.
c. On Figure LG 20.2 label the four valves that control blood flow through the heart.
d. On Figure LG 20.3 label the *ligamentum arteriosum*. In fetal life this is the site of the ductus *(arteriosus? venosus?)*, which connects the pulmonary *(artery? vein?)* to the

_____, thereby permitting blood to bypass the lungs.

C. Heart blood supply (pages 588–589)

C1. Defend or dispute this statement: "The myocardium receives all of the oxygen and nutrients it needs from blood that is passing through its four chambers."

■ **C2.** Refer to Figure LG 20.3 and complete this checkpoint.

    a. Using the color code ovals on the figure, color the two coronary arteries and their main branches.

    b. Label each of the vessels, using the list of labels on the figure.

■ **C3.** The coronary sinus functions as *(an artery? a vein?)*. It collects blood that has passed

through coronary arteries and capillaries into _____ veins. The coronary
sinus finally empties this blood into the *(right? left?) (atrium? ventricle?)*.

C4. Define *ischemia* and explain how it is related to *angina pectoris*.

■ **C5.** *A clinical challenge.* Do this exercise about "heart attacks."

    a. "Heart attack" is a common name for a myocardial _____ (MI).

        What does the term *infarction* mean? _____.

        In what other areas of the body might infarctions also occur? Cerebral infarction: _____;

        pulmonary infarction: _____.

    b. List two or more immediate causes of an MI.

    c. Define *reperfusion*.

    d. Explain possible damage caused by reperfusion of the heart, including the role of *free radicals*.

    e. List three or more other diseases or conditions that may be related to free radicals.

D. Conduction system; cardiac muscle physiology; electrocardiogram   (pages 589–594)

■ **D1.** In this exercise describe how the heart beats regularly and continuously.

    a. In embryonic life about _____% of cardiac muscle fibers become *autorhythmic*.
        State two functions of these specialized cells of the heart.

    b. Label the parts of the conduction system on Figure LG 20.2.

c. The normal pacemaker of the heart is the *(SA? AV?)* node. Each time the SA node "fires," impulses travel via the conduction system and also by the

_____ junctions in _____ discs of cardiac muscle.

d. What structural feature of the heart makes the AV node and AV bundle necessary for conduction from atria to ventricles?

e. Through which part of the conduction system do impulses pass most slowly? *(SA node? AV node? AV bundle [of His]?).* Based on the anatomy of this tissue, why does the rate of impulse conduction slow down?

Of what advantage is this slowing?

■ **D2.** Do this exercise about the pacemaker(s) of the heart.

a. On its own, the SA node, located in the *(left? right?)* atrium, normally fires at about

_____ to _____ times per minute. This rate is *(a little faster than? a little slower than? about the same as?)* a typical resting heart rate. At rest, *(acetylcholine? epinephrine?)* released from *(sympathetic? parasympathetic?)* nerves normally slows the pace of the SA node to modify the heart rate to about _____ beats per minute.

b. How might this "normal pacemaker" be damaged? _____
If this occurs, then responsibility for setting the pace of the heart may be passed on to

the _____ node, which fires at _____ to _____ times per minute.

c. If both the SA and AV nodes fail, then autorhythmic fibers in _____

may take over with a rate of only _____ to _____ beats per minute. Pacemakers at

"other than the normal site" are known as _____ pacemakers
and tend to be *(more? less?)* effective than the SA node in pacing the heart.

■ **D3.** Describe the physiology of cardiac muscle contraction in this activity.

a. Contractile fibers of the normal heart have a resting membrane potential of *(−70? −90?)* mV.

b. Arrange in correct sequence the events in an action potential of cardiac muscle. The blank parentheses are for Activity D3c. One is done for you.

<u>C (RD)</u> → _____ (_____) → _____ or _____ (_____)

→ _____ (_____) → _____ (_____)

A. *Slow Ca^{2+} channels* open so Ca^{2+} enters muscle fibers. ___**P**___

B. Combined flow of Na^+ and Ca^{2+} maintains depolarization for about 250 msec. _____

C. Na^+ enters through *fast Na^+ channels;* voltage rises rapidly to about +20 mV. _____

D. A second contraction cannot be triggered during this period, which is longer than

for skeletal muscle. _____

E. The presence of Ca^{2+} binding to troponin permits myocardial contraction via sliding

of actin filaments next to myosin filaments. _____

F. K^+ channels open so K^+ ions leave the fiber; meanwhile less Ca^{2+} enters as those

channels are closing; voltage returns to resting level. _____

c. Now label events A–F above by filling in parentheses. One is done for you.

P. Plateau	Rf. Refractory
RD. Rapid depolarization	Rp. Repolarization

(For help refer to Figure 20.9, page 592 in your text.) Use these answers:

d. *A clinical challenge.* Calcium channel blockers [such as verapamil (Procardia)] tend to *(in? de?)*-crease contraction of the cardiac muscle (and of smooth muscle of coronary arteries). Write one or more condition(s) for which such a medication might be prescribed.

Now name one chemical that enhances flow of Ca^{2+} through these channels to

strengthen contraction of the heart. _____

■ **D4.** Describe an ECG (or EKG) in this activity.

a. What do the letters ECG (or EKG) stand for?

An ECG is a recording of *(electrical changes associated with impulse conduction? muscle contractions?)* of the heart.

b. On what parts of the body are leads (electrodes) placed for a 12-lead ECG?

c. In Figure LG 20.4b, an ECG tracing using lead *(I? II? III?)* is shown. Label the following parts of that ECG on the figure: *P wave, P-R interval, QRS wave (complex), S-T segment, T wave.*

d. *A clinical challenge.* Answer the following questions with names of parts of an ECG.

1. Indicates atrial depolarization leading to atrial contraction _____
2. Prolonged if the AV node is damaged, for example, by rheumatic fever, or in the

condition described as "bundle (of His) block" _____

3. Elevated in acute MI _____

4. Elevated in hyperkalemia (high blood level of K^+) _____

5. Flatter than normal in coronary artery disease _____

6. Represents ventricular repolarization _____

7. Represents ventricular depolarization _____

8. Represents atrial repolarization _____

E. Cardiac cycle   (pages 594–597)

■ **E1.** Complete the following overview of movement of blood through the heart.

   a. Blood moves through the heart as a result of _____ and _____

     of cardiac muscle, as well as _____ and _____ of valves.

     Valves open or close due to _____ within the heart.

   b. Contraction of heart muscle is known as _____, whereas relaxation of

     myocardium is called _____.

■ **E2.** In this exercise, we will consider details of the cardiac cycle. Refer alternately to
Figure LG 20.4 and Figure LG 20.5. Figure LG 20.4 will allow you to correlate pressure
and volume changes within the heart to an ECG recording and to heart sounds. Figure LG
20.5 will help you to visualize events as they occur in a one-cycle "clock."

   a. If the heart beats at 75 beats per minute, then *(54? 60? 75?)* complete cardiac cycles
occur per 60-sec period. In other words, if your pulse is 75 (beats per minute), the

     duration of one cardiac cycle ("heartbeat") within you is _____ sec.

   b. Now look at Figure LG 20.5 which represents one complete 0.8-sec cardiac cycle on a
special 0.8-sec "clock." The clock is divided into eight sectors, or time periods, each of

     which is _____ sec in duration.
     *Note: In the following activities, it is critical to remember that the total time elapsed
during one cardiac (clock) cycle is only 0.8 sec.*

   c. We could examine the cycle shown in Figure LG 20.5 at any point on the "clock."
We will, however, begin our study with an overview of the cycle. Beginning at the top
of the clock, the "12 o'clock" position, identify the three distinct phases—1, 2, and 3—
that are bracketed on Figures LG 20.4 and LG 20.5.
     1. Phase 1 is a period in which the atria and ventricles are both *(contracting?
relaxing?)*. In other words, they are both in *(systole? diastole?)*.
     2. Throughout phase 2 the ventricles *(eject? fill with?)* blood as they continue in

       (systole? diastole?) while atria go into _____ toward the end
of Phase 2.
     3. Now the filled ventricles are ready to be stimulated (indicated by the _____
wave of the ECG) to contract and then eject the blood. That is exactly what happens

       in phase _____.
     At the bottom of Figure 20.4, label these three bracketed phases.

   d. Now let's look at details. Begin again at "12 o'clock" on Figure LG 20.5 and at the left
side of LG Figure LG 20.4. As we come on to the scene, ventricles have just stopped
contracting in the last cycle. Therefore ventricular pressure is *(rising? dropping?)*.
When pressure within ventricles dips below that in the great arteries, blood fills

     semilunar valves and closes them (point _____). Notice that pressure within
ventricles drops at a rapid rate now as ventricles continue to relax.

   e. When intraventricular pressure becomes lower than that in atria, the force of blood within

     atria causes (atrioventricular) AV valves to *(open? close?)*. This happens at point _____.

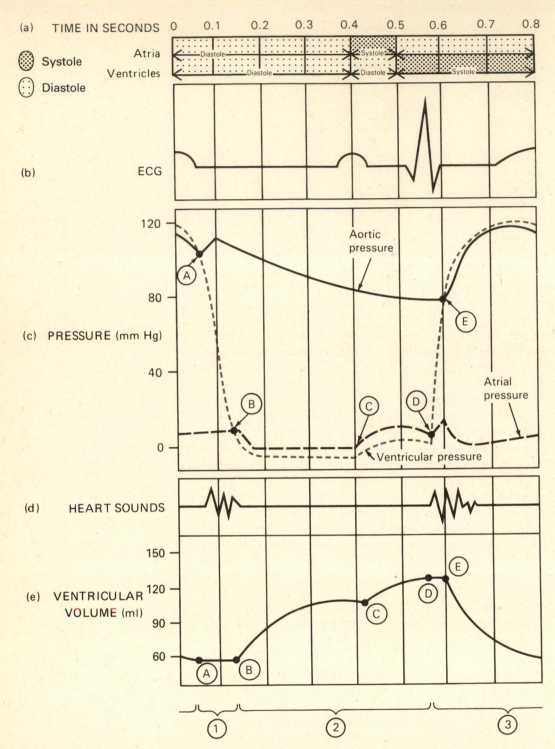

Figure LG 20.4 Cardiac cycle. (a) Systole and diastole of atria and ventricles related to time. (b) ECG related to cardiac cycle. (c) Pressures in the aorta, atria, and ventricles. (d) Heart sounds related to cardiac cycle. (e) Volume of blood in ventricles. Label as directed in Checkpoints D4c and E2 to E4

   f. Note (Figure LG 20.5) that AV valves now remain open all the way to point D, in other words, throughout the time of ventricular *(ejection? filling?).* Trace a pencil lightly along the filling curve shown on Figure LG 20.4e, starting at point B. Note that the curve is steepest just *(after B? before C?).* In other words, the first part of ventricular diastole (just after the AV "floodgates" open) is the time when ventricles fill *(rapidly?*

   slowly?), whereas slower filling, known as _____, occurs later (just before point C).

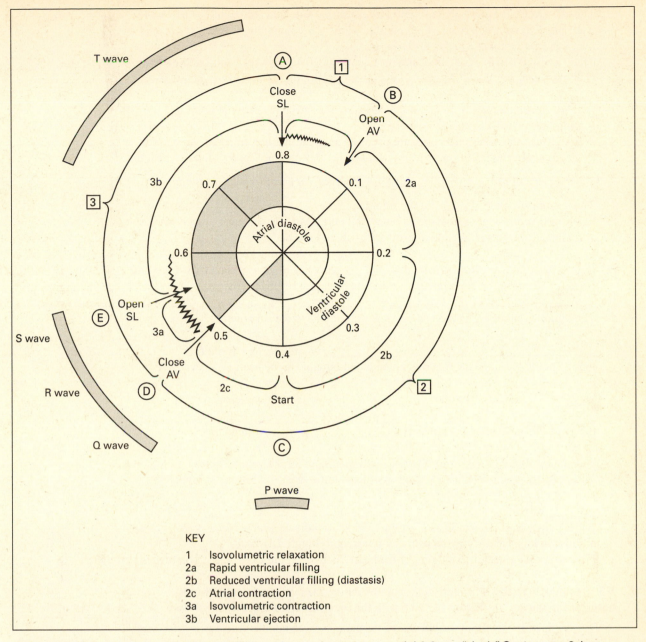

Figure LG 20.5 Cardiac cycle. Events of this cycle are shown on a special 0.8-sec "clock." Sectors are 0.1 sec each. Atrial activity is represented within inner circle and ventricular activity within outer circle. Shaded areas: systole; white areas of circle: diastole. Sounds are represented as ᴡᴡᴡ. Valve openings and closings and ECG correlations are also included. Brackets with key numbers 1 to 3b define important periods within the cycle. Refer to Checkpoints E2 and E3.

g. Which wave of the ECG signals atrial systole? *(P? QRS? T?)* Duration of atrial systole is about *(0.1? 0.3? 0.5?)* sec. As a result of atrial contraction, pressure within the atria *(increases? decreases?)* (point C in Figure LG 20.4c), forcing the last 20–30% of all the blood that will enter ventricles into those chambers.

h. Point D marks the end of atrial systole, as well as the end of ventricular *(diastole? systole?)*. Ventricles are now filled to their maximum volume

(see Figure LG 20.4e), which is about _____ ml of blood. This volume is known as

the _____-_____ volume (EDV).
Write EDV next to point E on Figure LG 20.4e.

i. The *(P? QRS? T?)* wave signals ventricular systole. Note that although the QRS wave itself happens over a very short period, the ventricular contraction that follows occurs over a period of about *(0.1? 0.3? 0.5?)* sec.

j. Trace a pencil lightly along the curve in Figure LG 20.4c showing ventricular pressure changes following the QRS wave. Notice that pressure there *(increases? decreases?)*

(slightly? dramatically?). In fact it quickly surpasses atrial pressure (at point _____), forcing the AV valves *(open? closed?)*.

k. A brief time later ventricular pressure becomes so great that it even surpasses pressure in the great arteries (pulmonary artery and aorta). This pressure forces blood against the

undersurface of the _____ valves, *(opening? closing?)* them.

This occurs at point _____ on the figure.

l. Continued ventricular systole ejects blood from the heart into the great vessels. In the aorta this creates a systolic blood pressure of about *(15? 80? 120?)* mmHg.

m. Repolarization of ventricles (after the _____ wave) causes these chambers to go into *(systole? diastole?)*. Consequently, ventricular pressure begins to *(increase? decrease?)*. This completes one cardiac cycle.

n. The volume remaining in the ventricles at the end of systole is known as

_____-_____ volume (ESV). Write ESV next to point A to indicate that this much blood remains in the heart as a new cycle

begins. ESV is normally about _____ ml. The amount of blood ejected from

the heart during systole is known as _____ volume (SV)

and equals about _____ ml. Show the calculation of stroke volume here:
End-diastolic volume (EDV) − End-systolic volume (ESV) = Stroke volume (SV)

_____ ml   −   _____ ml   =   _____ ml

o. Label points A, B, D, and E on Figure LG 20.4c to indicate which valves open or close at those points. Check your answers by referring to Figure LG 20.5.

p. *For extra review.* Refer to a figure of the heart, such as Figure LG 20.2, as you review this activity.

■ **E3.** Summarize changes in ventricular volume during the cycle by relating events described in the key for Figure LG 20.5 with the curve in Figure LG 20.4e.

a. For example, the fact that the portion of the curve between points A and B is "flat" indicates that the volume of these chambers *(does? does not?)* change during this period. This is logical because the AV and SL valves are all *(open? closed?)* during this time.

This period (A to B) is known as the period of _____ relaxation.

b. Note that the "flat" portion of the curve in Figure LG 20.4e between points D and E is a time when ventricles are *(relaxing? contracting?)*. Which valves are closed during this

period? _____ In other words, ventricles have exerted enough pressure

to close *(AV? SL?)* valves but not enough pressure yet to open _____ valves. Name this

period: isovolumetric _____.

(*Note:* Remember that the reason that periods A–B and D–E are called *isovolumetric* (iso = same) is that no blood enters or leaves ventricles because both sets of valves are *closed*.)

E4. Do this exercise on heart sounds.

a. The first heart sound *(lubb? dupp?)* is produced by turbulence of blood at the *(opening?*

closing?) of the _____ valves. What causes the second sound?

b. Write *first* and *second* next to the parts of Figure LG 20.4d showing each of these sounds.

c. *A clinical challenge.* Using Figure 20.12, page 596 in your text, recognize why a stethoscope is placed at several points on the chest in order to best hear different heart sounds. If a stethoscope is placed at the level of about the fifth intercostal space (between ribs 5 and 6) on the left side of the sternum, the *(first? second?)* sound is better heard because turbulence from the *(mitral? aortic semilunar?)* valve is readily detected there. To best hear the second sound (related to SL valve closure), the stethoscope must then be moved in a more *(inferior? superior?)* direction on the chest.

E5. Explain what causes heart murmurs.

E6. *A clinical challenge.* Consider effects of a significant change in heart rate.

a. Recall that a heart beating at about 75 beats/min has a cardiac cycle of about 0.8 sec duration:

ventricles contract for _____ sec and they relax and fill with blood for _____ sec.

b. When the heart rate doubles, for example, to 150 beats/min, the duration of the cardiac cycle *(doubles? is halved?),* for example, to 0.4 sec. Which part of the cycle is especially affected? Ventricular *(systole? diastole?)* What are the consequences of this change?

F. Cardiac output; regulation of the heart   (pages 597–600)

F1. Determine the average cardiac output in a resting adult.

Cardiac output = stroke volume × heart rate

= _____ ml/stroke × _____ strokes/min

= _____ ml/min (_____ liter/min)

F2. At rest Dave has a cardiac output of 5 liters per minute. During a strenuous cross-country run, Dave's maximal cardiac output is 20 liters per minute. Calculate Dave's cardiac reserve.

$$\text{Cardiac reserve} = \frac{\text{maximal cardiac output}}{\text{cardiac output at rest}} =$$

Note that cardiac reserve is usually *(higher? lower?)* in trained athletes than in sedentary persons.

F3. The two major factors (shown in Activity F1) that control cardiac output are

_____ and _____.

F4. Do this activity about stroke volume (SV).

a. Which statement is true about stroke volume (SV)? *(Hint:* Refer to Checkpoint E2n.)

A. SV = ESV − EDV   B. SV = EDV − ESV   C. SV = $\dfrac{\text{ESV}}{\text{EDV}}$ D. SV = $\dfrac{\text{EDV}}{\text{ESV}}$

b. List the three factors that control stroke volume (SV):

_____ _____ _____

c. During exercise, skeletal muscles surrounding blood vessels squeeze *(more? less?)* blood back to the heart. This increase in venous return to the heart *(increases? decreases?)* end-diastolic volume (EDV). As a result, muscles of the heart are stretched *(more? less?)* so that preload (= stretching of the heart muscle) *(increases? decreases?)*.

d. Within limits, a stretched muscle contracts with *(greater? less?)* force than a muscle that is only slightly stretched. This is a statement of the

_____-_____ law of the heart. To get an idea of this, blow up a balloon slightly and then let it go. Then blow up a balloon quite full of air (much like the heart when stretched by a large venous return) and let that balloon go. In which stage do the walls of the balloon (heart) compress the air (blood) with greater force?

e. As a result of the Frank-Starling law, during exercise the ventricles of the normal heart contract *(more? less?)* forcefully, and stroke volume *(increases? decreases?)*.

f. State several reasons why venous return might be decreased, leading to reduction in stroke volume and cardiac output.

g. Elevated blood pressure or narrowing of blood vessels tends to _____-crease

afterload and therefore to _____-crease stroke volume and cardiac output. Identify two causes of increased afterload.

F5. *The Big Picture: Looking Back.* **Return to text references given below and answer** questions related to cardiac function.

a. In Exhibit 4.3, page 117 of the text, cardiac muscle cells appear *(parallel? branching?)*, which is a characteristic *(similar to? different from?)* skeletal muscle. What is the significance of this pattern of cardiac muscle tissue?

b. Refer to Exhibit 17.3, pages 497–498 of the text. Sympathetic nerves _____-crease heart rate and force of atrial and ventricular contraction and also *(dilate? constrict?)*

coronary vessels. Therefore sympathetic nerves tend to _____-crease stroke volume and cardiac output.

c. Figure 17.1, page 490 of the text, shows that the vagus nerve is a *(sympathetic? parasympathetic?)* nerve. Its effects upon heart muscle, coronary vessels, and cardiac output are *(similar? opposite?)* to those of sympathetic nerves.

d. Exhibit 18.7, page 534 of the text, states that epinephrine and norepinephrine have *(sympathomimetic? parasympathomimetic?)* effects. In other words, these hormones, produced by the adrenal *(cortex? medulla?)* have effects similar to those of the sympathetic division of the autonomic nervous system: they prepare the body for a

_____ response.

■ **F6.** Summarize factors affecting heart rate (HR), stroke volume (SV), and cardiac output (CO) by completing arrows: ↑ (for increase) or ↓ (for decrease). The first one is done for you.

    a. ↑ exercise → ↑ preload

    b. Moderate ↑ in preload → | SV

    c. ↑ SV (if heart rate is constant) → | CO

    d. ↑ in heart contractility → | SV and CO

    e. Positively inotropic medication such as digoxin (Lanoxin) → | SV and CO

    f. ↑ afterload (for example, due to hypertension) → | SV and CO

    g. Hypothermia → | heart rate and CO

    h. ↑ heart rate (if SV is constant) → | CO

    i. ↑ stimulation of the cardioaccelerator center → | heart rate

    j. ↑ sympathetic nerve impulses → | heart strength (and SV) and | heart rate

    k. ↑ vagal nerve impulses → | heart rate

    l. Hyperthyroidism → | heart rate

    m. ↑ Ca^{2+} → | heart strength (and SV) and | heart rate

    n. ↓ SV (as in CHF) → | ESV → excessive EDV and preload → | SV and CO

■ **F7.** *A clinical challenge.* Do this exercise about Ms. S, who has congestive heart failure (CHF).

    a. Her weakened heart becomes overstretched much like a balloon (except much thicker!) that has been expanded 5000 times. Now her heart myofibers are stretched beyond the optimum length according to the Frank-Starling law. As a result, the force of her heart is

    _____-creased, and its stroke volume _____-creases (as shown in Activity F6n).

    b. Ms. S has right-sided heart failure, so her blood is likely to back up, distending

    vessels in *(lungs? systemic regions, such as in neck and ankles?);* thus _____ edema results, with signs such as swollen hands and feet.

    c. Write a sign or symptom of left-sided heart failure.

    d. An intra-aortic balloon pump (IABP) is especially helpful for *(right? left?)*-sided

    CHF because the IABP _____-creases afterload and therefore can _____-crease stroke volume.

G. Risk factors in heart disease   (pages 600–603)

    G1. List eight risk factors for heart disease. Circle five of those that can be modified by a healthy lifestyle.

    a._____   e._____

    b._____   f._____

    c._____   g._____

    d._____   h._____

■ **G2.** One risk factor, obesity, is associated with the fact that the heart must work harder to

    pump blood into an estimated _____ km (_____ miles) of blood vessels for each extra pound of fat.

■ **G3.** Distinguish classes of chemicals related to atherosclerosis by doing this exercise.

a. HDL and LDL both consist of complexes of _____

combined with _____. HDL refers to *(high? low?)*-density

lipoprotein, whereas LDL means _____.

b. *(HDLs? LDLs?)* are considered culprits in causing atherosclerosis because they contain large amounts of cholesterol and deposit some of this lipid in walls of arteries. So a *(high? low?)* blood level of LDL is desirable.

c. High levels of LDL may occur in blood of persons who have too *(many? few?)* LDL receptors on their cells. Consequently, LDL remains in blood and may travel to blood vessel walls and take up residence there.

d. A high level of HDL tends to *(lead to? prevent?)* atherosclerosis. List two ways that HDLs may be increased in your own blood.

■ **G4.** Circle the more ideal value in each pair in order to lower risk of heart attack.

a. Total cholesterol (TC):   180   260

b. High-density lipoproteins (HDL):   30   50

c. Low-density lipoproteins (LDL):   129   169

d. Ratio of TC to HDL: 2.8   5.1

■ **G5.** Circle answers that indicate benefits of physical conditioning, such as three to five aerobic workouts each week.

a. _____-crease in maximal cardiac output   f. _____-crease in resting heart rate

b. _____-crease in hemoglobin level of blood   g. _____-crease in blood pressure

c. _____-crease in capillary networks in   h. _____-crease in fibrinolytic activity
skeletal muscles

d. _____-crease in HDLs   i. _____-crease in endorphins

e. _____-crease in triglycerides

H. Developmental anatomy   (page 604)

■ **H1.** The heart is derived from _____-derm. The heart begins to develop during the *(third? fifth? seventh?)* week. Its initial formation consists of two endothelial

tubes which unite to form the _____ tube.

■ **H2.** Match the regions of the primitive heart below with the related parts of a mature heart listed below.

A. Atria	SV. Sinus venosus
BCTA. Bulbus cordis and truncus arteriosus	V. Ventricle

_____ a. Superior and inferior vena cava   _____ c. Parts of the fetal heart connected by foramen ovale

_____ b. Right and left ventricles   _____ d. Aorta and pulmonary trunk

I. Disorders, medical terminology   (pages 605–608)

■ **I1.** Outline the sequence of changes in atherosclerosis in this learning activity.

   a. Atherosclerosis involves damage to the walls of arteries. State several factors that may lead to damage of this lining.

   b. The resulting fatty lesion in the arterial lining is known as an atherosclerotic

   _____. The rough surface of plaque may snag platelets.

   These may then cause _____ formation at the site, possibly leading to unwanted emboli. Platelets may also release a chemical

   called _____.

   c. Name two other types of cells that release PDGF: _____ and

   _____. What is the effect of this chemical?

I2. List several factors that may lead to a coronary artery spasm.

■ **I3.** Match diagnostic techniques and treatments in the box with descriptions below.

C. Catheterization	CABG. Coronary artery bypass graft
CA. Cardiac angiography	PTCA. Percutaneous transluminal coronary angioplasty

_____ a. Invasive procedure in which a long, slender tube is inserted into an artery or vein under x-ray observation

_____ b. Catheterization with use of contrast dye for visualization of coronary blood vessels or heart chambers

_____ c. Nonsurgical technique to increase blood supply to the heart as arterial plaque is compressed

_____ d. Surgical procedure utilizing a vessel from another part of the body (such as a vein from the thigh) to reroute blood around a blocked portion of a coronary artery

■ **I4.** Match each congenital heart disease in the box with the related description below.

C. Coarctation of aorta.	PDA. Patent ductus arteriosus
IASD. Interartrial septal defect	VS. Valvular stenosis
IVSD. Interventricular septal defect	

_____ a. Connection between aorta and pulmonary artery is retained after birth, allowing backflow of blood to right ventricle.

_____ b. Foramen ovale fails to close.

_____ c. Septum between ventricles does not develop properly.

_____ d. Valve is narrowed, limiting forward flow and causing increase in workload of heart.

_____ e. Aorta is abnormally narrow.

■ **I5.** Do this exercise on arrhythmias.

a. *(All? Not all?)* arrhythmias are serious.

b. Conduction failure across the AV node results in an arrhythmia known as

_____.

c. Which is more serious? *(Atrial? Ventricular?)* fibrillation. Explain why.

d. Excitation of a part of the heart other than the normal pacemaker (SA node) is known as

a(n) _____ focus. Such excitations, which may be caused by
ingestion of caffeine or by lack of sleep or by more serious problems such as ischemia,
may lead to an extra early ventricular systole; these are known as _____.

■ **I6.** Match disorders in the box with descriptions below.

CHF. Congestive heart failure	CP. Cor pulmonale	
CM. Cardiomegaly	PT. Paroxysmal tachycardia	
Com. Compliance		

_____ a. Enlarged heart

_____ b. Right ventricular hypertrophy resulting
from lung disorders that create resistance
to blood flow through pulmonary vessels

_____ c. Ease of movement (lack of stiffness) of heart
or lungs

_____ d. Sudden period of rapid heart rate

_____ e. Inability of the heart to function as an
effective pump; blood backs up leading
to pulmonary or peripheral edema

A1. (a) It is a pump. (b) 4 (one gallon = 4 quarts or 3.86 liters). (b1) 237. (b2) 5 (1.3 gallons = 5 liters). (b3) 7200 (1.3 gallons/min × 1440 min/day = 1800 gallons/day = over 7000 liters). (c) Perhaps: "Thanks!" "Great job!" "Amazing!" or "You deserve the best of care, and I intend to see that you get it."

A2. (a) 1, aorta; 2, small artery; 3, arteriole; 4, capillary; 5, venule; 6, small vein; 7, inferior vena cava. (b) See Figure 21.17 (page 633 of the text. (c) Closed; blood is carried from the heart to tissues and back to the heart within a closed system of tubes that contributes to hemodynamics, for example, by large arteries converting stored energy into kinetic energy of blood so that blood continues to move on through smaller vessels.

A4. (a) Mediastinal, lungs; two-thirds. (b) Fist. (c) Apex, fifth, 8, 3. (d) Ventricle.

B1. FP PP PC VP M E

B2. Pericarditis; compression of heart due to accumulation of fluid in the pericardial space.

B3. See Fig. LG 20.2A

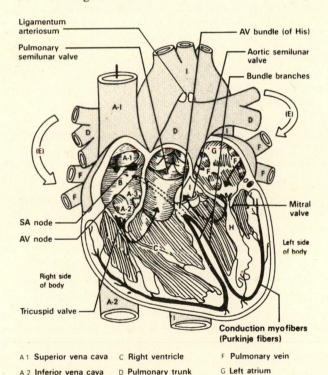

A1 Superior vena cava   C Right ventricle   F Pulmonary vein
A2 Inferior vena cava   D Pulmonary trunk   G Left atrium
A3 Coronary sinus   and arteries   H Left ventricle
B Right atrium   E Vessels in lungs I Aorta

Figure LG 20.2A   Diagram of a frontal section of the heart.

B4. (a) A. (b) PV. (c) IAS. (d) IVS. (e) V. (f) CS. (g) TC. (h) CT.

B5. (a) B. (b) T. (c) P. (d) B. (e) A, P. (f) B, T.

B6. Group A, β-hemolytic Streptococcus pyogenes; streptococcal sore throat ("strep throat"); valves, particularly the bicuspid (mitral) and the aortic semilunar. These two valves control blood flow through the left side of the heart, which is a higher-pressure system than the right side; therefore, valve damage is more likely, making these valves more vulnerable to effects of RF.

B7. (a) See Figure LG 20.2A. (b) Blue: A–D and first part of E; red: last part of E and F–I. (c) See Figure LG 20.2A. (d) See Figure 20.4, page 583 of the text; arteriosus, artery, aorta.

C2. (a) Left coronary artery: 4–6; right coronary artery: 1–3. (b) 1, right coronary artery; 2, posterior interventricular branch; 3, marginal branch; 4, left coronary artery; 5, circumflex branch; 6, anterior interventricular branch.

C3. Vein; cardiac; right atrium.

C5. Infarction; death of tissue because of lack of blood flow to that tissue; brain, lung. (b) Thrombus (clot), embolus (mobile clot), spasm of coronary artery. (c) Reestablishing blood flow (with oxygen supply) after tissue has been deprived of blood flow (ischemia). (d) An oxygen free radical borrows an electron from another molecule. As a result, a chain reaction of "borrowing electrons" may occur causing damage to heart tissue. (e) Alzheimer's disease, Parkinson's disease, cancer, cataracts, rheumatoid arthritis, and aging.

D1. (a) One; pacemaker and impulse conduction. (b) See Figure LG 20.2A. (c) SA; gap, intercalated. (d) The fibrous ring (including AV valves) that completely separates atria from ventricles. (e) AV node; fibers have smaller diameter; affords time for ventricles to fill more completely before they contract.

D2. (a) Right, 90, 100; a little faster than; acetylcholine, parasympathetic, 75. (b) Ischemia as in myocardial infarction (MI); AV, 40, 50. (c) Conduction myofibers (Purkinje fibers), 20, 40; ectopic, less.

D3. (a) − 90. (b,c) C (RD) → A (P) → B or E (both P) → F (Rp) → D (Rf). (d) De; coronary artery disease (CAD), high blood pressure (hypertension), rapid heart rate (tachycardia); epinephrine or other catecholamines.

D4. (a) An electrocardiogram (the recording) or electrocardiograph (the instrument); electrical changes associated with impulse conduction. (b) Arms, legs, and chest. (c) II; refer to Figures 20.10 and 20.11 pages 593 and 595 of the text. (d1) P wave; (d2) P-R interval; (d3) S-T segment; (d4–d6) T wave; (d7) QRS complex; (d8) No wave because masked by QRS complex.

E1. (a) Contraction, relaxation, opening, closing; pressure changes. (b) Systole, diastole.

E2. (a) 75; 0.8 (60 sec/75 cardiac cycles = 0.8 sec/cardiac cycle). (b) 0.1. (c1) Relaxing; diastole; (c2) fill with, diastole, systole; (c3) QRS; 3; phase 1: Atrial and ventricular diastole (or isovolumetric relaxation); phase 2: Ventricular diastole (and filling) with atrial contraction toward the end of this phase; phase 3: ventricular systole and atrial diastole. (d) Dropping; A. (e) Open; B. (f) Filling; after B; rapidly, diastasis. (g) P; 0.1; increases. (h) Diastole; 130; end-diastolic. (i) QRS; 0.3. (j) Increases dramatically; D, closed. (k) Semilunar, opening; E. (1) 120. (m) T, diastole; decrease. (n) End-systolic; 60; stroke, 70; 130 − 60 = 70. (o) A, closing of SL valves; B, opening of AV valves; D, closing of AV valves; E, opening of SL valves.

E3. (a) Does not; closed; isovolumetric. (b) Contracting; AV and SL; AV, SL; contraction.

E4. (a) Lubb, closing, A–V; turbulence due to closing of SL valves. (b) Refer to Figure 20.11 page 595 in the text. (c) First, mitral; superior.

E6. (a) 0.3, 0.5. (b) Is halved, diastole; because the heart cannot adequately fill with blood during a very brief ventricular diastole, only a very small volume of blood is ejected from the heart during each cardiac cycle.

F1. CO = 70 ml/stroke × 75 strokes (beats)/min = 5250 ml/min = 5.25 liters/min.

F2. Four; 20 liters per min/5 liters per min = 4; higher.

F3. Stroke volume (SV), heart rate (HR).

F4. (a) B. (b) Preload, contractility, afterload. (c) More; increases; more, increases. (d) Greater; Frank-Starling; the more expanded balloon (heart) exerts greater force on air (blood). (e) More, increases. (f) Lack of exercise, loss of blood (hemorrhage), heart attack (myocardial infarction), or rapid heart rate (shorter diastole: see Checkpoint E6). (g) In, de; vasoconstriction of blood vessels, atherosclerosis.

F5. (a) Branching, different from; allows for more effective spreading of action potentials and contraction of the heart muscle as a unit, rather than as discrete muscle fibers. (b) In, dilate; in. (c) Parasympathetic; opposite. (d) Sympathomimetic; medulla, fight-or-flight.

F6. (a) ↑ exercise → ↑ preload. (b) Moderate ↑ in preload → ↑ SV. (c) ↑ SV (if heart rate is constant) → ↑ CO. (d) ↑ in heart contractility → ↑ SV and CO. (e) Positively inotropic medication such as digoxin (Lanoxin) → ↑ SV and CO. (f) ↑ afterload (for example, due to hypertension) → ↓ SV and CO. (g) Hypothermia → ↓ heart rate and CO. (h) ↑ heart rate (if SV is constant) → ↑ CO. (i) ↑ stimulation of the cardioaccelerator center → ↑ heart rate. (j) ↑ sympathetic nerve impulses → ↑ heart strength (and SV) and ↑ heart rate. (k) ↑ vagal nerve impulses → ↓ heart rate. (l) Hyperthyroidism → ↑ heart rate. (m) ↑ Ca^{2+} → ↑ heart strength (and SV) and ↑ heart rate. (n) ↓ SV (due to CHF) → ↑ ESV → excessive EDV and preload → ↓ SV and CO.

F7. (a) De, de. (b) Systemic regions, such as in neck and ankles; peripheral or systemic. (c) Difficulty breathing or shortness of breath (dyspnea), which is a sign of pulmonary edema. (d) Left, de, in.

G2. 300 km (200 miles) per pound of fat.

G3. (a) Lipids (like cholesterol and triglycerides), proteins; high, low density lipoproteins. (b) LDLs; low. (c) Few. (d) Prevent; exercise and a diet low in total fat, saturated fat, and cholesterol.

G4. (a) 180. (b) 50. (c) 129. (d) 2.8.

G5. (a–d) In. (e–g) De. (h–i) In.

H1. Meso; third; primitive heart.

H2. (a) SV. (b) V. (c) A. (d) BCTA.

I1. (a) Cytomegalovirus, carbon monoxide from smoking, diabetes mellitus, prolonged hypertension, and fatty diet. (b) Plaque; clot (thrombus); platelet-derived growth factor (PDGF). (c) Macrophages, endothelial cells; promotes growth of smooth muscle cells, thus narrowing arterial lumen.

I3. (a) C. (b) CA. (c) PTCA. (d) CABG.

I4. (a) PDA. (b) IASD. (c) IVSD. (d) VS. (e) C.

I5. (a) Not all. (b) Heart block. (c) Ventricular; atrial fibrillation reduces effectiveness of the atria by only about 20–30%, but ventricular fibrillation causes the heart to fail as a pump. (d) Ectopic; premature ventricular contractions (PVCs) or ventricular premature contractions (VPCs).

I6. (a) CM. (b) CP. (c) Com. (d) PT. (e) CHF.

WRITING ACROSS THE CURRICULUM: CHAPTER 20

1. Relate opening and closing of heart valves to systole and diastole of atria and ventricles.
2. Describe components of a wellness lifestyle that is likely to enhance heart health.
3. Contrast *pulmonary edema* with *peripheral edema.* Define each condition and relate them to different types of congestive heart failure (CHF). Also state two or more prominent signs or symptoms.

MASTERY TEST: CHAPTER 20

Questions 1–3: Arrange the answers in correct sequence.

_____ _____ _____ _____ 1. Pathway of the conduction system of the heart:
    A. AV node
    B. AV bundle and bundle branches
    C. SA node
    D. Conduction myofibers (Purkinje fibers)

_____ _____ _____ _____ _____ 2. Route of a red blood cell supplying oxygen to myocardium of left atrium and returning to right atrium:
    A. Arteriole, capillary, and venule within myocardium
    B. Branch of great cardiac vein
    C. Coronary sinus leading to right atrium
    D. Left coronary artery
    E. Circumflex artery

_____ _____ _____ _____ _____ 3. Route of a red blood cell now in the right atrium:
    A. Left atrium
    B. Left ventricle
    C. Right ventricle
    D. Pulmonary artery
    E. Pulmonary vein

Questions 4–15: Circle the letter preceding the one best answer to each question.

4. All of the following are correctly matched *except:*
    A. Myocardium—heart muscle
    B. Visceral pericardium—epicardium
    C. Endocardium—forms heart valves which are especially affected by rheumatic fever
    D. Pericardial cavity—space between fibrous pericardium and parietal layer of serous pericardium

5. Which of the following factors will tend to decrease heart rate?
    A. Stimulation by cardiac nerves
    B. Release of the transmitter substance norepinephrine in the heart
    C. Activation of neurons in the cardioaccelerator center
    D. Increase of vagal nerve impulses
    E. Increase of sympathetic nerve impulses

6. The average cardiac output for a resting adult is about _____ per minute.
    A. 1 quart
    B. 5 pint
    C. 1.25 gallons (5 liters)
    D. 0.5 liter
    E. 2000 ml

7. When ventricular pressure exceeds atrial pressure, what event occurs first?
    A. AV valves open
    B. AV valves close
    C. Semilunar valves open
    D. Semilunar valves close

8. The second heart sound is due to turbulence of blood flow as a result of what event?
    A. AV valves opening
    B. AV valves closing
    C. Semilunar valves opening
    D. Semilunar valves closing

9. Choose the *false* statement about the circumflex artery.
    A. It is a branch of the left coronary artery.
    B. It provides the major blood supply to the right ventricle.
    C. It lies in a groove between the left atrium and left ventricle.
    D. Damage to this vessel would leave the left atrium with virtually no blood supply.

10. Choose the *false* statement about heart structure.
    A. The heart chamber with the thickest wall is the left ventricle.
    B. The apex of the heart is more superior in location than the base.
    C. The heart has four chambers.
    D. The left ventricle forms the apex and most of the left border of the heart.
11. All of the following are defects involved in Tetralogy of Fallot *except:*
    A. Ventricular septal defect
    B. Right ventricular hypertrophy
    C. Stenosed mitral valve
    D. Aorta emerging from both ventricles
12. Choose the *true* statement.
    A. The T wave is associated with atrial depolarization.
    B. The normal P-R interval is about 0.4 sec.
    C. Myocardial infarction means strengthening of heart muscle.
    D. Myocardial infarction is commonly known as a "heart attack" or a "coronary."

13. Choose the *false* statement.
    A. Pressure within the atria is known as intraarterial pressure.
    B. During most of ventricular diastole the semilunar valves are closed.
    C. During most of ventricular systole the semilunar valves are open.
    D. Diastole is another name for relaxation of heart muscle.
14. Which of the following structures are located in ventricles?
    A. Papillary muscles
    B. Fossa ovalis
    C. Ligamentum arteriosum
    D. Pectinate muscles
15. The heart is composed mostly of:
    A. Epithelium
    B. Muscle
    C. Dense connective tissue

Questions 16–25: Circle T (true) or F (false). If the statement is false, change the underlined word or phrase so that the statement is correct.

T F 16. The blood in the left chambers of the heart contains <u>higher</u> oxygen content than blood in the right chambers.

T F 17. At the point when intraarterial pressure surpasses ventricular pressure, semilunar valves <u>open.</u>

T F 18. The pulmonary <u>artery carries</u> blood from the lungs to the left atrium.

T F 19. The normal cardiac cycle <u>does not</u> require direct stimulation by the autonomic nervous system.

T F 20. During <u>about half</u> of the cardiac cycle, atria and ventricles are contracting simultaneously.

Questions 21–25: Fill-ins. Complete each sentence with the word or phrase that best fits.

_____ 21. Most ventricular filling occurs during atrial _____.

_____ 22. _____ is a cardiovascular disorder that is the leading cause of death in the U.S.

_____ 23. An ECG is a recording of _____ of the heart.

_____ 24. _____ is a term that means abnormality or irregularity of heart rhythm.

_____ 25. Are atrioventricular and semilunar valves ever open at the same time during the cardiac cycle? _____

ANSWERS TO MASTERY TEST: ★ CHAPTER 20

Arrange
1. C A B D
2. D E A B C
3. C D E A B

Multiple Choice
4. D
5. D
6. C
7. B
8. D
9. B
10. B
11. C
12. D
13. A
14. A
15. B

True–False
16. T
17. F. Close
18. F. Veins carry
19. T
20. F. No part

Fill-ins
21. Diastole
22. Coronary artery disease
23. Electrical changes or currents that precede myocardial contractions
24. Arrhythmia or dysrhythmia
25. No

FRAMEWORK 21

The Cardiovascular System: Blood Vessels and Hemodynamics

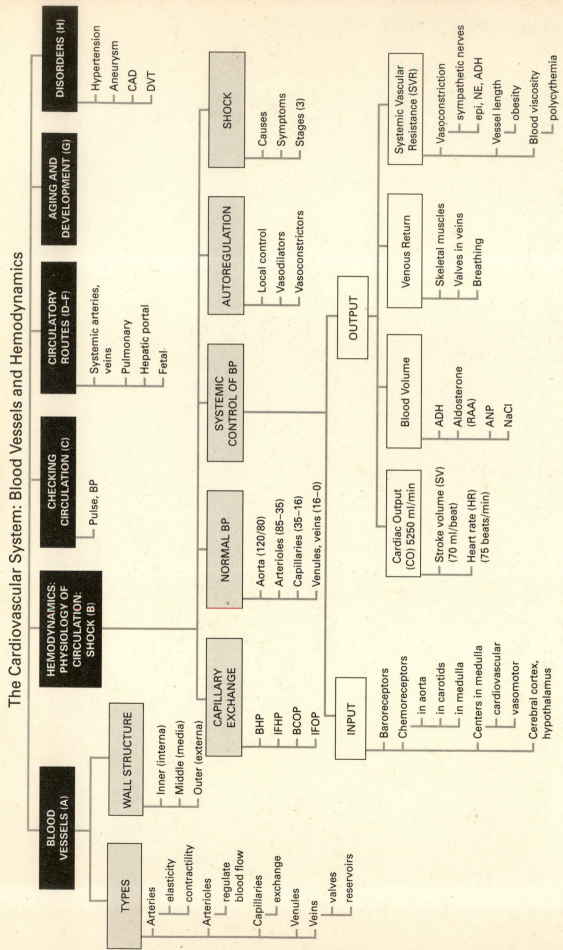

The Cardiovascular System: Blood Vessels and Hemodynamics

CHAPTER 21

Blood is pumped by heart action to all regions of the body. The network of vessels is organized much like highways, major roads, side streets, and tiny alleys to provide interconnecting routes through all sections of the body. In city traffic, tension mounts when traffic is heavy or side streets are blocked. Similarly, pressure builds (arterial blood pressure) when volume of blood increases or when constriction of small vessels (arterioles) occurs. In a city, traffic may be backed up for hours, with vehicles unable to deliver passengers to destinations. A weakened heart may lead to similar and disastrous effects, backing up fluid into tissues (edema) and preventing transport of oxygen to tissues (ischemia, shock, and death of tissues).

Careful map study leads to trouble-free travel along the major arteries of a city. A similar examination of the major arteries and veins of the human body facilitates an understanding of the normal path of blood as well as clinical applications, such as the exact placement of a blood pressure cuff over the brachial artery or the expected route of a deep venous thrombus (pulmonary embolism) to the lungs.

As you continue your study of the cardiovascular system, carefully examine the Chapter 21 Topic Outline and Objectives; check off each one as you complete it. To organize your study of blood vessels and hemodynamics, glance over the Chapter 21 Framework now. Be sure to refer to the Framework frequently and note relationships among key terms in each section.

TOPIC OUTLINE AND OBJECTIVES

A. Anatomy of blood vessels and capillary exchange

☐ 1. Contrast the structure and function of the various types of blood vessels.

B. Hemodynamics: physiology of circulation; shock

☐ 2. Discuss the various pressures involved in the movement of fluids between capillaries and interstitial spaces.

☐ 3. Explain the factors that regulate the velocity and volume of blood flow.

☐ 4. Explain how blood pressure changes throughout the cardiovascular system and describe the factors that determine mean arterial blood pressure.

☐ 5. Describe the factors that determine systemic vascular resistance and explain how the return of venous blood to the heart is accomplished.

☐ 6. Describe how blood pressure is regulated.

☐ 7. Define the three stages of shock.

C. Checking circulation

☐ 8. Define pulse and blood pressure (BP) and contrast the clinical significance of systolic, diastolic, and pulse pressure.

D. Circulatory routes: systemic arteries

E. Circulatory routes: systemic veins

F. Circulatory routes: hepatic portal, pulmonary, and fetal circulation

☐ 9. Identify the principal arteries and veins and describe the flow of blood through the systemic, hepatic portal, pulmonary, and fetal circulations.

developmental anatomy

...n the effects of aging on the
...iovascular system.
...escribe the development of blood vessels
and blood.

H. Disorders, medical terminology

☐ 12. List the causes and symptoms of hypertension, aneurysm, coronary artery disease (CAD), and deep-venous thrombosis (DTV).

☐ 13. Define medical terminology associated with blood vessals.

WORDBYTES

Now become familiar with the language of this chapter by studying each wordbyte, its meaning, and an example of its use within a term. After you study the entire list, self-check your understanding by writing the meaning of each wordbyte on the line. As you continue through the *Learning Guide,* identify (and fill in) additional terms that contain the same wordbyte.

Wordbyte	Self-check	Meaning	Example(s)
hepat-	_____	liver	*hepat*ic artery
med-	_____	middle	tunica *med*ia
phleb-	_____	vein	*phleb*itis
port-	_____	carry	*port*al vein
retro-	_____	backward, behind	*retro*peritoneal
sphygmo-	_____	pulse	*sphygmo*manometer
tunica	_____	sheathe	*tunica* intima
vaso-	_____	vessel	*vaso*constriction
ven-	_____	vein	*ven*ipuncture

CHECKPOINTS

A. Anatomy of blood vessels and capillary exchange (pages 612–620)

A1. Review types of vessels in the human circulatory system by referring to Figure LG 20.1 and Chapter 20, Checkpoint A2, pages LG 425–426.

■ **A2.** Check your understanding of the structure of blood vessels by matching terms in the box to related descriptions.

A. Anastomosis	TI. Tunica intima
L. Lumen	TM. Tunica media
TE. Tunica externa	VV. Vasa vasorum

_____ a. The opening in blood vessels through which blood flows; narrows in atherosclerosis

_____ b. Formed largely of endothelium and basement membrane, it lines lumen; the only layer in capillaries.

_____ c. Junction of two or more vessels supplying one region of the body, such as the cerebral arterial circle (of Willis) to the brain

_____ d. Composed of muscle and elastic tissue; contraction permits vasoconstriction

_____ e. Strong outer layer composed mainly of elastic and collagenous fibers

_____ f. Tiny blood vessels that carry nutrients to walls of blood vessels

■ **A3.** Match the names of vessels in the box with descriptions below. Answers may be used once, more than once, or not at all.

Aorta and other large arteries	Medium-sized arteries
Arterioles	Small veins
Capillaries	Venules
Inferior vena cava (large vein)	

a. Sites of gas, nutrient, and waste exchange with tissues: _____

b. Play primary role in regulating moment-to-moment distribution of blood and in

   regulating blood pressure: _____

c. Sinusoids of liver are wider, more leaky versions of this type of vessel: _____

d. Reservoirs for about 60% of the volume of blood in the body; vasoconstriction (due to sympathetic impulses) permits redistribution of blood stored here (three

   answers): _____ _____ _____

e. Elastic (conducting) arteries that function collectively as a pressure reservoir: _____

f. Flow of blood here is called microcirculation: _____

A4. Contrast terms in the following pairs:

a. Structure of the tunica media of the following vessels:
   conducting arteries/distributing arteries

   artery/vein

b. Functions of *anatomosing arteries/end arteries*

c. Number of capillaries in *muscles and liver/tendons and ligaments*

d. Functions of *thoroughfare channel/true capillary*

e. Structure of *continuous capillary/fenestrated capillary*

■ **A5.** Refer to Figure 21.4, page 616 in your text, and identify structural features of different types of capillaries and sinusoids that permit passage of materials across the vessel wall by four different routes. Select answers from the box.

Direct movement across endothelial membranes	Intercellular clefts
Fenestrations	Pinocytic vesicles

a. Structural features of very thin endothelial membrane with thin, incomplete, or even

absent basement membrane, for example, in sinusoids: _____

b. Gaps known as _____ between neighboring endothelial cells;
   may be large, for example, to permit passage of large proteins made in
   liver cells into the blood in liver sinusoids.

c. Pores known as _____ in plasma membranes of endothelial cells,
   for example, in cells of kidneys, intestinal villi, and choroid plexuses of the brain.
   The tight seal of the blood–brain barrier is attributed partly to lack of such pores.

d. _____ permitting endocytosis and exocytosis for
   transport of large molecules that cannot cross capillary walls in any other way.

■ **A6.** Complete this exercise about the structure of veins.

a. Veins have *(thicker? thinner?)* walls than arteries. This structural feature relates to the
   fact that the pressure in veins is *(more? less?)* than in arteries. The pressure difference
   is demonstrated when a vein is cut; blood leaves a cut vein in *(rapid spurts?
   an even flow?).*

b. Gravity exerts back pressure on blood in veins located inferior to the heart.

   To counteract this, veins contain _____.

c. When valves weaken, veins become enlarged and twisted. This condition is called

   _____. This occurs more often in *(superficial? deep?)* veins. Why?

d. A vascular (venous) sinus has _____

   _____

   replacing the tunica media and tunica externa. So sinuses have the *(structure but not
   function? function but not structure?)* of veins. List two places where such sinuses
   are found.

■ **A7.** Complete this Checkpoint on distribution of blood. It may help to refer to Figure 21.6, page 617 in your text.

a. At rest, *(arteries and arterioles? capillaries? venules and veins?)* contain most of the blood in the body. In which vessels is the smallest percentage of blood found in the resting person? *(Arteries and arterioles? Capillaries? Venules and veins?)*

b. *(Sympathetic? Parasympathetic?)* nerve impulses to these vessels cause vasoconstriction and release of blood from these vessels. Name two locations of venous "reservoirs" that can be activated to release blood when needed.

c. State two examples of circumstances that might activate distribution of reservoir blood.

_____ _____

■ **A8.** Identify the types of transport mechanisms involved in capillary exchange by doing this exercise. It may help to refer back to Checkpoint A5. Use these answers:

B. Bulk flow	V. Vesicular transport
D. Diffusion	

_____ a. Passage of lipid-soluble chemicals, such as oxygen, carbon dioxide, and steroid hormones, through the phospholipid bilayer of capillary membrane endothelial cells

_____ b. Transport of water-soluble substances, such as glucose or amino acids, through fenestrations or intercellular clefts

_____ c. Movement of large plasma proteins, such as fibrinogen or albumin, made in liver cells through intercellular clefts in sinusoids to reach plasma

_____ d. Movement of antibody proteins from maternal blood to fetal blood

_____ e. Passive movements of large numbers of ions, molecules, and particles in the same direction, largely due to opposing forces of blood pressure and osmotic pressure

■ **A9.** Recall that pressure is the principal factor causing blood to move through the vessels of the body. Pressure is also the major factor determining movement of substances across capillary membranes. Refer to Figure LG 21.1 and do this activity.

a. Color arrows A–D on Figure LG 21.1 to identify the four pressures involved in capillary exchange.

b. These pressures interact in a kind of tug-of-war that results in a net filtration pressure

(NFP) of _____ mm Hg. In other words, at the arterial end of the capillary, substances tend to move *(into? out of?)* the vessel.

c. Which value is substantially different at the venous end of capillaries? _____

Calculate the NFP there: _____ mm Hg

d. The near-equilibrium at the two ends of capillaries is based on _____'s law of the capillaries. Which system "mops up" the slight amount of fluid that is not

drawn back into veins? _____

e. Excessive amounts of fluid accumulating in interstitial spaces is the condition known as

_____. This may result from high blood pressure in which *(BCOP? BHP?)* is increased, loss of plasma protein in which *(BCOP? BHP?)* is decreased, or *(increase? decrease?)* in capillary permeability.

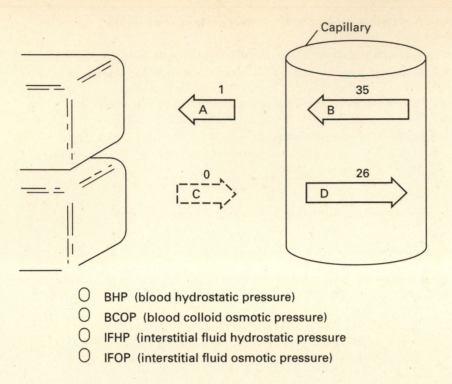

○ BHP (blood hydrostatic pressure)
○ BCOP (blood colloid osmotic pressure)
○ IFHP (interstitial fluid hydrostatic pressure
○ IFOP (interstitial fluid osmotic pressure)

Figure LG 21.1 Practice diagram showing forces controlling capillary exchange and for calculation of NFP at arterial end of capillary. Numbers are pressure values in mm Hg. Color arrows as directed in Checkpoint A9.

■ **A10.** *A clinical challenge.* On a random review of patient charts from a hospital, five clients were noted as having edema. Identify the physiological cause that was most likely the cause of each case of edema. Use these answers:

> ↑ BHP. Increased blood hydrostatic pressure
> ↑ PC. Increased permeability of capillaries
> ↑ ECV. Increased extracellular volume
> ↓ PP. Decreased concentration of plasma proteins
> LB. Lymphatic blockage

_____a. Maureen S, age 40, had axillary lymph nodes removed for biopsy following right-sided mastectomy; she stated that her "right arm seems more swollen" than her left.

_____b. Paul P, age 72, has a diagnosis of congestive heart failure and experiences what he describes as "spells when I just can't get my breath."

_____c. Tommy R, age 7, received second-degree burns on his left leg, which is swollen and seeping fluids.

_____d. Elaine B, age 41, has kidney failure; she has gained 11 pounds since her last hemodialysis treatment 3 days ago. She reported that she "went overboard eating and drinking at a couple of parties over the weekend." (Two answers)

_____e. Tracey D, who reports that she is allergic to iodine, accidentally ate an entree containing some seafood. She is experiencing pharyngeal and esophageal edema.

B. Hemodynamics: physiology of circulation; shock   (pages 620–631)

■ **B1**. Check your understanding of circulation of blood in this Checkpoint.

a. The typical human body contains about _____ liters of blood. At rest, this same

amount of blood is pumped out by the heart each minute, as cardiac _____
(5 liters/minute).

b. Now name the two factors that determine how rapidly and where this blood is distributed, for example, rapid delivery to brain and muscles, or slow delivery to intestine and
skin. (See Checkpoint B4 for more on this topic.)

■ **B2.** Do this exercise about blood flow.

a. Figure 21.8 (page 620 of your text) shows that velocity of blood flow is greatest in

(arteries? capillaries? veins?). Flow is slowest in _____.
As a result, ample time is available for exchange between capillary blood and tissues.

b. As blood moves from capillaries into venules and veins, its velocity *(increases?
decreases?),* enhancing venous return.

c. Although each individual capillary has a cross-sectional area which is microscopic, the
body contains so many capillaries that their total cross-sectional area is enormous.
(For comparison, look at the eraser end on one pencil; you are seeing the cross-sectional
area of that one pencil. If you bind together 100 pencils with a strong rubber band and
observe their eraser ends as a group, you see that the individual pencils contribute to a
relatively large total cross-sectional area.) All of the capillaries in the body have a total

cross-sectional area of about _____ cm^2.

d. *(A direct? An indirect?)* relationship exists between cross-sectional area and velocity of
blood flow. This is evident in the aorta also. This vessel has a cross section of only

about _____ cm^2 and has a very *(high? low?)* velocity.

e. *A clinical challenge.* Mr. Fey has an aortic aneurysm. The diameter of his aorta has increased
from normal size to about 10 cm. The region of the aorta with the aneurysm therefore has
a *(larger? smaller?)* cross-sectional area. One would expect the velocity of Mr. Fey's aortic

blood flow to *(in? de?)-crease.* This change may put Mr. Fey at risk for _____.

f. On the average, *(1? 5? 10?)* minute(s) is normally required for a blood cell to travel one
complete circulatory circuit, for example, heart → lung → heart → foot → heart.

■ **B3.** Complete this activity on blood pressure (BP) and mean arterial blood pressure (MABP).

a. In the aorta and brachial artery, blood pressure (BP) is about 120 mm Hg immediately
following ventricular contraction. This is called *(systolic? diastolic?)* BP. As ventricles

relax (or go into _____), blood is no longer ejected into these arteries.
However, the normally *(elastic? rigid?)* walls of these vessels recoil against blood,

pressing it onward with a diastolic BP of _____ mm Hg.

b. In a blood pressure (BP) of 120/80, the average of the two pressures (systolic and
diastolic) is 100. However the MABP (average or mean blood pressure in arteries) for a

BP of 120/80, is actually slightly less than 100; it is about _____.

Reason: during the typical cardiac cycle (0.8 sec), the ventricles are in *(systole? diastole?)* for about two-thirds of the cycle and in systole for only about one-third of the cycle (see Figure LG 20.5, page LG 435). As a result, the MABP is always slightly closer to the value of the *(systolic? diastolic?)* BP.

c. Try calculating MABP for a person with an arterial BP of 120/80. First write the

diastolic BP: _____ mm Hg. Then find the difference between systolic BP and diastolic BP, a value known as *pulse pressure:* 120 mm Hg − 80 mm Hg =

_____ mm Hg. Finally, add one-third of pulse pressure to diastolic BP to arrive at MABP:

$$BP_{diastolic} + (1/3 \times pulse\ pressure) = MABP$$

_____ mm Hg + (1/3 × _____ mm Hg) = _____ mm Hg

■ **B4.** Recall that in Checkpoint B1 you named two factors that determine distribution (or circulation) of cardiac output to the body. Closely examine these two factors in this Checkpoint.

a. One factor is _____ (such as blood pressure or MABP). Define blood

pressure: the pressure exerted by _____.
Picture the heart pumping harder and faster, much like a dam permitting more water (cardiac output) to pass over it so the river below swells; water (blood) presses forward and laps up against river banks (BP increases). So if cardiac output increases, BP or MABP are

likely to _____-crease. Finish the arrow to show the correct relationship:

↑ CO → | MABP. On the other hand, if CO decreases (as in hemorrhage or shock),

then MABP is likely to _____-crease.

b. Now visualize the effects of the second factor (resistance) on distribution of blood. Imagine yourself inside a red blood cell (much like a boat) surging forward at high velocity through the aorta or other large arteries. BP is high there (MABP of 93) as blood flows along freely. But as you enter a small artery or arteriole, your RBC encounters *(more? less?)* resistance to flow as the RBC repeatedly bounces up against and snags the walls of this narrow-diameter vessel. In other words, BP in any given blood vessel is *(directly? inversely?)* related to resistance to flow, while *(directly? inversely?)* related to cardiac output:

$$BP = \frac{CO}{R}$$

c. Because CO stays the same within the body at any given time, we can conclude that if resistance to flow is high through a vessel (such as an arteriole or capillary), that BP *in that vessel* will be *(high? low?)*. Referring to Figure 21.9, page 621 in your text, write the normal ranges of BP values for the following types of vessels: arterioles

_____ mm Hg; capillaries, _____ mm Hg; venules and veins, _____ mm Hg;

venae cavae, _____ mm Hg.

d. On Figure 21.9 the steepest decline in the pressure curve occurs as blood passes through the *(aorta? arterioles? capillaries?)*. This indicates that the greatest resistance to flow is present in the *(aorta? arterioles? capillaries?)*.

■ **B5.** Do this exercise on two very important and different relationships between resistance (R) and blood pressure.

a. In Checkpoint B4b, you saw that increase in resistance (R) *within a given blood vessel*

causes _____-crease in *BP in that vessel.* Therefore, BP (within a vessel) and R (in that same vessel) are *inversely* related. Show this by completing the arrow.

$$\uparrow R_{\text{within a vessel}} \rightarrow |\ BP_{\text{in that same vessel.}}$$

For example, the RBC (boat) passing from an artery (wide river) into an arteriole (tributary) meets more resistance to flow in this narrow tributary, thereby

_____-creasing BP *within that vessel.* So the increased resistance (R) blood meets as it passes through arterioles causes BP there to decrease from about 85 to 35 mm Hg.

b. However, several factors contribute to changes in *BP in systemic arteries* (usually measured in the brachial artery), for example, increase in BP from a normal

_____/_____ to a high BP of 162/98. One of these factors altering systemic BP is resistance. But here the resistance refers to inhibition of flow through many, many small systemic vessels, for example, by sympathetic nerve stimulation of arterioles causing

them to narrow. So the term SVR (_____ vascular _____) is used to describe this resistance. Because less blood is engaged in runoff into arterioles, more blood stays in main arteries, thereby increasing systemic BP. Complete arrows to show this:

$$\uparrow R_{\text{within many systemic arterioles, as in skin or abdomen}} \rightarrow |\ BP_{\text{in systemic arteries}}$$

or

$$\uparrow SVR \rightarrow |\ BP_{\text{systemic arterial}}$$

Consider this analogy: if blood (water) cannot pass from an artery (wide river) into many arterioles (tributaries) because entranceways are blocked by vasoconstriction of arterioles (tributaries are clogged with silt and debris), then blood (water) will remain in the

main artery (wide river), _____-creasing BP *in main systemic arteries* (wide rivers).

c. In fact, systemic BP tends to increase if either cardiac output or systemic vascular resistance (SVR) or both of those factors increase. Show this by completing the equation:

$$BP_{\text{systemic arterial}} = \underline{\hspace{1.5cm}} \times \underline{\hspace{1.5cm}}$$

or

$$MABP = \text{cardiac output} \times \text{systemic vascular resistance}$$

d. Describe three factors that may increase SVR that can raise systemic BP:

1. _____-crease in blood viscosity, for example, due to _____

2. _____-crease in total length of blood vessels in the body, for example,

   due to _____

3. _____-crease in diameter (or radius) of blood vessels, for example, by

   _____ of arterioles

e. SVR is also called total _____ resistance because the vessels that are targeted to vasoconstrict (decrease radius) and create that resistance are those in

regions of the body (such as skin and _____)
that could be considered more "peripheral," or less crucial, to survival.

■ **B6.** *A clinical challenge.* Do this exercise about Mr. Tyler, who has been depressed since his wife died two years ago and has gained 50 pounds during that time.

a. Which one of the following factors that increase systemic vascular resistance is most likely to be higher in Mr. Tyler?
    A. Blood viscosity
    B. Blood vessel length
    C. Blood vessel radius
    This factor *(does? does not?)* put Mr. Tyler at higher risk for hypertension.

b. Because Mr. Tyler's stress level is often high, his *(sympathetic? parasympathetic?)* nerves are likely to be more active. Which of the three factors (A–C) above is most likely to be affected? *(A? B? C?)* If his arterioles decrease in diameter by 50%, the resistance to

flow increases _____ times. This factor is likely to _____-crease his blood pressure.

B7. Explain how these two factors increase venous return during exercise.

a. Skeletal muscle pump (For help, refer to Chapter 20, Checkpoint F4c, page 438.)

b. Respiratory pump

■ **B8.** Imagine that your blood pressure (BP) is slightly lower than your body needs right now. Do this exercise to see how your body is likely to respond to increase your blood pressure.

a. Refer to Figure LG 21.2 and fill in box 1 to describe types of receptors that trigger changes in blood pressure. Baroreceptors are sensitive to changes in

_____ _____, and chemoceptors are

activated by chemicals such as _____. These receptors are located

in three locations: _____, _____,

and _____.

b. Nerve messages from the carotid sinuses travel to the brain via cranial nerves

_____, whereas messages from the aorta and right atrium travel by cranial

nerves _____. Write in these cranial nerves (C.N.) on the figure.

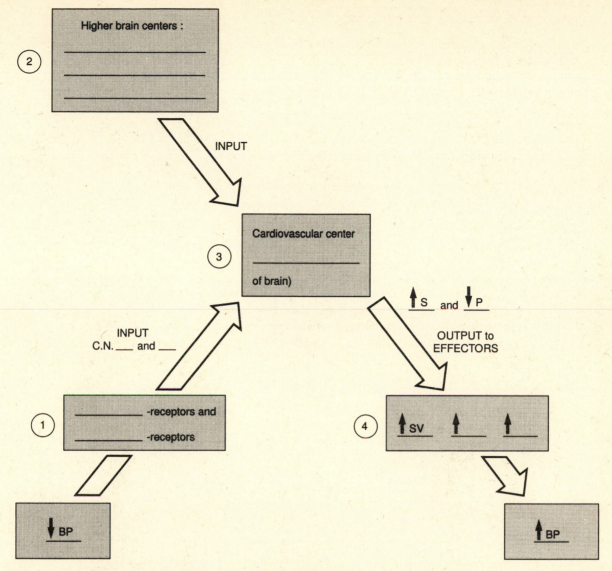

Figure LG 21.2 Overview of input and output regulating blood pressure. Complete as directed in Checkpoint B8.

Within the figure:

- Box ② Higher brain centers : _____ _____ _____
- INPUT (arrow down)
- Box ③ Cardiovascular center _____ of brain)
- ↑ S and ↓ P
- INPUT C.N. ___ and ___
- Box ① _____ -receptors and _____ -receptors
- ↓ BP
- OUTPUT to EFFECTORS
- Box ④ ↑ SV ↑ ___ ↑ ___
- ↑ BP

c. This *input* reaches the cardiovascular center located within the _____ of the brain. Fill in box 3 to indicate this region of the brain. This center receives input from other parts of the brain also. Note three of these brain parts in box 2 in the figure.

d. Consider the cardiovascular center as comparable to the control center of an automobile. The center has an accelerator portion (like a gas pedal) and an inhibitor portion (like the

_____ of a car). When baroreceptors report that BP is too low (as if the car is running too slowly), the *(cardiostimulatory? cardioinhibitory?)* center must be activated; in the car analogy, you would *(step on the gas pedal? step on the brakes?)*. At the same time, the cardioinhibitory center must be deactivated; continuing

with the car analogy, what response do you make? _____
What is the outcome for the car/heart?

e. In addition, a vasomotor portion of the cardiovascular center responds to lower blood pressure by *(stimulating? inhibiting?)* smooth muscle of blood vessels, resulting in

(vasoconstriction? vasodilatation?). This helps to increase BP by _____-creasing systemic vascular resistance (SVR).

f. Summarize nervous *output* to *effectors* from the cardiovascular center by writing in box 4 in Figure LG 21.2 the three factors that are altered. The result of this output is an

attempt to restore homeostasis by _____-creasing your blood pressure.

g. Overall, Figure LG 21.2 demonstrates *(positive? negative?)* feedback mechanisms for regulating BP, because a decrease in BP initiates factors to increase BP.

■ **B9.** On Figure LG 21.3 fill in blanks and complete arrows to show details of all of the nervous output, hormonal action, and other factors that increase blood pressure.

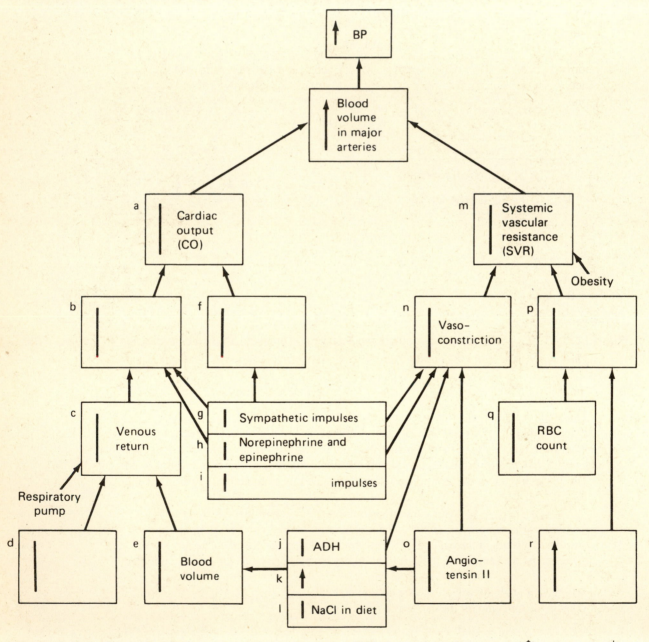

Figure LG 21.3 Summary of factors controlling blood pressure (BP). Arrows indicate increase (↑) or decrease (↓). Complete as directed in Checkpoint B9.

■ **B10.** *A clinical challenge.* Complete this exercise about blood pressure disorders.

a. Hypertension means *(high? low?)* blood pressure. It may be caused by an excess of any of the factors shown in Figure LG 21.3. If directions of arrows in Figure LG 21.3 are reversed, blood pressure can be *(increased? decreased?).*

b. Circle all the factors listed below that tend to decrease blood pressure.
   A. Increase in cardiac output, as by increased heart rate or stroke volume
   B. Increase in (vagal) impulses from cardioinhibitory center
   C. Decrease in blood volume, as following hemorrhage
   D. Increase in blood volume, by excess salt intake and water retention
   E. Increased systemic vascular resistance due to vasoconstriction of arterioles (blood prevented from leaving large arteries and entering small vessels)
   F. Decrease in sympathetic impulses to the smooth muscle of arterioles
   G. Decreased viscosity of blood via loss of blood protein or red blood cells, as by hemorrhage
   H. Use of medications called dilators because they dilate arterioles, especially in areas such as abdomen and skin

■ **B11.** *For extra review.* Do the following Checkpoint on additional details on blood pressure regulation.

a. Vasomotor _____ refers to a moderate level of vasoconstriction normally maintained by the medulla and sympathetic nerves. A sudden loss of sympathetic nerve output, for example, by head trauma or emotional trauma can reduce this tone and cause BP to *(skyrocket? drop precipitously?).* This may lead to

_____ (fainting) or even to shock.

b. Smooth muscle of most small arteries and arterioles contains *(alpha? beta?)* receptors;

in response to the neurotransmitter _____, these vessels vaso-*(constrict? dilate?).* Sympathetic stimulation of most veins causes them to vaso-

(constrict? dilate?), thereby _____-creasing venous return.

c. Vasodilator medications therefore tend to _____-crease BP as blood pools in nonessential areas (such as GI tract or skin). Name two vasodilator chemicals that

may be produced by the body: _____ from cells of the atria, and

_____ from mast cells during inflammation.

d. *(The Bainbridge? Marey's?)* reflex helps to increase heart rate and force of contraction

when receptors in the _____ detect increased return of blood to the heart, for example, during exercise. This is therefore a *(positive? negative?)* feedback mechanism.

e. Baroreceptors involved with the aortic reflex primarily maintain *(cerebral? general systemic?)* BP, whereas receptors that trigger the carotid sinus reflex primarily control

_____ BP.

f. Carotid bodies and aortic bodies are sites of *(baro? chemo?)*-receptors that provide

input for control of BP. Low oxygen (known as _____) or high

carbon dioxide (called _____) will tend to cause *(vasoconstriction?*

vasodilatation?) and consequently _____-crease BP. As a result, blood can move more rapidly to the lungs to pick up oxygen and give up carbon dioxide.

g. Write ↑ or ↓ next to each hormone or other chemical listed below to indicate whether the chemical increases or decreases blood pressure.

_____ 1. ADH   _____ 5. ANP

_____ 2. Angiotensin II   _____ 6. Histamine

_____ 3. Calcitriol   _____ 7. Kinins

_____ 4. Epinephrine of NE   _____ 8. PTH

■ **B12.** Describe the process of autoregulation in this activity.

a. Autoregulation is a mechanism for regulating blood flow through specific tissue areas. Unlike cardiac and vasomotor mechanisms, autoregulation occurs by *(local? autonomic nervous system?)* control.

b. In other words, when a tissue (such as an active muscle) is hypoxic, that is, has a *(high? low?)* oxygen level, the cells of that tissue release *(vasoconstrictor?*

vasodilator?) substances. These may include _____ acid, built up by active muscle. Other products of metabolism that serve as dilator substance include:

c. The fact that warming of an area causes: *(vasoconstriction? vasodilation?)* is the

principle behind application of heat to _____-increase local blood flow for healing.

d. Recall from Chapter 19 that the initial step in hemostasis is *(vasodilation? vasospasm?)*. Name two chemicals that produce this effect.

■ **B13.** *A clinical challenge.* Write arrows to indicate signs and symptoms of shock.
(↑ = increase; ↓ = decrease)

a. Sweat _____   d. Blood pH _____

b. Heart rate (pulse) _____   e. Mental status _____

c. BP _____   f. Urine volume _____

■ **B14.** Describe cause of shock and compensation for shock in this activity.

a. Shock is said to occur when tissues receive inadequate _____ supply to meet their demands for oxygen, nutrients, and waste removal. Shock resulting from decrease in circulating blood volume is known as

_____ shock. Write two or more causes of this type of shock.

b. Compensatory mechanisms activate *(sympathetic? parasympathetic)* nerves that

_____-crease both heart rate and systemic vascular resistance. These attempts to increase

BP normally work for blood losses up to _____% of total blood volume. One common indicator of this compensatory mechanism is *(warm, pink, and dry? cool, pale, and clammy?)*

skin related to vaso-_____ and _____-creased sweating.

c. Chemical vasoconstrictors include _____, released from the adrenal medulla;

angiotensin II, formed when _____ is secreted by kidneys; _____

made by the adrenal cortex; and the posterior pituitary hormone, _____

d. Which hormones cause water retention which helps to increase BP in shock?

e. How does hypoxia help to compensate for shock?

B15. Explain what is likely to happen in severe, prolonged shock:

a. Stage II

b. Stage III

C. Checking circulation (pages 631–632)

C1. Define *pulse.*

C2. In general, where may pulse best be felt? List six places where pulse can readily be palpated. Try to locate pulses at these points on yourself or on a friend.

■ **C3.** Normal pulse rate is about _____ beats per minute. *Tachycardia* means *(rapid? slow?)*

pulse rate; _____ means slow pulse rate.

■ **C4.** Describe the method commonly used for taking blood pressure by answering these questions about the procedure.

a. Name the instrument used to check BP. _____

b. The cuff is usually placed over the _____ artery. How can you tell that this artery is compressed after the bulb of the sphygmomanometer is squeezed?

c. As the cuff is deflated, the first sound heard indicates _____ pressure because reduction of pressure from the cuff below systolic pressure permits blood to begin spurting through the artery during diastole.

d. Diastolic pressure is indicated by the point at which the sounds

_____.
At this time blood can once again flow freely through the artery.

e. Sounds heard while taking blood pressure are known as _____ sounds.

■ **C5.** *A clinical challenge.* Mrs. Yesberger has a systolic pressure of 128 mm Hg and diastolic of 78 mm Hg. Write her blood pressure in the form in which BP is usually

expressed: _____. Determine her pulse pressure.

Would this be considered a normal pulse pressure? _____

D. Circulatory routes: systemic arteries   (pages 632–647)

D1. On Figure LG 20.1 (page 426), label the *pulmonary artery, pulmonary capillaries,* and a *pulmonary vein.* Briefly contrast pulmonary and systemic vessels.

■ **D2.** Use Figure LG 21.4 to do this learning activity about systemic arteries.

a. The major artery from which all systemic arteries branch is the _____. It exits from the chamber of the heart known as the *(right? left?)* ventricle. The aorta can be

divided into three portions. They are the _____ (N1 on the figure),

the _____ (N2), and the _____ (N3).

b. The first arteries to branch off the aorta are the _____ arteries. Label these

(at O) on the figure. These vessels supply blood to the _____.

c. Three major arteries branch from the aortic arch. Write the names of these (at K, L, and M) on the figure, beginning with the vessel closest to the heart.

d. Locate the two branches of the brachiocephalic artery. The right subclavian artery (letter _____)

supplies blood to _____

Vessel E is named the *(right? left?)* _____ artery. It supplies the right side of the head and neck.

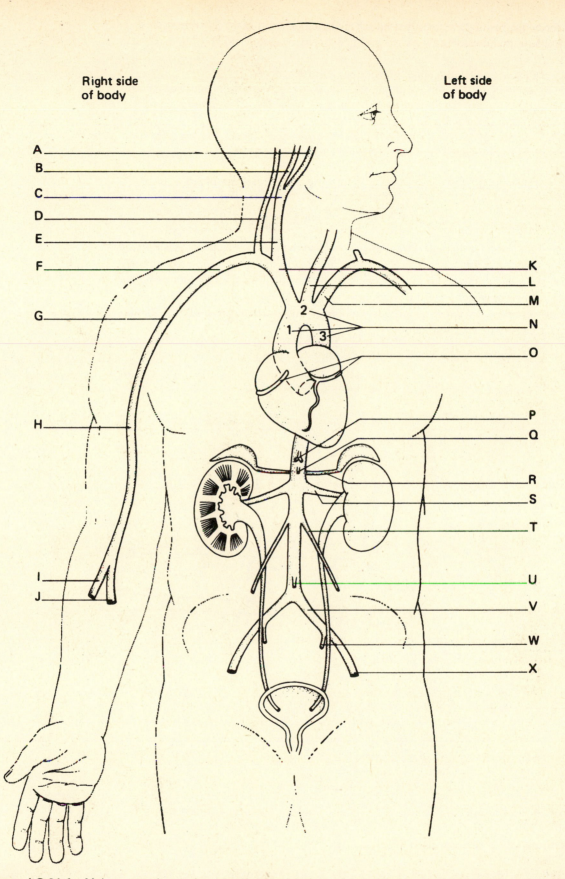

Right side
of body

Left side
of body

A
B
C
D
E
F

G

H

I
J

K
L
M

N

O

P
Q

R
S

T

U

V

W

X

2
1 3

Figure LG 21.4 Major systemic arteries. Identify letter labels as directed in Checkpoints D2, D3, and D6–D8.

e. As the subclavian artery continues into the axilla and arm, it bears different names (much as a road does when it passes into new towns). Label G and H. Note that H is

the vessel you studied as a common site for measurement of _____

f. Label vessels I and J, and then continue the drawing of these vessels into the forearm and hand. Notice that they anastomose in the hand. Vessel (*I? J?*) is often used for checking pulse in the wrist.

g. Besides supplying the arms, subclavian arteries each send a branch that ascends the neck through foramina in cervical vertebrae. This is the _____

artery. The right and left vertebral arteries join at the base of the brain to form

the _____ artery. (Refer to D and C on Figure LG 21.5.)

■ **D3.** The right and left common carotid arteries ascend the neck in positions considerably (*anterior? posterior?*) to the vertebral arteries. Find a pulse in one of your own carotids. These vessels each bifurcate (divide into two vessels) at about the level of your mandible. Identify the two branches (A and B) on Figure LG 21.4. Which supplies the brain and eye? (*Internal? External?*) What structures are supplied by the external carotids? (See Exhibit 21.4, page 638 in the text, for help.)

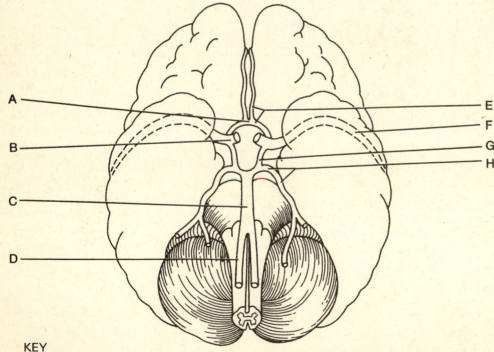

KEY

A. Anterior communicating cerebral artery (1)
B. Internal carotid arteries (cut) (2)
C. Basilar artery (1)
D. Vertebral arteries (2)

E. Anterior cerebral arteries (2)
F. Middle cerebral arteries (2)
G. Posterior communicating cerebral arteries (2)
H. Posterior cerebral arteries (2)

Figure LG 21.5 Arterial blood supply to the brain including vessels of the cerebral arterial circle (of Willis). View is from the undersurface of the brain. Numbers in parentheses indicate whether the arteries are single (1) or paired (2). Refer to Checkpoints D2, D4, and D5.

■ **D4.** Refer to Figure LG 21.5 and answer these questions about the vital blood supply to the brain.

a. An anastomosis, called the cerebral arterial circle of _____ is located just inferior to the brain. This circle closely surrounds the pituitary gland.

b. The circle is supplied with blood by two pairs of vessels: anteriorly, the *(internal? external?)* carotid

arteries and, posteriorly, the _____ arteries (joined to form the basilar artery).

c. The circle supplies blood to the brain by three pairs of cerebral vessels, identified on

Figure LG 21.5 by letter labels _____, _____, and _____. Vessels A and G

complete the circle; they are known as anterior and posterior _____ arteries. Now cover the key and correctly label each of the lettered vessels.

■ **D5.** *A clinical challenge.* Relate symptoms to the interruption of normal blood supply to the brain as described in each case. Explain your reasons.

a. A patient's basilar artery is sclerosed. Which part of the brain is most likely to be deprived of normal blood supply? *(Frontal lobes of cerebrum? Occipital lobes of cerebrum, as well as cerebellum and pons?)*

b. Paralysis and loss of some sensation are noted on a patient's *right* side. Angiography points to sclerosis in a common carotid artery. Which one is more likely to be sclerosed? *(Right? Left?)* Explain.

c. The middle cerebral artery which passes between temporal and frontal lobe (see broken lines on Figure LG 21.5) is blocked. Which sense is more likely to be affected? *(Hearing? Vision?)* Explain.

d. Anterior cerebral arteries pass anteriorly from the circle of Willis and then arch backward along the medial aspect of each cerebral hemisphere. Disruption of blood flow in the right anterior cerebral artery will be most likely to cause effects in the *(right? left?)* *(hand? foot?)*. Hint: The affected brain area would include region Z in Figure LG 15.1 and/or region R in Figure LG 15.2, pages LG 300 and 303. Note that these regions are on the medial aspect of the right hemisphere.)

■ **D6.** Complete the table describing visceral branches of the abdominal aorta. Identify these on Figure LG 21.4 (vessels P to U). Refer also to Figure LG 24.4, page LG 547.

Artery	Letter	Structures Supplied
a. Celiac (three major branches) 1. 2. 3.		1. Liver 2. 3.
b.	Q	Small intestine and part of large intestine and pancreas
c.		Adrenal (suprarenal) glands
d. Renal	S	
e.		Testes (or ovaries)
f. Inferior mesenteric		

■ **D7.** The aorta ends at about level _____ vertebra by dividing into right and left

_____ arteries. Label these at V on Figure LG 21.4.

D8. Label the two branches (W and X on Figure LG 21.4) of each common iliac artery. Name structures supplied by each.

a. Internal iliac artery

b. External iliac artery

■ **D9.** *For extra review.* Suppose that the right femoral artery were completely occluded. Describe an alternate (or collateral) route that would enable blood to pass from the right common iliac artery to the little toe of the right foot.

Right common iliac artery →

E. Circulatory routes: systemic veins   (pages 648–658)

■ **E1.** Name the three main vessels that empty venous blood into the right atrium of the

heart. _____ _____ _____

■ **E2.** Do this exercise about venous return from the head.

a. Venous blood from the brain drains into vessels known as _____.
    These are *(structurally? functionally?)* like veins.
b. Name the sinus that lies directly beneath the sagittal suture: _____
c. Blood from all of the cranial vascular sinuses eventually drains into the

    _____ veins, which descend in the neck. These veins are

    positioned close to the _____ arteries.
d. Where are the jugular arteries and carotid veins located?

■ **E3.** Imagine yourself as a red blood cell currently located in the *left brachial vein.* Answer these questions.

a. In order for you to flow along in the bloodstream to the *left brachial artery,* which of these must you pass through? *(A brachial capillary? The heart and a lung?)*

b. In order for you to flow from the left brachial vein to the *right arm,* must you pass through the heart? *(Yes? No?)* Both sides of the heart, that is, right and left? *(Yes? No?)* One lung? *(Yes? No?)*

■ **E4.** Once you are familiar with the pathways of systemic arteries, you know much about the veins that accompany these arteries. Demonstrate this by tracing routes of a drop of blood along the following deep vein pathways.

a. From the thumb side of the left forearm to the right atrium

b. From the medial aspect of the left knee to the right atrium

c. From the right kidney to the right atrium

■ **E5.** Contrast veins with arteries in this exercise.

a. Veins have *(higher? lower?)* blood pressure than arteries and a *(faster? slower?)* rate of blood flow.

b. Veins are therefore *(more? less?)* numerous, in order to compensate for slower blood flow in each vein. The "extra" veins consist of vessels located just under the skin

(in subcutaneous fascia); these are known as _____ veins.

■ **E6.** On text Figures 21.25 to 21.27 (pages 653–657), locate the following veins. Identify the location at which each empties into a deep vein.

a. Basilic _____   c. Great saphenous _____

b. Median cubital _____   d. Small saphenous _____

E7. Veins connecting the superior and inferior venae cavae are named _____ veins. Identify the azygos system of veins on Figure 21.26 (page 655) in the text. What parts of the thorax do they drain? Explain how these veins may help to drain blood from the lower part of the body if the inferior vena cava is obstructed.

E8. Name the visceral veins that empty into the abdominal portion of the inferior vena cava. Then color them blue on Figure LG 26.1 (page LG 594). Contrast accompanying systemic arteries by coloring them red on that figure. Note on Figure LG 26.1 that the inferior vena cava normally lies on the *(right? left?)* side of the aorta. One way to remember this is that the inferior vena cava is returning blood to the *(right? left?)* side of the heart.

F. Circulatory routes: hepatic portal, pulmonary, and fetal circulation (pages 658–664)

F1. Refer to Figure LG 20.1 (page 426) and do the following activity.

a. Label the following vessels: *gastric, splenic,* and *mesenteric* vessels. Note that all of these vessels drain directly into the vessel named *(7, inferior vena cava? H, portal vein to the liver?).* Label that vessel on Figure LG 20.1.

b. Label the vessel that carries oxygenated blood into the liver.

c. In the liver, blood from both the portal vein and hepatic artery mixes as it passes

through tortuous capillary-like vessels known as _____.
On Figure LG 20.1 label a vessel that collects this blood and transports it to the inferior vena cava.

d. What functions are served by the special hepatic portal circulation?

F2. Contrast pulmonary circulation with systemic circulation in this activity.
Write P (pulmonary) or S (systemic).

_____a. Which is considered a lower resistance system? List structural features that explain this lower resistance.

_____b. Which arteries are likely to have a higher blood pressure (BP)? Write typical

BP values for these vessels: aorta _____ mm Hg; pulmonary artery

_____ mm Hg.

_____c. Which capillaries normally have a hydrostatic pressure (BP) of about 10 mm Hg? Write one consequence of increased pulmonary capillary BP.

_____d. Which system has its blood vessels constrict in response to hypoxia? Write one consequence of this fact.

F3. What aspects of fetal life require the fetus to possess special cardiovascular structures not needed after birth?

■ **F4.** Complete the following table about fetal structures. (Note that the structures are arranged in order of blood flow from fetal aorta back to fetal aorta.)

Fetal Structure	Structures Connected	Function	Fate of Structure After Birth
a.		Carries fetal blood low in oxygen and nutrients and high in wastes	
b. Placenta	(Omit)		
c.	Placenta to liver and ductus venosus		
d.		Branch of umbilical vein, bypasses liver	
e. Foramen ovale			
f.			Becomes ligamentum arteriosum

■ **F5.** For each of the following pairs of fetal vessels, circle the vessel with higher oxygen content.

a. Umbilical artery/umbilical vein

b. Femoral artery/femoral vein

c. Aorta/thoracic portion of inferior vena cava

G. Aging and developmental anatomy   (page 664)

■ **G1.** Draw arrows next to each factor listed below to indicate whether it increases (↑) or decreases (↓) with normal aging.

_____ a. Size and strength of cardiac muscle cells   _____ c. Total cholesterol and LDLs in blood

_____ b. Cardiac output (CO)   _____ d. Systolic blood pressure

G2. Write a paragraph describing embryonic development of blood vessels. Include the following key terms: *15 to 16 days, mesoderm, mesenchyme, blood islands, spaces, endothelial, tunica.*

G3. List four sites of blood formation in the embryo and fetus.

H. Disorders, medical terminology   (pages 665–666)

■ **H1.** Match types of hypertension listed in the box with descriptions below. One answer will be used twice; one answer will not be used.

Pri.	Primary	Sta–1.	Stage 1
Sec.	Secondary	Sta–2.	Stage 3
		Sta–4.	Stage 4

_____ a. Also known as essential or idiopathic (no known cause) hypertension

_____ b. Over 90% of cases of hypertension fit into this category

_____ c. Hypertension that has a known cause

_____ d. BP of 150/94 fits into this category

_____ e. BP of 200/110 fits into this category

■ **H2.** Match each organ listed in the box with the related disorder that may cause secondary hypertension.

AC.	Adrenal cortex	K.	Kidney
AM.	Adrenal medulla		

_____ a. Excessive production of renin, which catalyzes formation of angiotensin II, a powerful vasoconstrictor

_____ b. Pheochromocytoma, a tumor that releases large amounts of norepinephrine and epinephrine

_____ c. Release of excessive amounts of aldosterone, which promotes salt and water retention

H3. Describe possible effects of hypertension upon the following body structures:

a.   Heart

b.   Cerebral blood vessels

c.   Kidneys

H4. Explain how each of the following therapeutic measures can help control hypertension.

a. Sodium restriction

b. Cessation of smoking

c. Exercise

d. Stress reduction

e. Vasodilators

f. A.C.E. inhibitors

g. Diuretics

H5. Define *aneurysm* and list four possible causes of aneurysms.

■ **H6.** *A clinical challenge.* Explain how a deep vein thrombosis (DVT) of the left femoral vein can lead to a pulmonary embolism.

■ **H7.** Match terms related to blood vessels with descriptions below. Use each answer once.

Angio.	Angiogenesis	Ortho.	Orthostatic hypotension
Art.	Arteritis	Phleb.	Phlebitis
C.	Claudication	WCH.	White coat hypertension
Occ.	Occlusion		

_____ a. Inflammation of a vein

_____ b. Inflammation of an artery

_____ c. Obstruction of a vessel, such as an artery, due to atherosclerotic plaque

_____ d. Pain upon exercise due to impaired circulation in limbs

_____ e. Significant drop in blood pressure upon standing up

_____ f. Growth of new blood vessels, for example with weight gain, pregnancy, or to increase blood supply to tumor

_____ g. Temporary elevation of blood pressure associated with stress of having BP measured by health-care personnel

A2. (a) L. (b) TI. (c) A. (d) TM. (e) TE. (f) VV.

A3. (a) Capillaries. (b) Arterioles (c) Capillaries. (d) Venules, small veins, inferior vena cava (actually all veins and venules). (e) Aorta and other large arteries. (f) Capillaries.

A5. (a) Direct movement across endothelial membranes. (b) Intercellular clefts. (c) Fenestrations. (d) Pinocytic vesicles.

A6. (a) Thinner; less; an even flow. (b) Valves. (c) Varicose veins; superficial; skeletal muscles around deep veins limit overstretching. (d) Surrounding dense connective tissue (such as the dura mater); function, but not structure; intracranial (such as superior sagittal venous sinus) and coronary sinus.

A7. (a) Venules and veins; capillaries. (b) Sympathetic; veins of skin and of abdominal organs such as liver and spleen. (c) Exercise, hemorrhage.

A8. (a–c) D. (d) V. (e) B.

A9. (a) A, IFOP; B, BHP; C, IFHP; D, BCOP. (b) +10; out of. (c) BHP is about 16 mm Hg; –9. (d) Starling; lymphatic. (e) Edema; BHP, BCOP, increase.

A10. (a) LB. (b) $\uparrow$ BHP (pulmonary hypertension due to backlog of blood from a failing left ventricle. (c) $\downarrow$ PP (as plasma proteins leaked out of damaged vessels); also $\uparrow$ PC. (d) $\uparrow$ ECV and $\uparrow$ BHP. (e) $\uparrow$ PC (related to release of histamine in allergic reaction).

B1. (a) 5–6; output. (b) Pressure (such as blood pressure) and resistance to flow.

B2. (a) Arteries; capillaries. (b) Increases. (c) 4500–6000. (d) An indirect; 3 to 5, high. (e) Larger; de; clot formation (also shock because pooled blood there does not circulate to tissues). (f) 1.

B3. (a) Systolic; diastole; elastic, 70–80. (b) 93; diastole; diastolic. (c) 80; 120–80 = 40; 80 + $(1/3 \times 40 = 13)$ = 93.

B4. (a) Pressure; blood on the wall of a blood vessel; in; $\uparrow$ CO $\rightarrow \uparrow$ MABP; de. (b) More; inversely, directly. (c) Low; 85–35; 35–16; 16–0; close to 0. (d) Arterioles; arterioles.

B5. (a) De; $\uparrow$ $R_{within\ a\ vessel} \rightarrow \downarrow$ $BP_{in\ that\ same\ vessel}$; de. (b) About 120/80; systemic (vascular) resistance; $\uparrow$ $R_{(systemic\ arterioles:\ skin\ or\ abdomen)} \rightarrow \uparrow$ $BP_{in\ systemic\ arteries}$; or $\uparrow$ SVR $\rightarrow \uparrow$ $BP_{systemic\ arterial}$; in. (c) $BP_{systemic\ arterial}$ = CO × SVR. (d1) In, dehydration or polycythemia. (d2) In, weight gain (or obesity), growth, or pregnancy. (d3) De, vasoconstriction. (e) Peripheral; digestive organs such as liver, spleen, stomach, intestine, as well as kidneys.

B6. (a) B; does. Recall that the body grows an average of about 300 km (200 miles) of new blood vessels for each extra pound of weight. (b) Sympathetic; C; 16; in.

B8. (a) Box 1: baro, chemo. Blood pressure; H^+, CO_2, and O_2; carotid arteries, aorta, and right atrium. (b) IX, X. (c) Box 3: medulla. Box 2: cerebral cortex, limbic system, and hypothalamus. (d) Brakes; cardiostimulatory, step on the gas pedal; take your foot off the gas; car goes faster/heart beats faster and with greater force of contraction. (e) Stimulating, vasoconstriction; in. (f) $\uparrow$ stroke volume (SV), $\uparrow$ heart rate (HR), and $\uparrow$ systemic vascular resistance (SVR); in. (g) Negative.

B9. (a) $\uparrow$ CO. (b) $\uparrow$ Stroke volume. (c) $\uparrow$ Venous return. (d) $\uparrow$ Exercise. (e) $\uparrow$ Blood volume. (f) $\uparrow$ Heart rate. (g) $\uparrow$ Sympathetic impulses. (h) $\uparrow$ Norepinephrine and epinephrine. (i) $\downarrow$ Vagal impulses. (j) $\uparrow$ ADH. (k) $\uparrow$ Aldosterone. (l) $\uparrow$ NaCl in diet. (m) $\uparrow$ Systemic vascular resistance (SVR). (n) $\uparrow$ Vasoconstriction. (o) $\uparrow$ Angiotensin II. (p) $\uparrow$ Viscosity of blood. (q) $\uparrow$ RBC count. (r) $\uparrow$ Plasma proteins.

B10. (a) High; decreased. (b) B C F G H.

B11. (a) Tone; drop precipitously; syncope. (b) Alpha, norepinephrine, constrict; constrict, in. (c) De; atrial natriuretic peptide (ANP), histamine. (d) The Bainbridge, right atrium; positive. (e) General systemic, cerebral. (f) Chemo; hypoxia, hypercapnia, vasconstriction, in. (g1–4) $\uparrow$; (g5–8) $\downarrow$.

B12. (a) Local. (b) Low, vasodilator; lactic; EDRF (nitric oxide), K^+, H^+, and adenosine. (c) Vasodilation, in. (d) Vasospasm; eicosanoids such as thromboxane A_2 and prostaglandin F_{2a}.

B13. (a–b) $\uparrow$. (c–f) $\downarrow$.

B14. (a) Blood; hypovolemic; loss of blood or other body fluids, or shifting of fluids into interstitial spaces (edema) as with burns. (b) Sympathetic, in; 10; cool, pale, and clammy, constriction, in. (c) Epinephrine, renin, aldosterone, ADH (vasopressin). (d) Aldosterone and ADH. (e) It causes local vasodilation.

C3. 60–100; rapid, bradycardia.

C4. (a) Sphygmomanometer. (b) Brachial; no pulse is felt or no sound is heard via a stethoscope. (c) Systolic. (d) Change or cease. (e) Korotkoff sounds.

C5. 128/78, 50 (128–78); yes, it is in the normal range.

D2. (a) Aorta; left; ascending aorta, arch of the aorta, descending aorta. (b) Coronary; heart.

(c) K, brachiocephalic; L, left common carotid; M, left subclavian. (d) F; in general, the right extremity and right side of thorax, neck, and head; right common carotid. (e) G, axillary; H, brachial; blood pressure. (f) I, radial; J, ulnar; I. (g) Vertebral; basilar.

D3. Anterior; internal; external portions of head (face, tongue, ear, scalp) and neck (throat, thyroid).

D4. (a) Willis. (b) Internal, vertebral. (c) E, F, H; communicating.

D5. (a) Occipital lobes of cerebrum, as well as cerebellum and pons. (b) Left, because this artery supplies the left side of the brain. (Remember that neurons in tracts cross to the opposite side of the body, but arteries do not.) (c) Hearing, because it is perceived in the temporal lobe. (Visual areas are in occipital lobes). (d) Left foot.

D6.

Artery	Letter	Structures Supplied
a. Celiac (three major branches) 　1. Hepatic 　2. Gastric 　3. Splenic	P	1. Liver, **gallbladder and parts of stomach, duodenum, pancreas** 2. **Stomach and esophagus** 3. **Spleen, pancreas, and stomach**
b. **Superior mesenteric**	Q	Small intestine and part of large intestine and pancreas
c. **Adrenal (suprarenal)**	R	Adrenal (suprarenal) glands
d. Renal	S	**Kidneys**
e. **Gonadal (testicular or ovarian)**	T	Testes (or ovaries)
f. Inferior mesenteric	U	**Parts of colon and rectum**

D7. L-4, common iliac.

D9. Right external iliac artery → right lateral circumflex artery (right descending branch) → right anterior tibial artery → right dorsalis pedis artery → to arch and digital arteries in foot.

E1. Superior vena cava, inferior vena cava, coronary sinus.

E2. (a) Intracranial vascular (venous) sinuses; functionally. (b) Superior sagittal. (c) Internal jugular; common carotid. (d) There are no such vessels.

E3. (a) The heart and a lung. (b) Yes; yes; yes.

E4. (a) Left radial vein → left brachial vein → left axillary vein → left subclavian vein → left brachiocephalic vein → superior vena cava. (Note that there are left and right brachiocephalic veins, but only one artery of that name.) (b) Small veins in medial aspect of left knee → left popliteal vein → left femoral vein → left external iliac vein → left common iliac vein → inferior vena cava. (c) Right renal vein → inferior vena cava.

E5. (a) Lower, slower. (b) More; superficial.

E6. (a) Axillary. (b) Axillary. (c) Femoral. (d) Popliteal.

E8. See key, page 592 of the Guide. Right, right.

F1. (a) E, gastric; F, splenic; I, mesenteric; H, portal vein to the liver. (b) D, hepatic artery. (c) Sinusoids; G, hepatic vein (there are two but only one is shown on Figure LG 20.1). (d) While in liver sinusoids, blood is cleaned, modified, detoxified. Ingested nutrients and other chemicals are metabolized and stored.

F2. (a) P; pulmonary arteries have larger diameters and shorter length (a shorter circuit), thinner walls, less elastic tissue. (b) S; 120/80; 25/8 (pulmonary BP is about one-fifth of systemic BP). (c) P; pulmonary edema with shortness of breath and impaired gas exchange. (d) P; an advantage is the diversion of blood to well-ventilated regions. One disadvantage is that chronic vasoconstriction in persons with chronic hypoxia (as in emphysema or chronic bronchitis) may lead to cor pulmonale (see page 608 of your text).

F4.

Fetal Structure	Structures Connected	Function	Fate of Structure After Birth
a. Umbilical arteries (2)	**Fetal Internal iliac arteries to placenta**	Carries fetal blood low in oxygen and nutrients and high in wastes	**Medial umbilical ligaments**
b. Placenta	(Omit)	**Site where maternal and fetal blood exchange gases, nutrients, and wastes**	**Delivered as "afterbirth"**
c. Umbilical vein (1)	Placenta to liver and ductus venosus	**Carries blood high in oxygen and nutrients, low in wastes**	**Round ligament of the liver (ligamentum teres)**
d. Ductus venosus	**Umbilical vein to inferior vena cava**	Branch of umbilical vein, bypasses liver	**Ligamentum venosum**
e. Foramen ovale	**Right and left atria**	**Bypasses lungs**	**Fossa ovalis**
f. Ductus arteriosus	**Pulmonary artery to aorta**	**Bypasses lungs**	Becomes ligamentum arteriosum

F5. (a) Umbilical vein. (b) Femoral artery. (c) Thoracic portion of inferior vena cava (because umbilical vein blood enters it via ductus venosus).

G1. (a) ↓. (b) ↓. (c) ↑. (d) ↑.

H1. (a–b) Pri. (c) Sec. (d) Sta-1. (e) Sta-3.

H2. (a) K. (b) AM. (c) AC.

H6. Embolus (a "clot-on-the-run") travels through femoral vein to external and common iliac veins to inferior vena cava to right side of heart. It can lodge in pulmonary arterial branches, blocking further blood flow into the pulmonary vessels.

H7. (a) Phleb. (b) Art. (c) Occ. (d) C. (e) Ortho. (f) Angio. (g) WCH.

WRITING ACROSS THE CURRICULUM: CHAPTER 21

1. Explain how the structure of the following types of blood vessels is admirably suited to their functions: elastic arteries, muscular arteries, arterioles, metarterioles, and capillaries.
2. Discuss advantages to health offered by collateral circulation. Include the terms anastomoses and end arteries in your discussion.
3. The typical human body contains about 5–6 liters of blood. Describe mechanisms that regulate where this blood goes at any given moment, for example, at rest versus during a period of vigorous exercise. Be sure to include the following terms in your discussion: metarteriole, precapillary sphincter, sympathetic, and vasoconstriction.
4. In most cases, arteries and veins are parallel in both name and locations, such as the brachial artery and brachial vein lying close together in the arm. Identify major arteries and veins that differ in names and locations.

MASTERY TEST: CHAPTER 21

Questions 1–2: Circle the letters preceding all correct answers to each question.

1. In the most direct route from the left leg to the left arm of an adult, blood must pass through all of these structures:
   A. Inferior vena cava
   B. Brachiocephalic artery
   C. Capillaries in lung
   D. Hepatic portal vein
   E. Left subclavian artery
   F. Right ventricle of heart
   G. Left external iliac vein

2. In the most direct route from the fetal right ventricle to the fetal left leg, blood must pass through all of these structures:
   A. Aorta
   B. Umbilical artery
   C. Lung
   D. Ductus arteriosus
   E. Ductus venosus
   F. Left ventricle
   G. Left common iliac artery

Questions 3–9: Circle the letter preceding the one best answer to each question. Note whether the underlined word or phrase makes the statement true or false.

3. Choose the *false* statement.
   A. In order for blood to pass from a vein to an artery, it must pass through chambers of the <u>heart.</u>
   B. In its passage from an artery to a vein a red blood cell must ordinarily travel through a <u>capillary.</u>
   C. The wall of the femoral artery is <u>thicker</u> than the wall of the femoral vein.
   D. Most of the smooth muscle in arteries is in the <u>tunica interna.</u>

4. Choose the *false* statement.
   A. Arteries <u>contain valves, but veins do not.</u>
   B. Decrease in the size of the lumen of a blood vessel by contraction of smooth muscle is called <u>vasoconstriction.</u>
   C. Most capillary exchange occurs by <u>simple diffusion.</u>
   D. Sinusoids are <u>wider and more tortuous (winding)</u> than capillaries.

5. Choose the *false* statement.
   A. Cool, clammy skin is a sign of shock that results from <u>sympathetic stimulation of blood vessels and sweat glands.</u>
   B. Nicotine is a <u>vasodilator that helps control</u> hypertension.
   C. <u>Orthostatic hypotension</u> refers to a sudden, dramatic drop in blood pressure upon standing or sitting up straight.
   D. Phlebitis is an <u>inflammation of a vein.</u>

6. Choose the *false* statement.
   A. The vessels that act as the major regulators of blood pressure are <u>arterioles.</u>
   B. The only blood vessels that carry out exchange of nutrients, oxygen, and wastes are <u>capillaries.</u>
   C. Blood in the umbilical artery is normally <u>more highly</u> oxygenated than blood in the umbilical vein.
   D. At any given moment more than 50% of the blood in the body is in the <u>veins.</u>

7. Choose the *true* statement.
   A. The basilic vein is located <u>lateral</u> to the cephalic vein.
   B. The great saphenous vein runs along the <u>lateral</u> aspect of the leg and thigh.
   C. The vein in the body most likely to become varicosed is the <u>femoral.</u>
   D. The hemiazygos and accessory hemiazygos veins lie to the <u>left</u> of the azygos.

8. Choose the *true* statement.
   A. A normal blood pressure for the average adult is <u>160/100.</u>
   B. During exercise, blood pressure will tend to <u>increase.</u>
   C. Sympathetic impulses to the heart and to arterioles tend to <u>decrease blood pressure.</u>
   D. Decreased cardiac output causes <u>increased</u> blood pressure.

9. Choose the *true* statement.
   A. Cranial venous sinuses are composed of the <u>typical three layers</u> found in walls of all veins.
   B. Cranial venous sinuses all eventually empty into the <u>external jugular veins.</u>
   C. Internal and external jugular veins <u>do</u> unite to form common jugular veins.
   D. Most parts of the body supplied by the internal carotid artery are ultimately drained by the <u>internal jugular vein.</u>

Questions 10–14: Circle the letter preceding the one best answer to each question.

10. Hepatic portal circulation carries blood from:
    A. Kidneys to liver
    B. Liver to heart
    C. Stomach, intestine, and spleen to liver
    D. Stomach, intestine, and spleen to kidneys
    E. Heart to lungs
    F. Pulmonary artery to aorta

11. Which of the following is a parietal (rather than a visceral) branch of the aorta?
    A. Superior mesenteric artery
    B. Bronchial artery
    C. Celiac artery
    D. Lumbar artery
    E. Esophageal artery

12. All of these vessels are in the leg or foot *except:*
    A. Saphenous vein
    B. Azygos vein
    C. Peroneal artery
    D. Dorsalis pedis artery
    E. Popliteal artery

13. All of these are superficial veins *except:*
    A. Median cubital
    B. Basilic
    C. Cephalic
    D. Great saphenous
    E. Brachial

14. Vasomotion refers to:
    A. Local control of blood flow to active tissues
    B. The Doppler method of checking blood flow
    C. Intermittent contraction of small vessels resulting in discontinuous blood flow through capillary networks
    D. A new aerobic dance sensation

Questions 15–20: Arrange the answers in correct sequence.

_____ _____ _____ _____ 15. Route of a drop of blood from the right side of the heart to the left side of the heart:
   A. Pulmonary artery
   B. Arterioles in lungs
   C. Capillaries in lungs
   D. Venules and veins in lungs

Questions 16–18, use the following answers:

A. Aorta and other arteries	C. Capillaries
B. Arterioles	D. Venules and veins

_____ _____ _____ _____16. Blood pressure in vessels, from highest to lowest:

_____ _____ _____ _____17. Total cross-sectional areas of vessels, from highest to lowest:

_____ _____ _____ _____18. Velocity of blood in vessels, from highest to lowest:

_____ _____ _____ _____ _____19. Route of a drop of blood from small intestine to heart:
    A. Superior mesenteric vein
    B. Hepatic portal vein
    C. Small vessels within the liver
    D. Hepatic vein
    E. Inferior vena cava

_____ _____ _____ 20. Layers of blood vessels, from most superficial to deepest:
    A. Tunica interna
    B. Tunica externa
    C. Tunica media

Questions 21–25: Fill-ins. Complete each sentence or answer the question with the word(s) or phrase that best fits.

_____21. A weakened section of a blood vessel forming a balloonlike sac is known as a(n) _____.

_____22. Name the major factor that creates blood osmotic pressure (BCOP) that prevents excessive flow of fluid from blood to interstitial areas.

_____23. Name the two blood vessels that are locations of baroreceptors and chemoreceptors for regulation of blood pressure.

_____24. In taking a blood pressure, the cuff is first inflated over an artery. As the cuff is then slowly deflated, the first sounds heard indicate the level of _____ blood pressure.

_____25. Most arteries are paired (one on the right side of the body, one on the left). Name five or more vessels that are unpaired.

Multiple Answers

1. A C E F G
2. A D G

Multiple Choice

3. D	
4. A	9. D
5. B	10. C
6. C	11. D
7. D	12. B
8. B	13. E
	14. C

Arrange

15 A B C D
16. A B C D
17. C D B A
18. A B D C
19. A B C D E
20. B C A

Fill-ins

21. Aneurysm
22. Plasma protein such as albumin
23. Aorta and carotids
24. Systolic
25. Aorta, anterior communicating cerebral, basilar, brachiocephalic, celiac, gastric, hepatic, splenic, superior and inferior mesenteric, middle sacral

FRAMEWORK 22
Lymphatic System & Immunity

FRAMEWORK 22 — Lymphatic System & Immunity

- LYMPHATIC SYSTEM
  - ORGANIZATION (A)
    - Lymphatic vessels
    - Lymph circulation
  - LYMPHATIC TISSUE (B)
    - Nodes
    - Tonsils
    - Spleen
    - Thymus
  - DEVELOPMENT (C)

- RESISTANCE TO DISEASE
  - NONSPECIFIC RESISTANCE (D)
    - Skin, mucosa
      - Mechanical factors
      - Chemical factors
    - NK cells
    - Antimicrobial substances
      - Transferrins
      - Interferon
      - Complement
      - Properdin
    - Phagocytosis
      - Chemotaxis
      - Opsonization
      - Adherence
      - Ingestion
      - Killing
    - Inflammation
      - Pain, redness, swelling, heat
      - Stages
    - Fever
  - IMMUNITY: SPECIFIC RESISTANCE (E)
    - Antigens
      - Immunogenicity
      - Reactivity
      - MHC antigens
      - APCs
    - Cytokines
      - Interleukins
      - Interferons
      - Lymphotoxin
      - Perforin
    - Cell-mediated immunity (CMI)
      - T cells
        - cytotoxic (T_C)
        - helper (T_H)
        - 3 others
    - Antibody-mediated immunity (AMI)
      - B cells
      - Plasma cells
      - Antibodies
        - Ig
          - Active
          - Passive

- AGING AND IMMUNITY (F)

- DISORDERS (G)
  - Cancer
  - AIDS
  - Autoimmune
  - Hypersensitivity
    - allergy
    - tissue rejection

The Lymphatic System, Nonspecific Resistance to Disease, and Immunity

CHAPTER 22

The human body is continually exposed to foreign (nonself) entities: bacteria or viruses, pollens, chemicals in a mosquito bite, or foods or medications that provoke allergic responses. Resistance to specific invaders is normally provided by a healthy immune system made up of antibodies and battalions of T lymphocytes.

Nonspecific resistance is offered by a variety of structures and mechanisms, such as skin, mucus, white blood cells assisting in inflammation, and even a rise in temperature (fever) that limits the survival of invading microbes. The lymphatic system provides sentinels at points of entry (as in tonsils) and at specific sites (lymph nodes) near entry to body cavities (inguinal and axillary regions). The lymphatic system also wards off the detrimental effects of edema by returning proteins and fluids to blood vessels.

As you begin your study of this system, carefully examine the Chapter 22 Topic Outline and Objectives; check off each one as you complete it. To organize your study of the lymphatic system, nonspecific resistance to disease, and immunity, glance over the Chapter 22 Framework now. Be sure to refer to the Framework frequently and note relationships among key terms in each section.

TOPIC OUTLINE AND OBJECTIVES

A. Lymphatic vessels and lymph circulation

- ☐ 1. Describe the general components of the lymphatic system and list its functions.
- ☐ 2. Describe the formation and flow of lymph.

B. Lymphatic tissue

- ☐ 3. List and describe the primary and secondary lympatic organs of the body.

C. Developmental anatomy of the lymphatic system

- ☐ 4. Describe the development of the lymphatic system.

D. Nonspecific resistance to disease

- ☐ 5. Discuss the roles of the skin and mucous membranes, antimicrobial substances, phagocytosis, inflammation, and fever in nonspecific resistance to disease.

E. Immunity (specific resistance to disease)

- ☐ 6. Define immunity and describe how T cells and B cells arise.
- ☐ 7. Explain the relationship between an antigen (Ag) and an antibody (Ab).
- ☐ 8. Describe the roles of antigen presenting cells, T cells, and B cells in cell-mediated and antibody-mediated immunity.
- ☐ 9. Explain how self-tolerance occurs.
- ☐ 10. Discuss the relationship of immunology to cancer.

F. Aging and the immune system

G. Disorders, medical terminology

- ☐ 11. Describe the clinical symptoms of the following disorders: acquired immunodeficiency syndrome (AIDS), autoimmune diseases, systemic lupus erythematosus (SLE), chronic fatigue syndrome, severe combined immunodeficiency (SCID), hypersensitivity (allergy), tissue rejection, and Hodgkin's disease (HD).
- ☐ 12. Define medical terminology associated with the lymphatic system.

WORDBYTES

Now become familiar with the language of this chapter by studying each wordbyte, its meaning, and an example of its use within a term. After you study the entire list, self-check your understanding by writing the meaning of each wordbyte on the line. As you continue through the *Learning Guide,* identify (and fill in) additional terms that contain the same wordbyte.

Wordbyte	Self-check	Meaning	Example(s)
anti-	_____	against	*anti*body
auto-	_____	self	*auto*immune
axilla-	_____	armpit	*axilla*ry nodes
chyl-	_____	juice	cisterna *chyl*i
gen-	_____	to produce	anti*gen,* *gen*etic
hetero-	_____	other	*hetero*graft
iso-	_____	same	*iso*graft
lact-	_____	milk	*lact*eals
meta-	_____	beyond	*meta*stasis
xeno-	_____	foreign	*xeno*graft

CHECKPOINTS

A. Lymphatic vessels and lymph circulation (pages 671–675)

A1. Summarize the body's defense mechanisms against invading microorganisms and other potentially harmful substances. Be sure to use these terms in your paragraph: *pathogens, resistance, susceptibility.*

■ **A2.** Write S next to specific immune responses and NS next to nonspecific responses.

_____ a. Barriers to infection, such as skin and mucous membranes

_____ b. Inflammation and phagocytosis

_____ c. Chemicals such as gastric or vaginal secretions, which are both acidic

_____ d. Antigen-antibody responses.

_____ e. Destruction of an intruder by a cytotoxic T lymphocyte

_____ f. Antimicrobial substances such as interferons (IFNs) or complement

■ **A3.** Describe the functions of the lymphatic system in this exercise.

a. Lymphatic vessels drain from tissue spaces _____ that are too large to

pass readily into blood capillaries. They also transport _____ and
(water? fat?)-soluble vitamins from the gastrointestinal (GI) tract to the bloodstream.
b. The defensive functions of the lymphatic system are carried out largely by white blood

cells named _____-cytes. *(B? T?)* lymphocytes destroy foreign

substances directly, whereas B lymphocytes lead to the production of _____.

■ **A4.** List the components of the lymphatic system.

A5. Describe relationships between terms in each pair.
a. *Lymphatic capillary epithelium/one-way valve*

b. *Anchoring filaments/edema*

c. *Lacteal/chyle*

A6. Describe the general pathway of lymph circulation. (*For extra review.* Refer to
Chapter 1, Checkpoint D2 and Figure LG 1.1, page 9 in the *Learning Guide.*)

■ **A7.** The lymphatic system does not have a separate heart for pumping lymph. Describe
two principal factors that are responsible for return of lymph from the entire body to major
blood vessels in the neck. (Note that these same factors also facilitate venous return.)

■ **A8.** Answer the following questions about the largest lymphatics.

  a. In general, the thoracic duct drains *(three-fourths? one-half? one-fourth?)* of the body's lymph. This vessel starts in the lumbar region as a dilation known as the

  _____. This cisterna chyli receives lymph from which areas of the body?

  b. In the neck the thoracic duct receives lymph from three trunks. Name these.

  c. One-fourth of the lymph of the body drains into the _____ duct. The regions of the body from which this vessel collects lymph are:

  d. The thoracic duct empties into the junction of the blood vessels named left

  _____ and left _____ veins. The right lymphatic duct empties into veins of the same name on the right side.

  e. In summary, lymph fluid starts out originally in plasma or cells and circulates as

  _____ fluid. Then as lymph it undergoes extensive cleaning as it passes through nodes. Finally, fluid and other substances in lymph are returned

  to _____ in veins of the neck.

■ **A9.** *A clinical challenge.* Explain why a person with a broken left clavicle might exhibit these two symptoms:

  a. Edema, especially of the left arm and both legs

  b. Excess fat in feces (steatorrhea)

B. Lymphatic tissue   (pages 675–679)

■ **B1.** Write P next to descriptions of primary lymphatic organs and S next to descriptions of secondary lymphatic organs.

_____ a. Origin of B cells and pre-T cells

_____ b. Sites where most immune responses occur

_____ c. Includes lymph nodes, spleen, and lymphatic nodules

_____ d. Includes red bone marrow and thymus

■ **B2.** Describe the thymus gland in this exercise.

    a. It is located in the *(neck? mediastinum?)*. Its position is just posterior to the

    _____ bone. Its size is relatively *(large? small?)* in childhood,

    _____-creasing in size after age 10 to 12 years.

    b. Like lymph nodes, the thymus contains a medulla surrounded by a

    _____. Within the cortex are densely packed

    _____-cytes. The thymus is the site of maturation of *(B? T?)*
    cells *(only in childhood? throughout life?)*.

    c. The *(cortex? medulla?)* consists mostly of epithelial cells. What do these produce?

■ **B3.** Label structures A–E on Figure LG 22.1. Then color each of those structures
according to descriptions below.
    a. ◯ Vessels for entrance of lymph into node
    b. ◯ Vessel for exit of lymph out of node
    c. ◯ Site of B lymphocyte proliferation into antibody-secreting plasma cells
    d. ◯ Channel for flow of lymph in the cortex of the node
    e. ◯ Dense strand of lymphocytes, macrophages, and plasma cells

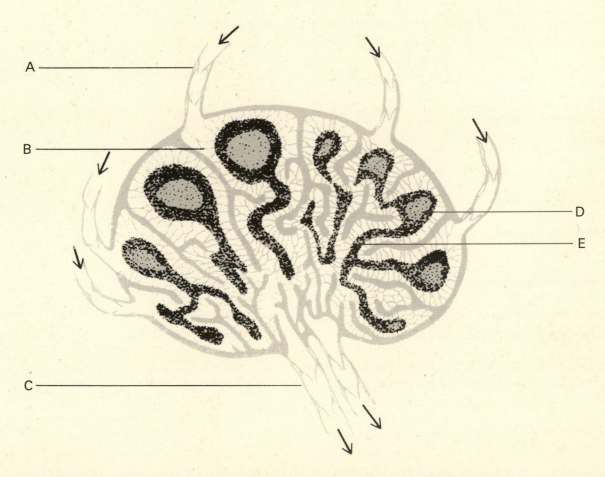

Figure LG 22.1   Structure of a lymph node. Color and label as directed in Checkpoint B3.

B4. Describe how lymph is "processed" as it circulates through lymph nodes. How does this function of lymph nodes explain the fact that nodes ("glands") become enlarged and tender during infection?

■ **B5.** Spread, or _____, of cancer may occur via the lymphatic system. Lymph nodes that have become cancerous tend to be enlarged, firm, and *(tender? nontender?)*.

■ **B6.** The spleen is located in the *(upper right? upper left? lower right? lower left?)* quadrant of the abdomen, immediately inferior to the _____. List three functions of the spleen:

_____   _____   _____

■ **B7.** Complete this exercise about lymphatic nodules.

a. The acronym MALT refers to M_____ A_____

L_____ T_____. Name four systems that contain MALT:

_____   _____   _____   _____

b. State advantages of the following locations of aggregations of lymphatic tissue:
   1. Peyer's patches

   2. Tonsils and adenoids

c. Which tonsils are most commonly removed in a tonsillectomy? _____

Which ones are called adenoids? _____

C. Developmental anatomy of the lymphatic system   (pages 679–680)

■ **C1.** The lymphatic system derives from _____-derm, beginning about the *(fifth? seventh? tenth?)* week of gestation.

■ **C2.** Lymphatic vessels arise from lymph sacs that form from early *(arteries? veins?)*. Name the embryonic lymph sac that develops vessels to each of the following regions of the body.

a. Pelvic area and lower limbs: _____ lymph sac

b. Upper limbs, thorax, head, and neck: _____ lymph sac. The left jugular lymph sac eventually develops into the upper portion of the main collecting

lymphatic vessel, the _____ duct.

D. Nonspecific resistance to disease   (pages 680–685)

D1. Briefly contrast the two types of resistance to disease: *specific* vs. *nonspecific*. (For help, refer back to Checkpoint A2.)

■ **D2.** *A clinical challenge.* Match the following conditions or factors with the nonspecific resistance that is lost in each case.

Nonspecific Resistance Lost

_____ a. Cleansing of oral mucosa

_____ b. Cleansing of vaginal mucosa

_____ c. Flushing of microbes from urinary tract

_____ d. Lacrimal fluid with lysozyme

_____ e. Respiratory mucosa with cilia

_____ f. Closely packed layers of keratinized cells

_____ g. HCl production; pH 1 or 2

_____ h. Sebum production

Condition/Factor

A. Long-term smoking
B. Excessive dryness of skin
C. Use of anticholinergic medications that decrease salivation
D. Vaginal atrophy related to normal aging
E. Dry eyes related to decreased production of tears, as in aging
F. Decreased urinary output, as in prostatic hypertrophy
G. Skin wound
H. Partial gastrectomy (removal of part of the stomach)

■ **D3.** *For extra review.* Describe nonspecific defenses more fully in this exercise.

a. Interferons (IFNs) are proteins produced by cells infected with _____,

such as cells named _____, _____, and _____.

IFNs then stimulate *(infected? uninfected?)* cells to produce _____ proteins.

b. Alpha-IFN is classified as a *(type I? type II?)* interferon. Name three or more virus-associated disorders combatted by alpha-IFN.

c. Besides their antiviral functions, what other roles do IFNs play?

d. NK cells, or _____ _____ cells, are a type of

(lymphocyte? neutrophil?). Type II (gamma) IFNs stimulate the _____

activity of these cells. NK cells are less functional in patients with _____.

e. The primary type of white blood cells that are involved in phagocytosis is _____.

Other leukocytes, known as _____, travel to infection sites and develop

into highly phagocytic _____ cells. Macrophages compose the reticulo-

endothelial, or _____ _____, system.

f. _____ consists of a group of proteins in blood plasma and on cell membranes. When activated, these enhance (or complement) immune, allergic, and inflammatory reactions. List the components of this system.

D4. Write a paragraph describing roles of the complement system. Include these key terms: *C3, histamine, opsonization,* and *membrane attack complex (MAC).*

■ **D5.** Match each of the steps in phagocytosis with the related description. Note that descriptions are listed in chronological sequence.

A. Adherence	I. Ingestion
C. Chemotaxis	K. Killing

_____ a. Phagocytic cells "sniff" the delectable fragrance of invading microbes and especially savor bacteria with the tasty *complement* coating. "Ah, dinner soon!"

_____ b. Phagocytes trap the invaders, possibly wedging them against a blood vessel or a clot, and then attach to the surface of the invader.

_____ c. Phagocytic tentacles called pseudopods surround the tasty morsels forming a culinary treat, the *phagocytic vesicle.* "Slurp!"

_____ d. Lysosomal enzymes, oxidants, and *defensins* pounce on the microbial meal, a has-been within a half hour. *Residual bodies,* like the pits of prunes, are discarded, and the phagocytic face smiles in contentment.

D6. Do this exercise on inflammation.

a. Defend or dispute this statement: "Inflammation is a process that can be both helpful and harmful."

b. List the four cardinal signs or symptoms of inflammation.

_____   _____

_____   _____

c. Briefly describe the three basic stages of inflammation.
   1.

   2.

   3.

■ **D7.** *For extra review.* Select the number that best fits each description of a phase of inflammation. Use these answers:

> 1. Vasodilation and increased capillary permeability
> 2. Phagocytic migration
> 3. Repair

_____ a. Brings defensive substances to the injured site and helps remove toxic wastes

_____ b. Histamine, prostaglandins, kinins, leukotrienes, and complement enhance this process

_____ c. Causes redness, warmth, and swelling of inflammation

_____ d. Localizes and traps invading organisms

_____ e. Occurs within an hour after initiation of inflammation; involves margination, diapedesis, chemotaxis, and leukocytosis

_____ f. White blood cell and debris formation that may lead to abscess or ulcer.

■ **D8.** Identify the chemicals that fit the descriptions below.

> C. Complement   LT. Leukotrienes
> H. Histamine   PG. Prostaglandins
> I. Interferon   T. Transferrins
> K. Kinins

_____ a. Produced by damaged or inflamed tissues; aspirin and ibuprofen neutralize these pain-inducing chemicals

_____ b. Produced by mast cells and basophils (two answers)

_____ c. Made by some leukocytes and fibroblasts; stimulates cells near an invading virus to produce antiviral proteins that interfere with survival of the virus

_____ d. Induces vasodilation, permeability, chemotaxis, and irritation of nerve endings (pain)

_____ e. Neutrophils are attracted by these chemicals present in the inflamed area (two answers)

_____ f. Proteins that inhibit bacterial growth by depriving microbes of iron.

_____ g. Group of 20 serum proteins that stimulate release of histamine and also make bacteria "tastier" to phagocytes by opsonization.

E. Immunity (specific resistance to disease) (pages 685–699)

■ **E1.** Complete this exercise about cells that carry out immune responses.

a. Name the two categories of cells that carry out immune responses. _____

and _____ Where do lymphocytes originate?

b. Some immature lymphocytes migrate to the thymus and become *(B? T?)* cells. Here T cells develop immunocompetence, meaning that these cells have the ability to

_____.

Some T cells become CD4+ cells and others become _____ cells, based on the type of

_____ in their plasma membranes.

c. Where do B cells mature into immune cells? _____

■ **E2.** Contrast two types of immunity by writing AMI before descriptions of *antibody-mediated immunity* and CMI before those describing *cell-mediated immunity*.

_____ a. Especially effective against microbes that enter cells, such as viruses and parasites

_____ b. Especially effective against bacteria present in extracellular fluids

_____ c. Involves plasma cells (derived from B cells) that produce antibodies

_____ d. Utilizes killer T cells (derived from CD8+ T cells) that directly attack the antigen

_____ e. Facilitated by helper T (CD4+) cells

■ **E3.** Do this exercise on antigens.

a. An antigen is defined as "any chemical substance which, when introduced into the body,"

_____

In general, antigens are *(parts of the body? foreign substances?)*.

b. Complete the definitions of the two properties of antigens.

1. Immunogenicity: ability to _____ specific antibodies or specific T cells

2. Reactivity: ability to _____ specific antibodies or specific T cells

c. An antigen with both of these characteristics is called a(n) _____.

d. A partial antigen is known as a _____. It displays *(immunogenicity?*

reactivity?) but not _____. For example, for the hapten penicillin to evoke an immune response (immunogenicity), it must form a complete antigen by

combining with a _____. Persons who have this particular protein

are said to be _____ to penicillin.

e. Describe the chemical nature of antigens.

f. Explain why plastics used for valves or joints are not likely to initiate an allergic response and be rejected.

g. Can an entire microbe serve as an antigen? *(Yes? No?)* List the parts of microbes that may be antigenic.

h. If you are allergic to pollen in the spring or fall or to certain foods, the pollen or foods serve as *(antigens? antibodies?)* to you.

i. Antibodies or specific T cells form against *(the entire? only a specific region of the?)*

antigen. This region is known as the _____. Most antigens have *(only one? only two? a number of?)* antigenic determinant sites.

E4. Explain how humans have the ability to recognize, bind to, and then evoke an immune response against over a billion different antigenic determinants. Describe two aspects.

a. Genetic recombination

b. Somatic mutations

■ **E5.** Describe roles of major histocompatibility complex (MHC) antigens in this exercise.

a. Describe the "good news" (how they help you) and the "bad news" (how they may be harmful) about MHC antigens.

b. Contrast two classes of MHC antigens according to types of cells incorporating the MHC into plasma membrane and the lengths of amino acid fragments picked up by each MHC.
MHC-I

MHC-II

c. Which of the following do T cells normally ignore?
   A. MHC with peptide fragment from foreign protein
   B. MHC with peptide fragment from self-protein

■ **E6.** Describe the processing and presenting of antigens in this activity.

a. For an immune response to occur, either _____ or _____ cells must recognize the presence of a foreign antigen. *(B? T?)* cells can recognize antigens located in extracellular fluid (ECF) and not attached to any cells. However, for *(B? T?)* cells to recognize antigens, the antigens must have been processed by a cell, and then they must be presented in association with MHC-I or MHC-II self-antigens (as described in Checkpoint E5b).

b. Define *exogenous antigens*.

Cells that present these antigenic proteins are called APCs, or a_____

p_____ c_____. These cells are generally found *(in lymph nodes? at sites where antigens are likely to enter the body, such as skin or mucous membranes?)*. Name three types of APCs.

c. Place in correct sequence the six boxed codes that describe the steps APCs or other body cells take to initiate an immune response to intruder antigens. Use the lines provided.

Bind to MHC–II. Bind peptide fragment to MHC–II
Exo/insert. Exocytosis and insertion of antigen fragment-MHC–II into APC plasma membrane.
Fusion. Fusion of vesicles of peptide fragments with MHC–II
Partial dig. Partial digestion of antigen into peptide fragments
Phago/endo. Phagocytosis or endocytosis of antigen
Present to T. Present antigen to T cell in lymphatic tissue

---

1. _____ → 2. _____ → 3. _____ →

4. _____ → 5. _____ → 6. _____

d. Although only selected cells can present exogenous antigens, most cells of the body

can present _____-genous antigens. State an example of endogenous antigens.

■ **E7.** Match the name of each cytokine with the description that best fits.

Gamma-IFN. Gamma interferon
IL–1. Interleukin-1
IL–2. Interleukin-2
IL–4. Interleukin-4
LT. Lymphotoxin
M. Monokine
MMIF. Macrophage migration inhibiting factor
P. Perforin
TNF. Tumor necrosis factor

_____ a. General name for a cytokine made by a
monocyte

_____ b. Made by activated helper T cells;
stimulates B cells, leading to IgE
production

_____ c. Known as T cell growth factor, it is made
by helper T cells; needed for almost all
immune responses; causes proliferation
of cytotoxic T cells (as well as B cells)
and activates NK cells

_____ d. Formerly called *macrophage activating
factor* (MAF), stimulates phagocytosis in
macrophages and neutrophils; enhances
AMI and CMI responses

_____ e. Prevents macrophages from leaving the
site of infection

_____ f. Destroys target cells by perforating their
cell membranes

_____ g. Activates enzymes in the target cell that it
kills by destruction of cell's DNA

_____ h. Produced largely by macrophages;
induces fever (two answers)

■ **E8.** Refer to Figure LG 22.2 and check your understanding of antibody structure in this
Checkpoint.

a. Chemically, all antibodies are composed of glycoproteins named _____.
They consist of *(2? 4? 8?)* polypeptide chains. The two heavy chains contain about

_____ amino acids; light chains contain about _____ amino acids. Crosshatch the
heavy chains on the figure.

b. *(Constant? Variable?)* portions are diverse for different antibodies because these serve
as binding sites for different antigens. *(Constant? Variable?)* portions *(differ? are
identical?)* for all antibodies within the same class. Color constant and variable portions
of both chains as indicated by color code ovals on the figure. Also label antibody *hinge
region, binding sites, disulfide bonds,* and *carbohydrates.*

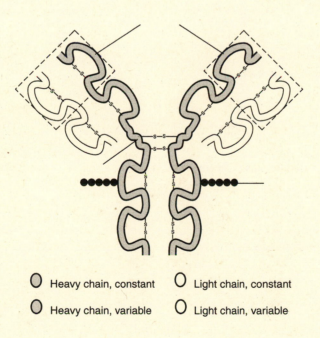

O Heavy chain, constant   O Light chain, constant

O Heavy chain, variable   O Light chain, variable

Figure LG 22.2   Diagram of an antibody molecule. Color and label as directed in Checkpoint E8.

■ **E9.** Because glycoproteins forming antibodies are involved in immunity, they are called

immunoglobulins, abbreviated _____. The different classes of Ig's are distinguishable by the *(constant? variable?)* portions of the antibody structure. Write the name of the related class of Ig next to each description.

_____ a. The first antibodies to be secreted after initial exposure to antigen, they are short lived; they destroy invading microbes by agglutination and lysis.

_____ b. The only type of Ig to cross the placenta, it provides specific resistance of newborns. Also significantly enhances phagocytosis and neutralizing of toxins in persons of all ages because it is the most abundant type of antibody.

_____ c. Found in secretions such as mucus, saliva, and tears and so protects against oral, vaginal, and respiratory infections.

_____ d. Located on mast and basophil cells and involved in allergic reactions, for example, to certain foods, pollen, or bee venom.

_____ e. Activate B cells to produce antibodies

E10. Describe the stages of cell-mediated immunity (CMI) by completing the following outline that provides key terms. *(Suggestion:* Refer to Figure 22.16, page 693 in the text.)

a. Antigen recognition (car-starting analogy) and activation
   1. T cell receptors (TCRs)

   2. Costimulators

   3. (What is the meaning of *anergy?*)

b. Proliferation and differentiation of effector cells
   1. Sensitized T cells

   2. Clones of five different T cells (See also Checkpoint E11.)

c. Elimination of the intruder by chemicals from T_c cells
   1. Perforin

   2. Lymphotoxin

   3. Gamma-IFN

■ **E11.** Match T cells to descriptions of their roles. (*Suggestion:* Refer to Checkpoint E7 above.)

> DTH. Delayed type hypersensitivity cells T_H. Helper T cells
> M. Memory T cells T_s. Suppressor T cells
> T_c. Cytotoxic T cells

_____ a. Cells that may produce chemicals that inhibit production of B cells and T cells

_____ b. Programmed to recognize the original invader; can initiate dramatic responses to reappearance of the intruder

_____ c. Called T8 cells because developed from cells with CD8 protein; to become cytolytic, must be costimulated by IL-2 or other cytokines from T_H cells

_____ d. Called T4 cells because developed from cells with CD4 protein; activated by APCs and costimulated by IL-1 and IL-2; produce IL-2, IL-4, and IL-5

_____ e. Mediate allergic responses by secreting cytokines (such as gamma-IFN) that activate macrophages

E12. Describe antibody-mediated immunity (AMI) in this activity.

a. Where do B cells perform their functions? (*At sites where invaders enter the body? In lymphatic tissue?*)

b. Now describe the steps in AMI, using these key terms or phrases as a guide.

1. Activation of B cells

2. Follicular dendritic cells

3. Antigen-MHC-II

4. Costimulation

5. Plasma cells

6. Four or five days

7. Memory B cells

8. Antibodies with identical structure to original antigen receptors

E13. List six mechanisms that antibodies may use to attack and inactivate antigens.

E14. *A clinical challenge.* Define monoclonal antibodies (MAb's). These antibodies are

produced by _____ cells. Explain how a *hybridoma* cell is produced.

Describe three or more clinical uses of monoclonal antibodies.

■ **E15.** Do this exercise on immunizations.

a. Once a specific antigen has initiated an immune response, either by infection or by a(n)
   (initial? booster dose?) immunization, the person produces some long-lived B and

   T cells called _____.

b. The level of antibodies rises *(slowly? rapidly?)* after the initial immunization. This level

   measured in serum is known as the antibody _____. Which antibody

   titer rises first? *(IgA? IgG? IgM?)* Which rises next? _____

c. Upon subsequent exposure to the same antigen, such as during another infection or by a

   _____ dose of vaccine, memory cells provide a *(more? less?)* intense
   response. Antibodies produced by this secondary response, also known as

   _____ _____, have a *(higher? lower?)* affinity for
   the antigen compared to antibodies produced in the primary response.

■ **E16.** *A clinical challenge.* Identify types of immunity described below by selecting from
these answers:

AAAI. Artificially acquired active immunity	NAAI. Naturally acquired active immunity
AAPI. Artificially acquired passive immunity	NAPI. Naturally acquired passive immunity

_____ a. As 31-year-old Bud tore apart an old
shed, a rusty nail entered his left hand.
Bud, who reported that he had not had a
tetanus shot since he was "about 14,"
received a shot of tetanus immunoglobulins at the emergency room.

_____ b. Kelly has provided her 3-month-old baby
Crystal with temporary immunity by the
antibodies that crossed over the placenta
during Kelly's pregnancy and also by
antibodies in milk from breast feeding.

_____ c. Kim took her baby Jamie to the clinic for
Jamie's regularly scheduled MMR
(measles-mumps-rubella) immunization.

■ **E17.** The ability of your T cells to avoid reacting with your own body proteins is known as

immunological _____. B cells *(also? do not?)* display this tolerance.

Loss of such tolerance results in _____ disorders.

E18. Describe each of the processes listed below, which are steps in development of immunological tolerance involving T cells.

a. Positive selection

b. Deletion

c. Anergy

Which of the above steps *(a? b? c?)* appears to be the main mechanism that B cells use to

prevent responses to self-proteins? _____

E19. "The ability to recognize tumor antigens as nonself" is a definition of the characteristic of immunological *(surveillance? tolerance?).* Describe how each of the following forms of therapy may help cancer patients whose own immunological surveillance is failing.

a. Adoptive cellular immunotherapy

b. Cytokine therapy

F. Aging and the immune system (page 699)

■ **F1.** Complete arrows to indicate immune changes that usually accompany aging.

ǀ a. Number of helper T cells

ǀ b. Risk of autoimmune conditions by production of antibodies against self

ǀ c. Response to vaccines

F2. Write several health practices that can help you to maximize your resistance against disease.

G. Disorders, medical terminology (pages 699–704)

■ **G1.** Answer these questions about AIDS.

a. AIDS refers to a _____ i_____ -

d_____ s_____. This condition was first
recognized in the United States by the Centers for Disease Control (CDC) in the year

19_____ following the increase of two relatively rare conditions in AIDS patients.

Name those conditions: p_____ c_____

p_____ (PCP), a respiratory infection, and K_____

s_____ (KS), a skin cancer.

b. The causative agent for AIDS has been identified as the h_____

i_____ v_____. What is the major effect of this virus?

c. List the groups of persons that have been most infected in the United States.

In the United States, almost 9 out of 10 AIDS patients are *(female? male?)*. Worldwide,
75% of persons with AIDS are thought to have gotten AIDS through *(heterosexual?
homosexual?)* contacts.

d. The incubation period (from HIV infection to full-blown AIDS) is about 10 *(days?
months? years?)*.

e. Like other viruses, the HIV virus *(does? does not?)* depend on host cells for replication.

The HIV virus is a _____-virus because its genetic code is
carried in *(DNA? RNA?)*. What role does reverse transcriptase play in a retrovirus?

Name AIDS drugs that work by inhibiting this enzyme.

f. HIV primarily affects *(B? cytotoxic T? helper T?)* lymphocytes because these cells display

the receptor or "_____ protein," known as the _____ molecule.

A normal CD4 (helper T) lymphocyte count is _____/mm³; a count of under

_____/mm³ was designated in 1992 as diagnostic of AIDS.

g. Name several other cells that may be attacked by the HIV virus.

h. Formation of antibodies in the infected person usually occurs within _____
weeks after viral exposure. The presence of HIV *(antigens? antibodies?)* is commonly
used to diagnose AIDS. If a person was infected with AIDS 48 days ago, the person is
likely to test *(positive? negative?)* for AIDS but *(is? is not?)* likely to be able to transmit
the virus.

i. Circle the four fluids through which the HIV virus has been found to be transmitted in
sufficient quantities to be infective.

Blood	Breast milk	Mosquito venom	Saliva
Semen	Tears	Vaginal fluids	

Health care personnel who use barrier precautions *(are? are not?)* at high risk for
contracting AIDS from their HIV-positive clients.

G2. Discuss the following aspects of AIDS.

a. Describe the stages of AIDS, including major infections, signs, and symptoms that
commonly occur.

b. Discuss development of an AIDS vaccine.

c. List several ways of avoiding contracting AIDS.

G3. Do this exercise on autoimmune disease.

a. Normally the body *(does? does not?)* produce antibodies and T cells against its own

   tissues. Such self-recognition is called _____.

b. Describe possible mechanisms by which autoimmune disease may develop.

c. List five or more autoimmune diseases.

■ **G4.** Match the four types of allergic responses in the box with the correct descriptions below.

> I. Type I
> II. Type II
> III. Type III
> IV. Type IV

_____ a. Known as *cell-mediated* reactions, or delayed-type-hypersensitivity reactions, these lead to responses such as those that occur after exposure to poison ivy or after tuberculin testing (Mantoux test)

_____ b. Antigen-antibody *immune complexes* trapped in tissues activate complement and lead to inflammation, as in lupus (SLE) or rheumatoid arthritis (RA)

_____ c. Caused by IgG or IgM antibodies directed against RBCs or other cells, as in, for example, response to an incompatible transfusion; called a *cytotoxic* reaction

_____ d. Involves IgEs produced by mast cells and basophils; effects may be localized (such as swelling of the lips) or systemic, such as acute *anaphylaxis*

■ **G5.** Transplantation is likely to lead to tissue rejection because the transplanted organ or tissue serves as an *(antigen? antibody?)*. The body tries to reject this foreign tissue by

producing _____.

■ **G6.** Match types of transplants in the box with descriptions below.

> A. Allograft   I. Isograft   X. Xenograft

_____ a. Between animals of different species

_____ b. The most successful type of transplant

_____ c. Between individuals of same species, but of different genetic backgrounds, such as a mother's kidney donated to her daughter or a blood transfusion

■ **G7.** Name one drug often used for transplant patients because it is selectively immuno-

suppressant. _____

■ **G8.** Match the condition in the box with the description below.

A. Autoimmune disease	SCID. Severe combined immune deficiency
H. Hodgkin's disease	SLE. Systemic lupus erythematosus
L. Lymphangioma	SP. Splenomegaly

_____ a. A curable malignancy usually arising in lymph nodes

_____ b. Multiple sclerosis (MS), rheumatoid arthritis (RA), and systemic lupus erythematosus (SLE) are examples

_____ c. Benign tumor of lymph vessels

_____ d. Characterized by lack of both B and T cells

_____ e. Autoimmune, inflammatory disease with skin and joint changes and possibly serious effects on organs such as kidneys; skin lesions may resemble wolf bites, and part of name is the Latin word for "wolf"

_____ f. Enlarged spleen

ANSWERS TO SELECTED CHECKPOINTS: CHAPTER 22

A2. (a–c) NS. (d–e) S. (f) NS.

A3. (a) Proteins; lipids (or fats), fat. (b) Lympho; T, antibodies.

A4. Lymph, lymph vessels, and lymph organs (such as tonsils, spleen, and lymph nodes).

A7. Skeletal muscle contraction squeezing lymphatics which contain valves that direct flow of lymph. Respiratory movements.

A8. (a) Three-fourths; cisterna chyli; digestive and other abdominal organs and both lower limbs. (b) Left jugular, left subclavian, and left bronchomediastinal. (c) Right lymphatic; right upper limb and right side of thorax, neck, and head. (d) Internal jugular, subclavian. (e) Interstitial; plasma.

A9. Fracture of this bone might block the thoracic duct, which enters veins close to the left clavicle. Lymph therefore backs up with these results: (a) Extra fluid remains in tissue spaces normally drained by vessels leading to the thoracic duct. (b) Lymph capillaries in the intestine normally absorb fat from foods. Slow lymph flow can decrease such absorption leaving fat in digestive wastes.

B1. (a) P. (b) S. (c) S. (d) P.

B2. (a) Mediastinum; sternum; large, de. (b) Cortex; lympho; T, throughout life. (c) Medulla; thymic hormones.

B3. (a) A, afferent lymphatic vessel. (b) C, efferent lymphatic vessel. (c) D, germinal center. (d) B, cortical sinus. (e) E, medullary cord.

B5. Metastasis; nontender.

B6. Upper left, diaphragm; B lymphocyte and antibody production, blood reservoir, phagocytosis of microbes, red blood cells, and platelets.

B7. (a) Mucosa-associated lymphoid tissue; digestive, respiratory, urinary, and reproductive. (b1) Located in the wall of the small intestine, this tissue can protect against microbes found in food or present in excessive numbers in the Gl tract; (b2) Because tonsils and adenoids surround the pharynx (throat), they protect against inhaled or ingested microbes. (c) Palatine; pharyngeals.

C1. Meso, fifth.

C2. Veins. (a) Posterior. (b) Jugular; thoracic.

D2. (a) C. (b) D. (c) F. (d) E. (e) A. (f) G. (g) H. (h) B.

D3. (a) Viruses; lymphocytes, macrophages, fibroblasts; uninfected, antiviral. (b) Type I; Kaposi's sarcoma, occurring almost exclusively in persons with AIDS (caused by the HIV virus), genital warts, and hepatitis B and C. (c) Enhance action of phagocytes and NK cells; inhibit tumor growth. (d) Natural killer, lymphocyte; cytolytic; AIDS and some forms of cancer. (e) Neutrophils; monocytes, macrophages; mononuclear phagocytic. (f) Complement; C1–C9, factors B, D, and P (properdin).

D5. (a) C. (b) A. (c) I. (d) K.

D7. (a–d) 1. (e) 2. (f) 3.

D8. (a) PG. (b) H, LT. (c) I. (d) K. (e) K, C. (f) T. (g) C.

E1. (a) B cells (or B lymphocytes), T cells (or T lymphocytes); from stem cells in bone marrow. (b) T; perform immune functions against specific antigens if properly stimulated; CD8+, antigen-receptor protein. (c) Bone marrow.

E2. (a) CMI. (b) AMI. (c) AMI. (d) CMI. (e) AMI and CMI

E3. (a) Is recognized as foreign and provokes an immune response; foreign substances. (b1) Stimulate production of; (b2) React with (and potentially be destroyed by). (c) Complete antigen, or immunogen. (d) Hapten; reactivity, immunogenicity; body protein; allergic. (e) Large, complex molecules, such as proteins, nucleoproteins, lipoproteins, glycoproteins, or complex polysaccharides. (f) They are made of simple, repeating subunits that are not likely to be antigenic. (g). Yes; flagella, capsules, cell walls, as well as toxins made by bacteria. (h) Antigens. (i) Only a specific region of the; antigenic determinant (epitope); a number of.

E5. (a) Good news: they help T cells to recognize foreign invaders because antigenic proteins must be processed and presented in association with MHC antigens before T cells can recognize the antigenic proteins; bad news: MHC antigens present transplanted tissue as "foreign" to the body, causing it to be rejected. (b) MHC-I molecules are in plasma membranes of all body cells except RBCs; MHC-I antigens pick up short (8–9 amino acid) peptide fragments to present to T cells; MHC-II antigens pick up longer (13–17 amino acid) peptides and are found on antigen-presenting cells (APCs), thymic cells, and T cells activated by previous exposure to the antigen. (c) B (basis of the principle of self-tolerance).

E6. (a) B or T; B; T. (b) Antigens from intruders outside of body cells, such as bacteria, pollen, foods, cat hair; antigen presenting cells; at sites where antigens are likely to enter the body, such as skin or mucous membranes; macrophages, B cells, and dendritic cells (including Langerhans cells in skin). (c) 1. Phago/endo. 2. Partial dig. 3. Fusion. 4. Bind to MHC-II. 5. Exo/insert. 6. Present to T. (d) Endo; viral proteins made after a virus infects a cell and takes over cell metabolism.

E7. (a) M. (b) IL-4. (c) IL-2. (d) Gamma-IFN. (e) MMIF. (f) P. (g) LT. (h) IL-1, TNF.

E8. (a) Globulins; 4; 450, 220. See Figure LG 22.2A. (b) Variable; constant, are identical. See Figure LG 22.2A.

E9. Ig's; constant. (a) M. (b) G. (c) A. (d) E. (e) D.

E11. (a) T_S. (b) M. (c) T_C. (d) T_H. (e) DTH.

E15. (a) Initial, memory cells. (b) Slowly; titer; IgM; IgG. (c) Booster, more; immunological memory, higher.

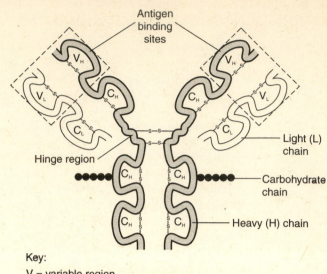

Figure LG 22.2A Diagram of an antibody molecule:
C = constant portion   V = variable portion

E16. (a) AAPI. (b) NAPI. (c) AAAI.

E17. Tolerance; also; autoimmune.

F1. (a) ↓. (b) ↑. (c) ↓.

G1. (a) Acquired immunodeficiency syndrome; 81; pneumocystis carinii pneumonia, Kaposi's sarcoma. (b) Human immunodeficiency virus; by destroying the immune system, the disease places the patient at high risk for opportunistic infections (normally harmless), such as PCP, and for other conditions such as KS. (c) Gay men who have not used safe sexual practices, IV drug users, hemophiliacs who received contaminated blood products before 1985, and heterosexual partners of HIV-infected persons; male; heterosexual. (d) Years. (e) Does; retro, RNA; directs DNA synthesis in host cells modeled on the viral RNA code; AZT, ddl, and ddC. (f) Helper T, docking, CD4; 1200, 200. (g) Macrophages and brain cells. (h) 3–20; antibodies; negative, is. (i) Blood, semen, vaginal fluids, and breast milk; are not.

G4. (a) IV. (b) III. (c) II. (d) I.

G5. Antigen; antibodies or T lymphocytes against the tissue.

G6. (a) X. (b) I. (c) A.

G7. Cyclosporine.

G8. (a) H. (b) A. (c) L. (d) SCID. (e) SLE. (f) Sp.

WRITING ACROSS THE CURRICULUM: CHAPTER 22.

1. Contrast lymphatic flow with blood circulation. Incorporate into your discussion the differences between (a) blood and lymph, (b) forces controlling flow of blood and of lymph, (c) lymph capillaries and blood capillaries, and (d) major lymph vessels and major blood vessels.

2. List the four cardinal signs or symptoms of inflammation. Then describe the processes that occur during inflammation that account for these signs or symptoms. Be sure to include roles of these chemicals: histamine, kinins, prostaglandins (PGs), leukotrienes (LTs), and complement. Also discuss these processes: vasodilation, emigration, and leukocytosis.

3. Contrast antigens and antibodies. Include several examples of each. Further distinguish endogenous from exogenous antigens, as well as complete and partial antigens.

4. Contrast antibody-mediated immunity (AMI) and cell-mediated immunity (CMI).

5. Describe the steps involved in processing and presenting antigens.

6. Contrast each of the following cytokines by describing types of cells producing the chemicals and also functions of each cytokine: interferons (alpha, beta, and gamma IFNs), interleukins (IL-1, IL-2, IL-4, IL-5), lymphotoxin, and perforin.

7. Differentiate the five classes of immunoglobulins.

8. Contrast helper T (T4) cells with cytotoxic T (T8) cells.

MASTERY TEST: CHAPTER 22

Questions 1–2: Arrange the answers in correct sequence.

_____ _____ _____ _____ 1. Flow of lymph through a lymph node:
   A. Afferent lymphatic vessel
   B. Medullary sinus
   C. Cortical sinus
   D. Efferent lymphatic vessel

_____ _____ _____ _____ _____ 2. Activities in humoral immunity, in chronological order:
   A. B cells develop in bone marrow or other part of the body.
   B. B cells differentiate and divide (clone), forming plasma cells.
   C. B cells migrate to lymphoid tissue.
   D. B cells are activated by specific antigen that is presented.
   E. Antibodies are released and are specific against the antigen that activated the B cell.

Questions 3–16: Circle the letter preceding the one best answer to each question. Take particular note of underlined terms.

3. Both the thoracic duct and the right lymphatic duct empty directly into:
   A. Axillary lymph nodes
   B. Superior vena cava
   C. Cisterna chyli
   D. Subclavian arteries
   E. Junction of internal jugular and subclavian veins

4. Which lymphatic trunk drains lymph primarily from the left upper limb?
   A. Left jugular trunk
   B. Left bronchiomediastinal trunk
   C. Left subclavian trunk
   D. Left lumbar trunk

5. All of these are examples of nonspecific defenses *except:*
   A. Antigens and antibodies
   B. Saliva
   C. Complement
   D. Interferon
   E. Skin
   F. Phagocytes

6. All of the following correctly match lymphocytes with their functions *except:*
   A. Helper T cells: stimulate B cells to divide and differentiate
   B. Cytotoxic (killer) T cells: produce lymphokines that attract and activate macrophages and lymphotoxins that directly destroy antigens
   C. Natural killer (NK) cells: suppress action of cytotoxic T cells and B cells
   D. B cells: become plasma cells that secrete antibodies

7. All of the following match types of allergic reactions with correct descriptions *except:*
   A. Type I: anaphylactic shock, for example, from exposure to iodine or bee venom
   B. Type II: transfusion reaction in which recipient's IgG or IgM antibodies attack donor's red blood cells
   C. Type III: Involves Ag-Ab-complement complexes that cause inflammations, as in lupus (SLE) or rheumatoid arthritis (RA)
   D. Type IV: cell-mediated reactions that involve immediate responses by B cells, as in the TB skin test.

8. Choose the one *false* statement about T cells.
   A. T_C cells are best activated by antigens associated with <u>both MHC-I and MHC-II</u> molecules.
   B. T_H cells to activated by antigens associated with <u>MHC-II</u> molecules.
   C. For T_C cells to become cytolytic, <u>they need costimulation by IL-2.</u>
   D. Perforin is a chemical released by <u>T_H cells.</u>

9. Choose the *false* statement about lymphatic vessels.
   A. Lymph capillaries are <u>more</u> permeable than blood capillaries.
   B. Lymphatics have <u>thinner</u> walls than veins.
   C. Like arteries, lymphatics contain <u>no</u> valves.
   D. Lymph vessels are <u>blind-ended.</u>

10. Choose the *false* statement about nonspecific defenses.
    A. Complement functions protectively by <u>being converted into histamine.</u>
    B. Histamine <u>increases</u> permeability of capillaries so that leukocytes can more readily reach the infection site.
    C. Complement and properdin are both <u>enzymes found in serum.</u>
    D. Opsonization <u>enhances</u> phagocytosis.

11. Choose the *false* statement about T cells.
    A. Some are called <u>memory cells.</u>
    B. They are called T cells because they are processed in the <u>thymus.</u>
    C. They are involved primarily in <u>antibody-mediated immunity (AMI).</u>
    D. Like B cells, they originate from <u>stem cells in bone marrow.</u>

12. Choose the *false* statement about lymph nodes.
    A. Lymphocytes <u>are</u> produced here.
    B. Lymph nodes are distributed <u>evenly</u> throughout the body, with <u>equal</u> numbers in all tissues.
    C. Lymph may pass through <u>several lymph nodes in a number of regions</u> before returning to blood.
    D. Lymph nodes are shaped roughly like <u>kidney (or lima) beans.</u>

13. Choose the *false* statement about lymphatic organs.
    A. The <u>palatine tonsils</u> are the ones most often removed in a tonsillectomy.
    B. The <u>spleen</u> is the largest lymphatic organ in the body.
    C. The thymus reaches its maximum size at age <u>40.</u>
    D. The spleen is located in the <u>upper left quadrant of the abdomen.</u>

14. Choose the *false* statement.
    A. Skeletal muscle contraction <u>aids</u> lymph flow.
    B. Skin is normally <u>more</u> effective than mucous membranes in preventing entrance of microbes into the body.
    C. Interferon is produced by <u>viruses.</u>
    D. An allergen is an <u>antigen,</u> not an antibody.

15. Choose the one *true* statement about AIDS.
    A. It appears to be caused by a virus that carries its genetic code in <u>DNA.</u>
    B. Persons at highest risk are <u>homosexual men and women.</u>
    C. Persons with AIDS experience a decline in <u>helper T cells.</u>
    D. AIDS is transmitted primarily through <u>semen and saliva.</u>

16. All of the following are cytokines known to be secreted by helper T cells *except:*
    A. Gamma interferon
    B. Interleukin-1
    C. Interleukin-2
    D. Interleukin-4

Questions 17–20: Circle T (true) or F (false). If the statement is false, change the underlined word or phrase so that the statement is correct.

T F 17. Antibodies are usually composed of <u>one light and one heavy</u> polypeptide <u>chain.</u>

T F 18. T_C cells are especially active against slowly growing bacterial diseases, some viruses, cancer cells associated with viral infections, and transplanted cells.

T F 19. Immunogenicity means the ability of an antigen to <u>react with</u> a specific antibody or T cell.

T F 20. A person with autoimmune disease produces <u>fewer than normal antibodies.</u>

Questions 21–25: Fill-ins. Complete each sentence with the word or phrase that best fits.

_____ 21. _____ lymphocytes provide antibody-mediated immunity, and _____ lymphocytes and macrophages offer cellular (cell-mediated) immunity.

_____ 22. Helper T cells are also known as _____ cells.

_____ 23. List the four fundamental signs or symptoms of inflammation.

_____ 24. The class of immunoglobins most associated with allergy are Ig _____.

_____ 25. Three or more examples of mechanical factors that provide non-specific resistance to disease are _____.

ANSWERS TO MASTERY TEST: ☆ CHAPTER 22

Arrange

1. A C B D
2. A C D B E

Multiple Choice

3. E	10. A
4. C	11. C
5. A	12. B
6. C	13. C
7. D	14. C
8. D	15. C
9. C	16 B

True–False

17. F. Two light and two heavy; chains
18. T
19. F. Stimulate formation of
20. F. Antibodies against the individual's own tissues

Fill-ins

21. B, T
22. T_H cells, T4 cells, and CD4 cells.
23. Redness, warmth, pain, and swelling
24. E
25. Skin, mucosa, cilia, epiglottis; also flushing by tears, saliva, and urine

FRAMEWORK 23
Respiratory System

PHYSIOLOGY

ANATOMY

UPPER RESPIRATORY (A)
- Nose
- Pharynx

LOWER RESPIRATORY (B)
- Larynx
- Trachea
- Bronchi
- Bronchioles
- Alveoli of lungs
- Alveolar-capillary membrane

CONTROL OF RESPIRATION (F)
- Control centers
  - medullary rhythmicity center
  - pons: pneumotaxic area
  - pons: apneustic area
- Regulation
  - cortical influence
  - inflation reflex
  - primary stimulus: $\uparrow H^+$ or $\uparrow CO_2$
  - secondary stimulus: $\downarrow O_2$
  - other factors: BP, pain, temperature, irritation

AGING, DEVELOPMENT (G)

DISORDERS (H)
- Lung cancer
- Asthma, bronchitis, emphysema
- Infections
- RDS, SIDS
- Pulmonary embolism
- Cor pulmonale
- Cystic fibrosis

PULMONARY VENTILATION (C)
(atmosphere–alveoli)
- Pressure changes
  - inspiration—active
  - expiration—passive
- Pneumothorax
- Surfactant and compliance
- Pulmonary volumes and capacities
- Gas laws: Charles', Dalton's, Henry's

EXTERNAL RESPIRATION (D)
(alveolar air–pulmonary blood)
- pO_2, pCO_2
- Thin a-c membrane, large surface area, pulmonary blood flow

INTERNAL RESPIRATION (D)
(systemic blood–tissues)

TRANSPORT OF OXYGEN AND CARBON DIOXIDE IN BLOOD (E)
- Oxygen
  - oxygen-Hb dissociation curve
  - Bohr effect
  - hypoxia
- CO_2
  - carbonic acid-bicarbonate
  - Haldane effect

The Respiratory System

CHAPTER
23

Oxygen is available all around us. But to get an oxygen molecule to a muscle cell in the stomach or a neuron of the brain requires coordinated efforts of the respiratory system with its transport adjunct, the cardiovascular system. Oxygen traverses a system of ever-narrowing and diverging airways to reach millions of air sacs (alveoli). Each alveolus is surrounded by a meshwork of pulmonary capillaries, much like a balloon encased in a nylon stocking. Here oxygen changes places with the carbon dioxide wastes in blood, and oxygen travels through the blood to reach the distant stomach or brain cells. Interference via weakened respiratory muscles (as in normal aging) or extremely narrowed airways (as in asthma or bronchitis) can limit oxygen delivery to lungs. Decreased diffusion of gases from alveoli to blood (as in emphysema, pneumonia, or pulmonary edema) can profoundly affect total body function and potentially lead to death.

Take a couple of deep breaths and, as you begin your study of the respiratory system, carefully examine the Chapter 23 Topic Outline and Objectives. Check off each one as you complete it. To further organize your study of the respiratory system, glance over the Chapter 23 Framework. Be sure to refer to the Framework frequently and note relationships among key terms in each section.

TOPIC OUTLINE AND OBJECTIVES

A. Upper respiratory passageways

☐ 1. Identify the organs of the respiratory system and describe their functions.

B. Lower respiratory passageways, lungs

☐ 2. Explain the structure of the alveolar-capillary (respiratory) membrane, and descibe its function in the diffusion of respiratory gases.

C. Physiology of respiration: pulmonary ventilation

☐ 3. Describe the events involved in inspiration and expiration.

D. Physiology of respiration: exchange of oxygen and carbon dioxide

E. Transport of oxygen and carbon dioxide in blood

☐ 4. Explain how respiratory gases are transported by blood.

F. Control of respiration

☐ 5. Describe the various factors that control the rate of respiration.
☐ 6. Describe the responses of the respiratory system to exercise.

G. Aging and development of the respiratory system

☐ 7. Describe the effects of aging on the respiratory system.

☐ 8. Describe the development of the respiratory system.

H. Disorders, medical terminology

☐ 9. Define asthma, bronchitis, emphysema, bronchogenic carcinoma, pneumonia, tuberculosis, respiratory distress syndrome, respiratory failure, sudden infant death syndrome, coryza, influenza, pulmonary embolism, pulmonary edema, cystic fibrosis, and smoke inhalation injury as disorders of the respiratory system.

☐ 10. Define medical terminology associated with the respiratory system.

WORDBYTES

Now become familiar with the language of this chapter by studying each wordbyte, its meaning, and an example of its use within a term. After you study the entire list, self-check your understanding by writing the meaning of each wordbyte on the line. As you continue through the *Learning Guide,* identify (and fill in) additional terms that contain the same wordbyte.

Wordbyte	Self-check	Meaning	Example(s)
atele-	_____	incomplete	*atele*ctasis
dia-	_____	through	*dia*phragm
dys-	_____	bad, difficult	*dys*pnea
-ectasis	_____	dilation	atel*ectasis*
ex-	_____	out	*ex*piration
in-	_____	in	*in*spiration
intra-	_____	within	*intra*pulmonic
-pnea	_____	breathe	a*pnea*
pneumo-	_____	air	*pneumo*thorax
pulmo-	_____	lung	*pulmo*nary
rhin-	_____	nose	*rhin*orrhea
spir-	_____	breathe	in*spir*ation

CHECKPOINTS

A. Upper respiratory passageways (pages 708–712)

A1. Explain how the respiratory and cardiovascular systems work together to accomplish gaseous exchange among the atmosphere, blood, and cells.

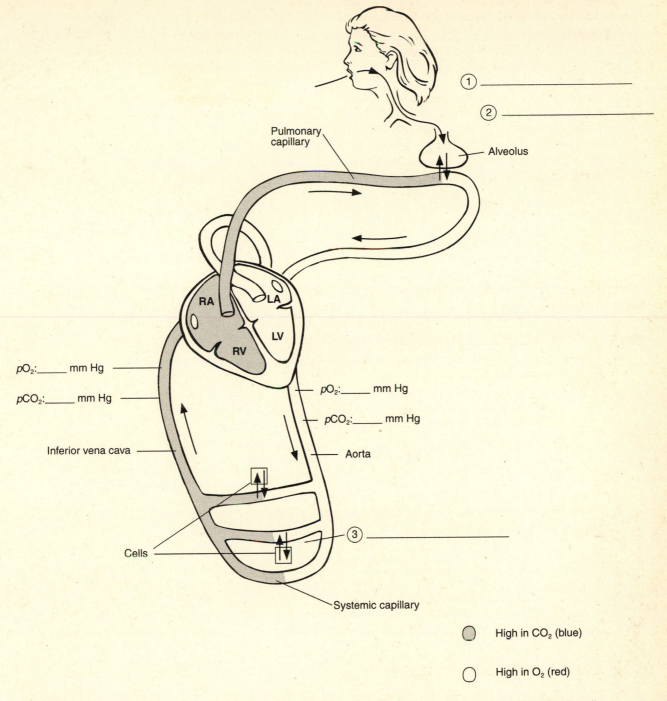

Figure LG 23.1 Phases of respiration and circulatory routes. Numbers refer to Checkpoint A2. Fill in short lines according to Checkpoint D3.

■ **A2.** Refer to Figure LG 23.1 and visualize the process of respiration in yourself as you do this exercise.

a. Label the first phase of respiration at (1) on the figure. This process involves transfer of air from your environment to *(alveoli? pulmonary capillaries?)* in your lungs.

Inflow of air rich in *(O$_2$? CO$_2$?)* is known as _____-spiration, whereas outflow of

air concentrated in *(O$_2$? CO$_2$?)* is known as _____-spiration.

b. Phase 2 of respiration involves exchange of gases between _____

in alveoli and _____ in pulmonary capillaries. This process is known as *(external? internal?)* respiration. Label phase 2 on the figure.

c. After oxygenated blood returns to your heart via your pulmonary *(artery? veins?),* blood is pumped to your tissues where gas exchange occurs between

_____ and _____. This process is known as *(external? internal?)* respiration. Label phase 3 on the figure.

d. Blood high in CO_2 then returns to the heart where it passes directly into the pulmonary *(artery? vein?)* to reach the lungs. Here diffusion of CO_2 from pulmonary capillaries to alveoli occurs; this is another aspect of *(external? internal?)* respiration (phase

_____). Finally, you exhale this air, a process that is part of *(external respiration?*

pulmonary ventilation?) (phase _____).

e. Which process takes place within tissue cells, utilizing O_2 and producing CO_2 as a waste product? *(Cellular? Internal?)* respiration.

f. *For extra review,* color all blood vessels on Figure LG 23.1 according to color code ovals on the figure.

■ **A3.** Identify parts of the respiratory system in this activity.

a. Arrange respiratory structures listed in the box in correct sequence from first to last in the pathway of inspiration.

ADA. Alveolar ducts and alveoli	NM. Mouth and nose
BB. Bronchi and bronchioles	P. Pharynx
L. Larynx	T. Trachea

_____ → _____ → _____ → _____ → _____ → _____

b. Write asterisks (∗) next to structures forming the upper respiratory system.

c. Most of the structures listed in the box above are parts of the *(conducting? respiratory?)* portion of the respiratory system.

■ **A4.** Match each term related to the nose with the description that best fits.

External nares	Rhinoplasty
Internal nares (choanae)	Vestibule
Paranasal sinuses and meatuses	

a. Mucous membrane-lined spaces and passageways: _____

b. Nostrils: _____

c. Anterior portion of nasal cavity: _____

d. Openings from posterior of nose into nasopharynx: _____

e. Repair of the nose: _____

■ **A5.** Name the structures of the nose that are designed to carry out each of the following functions.

a. Warm, moisten, and filter air

b. Sense smell

c. Assist in speech

■ **A6**. Write *LP* (laryngopharynx), *NP* (nasopharynx), or *OP* (oropharynx) to indicate the location of each of the following structures.

_____ a. Adenoids

_____ b. Palatine tonsils

_____ c. Lingual tonsils

_____ d. Opening (fauces) from oral cavity

_____ e. Opening into larynx and esophagus

A7. On Figure LG 23.2, cover the key and identify all structures associated with the nose, palate, pharynx, and larynx. Color structures with color code ovals.

B. Lower respiratory passageways, lungs (pages 712–723)

■ **B1.** Match parts of the larynx with the descriptions that best fit.

A. Arytenoid cartilages	RG. Rima glottidis
C. Cricoid cartilage	T. Thyroid cartilage
E. Epiglottis	

_____ a. Also known as the Adam's apple; longer in males than in females

_____ b. Includes a leaflike portion that covers the airway during swallowing

_____ c. A pair of triangular-shaped cartilages attached to muscles that control vocal cords

_____ d. Forms the inferior portion of the larynx

_____ e. Space between the vocal cords through which air passes during ventilation

B2. Explain how the larynx prevents food from entering the trachea. (*Hint:* Notice the arrow on Figure LG 23.2.)

B3. Tell how the larynx produces sound. Explain how pitch is controlled and what causes male pitch usually to be lower than female pitch.

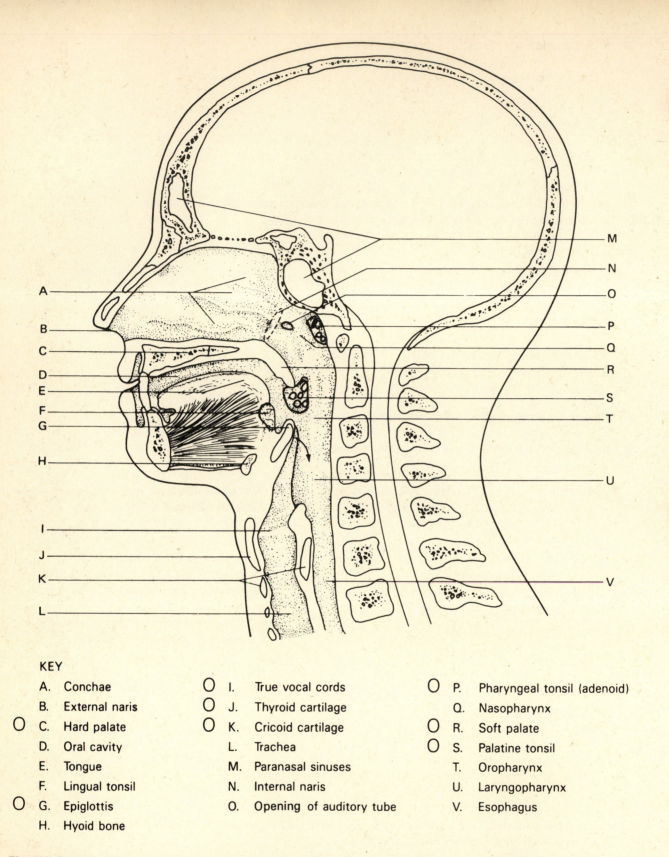

A ——————
B ——————
C ——————
D ——————
E ——————
F ——————
G ——————
H ——————
I ——————
J ——————
K ——————
L ——————

M ——————
N ——————
O ——————
P ——————
Q ——————
R ——————
S ——————
T ——————
U ——————
V ——————

KEY

A.	Conchae
B.	External naris
○ C.	Hard palate
D.	Oral cavity
E.	Tongue
F.	Lingual tonsil
○ G.	Epiglottis
H.	Hyoid bone

○ I.	True vocal cords
○ J.	Thyroid cartilage
○ K.	Cricoid cartilage
L.	Trachea
M.	Paranasal sinuses
N.	Internal naris
O.	Opening of auditory tube

○ P.	Pharyngeal tonsil (adenoid)
Q.	Nasopharynx
○ R.	Soft palate
○ S.	Palatine tonsil
T.	Oropharynx
U.	Laryngopharynx
V.	Esophagus

Figure LG 23.2 Sagittal section of the right side of the head with the nasal septum removed. Color and identify labeled structures as directed in Checkpoint A7. Arrow refers to Checkpoint B2.

■ **B4.** Complete this Checkpoint about the trachea.

a. The trachea is commonly known as the _____. It is located

(*anterior? posterior?*) to the esophagus. About _____ cm (_____ inches) long,

the trachea terminates at the _____, where the trachea leads to a
Y-shaped intersection with the primary bronchi.

b. The tracheal wall is lined with a _____ membrane and

strengthened by 16–18 C-shaped rings composed of _____.

■ **B5.** Describe the functional advantages provided by each of these parts of the trachea:
a. Pseudostratified ciliated epithelium lining the mucosa

b. C-shaped cartilage rings with the trachealis muscle completing the "ring" posteriorly

c. Sensitivity of the mucosa at the carina

■ **B6.** *A clinical challenge.* Anna has aspirated a small piece of candy. Dr. Lennon expects
to find it in the right bronchus rather than in the left. Why?

■ **B7.** Bronchioles (*do? do not?*) have cartilage rings. How is this fact significant during an
asthma attack?

B8. Refer to Figure LG 23.3 and do this activity. Use Figure 23.7 (page 718 of your text)
for help.
a. Identify tracheobronchial tree structures J–N, and color each structure indicated with a
color code oval.
b. Color the two layers of the pleura and the diaphragm according to color code ovals.
Also, cover the key and identify all parts of the diagram from A to I.

■ **B9.** Write the correct term for each of these conditions.

a. Inflammation of the pleura: _____

b. Air in the pleural cavity: _____

c. Blood in the pleural cavity: _____

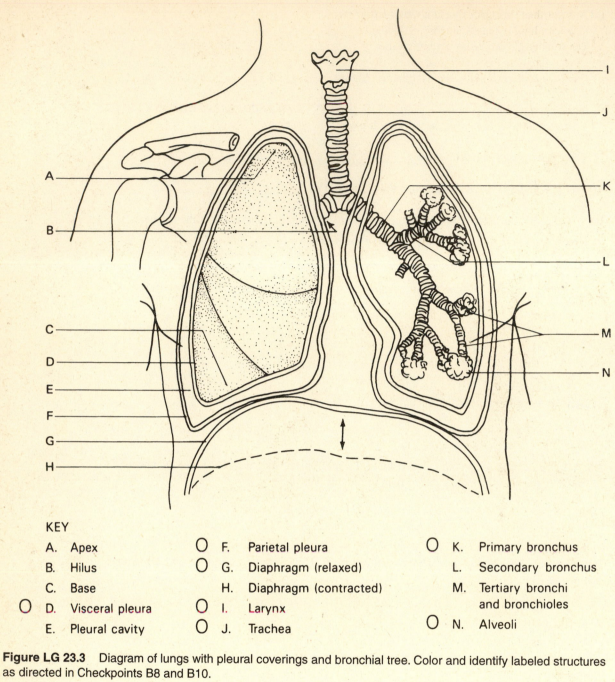

KEY

A. Apex
B. Hilus
C. Base
O D. Visceral pleura
E. Pleural cavity

O F. Parietal pleura
O G. Diaphragm (relaxed)
H. Diaphragm (contracted)
O I. Larynx
O J. Trachea

O K. Primary bronchus
L. Secondary bronchus
M. Tertiary bronchi
   and bronchioles
O N. Alveoli

Figure LG 23.3 Diagram of lungs with pleural coverings and bronchial tree. Color and identify labeled structures as directed in Checkpoints B8 and B10.

■ **B10.** Answer these questions about the lungs. (As you do the exercise, locate the parts of the lung on Figure LG 23.3.)

a. The broad, inferior portion of the lung that sits on the diaphragm is called the

_____. The upper narrow apex of each lung extends just superior to

the _____. The costal surfaces lie against the

_____.

b. Along the mediastinal surface is the _____ where the root of the lung

is located. This root consists of _____.

c. Answer these questions with *right* or *left*.
In which lung is the cardiac notch located? _____

Which lung has just two lobes and so only two lobar bronchi? _____

Which lung has a horizontal fissure? _____
Write S (superior), M (middle), and I (inferior) on the three lobes of the right lung.

■ **B11.** Describe the structures of a lobule in this exercise.

a. Arrange in order the structures through which air passes as it enters a lobule en route to

alveoli. _____ _____ _____

| A. Alveolar ducts   R. Respiratory bronchiole   T. Terminal bronchiole |

b. In order for air to pass from alveoli to blood in pulmonary capillaries, it must pass

through the _____-_____ (respiratory) mem-
brane. Identify structures in this pathway: A to E on Figure LG 23.4.

c. *For extra review.* Now label layers 1–6 of the alveolar-capillary membrane in the insert
in Figure LG 23.4.

d. *A clinical challenge.* Normally the respiratory membrane is extremely *(thick? thin?)*.
Suppose pulmonary capillary blood pressure rises dramatically, pushing extra fluid out
of blood. The presence of excess fluid in interstitial areas and ultimately in alveoli is

known as pulmonary _____. Diffusion occurs *(more? less?)* readily
through the fluid-filled membrane.

e. The alveolar epithelial layer of the respiratory membrane contains three types of cells.
Forming most of the layer are flat *(squamous? septal? macrophage?)* (Type I alveolar)
cells which are ideal for diffusion of gases. Septal (Type II alveolar) cells produce an

important phospholipid and lipoprotein substance called _____,
which reduces tendency for collapse of alveoli due to surface tension. Macrophage cells

help to remove _____ from lungs.

■ **B12.** *The Big Picture: Looking Back.* Return to earlier chapters in your text and *Guide* as
you contrast the dual blood supply to the lungs.

a. Which vessels supply oxygen-rich blood to the lungs? *(Bronchial? Pulmonary?)* arteries. These vessels carry blood directly from the *(aorta? right ventricle?)*. Review these
on Figure 21.18b and Exhibit 21.5, pages 636 and 641, in your text.

b. Which veins return blood carrying CO_2 from lung tissue toward the heart (Figure 21.26
and Exhibit 21.11, pages 654–655, in your text, and Chapter LG 21, Checkpoint E7,

page LG 469)? _____ or _____ veins.

c. Refer to Chapter 20 in your *Learning Guide,* Checkpoints A3 and B7 (pages LG 425
and 429). Which vessel contains more oxygenated blood? Pulmonary *(artery? vein?)*.

d. Return to Chapter LG 21, Checkpoint F2, page 470, and answer these questions.
1. Which system of vessels has higher blood pressure? *(Pulmonary? Systemic?)* One
reason for this difference in blood pressure is that the extensive (lengthy) network of
arterioles and capillaries in the *(pulmonary? systemic?)* circulation provides greater
resistance to blood flow.
2. Which vessels respond to hypoxia by vasoconstricting so that blood bypasses
hypoxic regions and more efficiently circulates to oxygen-rich regions? *(Pulmonary?
Systemic vessels, for example, those in skeletal muscle?)*

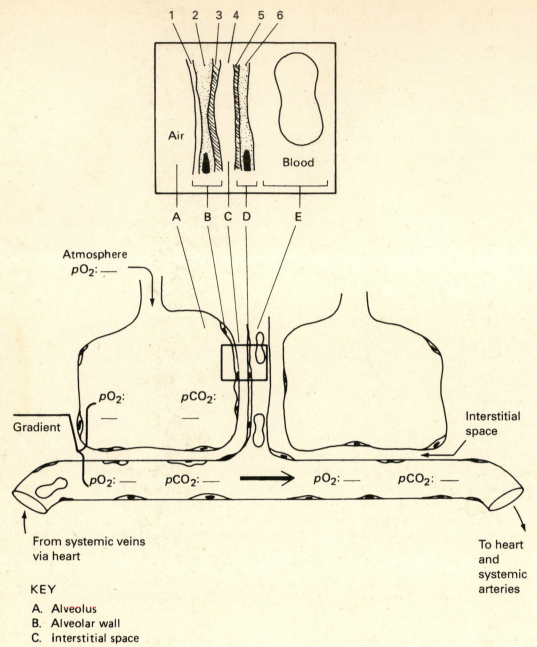

KEY

A. Alveolus
B. Alveolar wall
C. Interstitial space
D. Capillary wall
E. Blood plasma and blood cells

Figure LG 23.4 Diagram of alveoli and pulmonary capillary. Insert enlarges alveolar-capillary membrane. Numbers and letters refer to Checkpoint B11. Complete partial pressures (in mm Hg) as directed in Checkpoint D3.

C. Physiology of respiration: pulmonary ventilation (pages 723–729)

■ **C1.** Review the three phases of respiration in the box by matching them with the correct descriptions below.

E. External (pulmonary) respiration	PV. Pulmonary ventilation
I. Internal (tissue) respiration	

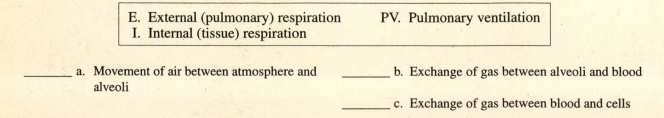

_____ a. Movement of air between atmosphere and alveoli

_____ b. Exchange of gas between alveoli and blood

_____ c. Exchange of gas between blood and cells

■ **C2.** Refer to Figure 23.14 (page 725 in the text) and describe the process of ventilation in this exercise.

a. In diagram (a), just before the start of inspiration, pressure within the lungs (called

_____) is _____ mm Hg. This is *(more than? less than? the same as?)* atmospheric pressure.

b. At the same time, pressure within the pleural cavity (called _____

pressure) is _____ mm Hg. This is *(more than? less than? the same as?)* alveolar and atmospheric pressure.

c. The first step in inspiration occurs as the muscles in the floor and walls of the thorax

contract. These are the _____ and _____ muscles. Note in diagram (b) (and also in Figure LG 23.3) that the size of the thorax *(increases? decreases?)*. Because the two layers of pleura tend to adhere to one another, the lungs will *(increase? decrease?)* in size also.

d. Increase in volume of a closed space such as the pleural cavity causes the pressure

there to *(increase? decrease?)* to _____ mm Hg. Because the lungs also increase in size (due to pleural cohesion), alveolar pressure also *(increases? decreases?)*

to _____ mm Hg. This inverse relationship between volume and pressure is a

statement of _____'s law.

e. A pressure gradient is now established. Air flows from high pressure area *(alveoli? atmosphere?)* to low pressure area *(alveoli? atmosphere?)*. So air flows *(into? out of?)* lungs. Thus *(inspiration? expiration?)* occurs. By the end of inspiration, sufficient air will have moved into the lungs to make pressure there equal to atmospheric pressure,

that is, _____ mmHg.

f. Inspiration is a(n) *(active? passive?)* process, whereas expiration is normally a(n)

(active? passive?) process resulting from _____ recoil. Identify two factors that contribute to this recoil.

During expiration, the diaphragm moves *(inf? sup?)*-eriorly and the sternum moves *(ant? post?)*-eriorly.

g. During labored breathing, accessory muscles help to force air out. Name two of these sets of muscles.

C3. Contrast these types of breathing patterns:

a. *Eupnea/dyspnea*

b. *Apnea/tachypnea*

c. *Costal breathing/diaphragmatic breathing*

■ **C4.** Answer these questions about compliance.

a. Imagine trying to blow up a new balloon. Initially, the balloon resists your efforts. Compliance is the ability of a substance to yield elastically to a force; in this case it is the ease with which a balloon can be inflated. So a new balloon has a *(high? low?)* level of compliance, whereas a balloon that has been inflated many times has *(high? low?)* compliance.

b. Similarly, alveoli that inflate easily have *(high? low?)* compliance. The presence of a

coating called _____ lining the inside of alveoli prevents alveolar walls from sticking together during ventilation and so *(increases? decreases?)* compliance.

c. Describe the chemical composition of surfactant.

Surfactant is made by type *(I? II? III?)* alveolar cells of the lungs, which produce this chemical especially during the final weeks before birth. A premature infant may lack

adequate production of surfactant; this condition is known as _____.

_____.

d. Collapse of all or part of a lung is known as _____. Such collapse may result from lack of surfactant or from other factors. Name several of these factors.

C5. Describe relationships between these pairs of terms:

a. *Surface tension/surfactant*

b. *Pneumothorax/atelectasis*

■ **C6.** Complete the arrows to indicate whether resistance to air flow increases (↑) or decreases (↓) in each case.

    a. Inhalation, causing dilation of bronchi and bronchioles: |

    b. Increase in sympathetic nerve impulses to relax smooth muscle of airways: |

    c. COPD such as bronchial asthma or emphysema, with obstruction or collapse or airways: |

C7. Refer to Exhibit 23.1 (page 728 in your text). Read carefully the descriptions of modified respiratory movements such as sobbing, yawning, sighing, or laughing while demonstrating those actions yourself.

C8. Each minute the average adult takes _____ breaths (respirations). Check your own respiratory rate and write it here: _____ breaths per minute.

■ **C9.** Maureen is breathing at the rate of 15 breaths per minute. She has a tidal volume of 480 ml per breath. Her *minute volume of respiration (MVR)* is _____.

■ **C10.** Of the total amount of air that enters the lungs with each breath, about *(99%? 70%? 30%? 5%?)* actually enters the alveoli. The remaining amount of air is much like the last portion of a crowd trying to rush into a store: it does not succeed in entering the alveoli during an inspiration but just reaches airways and then is quickly ushered out during the next expiration. Such air is known as anatomic _____ and constitutes about _____ ml of a typical breath.

■ **C11.** Match the lung volumes and capacities with the descriptions given. You may find it helpful to refer to Figure 23.16 (page 729 in the text).

ERV. Expiratory reserve volume		RV. Residual Volume
FEV_1. Forced expiratory volume in one second		TLC. Total lung capacity
FRC. Functional residual capacity		VC. Vital capacity
IRV. Inspiratory reserve volume		V_T. Tidal volume

_____ a. The amount of air taken in with each inspiration during normal breathing is called _____

_____ b. At the end of a normal expiration the volume of air left in the lungs is called _____. Emphysemics who have lost elastic recoil of their lungs cannot exhale adequately, so this volume will be large.

_____ c. Forced exhalation can remove some of the air in FRC. The maximum volume of air that can be expired beyond normal expiration is called _____. This volume will be small in emphysema patients.

_____ d. Even after the most strenuous expiratory effort, some air still remains in the lungs; this amount, which cannot be removed voluntarily, is called _____.

_____ e. The volume of air that represents a person's maximum breathing ability is called _____. This is the sum of ERV, TV, and IRV.

_____ f. Adding RV to VC gives _____.

_____ g. The excess air a person can take in after a normal inhalation is called _____.

_____ h. The amount of the vital capacity that can be forced out in 1 second is called _____.

■ **C12.** Indicate normal values for each of the following.

a. V_T = _____ ml (_____ liter) d. FRC = _____ ml (_____ liter)

b. TLC = _____ ml (_____ liter) e. RV = _____ ml (_____ liter)

c. VC = _____ ml (_____ liter)

D. Physiology of respiration: exchange of oxygen and carbon dioxide   (pages 729–732)

■ **D1.** Check your understanding of gas laws by matching the correct law with the condition that it explains.

┌──┐
│   B. Boyle's law   D. Dalton's law   H. Henry's law   │
└──┘

_____ a. If a patient breathes air highly concentrated in oxygen (as in a hyperbaric chamber), a higher percentage of oxygen will dissolve in the blood and tissues. This is the principle underlying the use of a hyperbaric chamber.

_____ b. The total atmospheric pressure (760 mm Hg) is due mostly to pressure caused by nitrogen, partly to pO_2 and slightly to pCO_2.

_____ c. Under high pN_2 more nitrogen dissolves in blood. The *bends* occurs when pressure decreases and nitrogen forms bubbles in tissue as it comes out of solution.

_____ d. As the size of the thorax increases, the pressure within it decreases.

■ **D2.** Refer to Figure LG 23.5 and do this exercise about Dalton's law.

a. The atmosphere contains enough gaseous molecules to exert pressure upon a column of

mercury to make it rise about _____ mm. The atmosphere is said to have a pressure of 760 mm Hg.

b. Air is about _____% nitrogen and _____% oxygen. To show this, circle the nitrogen molecules in *green* and oxygen molecules in *red* in Figure LG 23.5b. Note that

carbon dioxide is not represented in the figure because only about _____% of air is CO_2.

c. Dalton's law explains that of the total 760 mm of atmospheric pressure, a certain amount is due to each type of gas. Determine the portion of the total pressure due to nitrogen molecules:

$$79\% \times 760 \text{ mm Hg} \quad = \quad \text{_____ mm Hg}$$

This is the partial pressure of nitrogen, or _____.

d. Calculate the partial pressure of oxygen.

e. In Figure LG 23.5a, color the part of the Hg column due to N_2 pressure (up to the 600 mm Hg mark) *green*. Then color the portion of the Hg column due to O_2 pressure *red*.

■ **D3.** Answer these questions about external and internal respiration.

a. A primary factor in the diffusion of gas across a membrane is the difference in concentration of the gas (reflected by _____ pressures) on the two sides of

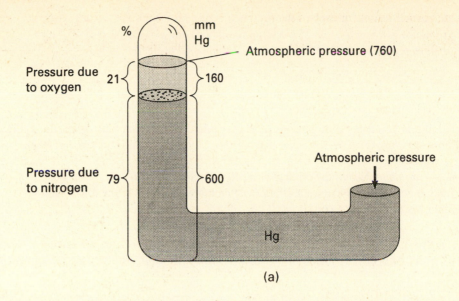

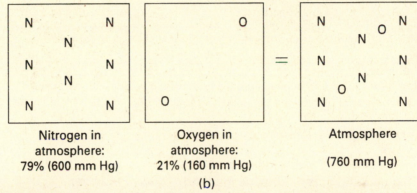

Figure LG 23.5 Atmospheric pressure. (a) Effect of gas molecules on column of mercury (Hg). (b) Partial pressures contributing to total atmospheric pressure. Color according to directions in Checkpoint D2.

the membrane. On the left side of Figure LG 23.4 write values for pO_2 (in mm Hg) in each of the following areas. (Refer to text Figure 23.17, page 731, for help.)

Atmospheric air (Recall this value from Figure LG 23.5.)

Alveolar air (Note that this value is lower than that for atmospheric pO_2 because some alveolar O_2 enters pulmonary blood.)

Blood entering lungs

b. Calculate the pO_2 difference (gradient) between alveolar air and blood entering lungs.

_____ mm Hg – _____ mm Hg = _____ mm Hg

c. Three other factors that increase exchange of gases between alveoli and blood are: *(large? small?)* surface area of lungs; *(thick? thin?)* respiratory membrane; and *(increased? decreased?)* blood flow through lungs (as in exercise).

d. By the time blood leaves the lungs to return to heart and systemic arteries, its pO_2 is normally *(greater than? about the same as? less than?)* pO_2 of alveoli. Write the correct value on Figure LG 23.4.

e. Now fill in all three pCO_2 values on Figure LG 23.4

f. *A clinical challenge.* Jenny, age 45, has had blood drawn from her radial artery to determine her arterial blood gases. Her pO_2 is 56 and her pCO_2 is 48. Are these typical values

for a healthy adult? _____

g. Fill in blood gas values (pO_2 and pCO_2) on the short lines on Figure LG 23.1.

■ **D4.** With increasing altitude the air is "thinner," that is, gas molecules are farther apart, so atmospheric pressure is lower. Atop a 25,000-foot mountain, this pressure is only 282 mm Hg. Oxygen still accounts for 21% of the pressure. What is pO_2 at that level?

From this calculation you can see limitations of life (or modifications that must be made) at high altitudes. If atmospheric pO_2 is 59.2, neither alveolar nor blood pO_2 could surpass that level.

■ **D5.** *A clinical challenge.* Match the clinical cases listed below with the factors (in box) affecting diffusion across the alveolar-capillary (a-c) membrane. On each line that follows, write a rationale. One is done for you.

DD. Diffusion distance	SA. Surface area for gas exchange
PP. Partial pressure of gases	SMW. Solubility and molecular weight of gases

a. __**PP**__ Morphine has been prescribed for Ms. Johnson to alleviate her pain.
Morphine slows the respiratory rate, so decreases alveolar pO_2.

b. _____ Mr. Schmidt has interstitial viral pneumonia, with excessive accumulation of

fluid in interstital spaces. Rationale: _____.

c. _____ Mrs. McLaughlin was diagnosed with emphysema nine years ago.

Rationale: _____.

d. _____ Rev. Lewis has a left-sided congestive heart failure (CHF) with pulmonary edema. Arterial blood gas (ABG) analysis indicates her pO_2 is 54 mm Hg and

pCO_2 is 43 mm Hg. Rationale: _____.

E. Transport of oxygen and carbon dioxide in blood (pages 732–736)

■ **E1.** Answer these questions about oxygen transport. Refer to Figure 23.18 (page 733 in your text).

a. One hundred ml of blood contains about _____ ml of oxygen. Of this, about 19.7 ml

is carried as _____. Only a small amount of oxygen is carried in the dissolved state because oxygen has a *(high? low?)* solubility in blood or water.

b. Oxygen is attached to the _____ atoms in hemoglobin. The chemical formula

for oxyhemoglobin is _____. When hemoglobin carries all of the oxygen it can hold, it is

said to be fully _____. High pO_2 in alveoli will tend to *(increase? decrease?)* oxygen saturation of hemoglobin.

c. Refer to Figure 23.19 (page 733 in your text). Note that arterial blood, with a pO_2 of about 105 mm Hg, has its hemoglobin _____% saturated with oxygen. (This may be

expressed as $SO_2 = $ _____%.)

d. List four factors that will enhance the dissociation of oxygen from hemoglobin so that oxygen can enter tissues. (*Hint:* Think of conditions within active muscle tissue.)

e. *For extra review.* Demonstrate the effect of temperature on the oxygen-hemoglobin dissociation curve. Draw a vertical line on Figure 23.21 (page 735 of the text) at pO_2 = 40 (the value for venous blood). Compare the percentage saturation of hemoglobin (SO_2) at these two body temperatures:

38°C (100.4°F)_____%SO_2 43°C(109.4°F)_____%SO_2

In other words, at a higher body temperature, for example in high fever, *(more? less?)* oxygen will be attached to hemoglobin, while *(more? less?)* oxygen will enter tissues to fuel metabolism.

f. By the time blood enters veins to return to the heart, its oxygen saturation of hemoglobin (SO_2) is about _____%. Note on Figure 23.19 (page 733 of your text) that although pO_2 drops from 105 in arterial blood to 40 in venous blood, oxygen saturation drops *(more? less?)* dramatically, that is, from 98% to 75%. Of what significance is this?

■ **E2.** Carbon monoxide has about _____ times the affinity that oxygen has for hemoglobin. State the significance of this fact.

■ **E3.** *A clinical challenge.* Identify types of hypoxia likely to be present in each situation. Use these answers: *anemic, histotoxic, hypoxic, stagnant.*

a. Cyanide poisoning: _____ hypoxia

b. Decreased circulation, as in circulatory shock: _____ hypoxia

c. Low hematocrit due to hemorrhage or iron-poor diet: _____ hypoxia

d. Chronic bronchitis or pneumonia: _____ hypoxia

■ **E4.** Complete this Checkpoint about transport of carbon dioxide.

a. Write the percentage of CO_2 normally carried in each of these forms: _____% is

present in bicarbonate ion (HCO_3^-); _____% is bound to the globin portion of hemo-

globin; _____% is dissolved in plasma.

b. Carbon dioxide (CO_2) produced by cells of your body diffuses into red blood cells

(RBCs) and combines with water to form _____.

c. Carbonic acid tends to dissociate into two products. One is H^+ which binds to

_____. The other product is _____ (bicarbonate), which is carried in *(RBCs? plasma?)* in exchange for a *(K^+? Cl^-?)* ion that shifts into the RBC.

d. Now write the entire sequence of reactions described in (b) and (c). Be sure to include the enzyme that catalyzes the first reaction. Notice that the reactions show that increase in CO_2 in the body (as when respiratory rate is slow) will tend to cause a buildup of acid (H^+) in the body.

$$CO_2 + \underline{\hspace{1cm}} \xrightarrow[\text{(enzyme)}]{} \underline{\hspace{1cm}} \rightarrow H^+ + \underline{\hspace{2cm}}$$

↓   ↓
Binds   Shifts to plasma
to Hb   in exchange for Cl⁻

E5. Note that as the red blood cells reach lung capillaries, the same reactions you just studied occur, but in reverse. Study Figure 23.23b (page 737 in the text) carefully. Then list the major steps that occur in the lungs so that CO_2 can be exhaled.

F. Control of respiration (pages 736–743)

F1. Complete the table about respiratory control areas. Indicate whether the area is located in the medulla (M) or pons (P).

F2. Answer these questions about respiratory control.

Name	M/P	Function
a.	M	Controls rhythm; consists of inspiratory and expiratory areas
b. Pneumotaxic		
c.		Prolongs inspiration and inhibits expiration

a. The main chemical change that stimulates respiration is increase in blood level of

_____, which is directly related to *(decrease in pO₂? increase in pCO₂?)* of blood.

b. Cells most sensitive to changes in blood CO_2 are located in the *(medulla? pons? aorta and carotid arteries?)*.

c. An increase in arterial blood pCO_2 is called _____. Write an

arterial pCO_2 value that is hypercapnic. _____ mm Hg *(Even slight? Only severe?)* hypercapnia will stimulate the respiratory system, leading to *(hyper? hypo?)*-ventilation.

d. State two locations of chemoreceptors sensitive to changes in pO_2.

(Even slight? Only large?) decreases in pO_2 level of blood will stimulate these chemo-receptors and lead to hyperventilation. Give an example of a pO_2 low enough to evoke

such a response. _____ mm Hg

e. Increase in body temperature (as in fever), as well as stretching of the anal sphincter,

will cause _____-crease in the respiratory rate.

f. Take a deep breath. Imagine the _____ receptors in your airways being stimulated. These will cause *(excitation? inhibition?)* of the inspiratory and apneustic areas, resulting in expiration. This reflex, known as the

_____ reflex, prevents overinflation of the lungs.

F3. Describe effects of each of the following upon the efficiency of the respiratory system.
   a. Exercise

   b. Smoking

G. Aging and development of the respiratory system   (page 743)

■ **G1.** Describe possible effects of these age-related changes.

   a. Chest wall becomes more rigid as bones and cartilage lose flexibility.

   b. Decreased macrophage and ciliary action of lining of respiratory tract.

■ **G2.** Describe development of the respiratory system in this exercise.

   a. The laryngotracheal bud is derived from _____-derm. List structures formed from this bud.

   b. Identify portions of the respiratory system derived from mesoderm.

H. Disorders, medical terminology (pages 744–748)

H1. A host of effects upon the respiratory system are associated with smoking. Describe these two.

a. Bronchogenic carcinoma (Include the roles of basal cells and excessive mucus production.)

b. Emphysema (Note that as walls of alveoli break down, surface area of the respiratory

   membrane _____-creases, so amount of gas diffused _____-creases also.)

■ **H2.** Match the condition with the correct description.

A. Asthma	P. Pneumonia
CB. Chronic bronchitis	RF. Respiratory failure
D. Dyspnea	TB. Tuberculosis
E. Emphysema	

_____ a. Permanent inflation of lungs due to loss of elasticity; rupture and merging of alveoli, followed by their replacement by fibrous tissue

_____ b. Inflammation of bronchi with excessive mucus production for at least three months per year for at least two consecutive years

_____ c. Acute infection or inflammation of alveoli, which fill with fluid

_____ d. Spasms of small passageways with wheezing and dyspnea

_____ e. Caused by a species of Mycobacterium; lung tissue is destroyed and replaced with inelastic connective tissue

_____ f. Difficult or painful breathing; shortness of breath

_____ g. May occur when pO_2 drops below 50 mm Hg and pCO_2 rises above 50 mm Hg

_____ h. A group of conditions known as chronic obstructive pulmonary disease (COPD) (three answers)

■ **H3.** Discuss respiratory distress syndrome (RDS) in this exercise.

a. Before birth fetal lungs are filled with *(air? fluid?)*. Inflation of lungs after birth depends

   largely on the presence of a chemical known as _____. This chemical
   is produced by *(basal cells of bronchi? septal cells of alveoli?)*.

b. Surfactant *(raises? lowers?)* surface tension so that alveolar walls are less likely to stick together. Thus surfactant *(facilitates? inhibits?)* lung inflation.

c. RDS especially targets *(premature? full-term?)* infants. Explain why.

H4. To what does SIDS refer? _____

What is the mortality rate of SIDS?

Discuss possible causes of SIDS.

H5. Explain how the abdominal thrust (Heimlich maneuver) helps to remove food that might otherwise cause death by choking.

■ **H6.** Write the ABCs of CPR.

ANSWERS TO SELECTED CHECKPOINTS: CHAPTER 23

A2. (a) Label (1) pulmonary ventilation; alveoli; O_2, in, CO_2, ex. (b) Air, blood; label (2) external respiration. (c) Veins, blood in systemic capillaries, tissue cells; label (3) internal respiration. (d) Artery, external, 2; pulmonary ventilation, 1. (e) Cellular. (f) Refer to Figure 21.17, p. 633 in your text.

A3. (a–b) NM* → P* → L → T → BB → ADA. (c) Conducting.

A4. (a) Paranasal sinuses and meatuses. (b) External nares. (c) Vestibule. (d) Internal nares (choanae). (e) Rhinoplasty.

A5. (a) Mucosa lining nose, septum, conchae, meati, and sinuses; lacrimal drainage; coarse hairs in vestibule. (b) Olfactory region lies superior to superior nasal conchae. (c) Sounds resonate in nose and paranasal sinuses.

A6. (a) NP. (b–d) OP. (e) LP.

B1. (a) T. (b) E. (c) A. (d) C. (d) RG.

B4. (a) Windpipe; anterior; 12 (4.5), carina. (b) Mucous, cartilage.

B5. (a) Protects against dust. (b) The open part of the cartilage ring faces posteriorly, permitting slight expansion of the esophagus during swallowing; the trachealis muscle provides a barrier preventing food from entering the trachea. (c) Promotes cough reflex, which helps deter food or foreign objects from moving from trachea into bronchi.

B6. The right bronchus is more vertical and slightly wider than the left.

B7. Do not; muscle spasms can collapse airways.

B9. (a) Pleurisy. (b) Pneumothorax. (c) Hemothorax.

B10. (a) Base; clavicle; ribs. (b) Hilus; bronchi, pulmonary vessels, and nerves. (c) Left; left; right; refer to Figure 23.9, page 720 of the text.

B11. (a) T R A. (b) Alveolar-capillary. See KEY to Figure LG 23.4A. (c) 1, Surfactant; 2, alveolar epithelium, 3, epithelial basement membrane; 4, interstitial space; 5, capillary basement membrane; 6, capillary endothelium. (d) Thin; edema; less. (e) Squamous; surfactant; debris.

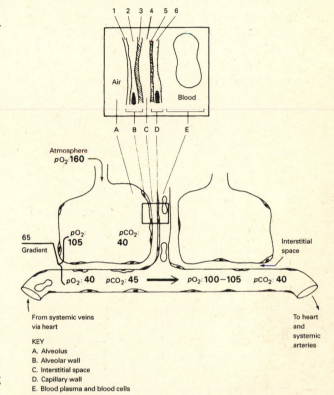

Figure LG 23.4A Diagram of alveoli and pulmonary capillary. Insert enlarges alveolar-capillary membrane.

B12. (a) Bronchial; aorta. (b) Azygos, accessory hemiazygos. (c) Vein. (d1) Systemic; systemic; (d2) pulmonary.

C1. (a) PV. (b) E. (c) I.

C2. (a) Alveolar or intrapulmonic, 760; the same as. (b) Intrapleural, 756; less than. (c) Diaphragm, external intercostal; increases; increase. (d) Decrease, 754; decreases, 758; Boyle. (e) Atmosphere, alveoli; into; inspiration; 760. (f) Active, passive, elastic; recoil of elastic fibers stretched during inspiration and inward pull of surface tension; sup, post. (g) Abdominal muscles and internal intercostals.

C4. (a) Low, high. (b) High; surfactant, increases. (c) Mixture of phospholipids and lipoproteins; II; respiratory distress syndrome (RDS) or hyaline membrane disease. (d) Atelectasis; examples: conditions creating external pressure on lungs, such as air, blood, or pleural effusions in the intrapleural space; conditions that prevent alveoli from reinflating with air, such as bronchitis, pneumonia, or pulmonary edema; or conditions that impede lung expansion, such as stiffening of lung tissue (as by tuberculosis or fibrosis) or paralysis of respiratory muscles.

C6. (a) ↓. (b) ↓. (c) ↑.

C9. 7,200 ml/min (= 15 breaths/min × 480 breaths/min).

C10. 70% (350/500 ml); dead space (V_D), 150 (or weight of the person).

C11. (a) V_T. (b) FRC. (c) ERV. (d) RV. (e) VC. (f) TLC. (g) IRV. (h) FEV_1.

C12. (a) 500 (0.5). (b) 6,000 (6). (c) 4,800 (4.8). (d) 2,400 (2.4). (e) 1,200 (1.2).

D1. (a) H. (b) D. (c) H. (d) B.

D2. (a) 760. (b) 79, 21; 0.04. (c) 600.4; pN_2. (d) 21% × 760 mm Hg = about 160 mm Hg = pO_2.

D3. (a) Partial; see Figure LG 23.4A. (b) 105 − 40 = 65. (c) Large, thin, increased. (d) About the same as; 105. (e) See Figure LG 23.4A. (f) No. Typical radial arterial values are same as for alveoli or blood leaving lungs: pO_2 = 100, pCO_2 = 40. Jenny's values indicate inadequate gas exchange. (g) Labels for Figure LG 23.1: inferior vena cava, pO_2 = 40 mm Hg and pCO_2 = 45 mm Hg; aorta, pO_2 = 100–105 mm Hg and pCO_2 = 40 mm Hg.

D4. 282 × 21% = 59.2 mm Hg.

D5. (b) DD; the excess fluid increases the thickness (diffusion distance) of the a-c membrane. (c) SA; emphysema destroys alveolar walls. (d) SMW. Because the a-c membrane is considerably more permeable to CO_2, she can exhale CO_2; so her pCO_2 is only slightly elevated above a normal of 40. Because O_2 exhibits lower solubility in the a-c membrane, her pO_2 (54) is considerably lower than normal (about 100).

E1. (a) 20; oxyhemoglobin; low. (b) Iron; HbO_2, saturated; increase. (c) 98; 98. (d) Increase in temperature, pCO_2, acidity, and BPG. (e) 62; 38; less, more. (f) 75; less. In the event that respiration is temporarily halted, even venous blood has much oxygen attached to hemoglobin and available to tissues.

E2. 200; oxygen-carrying capacity of hemoglobin is drastically reduced in carbon monoxide poisoning.

E3. (a) Histotoxic. (b) Stagnant. (c) Anemic. (d) Hypoxic.

E4. (a) 70, 23, 7. (b) H_2CO_3 (carbonic acid). (c) Hemoglobin (as H•Hb); HCO_3^-, plasma, Cl^-. (d) CO_2 + H_2O $\xrightarrow{\text{carbonic anhydrase}}$ H_2CO_3 (carbonic acid) → H^+ + HCO_3^- (bicarbonate).

F1.

Name	M/P	Function
a. **Medullary rhythmicity**	M	Controls rhythm; consists of inspiratory and expiratory areas
b. Pneumotaxic	**P**	**Limits inspiration and facilitates expiration**
c. **Apneustic**	**P**	Prolongs inspiration and inhibits expiration

F2. (a) H^+, increase in pCO_2. (b) Medulla. (c) Hypercapnia; any value higher than 40; even slight, hyper. (d) Aortic and carotid bodies, only large; usually below 60. (e) In. (f) Stretch; inhibition; inflation (Hering-Breuer).

G1. (a) Decreased vital capacity, and so decreased pO_2. (b) Risk of pneumonia.

G2. (a) Endo; lining of larynx, trachea, bronchial tree, and alveoli. (b) Smooth muscle, cartilage, and other connective tissues of airways.

H2. (a) E. (b) CB. (c) P. (d) A. (e) TB. (f) D. (g) RF. (h) A, CB, E.

H3. (a) Fluid; surfactant; septal cells of alveoli. (b) Lowers; facilitates. (c) Premature. Septal cells do not produce adequate amounts of surfactant until between weeks 28 and 32 of the 39-week human gestational period. A baby born at 7 months (30 weeks), for example, would be at high risk for RDS.

H6. Establish Airway, ventilate (Breathing), and reestablish Circulation.

WRITING ACROSS THE CURRICULUM: CHAPTER 23

1. After a larynx is removed (for example, due to cancer), what other structures would help the laryngectomee to speak?

2. Contrast location, structure, and function of each of these parts of the respiratory system: *vestibule/conchae, nasopharynx/oropharynx, epiglottis/thyroid cartilage, primary bronchi/bronchioles, parietal pleura/visceral pleura.*

3. Define and state one or more examples of reasons for the following procedures: bronchography, bronchoscopy, intubation, nebulization, tracheostomy.

4. Define and state one or more examples of causes of each of the following clinical conditions: laryngitis, atelectasis, pleural effusion.

MASTERY TEST: CHAPTER 23

Questions 1–4: Arrange the answers in correct sequence.

_____ _____ _____ 1. From first to last, the steps involved in inspiration:
  A. Diaphragm and intercostal muscles contract.
  B. Thoracic cavity and lungs increase in size.
  C. Alveolar pressure decreases to 758 mm Hg.

_____ _____ _____ 2. From most superficial to deepest:
  A. Parietal pleura
  B. Visceral pleura
  C. Pleural cavity

_____ _____ _____ _____ _____ 3. From superior to inferior:
  A. Bronchioles
  B. Bronchi
  C. Larynx
  D. Pharynx
  E. Trachea

_____ _____ _____ _____ _____ _____ 4. Pathway of inspired air:
  A. Alveolar ducts
  B. Bronchioles
  C. Lobar bronchi
  D. Primary bronchi
  E. Segmental bronchi
  F. Alveoli

Questions 5–9: Circle the letter preceding the one best answer to each question.

5. All of the following terms are matched correctly with descriptions *except:*
  A. Internal nares: choanae
  B. External nares: nostrils
  C. Posterior of nose: vestibule
  D. Pharyngeal tonsils: adenoids

6. Which of these values (in mm Hg) would be normal for pO_2 of blood in the femoral artery?
  A. 40
  B. 45
  C. 100
  D. 160
  E. 760
  F. 0

7. Pressure and volume in a closed space are inversely related, as described by _____ law.
  A. Boyle's
  B. Starling's
  C. Dalton's
  D. Henry's

8. Choose the correct formula for carbonic acid:
  A. HCO_3^-
  B. H_3CO_2
  C. H_2CO_3
  D. HO_3C_2
  E. H_2C_3O

9. A procedure in which an incision is made in the trachea and a tube inserted into the trachea is known as a(n):
  A. Tracheostomy
  B. Bronchogram
  C. Intubation
  D. Pneumothorax

Questions 10–20: Circle T (true) or F (false). If the statement is false, change the underlined word or phrase so that the statement is correct.

T F 10. In the chloride shift Cl⁻ moves into red blood cells in exchange for $\underline{H^+}$.

T F 11. When chemoreceptors sense <u>increase in pCO₂ or increase in acidity of blood H⁺</u> respiratory rate will normally be stimulated.

T F 12. Both increased temperature and increased acid content tend to cause oxygen to <u>bind more tightly</u> to hemoglobin.

T F 13. Under normal circumstances intrapleural pressure is <u>always less than atmospheric</u>.

T F 14. The pneumotaxic and apneustic areas controlling respiration are located in the <u>pons</u>.

T F 15. Fetal hemoglobin has a <u>lower</u> affinity for oxygen than maternal hemoglobin does.

T F 16. Most CO_2 is carried in the blood in the form of <u>bicarbonate</u>.

T F 17. The alveolar wall <u>does</u> contain macrophages that remove debris from the area.

T F 18. Intrapulmonic pressure means the same thing as <u>intrapleural</u> pressure.

T F 19. Inspiratory reserve volume is normally <u>larger than</u> expiratory reserve volume.

T F 20. The pO_2 and pCO_2 of blood leaving the lungs <u>are normally about the same as</u> pO_2 and pCO_2 of alveolar air.

Questions 21–25: Fill-ins. Complete each sentence with the word or phrase that best fits.

_____ 21. The process of exchange of gases between alveolar air and blood in pulmonary

capillaries is known as _____.

_____ 22. Take a normal breath and then let it out. The amount of air left in your lungs is the

capacity called _____ and it usually measures about _____ ml.

_____ 23. The Bohr effect states that when more H⁺ ions are bound to hemoglobin, less

_____ can be carried by hemoglobin.

_____ 24. The epiglottis, thyroid, and cricoid cartilages are all parts of the _____.

_____ 25. _____ is a chemical that lowers surface tension and therefore increases inflatability (compliance) of lungs.

ANSWERS TO MASTERY TEST: ☆ CHAPTER 23

Arrange

1. A B C
2. A C B
3. D C E B A
4. D C E B A F

Multiple Choice

5. C
6. C
7. A
8. C
9. A

True–False

10. F. HCO_3^-
11. T
12. F. Dissociate from
13. T
14. T
15. F. Higher
16. T
17. T
18. F. Alveolar
19. T
20. T

Fill-ins

21. External respiration (or pulmonary respiration, or diffusion)
22. Functional residual capacity (FRC), 2400
23. Oxygen
24. Larynx
25. Surfactant

FRAMEWORK 24
Digestive System

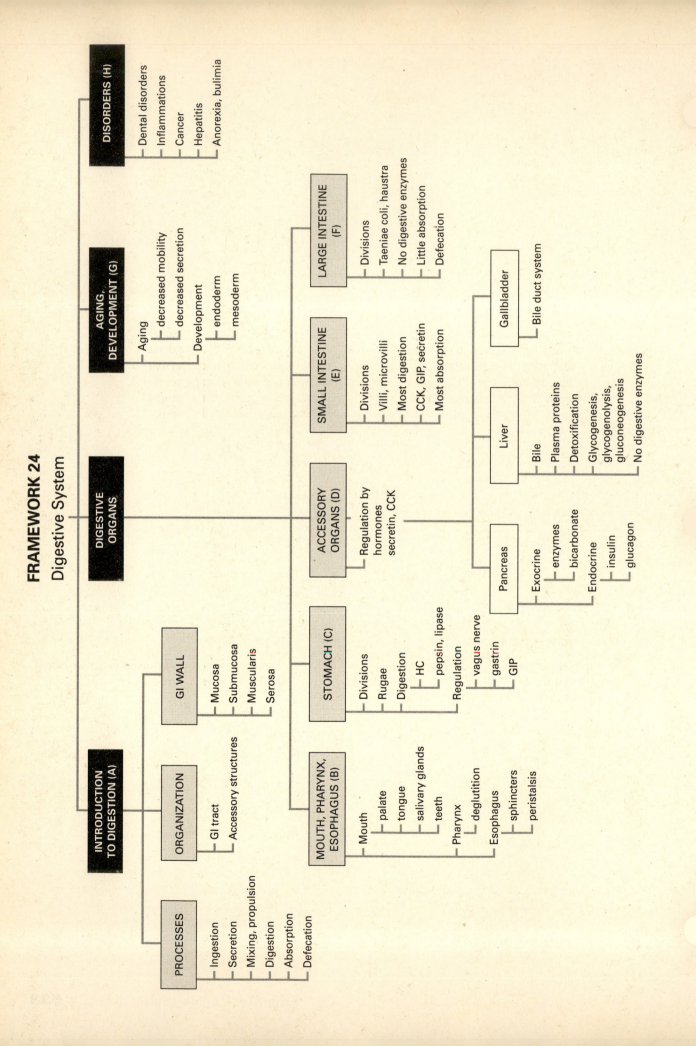

INTRODUCTION TO DIGESTION (A)

PROCESSES
- Ingestion
- Secretion
- Mixing, propulsion
- Digestion
- Absorption
- Defecation

ORGANIZATION
- GI tract
- Accessory structures

GI WALL
- Mucosa
- Submucosa
- Muscularis
- Serosa

DIGESTIVE ORGANS

MOUTH, PHARYNX, ESOPHAGUS (B)
- Mouth
  - palate
  - tongue
  - salivary glands
  - teeth
- Pharynx
  - deglutition
- Esophagus
  - sphincters
  - peristalsis

STOMACH (C)
- Divisions
- Rugae
- Digestion
  - HC
  - pepsin, lipase
- Regulation
  - vagus nerve
  - gastrin
  - GIP

ACCESSORY ORGANS (D)
- Regulation by hormones secretin, CCK

Pancreas
- Exocrine
  - enzymes
  - bicarbonate
- Endocrine
  - insulin
  - glucagon

Liver
- Bile
- Plasma proteins
- Detoxification
- Glycogenesis, glycogenolysis, gluconeogenesis
- No digestive enzymes

Gallbladder
- Bile duct system

SMALL INTESTINE (E)
- Divisions
- Villi, microvilli
- Most digestion
- CCK, GIP, secretin
- Most absorption

LARGE INTESTINE (F)
- Divisions
- Taeniae coli, haustra
- No digestive enzymes
- Little absorption
- Defecation

AGING, DEVELOPMENT (G)
- Aging
  - decreased mobility
  - decreased secretion
- Development
  - endoderm
  - mesoderm

DISORDERS (H)
- Dental disorders
- Inflammations
- Cancer
- Hepatitis
- Anorexia, bulimia

The Digestive System

CHAPTER 24

Food comes in very large pieces (like whole oranges, stalks of broccoli, and slices of bread) that must fit into very small spaces in the human body (like liver cells or brain cells). The digestive system makes such change possible. Foods are minced and enzymatically degraded so that absorption into blood en route to cells becomes a reality. The gastrointestinal (GI) tract provides a passageway complete with mucous glands to help food slide along, muscles that propel food forward and muscles (sphincters and valves) that regulate flow, and secretions that break apart foods or modify pH to suit local enzymes. Accessory structures such as teeth, tongue, pancreas, and liver lie outside the GI tract; yet each contributes to the mechanical and chemical dismantling of food from bite-sized to right-sized pieces. The ultimate fate of the absorbed products of digestion is the story line of metabolism (in Chapter 25).

As you begin your study of the digestive system, carefully examine the Chapter 24 Topic Outline and Objectives. Check off each one as you complete it. You will also find food for thought in the Chapter 24 Framework. Be sure to refer to the Framework frequently and note relationships among key terms in each section.

TOPIC OUTLINE AND OBJECTIVES

A. Introduction to digestion

☐ 1. Identify the organs of the gastrointestinal (GI) tract and the accessory organs of digestion and their functions in digestion.

B. Mouth, pharynx, and esophagus

C. Stomach

D. Accessory organs: pancreas, liver, gallbladder

E. Small intestine

☐ 2. Describe the mechanical movements of the gastrointestinal tract.

☐ 3. Explain how salivary secretion, gastric secretion, gastric emptying, pancreatic secretion, bile secretion, and small intestinal secretion are regulated.

☐ 4. Define absorption and explain how the end products of digestion are absorbed.

F. Large intestine

☐ 5. Define the processes involved in the formation of feces and defecation.

G. Aging and development anatomy of the digestive system.

☐ 6. Describe the effects of aging on the digestive system.

☐ 7. Describe the development of the digestive system.

H. Disorders, medical terminology

☐ 8. Describe the clinical symptoms of the following disorders: dental caries, periodontal disease, peptic ulcer disease (PUD), gastrointestinal tumors, cirrhosis, hepatitis, gallstones, anorexia nervosa, and bulimia.

☐ 9. Define medical terminology associated with the digestive system.

WORDBYTES

Now become familiar with the language of this chapter by studying each wordbyte, its meaning, and an example of its use within a term. After you study the entire list, self-check your understanding by writing the meaning of each wordbyte on the line. As you continue through the *Learning Guide,* identify (and fill in) additional terms that contain the same wordbyte.

Wordbyte	Self-check	Meaning	Example(s)
amyl-	_____	starch	*amyl*ase
-ase	_____	enzyme	malt*ase*
caec-, cec-	_____	blind	*cec*um
chole-, cholecyst-	_____	gallbladder	*chole*cystitis
chym-	_____	juice	*chym*otrypsin
dent-	_____	tooth	*dent*ures
-ectomy	_____	removal of	append*ectomy*
entero-	_____	intestine	*entero*kinase
gastro-	_____	stomach	*gastr*in
gingiv-	_____	gums	*gingiv*itis
ileo-	_____	ileum	*ileo*cecal valve
jejun-	_____	empty	*jejun*oileostomy
lith-	_____	stone	chole*lith*iasis
or-	_____	mouth	*or*al
-ose	_____	sugar	lact*ose*
retro-	_____	behind	*retro*peritoneal
-rhea	_____	to flow	diar*rhea*
stoma-	_____	mouth, opening	colo*stomy*
taen-	_____	ribbon	*taen*ia coli
vermi-	_____	worm	*vermi*form appendix

CHECKPOINTS

A. Introduction to digestion (pages 753–757)

A1. Explain why food is vital to life. Give three specific examples of uses of foods in the body.

A2. Cover the key and identify all digestive organs on Figure LG 24.1. Then color the structures with color code ovals; these five structures are all *(parts of the gastrointestinal tract? accessory structures?)*.

A3. List the six basic activities of the digestive system.

_____ _____ _____

_____ _____ _____

A4. _____ digestion occurs by action of enzymes (such as those in saliva) and intestinal secretion, whereas _____ digestion involves action of the teeth and muscles of the stomach and intestinal wall.

A5. Color layers of the GI wall indicated by color code ovals on Figure LG 24.2.

A6. Now match the names of layers of the GI wall with the correct description.

Muc. Mucosa	Ser. Serosa
Mus. Muscularis	Sub. Submucosa

_____ a. Also known as the peritoneum, it forms mesentery and omentum

_____ b. Consists of epithelium, lamina propria, and muscularis mucosae

_____ c. Connective tissue containing glands, nerves, blood and lymph vessels

_____ d. Consists of an inner circular layer and an outer longitudinal layer

A7. Describe the tissues listed below by completing this table. Identify the layer in which each tissue is located, type of tissue in each case, and functions of these tissues.

	Layer of GI Wall	Type of Tissue	Function(s)
a. Enteroendocrine tissue			Secretes hormones
b. MALT tissue		Lymphatic	
c. Submucosal plexus (of Meissner)	Submucosa and mucosa		
d. Myenteric plexus (of Auerbach)		ANS nerve tissue	

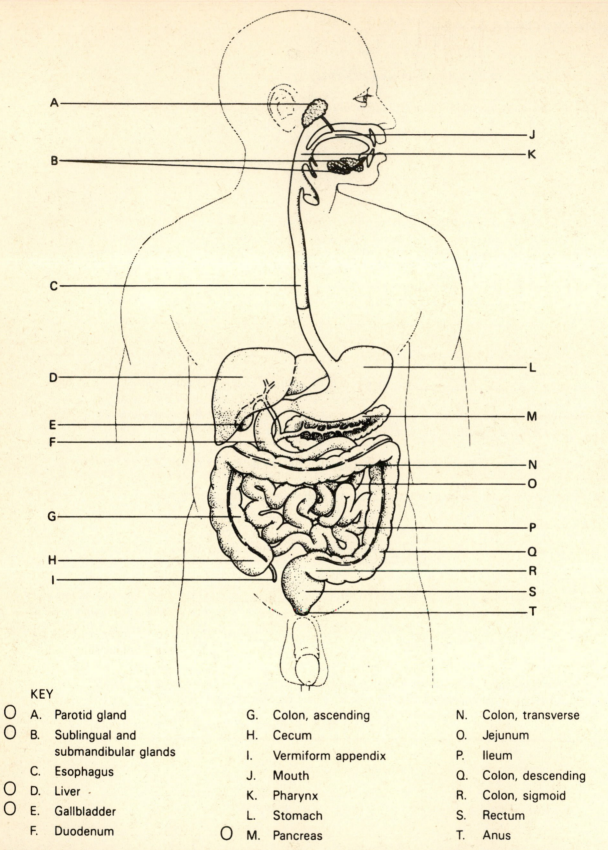

KEY

- A. Parotid gland
- B. Sublingual and submandibular glands
- C. Esophagus
- D. Liver
- E. Gallbladder
- F. Duodenum
- G. Colon, ascending
- H. Cecum
- I. Vermiform appendix
- J. Mouth
- K. Pharynx
- L. Stomach
- M. Pancreas
- N. Colon, transverse
- O. Jejunum
- P. Ileum
- Q. Colon, descending
- R. Colon, sigmoid
- S. Rectum
- T. Anus

Figure LG 24.1   Organs of the digestive system. Color, draw arrows, and identify labeled structures as directed in Checkpoints A2, B3, and F3.

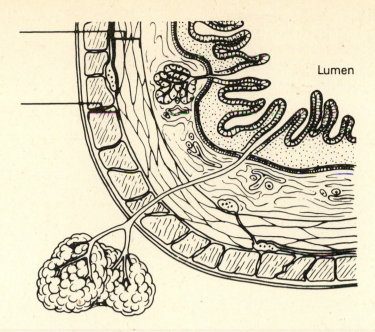

Lumen

Figure LG 24.2   Gastrointestinal (GI) tract and related gland seen in cross section. Refer to Checkpoint A5 and color layers of the wall according to color code ovals. Label muscularis layers.

■   **A8.**   Match the names of these peritoneal extensions with the correct descriptions.

F. Falciform ligament	M. Mesentery
G. Greater omentum	Meso. Mesocolon
L. Lesser omentum	

_____a. Attaches liver to anterior abdominal wall

_____b. Binds intestine to posterior abdominal wall; provides route for blood and lymph vessels and nerves to reach small intestine

_____c. Binds part of large intestine to posterior abdominal wall

_____d. "Fatty apron"; covers and helps prevent infection in small intestine

_____e. Suspends stomach and duodenum from liver

A9.   *A clinical challenge.* Define the following terms.

a. Ascites

b. Peritonitis

B. Mouth, pharynx, and esophagus (pages 758-767)

■ **B1.** Refer to Figure LG 24.3 and label the following structures.

Hard palate   Palatine tonsils   Uvula of soft palate
Lingual frenulum   Palatoglossal arch   Vestibule

B2. Contrast terms in each pair:

a. *Intrinsic muscles of the tongue/extrinsic muscles of the tongue*

b. *Fungiform papillae/filiform papillae*

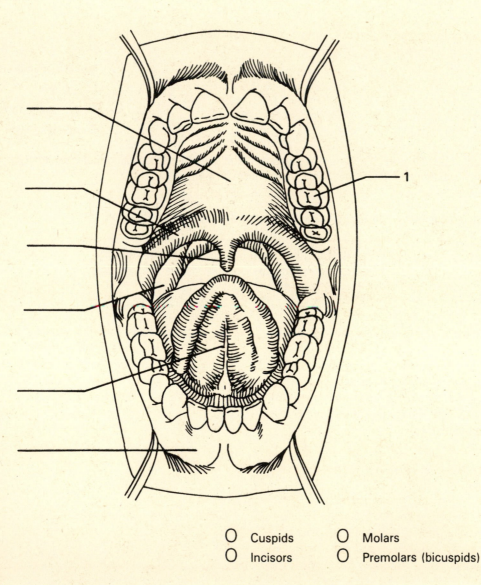

O Cuspids   O Molars
O Incisors   O Premolars (bicuspids)

Figure LG 24.3   Mouth (oral cavity). Complete as directed in Checkpoints B1 and B7.

B3. Identify the three salivary glands on Figure LG 24.1 and visualize their locations on yourself.

■ **B4.** Complete this exercise about salivary glands.

a. Which glands are largest? *(Parotid? Sublingual? Submandibular?)*

b. Which secrete the thickest secretion due to presence of much mucus?

_____

c. About 1 to 1½ *(tablespoons? cups? liters?)* of saliva are secreted daily.

d. State three or more functions of saliva.

e. The pH of the mouth is appropriate for action of salivary amylase. This is about pH *(2? 6.5? 9?)*.

f. List five or more chemicals that are in saliva.

g. List four or more factors that can stimulate increase in saliva production.

■ **B5.** Arrange the following in correct sequence:

_____ _____ _____ a. From most superficial to deepest:

> | A. Root | B. Neck | C. Crown |

_____ _____ _____ b. From most superficial to deepest within a tooth:

> | A. Enamel | B. Dentin | C. Pulp cavity |

_____ _____ _____ c. From hardest to softest:

> | A. Enamel | B. Dentin | C. Pulp cavity |

■ **B6.** List two functions of the periodontal ligament.

■ **B7.** Refer to Figure LG 24.3 and do this exercise on dentitions.

a. How many teeth (total) are in a complete set of deciduous (baby) teeth?

_____ How many are in a complete set of permanent teeth? _____

b. Color the teeth in the lower right of Figure LG 24.3 as indicated by color code ovals.

c. Number the teeth (1–8) on the upper right of the diagram according to order of eruption. One (the first one to erupt) is done for you.

d. Which teeth are most likely to become impacted? _____

■ **B8.** Check your understanding of digestion by completing this Checkpoint.

a. *(Deglutition? Mastication?)* is a term that means chewing. Food is formed into a soft mass known as a *(bolus? chyme?).*

b. Enzymes are used for *(chemical? mechanical?)* digestion of food. The main enzyme in

saliva is salivary *(amylase? lipase?).* This enzyme converts _____ into smaller carbohydrates. Most starch *(is? is not?)* broken down by the time food leaves the mouth. What inactivates amylase in the stomach?

c. Where is lingual lipase made? *(Salivary glands? Tongue? Tonsils?)* This enzyme starts

the breakdown of _____.

■ **B9.** The term *deglutition* means _____. Match each stage of deglutition with the correct description.

┌──┐
│ E. Esophageal   P. Pharyngeal   V. Voluntary │
└──┘

_____ a. Soft palate and epiglottis close off respiratory passageways.

_____ b. Tongue pushes food back into oropharynx.

_____ c. Peristaltic contractions push bolus from pharynx to stomach.

■ **B10.** Do this activity on the esophagus.

a. The esophagus is about _____ cm (_____ in.) long. It ends inferiorly in the

(stomach? duodenum?). In the condition known as _____

_____, the stomach protrudes superiorly through the diaphragm.

b. The muscularis of the superior one-third of the esophagus consists of *(smooth? striated?)* muscle, whereas the inferior one-third consists of *(smooth? striated?)* muscle similar to that found in the stomach.

c. Failure of the lower esophageal sphincter to relax is a condition known as

_____. Failure of the lower esophageal sphincter to close results in the stomach contents entering the esophagus and distending it and in irritation of the esophageal lining by the *(acidic? basic?)* contents of the stomach. The resulting

condition is known as _____. Pain associated with either of these

conditions may be confused with pain originating in the _____.

d. Forcible expulsion of the stomach contents into the esophagus is known as

_____. If this disorder is prolonged, what results may occur?

■ **B11.** *The Big Picture: Looking Back.* Refer to figures and exhibits in earlier chapters that relate to digestive activity in the mouth.

a. Exhibit 11.6 and Figure 11.7 (pages 284–285 in your text) demonstrate the genio-glossus, hyoglossus, and styloglossus. These are *(extrinsic? intrinsic?)* muscles of the tongue and are innervated by cranial nerve *(V? VII? XII?).*

b. Exhibit 17.3 (pages 497–498) demonstrates that increased secretion of saliva is a func-tion of the *(sympathetic? parasympathetic?)* division of the autonomic nervous system.

c. Exhibit 14.4 (pages 418–422) describes cranial nerves. Which ones carry sensory nerves for taste as well as the autonomic nerves stimulating salivary glands? Cranial nerves

_____ and _____ Pain in an upper tooth is transmitted by the *(mandibular?*

maxillary? ophthalmic?) branch of the trigeminal nerve (cranial nerve _____).

B12. Summarize digestion in the mouth, pharynx, and esophagus by completing parts a and b of Table LG 24.1 (next page).

C. Stomach   (pages 767–775)

■ **C1.**   Refer to Figure LG 24.1 and complete this Checkpoint.

a. Arrange these regions of the stomach according to the pathway of food from first to last.

Body	Fundus
Cardia	Pylorus

_____ → _____ →

_____ → _____

b. Match the terms in the box with the correct description.

GC. Greater curvature	LC. Lesser curvature

_____ More lateral and inferior in location

_____ Attached to lesser omentum

_____ Attached to greater omentum

c. In which condition does the pyloric sphincter fail to relax normally? *(Pyloric stenosis?*

Pylorospasm?) Write one sign or symptom of this condition. _____

Table LG 24.1   Summary of digestion.

Digestive Organs	Carbohydrate	Protein
a. Mouth, salivary glands	Salivary amylase: digests starch to maltose	
b. Pharynx, esophagus		
c. Stomach		
d. Pancreas		
e. Intestinal juices		
f. Liver	No enzymes for digestion of carbohydrates	
g. Large intestine	No enzymes for digestion of carbohydrates	No enzymes for digestion of proteins

Table LG 24.1 *(continued)*

Lipid	Mechanical	Other Functions
	Deglutition, peristalsis	
		1. Secretes intrinsic factor 2. Produces hormone stomach gastrin
Pancreatic lipase: digests about 80% of fats		
No enzymes for digestion of lipids		

■ **C2.** Complete this table about gastric secretions.

Name of Cell	Type of Secretion	Function of Secretion
a. Chief (zygomatic)		
b. Mucous		
c.	HCl	
d.	Gastrin	

■ **C3.** How does the structure of the stomach wall differ from that in other parts of the GI tract?

a. Mucosa

b. Muscularis

■ **C4.** Answer these questions about chemical digestion in the stomach.

a. Pepsin is most active at very *(acid? alkaline?)* pH.
b. State two factors that enable the stomach to digest protein without digesting its own cells (which are composed largely of protein).

c. If mucus fails to protect the gastric lining, the condition known as

_____ may result.
d. Another enzyme produced by the stomach is _____, which digests

_____. In adults it is quite *(effective? ineffective?)*. Why?

■ **C5.** Answer these questions about control of gastric secretion.

a. Name the three phases of gastric secretion: _____,

_____, and _____. Which of these causes gastric

secretion to begin when you smell or taste food? _____

b. Gastric glands are stimulated mainly by the _____ nerves. Their
fibers are *(sympathetic? parasympathetic?)*. Two stimuli that cause vagal impulses to

stimulate gastric activity are _____ and _____.
Three emotions that inhibit production of gastric secretions are

_____, _____, and _____.

c. Distension of the stomach triggers the _____ phase. Two effects of
this stimulus are *(increase? decrease?)* in parasympathetic impulses and release of the

hormone _____. What else triggers release of gastrin?

Gastrin travels through the blood to all parts of the body; its target cells are gastric
glands and gastric smooth muscle, which are *(stimulated? inhibited?)*, and the pyloric
and ileocecal sphincters, which are *(contracted? relaxed?)*.

d. *A clinical challenge.* Both acetylcholine, released by vagal nerves, and gastrin
(stimulate? inhibit?) HCl production. Their effect is *(greater? less?)* in the presence of

histamine, which is made by _____ cells in the stomach wall.
Describe the effect of H_2 blockers such as cimetidine (Tagamet) on the stomach.

e. When food (chyme) reaches the intestine, nerves initiate the _____
reflex, which *(stimulates? inhibits?)* further gastric secretion. Three hormones released
by the intestine also inhibit gastric secretion as well as gastric motility, so the effect of
these hormones is to *(stimulate? delay?)* gastric emptying. Name these three hormones:

_____, _____, and _____.

■ **C6.** Food stays in the stomach for about _____ hours. Which food type

leaves the stomach most quickly? _____ Which type stays in the

stomach longest? _____

■ **C7.** Do this exercise on alcohol absorption.

a. Is any alcohol absorbed in the stomach? *(Yes? No?)* Because of the difference of the
absorptive surface, alcohol is more readily absorbed into blood vessels in the wall of the
(stomach? small intestine?).

b. To decrease the rate of alcohol absorption, it is better to have alcohol enter the small intestine *(slowly? rapidly?)*. Identify types of food that could be eaten to slow the passage of food and alcohol into the small intestine.

c. Explain why a woman is more likely to experience a rapid rise in blood level of alcohol even if she drinks the same amount of alcohol as a man and has the same body size as the man.

C8. Complete part c of Table LG 24.1, describing the role of the stomach in digestion.

D. Accessory organs: pancreas, liver, gallbladder (pages 775–783)

D1. Refer to Figure LG 24.4 and color all of the structures with color code ovals according to colors indicated.

■ **D2.** Complete these statements about the pancreas.
a. The pancreas lies posterior to the _____.
b. The pancreas is shaped roughly like a fish, with its head in the curve of the

_____ and its tail nudging up next to the _____.
c. The pancreas contains two kinds of glands. Ninety-nine percent of its cells produce *(endocrine? exocrine?)* secretions. One type of these secretions is an *(acid? alkaline?)* fluid to neutralize the chyme entering the small intestine from the stomach.
d. One enzyme in the pancreatic secretions is trypsin; it digests *(fats? carbohydrates? proteins?)*. Trypsin is formed initially in the inactive form (trypsinogen) and is activated by *(HCl? NaHCO$_3$? enterokinase?)*. Trypsin itself serves as an activator in the formation of

the protease named _____ and the peptidase named

_____.

e. Most of the amylase and lipase produced in the body are secreted by the pancreas. Describe the functions of these two enzymes.

f. All exocrine secretions of the pancreas empty into ducts (_____ and _____ on

Figure LG 24.4). These empty into the _____.
g. *A clinical challenge.* In most persons the pancreatic duct also receives bile flowing

through the _____ (F on the figure). State one possible complication that may occur if a gallstone blocks the pancreatic duct (as at point H on the figure).

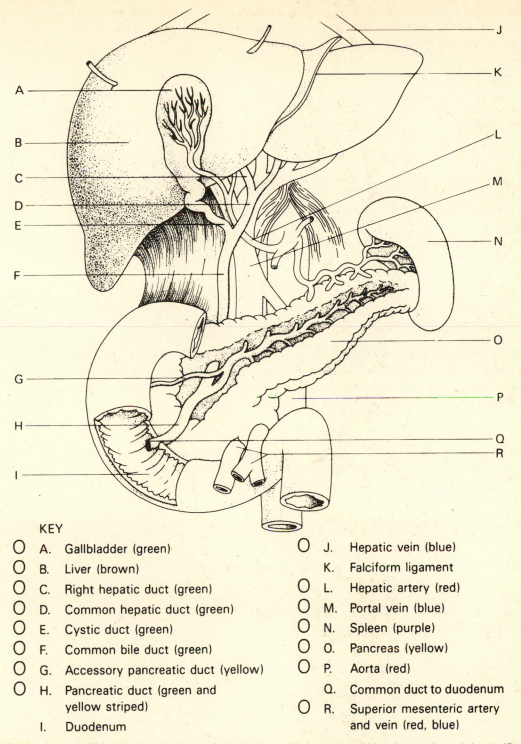

KEY

○ A. Gallbladder (green)
○ B. Liver (brown)
○ C. Right hepatic duct (green)
○ D. Common hepatic duct (green)
○ E. Cystic duct (green)
○ F. Common bile duct (green)
○ G. Accessory pancreatic duct (yellow)
○ H. Pancreatic duct (green and yellow striped)
  I. Duodenum

○ J. Hepatic vein (blue)
  K. Falciform ligament
○ L. Hepatic artery (red)
○ M. Portal vein (blue)
○ N. Spleen (purple)
○ O. Pancreas (yellow)
○ P. Aorta (red)
  Q. Common duct to duodenum
○ R. Superior mesenteric artery and vein (red, blue)

Figure LG 24.4 Liver, gallbladder, pancreas, and duodenum, with associated blood vessels and ducts. (Stomach has been removed). Refer to Checkpoints D1, D2 and D6; color and draw arrows as directed.

h. The endocrine portions of the pancreas are known as the _____

_____. List hormones they secrete. _____,

_____, _____, _____. Typical
of all hormones, these pass into *(ducts? blood vessels?)*, specifically into vessels that

empty into the _____ vein.

D3. Fill in part d of Table LG 24.1, describing the role of the pancreas in digestion.

■ **D4.** Describe regulation of the pancreas in this exercise.
a. The pancreas is stimulated by *(sympathetic? parasympathetic?)* nerves, specifically the

_____ nerves, and also by hormones produced by cells located in the

wall of the _____.
b. What is the chemical nature of chyme that stimulates release of CCK? Chyme that *(is
acidic? contains partially digested fats and proteins?)* CCK then activates the pancreas
to secrete fluid rich in *(HCO₃⁻? digestive enzymes such as trypsin, amylase, and
lipase?)*.
c. What is the chemical nature of chyme that stimulates release of secretion? Chyme that
(is acidic? contains partially digested fats and proteins?) CCK then activates the pancreas to secrete fluid rich in *(HCO₃⁻? digestive enzymes such as trypsin, amylase, and
lipase?)*.

■ **D5.** Answer these questions about the liver.
a. This organ weighs about _____ kg (_____ pounds). It lies in the

_____ quadrant of the abdomen. Of its _____ lobes, the

_____ is the largest.

b. The _____ ligament separates the right and left lobes. In the edge of
this ligament is the ligamentum teres (round ligament of the liver), which is the

obliterated _____ vein.

c. Blood enters the liver via vessels named _____ and

_____. In liver lobules, blood mixes in channels called

_____ before leaving the liver via vessels named

_____.

■ **D6.** Complete Figure LG 24.4 as directed.
a. Identify the pathway of bile from liver to intestine by following the structures you
colored green in that figure: C D F H. The general direction of bile is from *(superior
to inferior? inferior to superior?)*.
b. Now draw arrows to show the direction of blood through the liver. In order to reach the
inferior vena cava, blood must flow *(superiorly? inferiorly?)* through the liver.

■ **D7.** While blood is in liver sinusoids, hepatic cells have ample opportunity to act on this blood, modify it, and add new substances to it. Check your understanding of important functions of the liver in this Checkpoint.

a. Bile secreted from the liver is composed largely of the pigment named

_____, which is a breakdown product of _____ cells. Excessive amounts of this pigment give skin a yellowish color, a condition

known as _____. Two functions of bile are _____

of fats and _____ of fats (and fat-soluble vitamins).

b. Name several plasma proteins synthesized by the liver.

c. Identify two or more types of chemicals that are detoxified (metabolized) by the liver.

d. List three types of cells phagocytosed by the liver.

e. The liver can convert excess glucose to _____ and

_____ and also reverse that process. (More on this in Chapter 25.)

f. The liver stores the four fat-soluble vitamins named _____, _____, _____, and

_____. It also stores a vitamin necessary for erythropoiesis, namely vitamin

_____. The liver also cooperates with kidneys and skin to activate vitamin _____.

g. Note that the liver *(does? does not?)* secrete digestive enzymes.

D8. Complete part f of Table LG 24.1

■ **D9.** Check your understanding of accessory structures of the digestive system by naming the organ that best fits each description below. Select answers in the box.

| G. Gallbladder   L. Liver   P. Pancreas |

_____ a. Has both exocrine and endocrine functions

_____ b. Its primary functions are to store and concentrate bile

_____ c. Under influence of CCK, it contracts and ejects bile into the cystic duct

_____ d. Responds to CCK by secreting digestive enzymes

_____ e. Bilirubin, absorbed from worn-out red blood cells, is secreted into bile here

_____ f. Contains phagocytic cells known as stellate reticuloendothelial (Kupffer's) cells

_____ g. Stores iron in the form of ferritin

■ **D10.** Complete this activity about the "big four" hormones of the gut by selecting correct answers describing functions of each hormone. Complete arrows as ↑ if the hormone increases, stimulates, or promotes a certain activity; complete the arrow as ↓ if the hormone decreases, inhibits, or slows an activity.

a. Gastrin: | secretion of gastric juices, including HCl and pepsinogen; | gastric motility; | contraction of the lower esophageal sphincter; and | contraction of the pyloric sphincter. In summary, gastrin | digestive activity of the stomach.

b. GIP: | gastric secretions; | gastric emptying; and | pancreatic secretion of insulin. The overall effect of GIP (gastric inhibitory peptide) is to | digestive activity of the stomach.

c. CCK: | secretion of pancreatic juices rich in enzymes; | ejection of bile from gallbladder into bile duct system; | opening of hepatopancreatic ampulla (sphincter of Oddi) so that bile can flow into duodenum; | gastric emptying (by | contraction of pyloric sphincter and | satiety); | growth of pancreas; and | effects of secretin. The overall effect of CCK is to | digestive activity in the small intestine.

d. Secretin: | secretion of bicarbonate (in pancreatic juice and bile); | secretion of gastric juice; | effects of CCK; | growth of pancreas. The overall effect of CCK is to | digestive activity in the small intestine.

E. Small intestine   (pages 783–792)

■ **E1.** Describe the structure of the intestine in this Checkpoint.

a. The average diameter of the small intestine is:
   A. 2.5 cm (1 inch)   B. 5.0 cm (2 inches)

b. Its average length is about:
   A. 3 m (10 feet)   B. 6 m (20 feet)

c. Name its three main parts (in sequence from first to last):

_____ → _____ → _____

■ **E2.** Match the correct term related to the intestine to the description that fits.

ALF.	Aggregated lymphatic follicles (Peyer's patches)	MALT.	Mucosa-associated lymphoid tissue
BB.	Brush border	MV.	Microvilli
CF.	Circular folds (plicae circularis)	PC.	Paneth cells
DG.	Duodenal (Brunner's) glands	SKC.	S cells, K cells, and CCK cells
L.	Lacteal	V.	Villi

_____ a. Fingerlike projections up to 1 mm high that give the intestinal lining a velvety appearance and increase absorptive surface

_____ b. Fingerlike projections of plasma membrane

_____ c. Circular folds that further increase intestinal surface area

_____ d. Clusters of lymphatic follicles located primarily in the wall of the ileum

_____ e. Submucosal glands that secrete protective alkaline fluids

_____ f. Lymphatic nodules within the mucous membrane of the small intestine

_____ g. Fuzzy line marking the tips of microvilli; contains digestive enzymes

_____ h. Lymphatic capillary found within each villus

_____ i. Control level of microbes in intestine by phagocytosis and secretion of lysozyme

_____ j. Enteroendocrine cells that secrete the hormones secretin, GIP, and CCK

E3.   Name eight brush border enzymes.

These enzymes function primarily *(in the intestinal lumen? at the brush border?)*. Describe the action of one of these enzymes, α-*dextrinase.*

E4.   Contrast segmentation and peristalsis, the two types of movement of the small intestine.

■ **E5.**   Write the main steps in the digestion of each of the three major food types.

   a. *Carbohydrates*
      Polysaccharides   →   _____   →
      monosaccharides
   b. *Proteins*
      Proteins   →   _____   →   _____
   c. *Lipids*
      Neutral fats   →   emulsified fats   →   _____ _____

■ **E6.**   *A clinical challenge.* Persons with lactose intolerance lack the enzyme

   _____, which is necessary for digestion of lactose found in foods such

   as _____. Write a rationale for the usual symptoms of this condition.

E7.   Complete the description of functions of intestinal juices by filling in part e of Table LG 24.1. Then review digestion of each of the three major food groups by reading your columns vertically.

■ **E8.** *For extra review* of roles of GI organs in digestion, identify which chemicals in the list in the box are made by each of the following organs.

Alb. Albumin	HCl.	Hydrochloric acid
Amy. Amylase	IF.	Intrinsic factor
B. Bile	Ins.	Insulin
CCK. Cholecystokinin	KB.	Ketone bodies
Chol. Cholesterol	Lip.	Lipase
DA. Dipeptidases, aminopeptidases	Lys.	Lysozyme
E. Enterokinase	M.	Mucus
Ft. Ferritin	MLSD.	Maltase, lipase, sucrase, α-dextrinase
FP. Fibrinogen and prothrombin	NP.	Nucleosidases, phosphatases
Gas. Gastrin	P.	Pepsinogen
GIP. Gastric inhibitory peptide	RD.	Ribonuclease, deoxyribonuclease
Glu. Glucagon	S.	Secretin
Gly. Glycogen	TCP.	Trypsinogen, chymotrypsinogen, procarboxypeptidase

a. _____ a. Salivary glands and tongue (three answers)

b. _____ b. Pharynx and esophagus (one answer)

c. _____ c. Stomach (six answers)

d. _____ d. Pancreas (six answers)

e. _____ e. Small intestine (nine answers)

f. _____ f. Liver (seven answers)

■ **E9.** *For extra review* of secretions involved with digestion, match names of secretions listed for Checkpoint E8 with descriptions below. Blank lines follow some descriptions. On these indicate whether the secretion is classified as an enzyme (E), hormone (H), or neither of these (N). The first one is done for you.

CCK a. Causes contraction of gallbladder and relaxation of the sphincter of the hepatopancreatic ampulla so that bile enters duodenum __**H**__

_____ b. Stimulates production of alkaline pancreatic fluids and bile _____

_____ c. Stimulates pancreas to produce secretions rich in enzymes such as lipase and amylase_____

_____ d. Promote growth and maintenance of pancreas; enhance effects of each other (two answers)

_____ e. Increases gastric activity (secretion, motility, and growth) _____

_____ f. Inhibit stomach motility and/or secretions (three answers)

_____ g. Precursors to protein-digesting enzymes (two answers)

_____ h. Activates pepsinogen _____

_____ i. The active form of this enzyme starts protein digestion in the GI tract

_____ j. Activates the inactive precursor to form trypsin _____

_____ k. Plasma proteins (two answers)

_____ l. A storage form of carbohydrate

_____ m. Most effective of this type of enzyme is produced by the pancreas; digests fats

_____ n. Emulsifies fats before they can be digested effectively _____

_____ o. Starch-digesting enzymes secreted by salivary glands and pancreas

_____ p. Intestinal enzymes that complete carbohydrate breakdown, resulting in simple sugars

_____ q. Digest DNA and RNA_____

_____ r. Brush border enzymes that carry out digestion on the surface of villi (three answers)

_____ s. Complete digestion of proteins into amino acids

_____ t. Storage form of iron

_____ u. Bacteriocidal enzyme

■ **E10.** Describe the absorption of end products of digestion in this exercise.

a. Almost all absorption takes place in the *(large? small?)* intestine. Absorption is a two-step process. Products of digestion must first be absorbed into *(blood or lymph capillaries? epithelial cells lining the intestine?)* and then be transported into *(blood or lymph capillaries? epithelial cells lining the intestine?)*.

b. Glucose and some amino acids move into epithelial cells lining the intestine by

   (primary? secondary?) active transport coupled with active transport of _____. Fructose enters epithelial cells by *(diffusion? facilitated diffusion?)*. All simple sugars

   then enter capillaries by the process of _____.

c. Simple sugars, amino acids, and short-chain fatty acids are absorbed into *(blood?*

   lymph?) capillaries located in _____ in the intestinal wall. These

   vessels lead to the _____ vein and then to the

   _____ for storage or metabolism.

d. Long-chain fatty acids and monoglycerides first combine with _____ salts to form *(micelles? chylomicrons?)*. This enables fatty acids and monoglycerides to enter epithelial cells in the intestinal lining and soon enter lacteals leading to the *(portal vein? thoracic duct?)*. Most bile salts are ultimately *(eliminated in feces? recycled to the*

   liver?). This cycle is known as _____ circulation.

e. Aggregates of fats coated with _____ are known as chylomicrons.

   After traveling through lymph and blood, they reach the _____,

   where they may undergo conversions to form _____-proteins such

   as LDLs, VLDLs or _____. Name three types of fats found in lipoproteins.

f. About _____ liters (or _____ quarts) of fluids are ingested or secreted into the

   GI tract each day. Of this fluid, about _____ liters come from ingested food and about

   _____ liters derive from GI secretions. All but about 1.0 liter of the 9 liters is reabsorbed by *(facilitated diffusion? osmosis?)* into blood capillaries in the walls of the *(small? large?)* intestine. Several hundred milliliters of fluid are also reabsorbed each day into the *(small? large?)* intestine.

g. Normally about _____ ml of fluid exits each day in feces. When inadequate water

   reabsorption occurs, as in *(constipation? diarrhea?)*, then _____ such as Na+ and Cl− are also lost.

- **E11.** *The Big Picture: Looking Ahead* and *A clinical challenge.* Refer to Exhibit 25.6, pages 839–841 of the text, and answer these questions.
    a. Which vitamins are absorbed with the help of bile? *(Water-soluble? Fat-soluble?)* Circle the fat-soluble vitamins: A B$_{12}$ C D E K
    b. Inadequate bile production or obstruction of bile pathways may lead to signs or symptoms related to deficiencies of fat-soluble vitamins. List several.

F. Large intestine (pages 793–797)

- **F1.** The large intestine is so named based on its *(diameter? length?)* compared to that of

    the small intestine. The total length of the large intestine is about _____ m (_____ ft).

    More than 90% of its length consists of the part known as the _____.

- **F2.** Arrange the parts of the large intestine listed in the box in correct sequence in the pathway of wastes

AnC. Anal canal	R. Rectum	
AsC. Ascending colon	SC. Sigmoid colon	
C. Cecum	SF. Splenic flexure	
DC. Descending colon	TC. Transverse colon	
HF. Hepatic flexure		

    _____ → _____ → _____ → _____ → _____ →

    _____ → _____ → _____ → _____

F3. Identify the regions of the large intestine in Figure LG 24.1. Draw arrows to indicate direction of movement of intestinal contents.

- **F4.** Contrast different portions of the GI tract by identifying structures or functions associated with each. Use these answers:

LI. Large intestine	Sto. Stomach
SI. Small intestine	

_____ a. Has thickened bands of longitudinal muscle known as taeniae coli

_____ b. Pouches known as haustra give this structure its puckered appearance.

_____ c. Its fat-filled peritoneal attachments are known as epiploic appendages

_____ d. Bacteria here decompose bilirubin to urobilinogen, which gives feces its brown color

_____ e. Has rugae

_____ f. Has villi and microvilli

_____ g. (Vermiform) appendix is attached to the cecum here

_____ h. Diverticuli are outpouchings where the muscularis here has weakened

_____ i. Ileocecal valve is located here (two answers)

F5. Define each of these terms and discuss possible causes. Discuss the effects of high fiber diet on each condition.

a. Hemorrhoids

b. Diverticulosis

F6. Contrast terms in each pair:

a. *Gastroileal reflex/gastrocolic reflex*

b. *Haustral churning/mass peristalsis*

c. *Diarrhea/constipation*

■ **F7.** Complete this activity describing chemicals associated with digestive wastes.

a. Flatus (gas) in the colon due to the gases _____,

_____, _____; these are products of bacterial fermentation of *(carbohydrates? proteins? fats?)*.

b. Odors due to the chemicals _____ and _____; these are products of bacterial breakdown of *(carbohydrates? proteins and amino acids? fats?)*.

c. Brown color due to the chemical _____, a product of bacterial

decomposition of _____.

■ **F8.** Feces are formed by the time chyme has remained in the large intestine for about

_____ hours. List the chemical components of feces.

F9. Describe the process of defecation. Include these terms: *stretch receptors, rectal muscles, sphincters, diaphragm,* and *abdominal muscles.*

■ **F10.** Describe helpful effects of dietary fiber in this activity.

  a. *(Soluble? Insoluble?)* fibers tend to speed up passage of digestive wastes. Potential health benefits include lowering risk for a number of disorders, including

  _____ (enlarged and inflamed rectal veins), _____ (hard, dry stool), and colon cancer. Circle the two best sources of insoluble fiber in the following list.

  apples   broccoli   oats and oat bran   vegetable skins   wheat bran

  b. Write one health benefit of soluble fiber.

  c. Write three sources of soluble fiber from the list above.

F11. Complete part g of Table 24.1, describing the role of the large intestine in digestion.

G. Aging and developmental anatomy of the digestive system   (pages 797–799)

■ **G1.** *A clinical challenge.* List physical changes with aging that may lead to a decreased desire to eat among the elderly population.

  a. Related to the upper GI tract (to stomach)

  b. Related to the lower GI tract (stomach and beyond)

■ **G2.** Indicate which germ layer, endoderm (E) or mesoderm (M), gives rise to each of these structures.

  _____ a. Epithelial lining and digestive glands of the GI tract

  _____ b. Liver, gallbladder, and pancreas

  _____ c. Muscularis layer and connective tissue of submucosa

G3. List GI structures derived from each portion of the primitive gut.

a. Foregut

b. Midgut

c. Hindgut

H. Disorders, medical terminology (pages 799–802)

H1. Explain how tooth decay occurs. Include the roles of bacteria, dextran, plaque, and acid. List the most effective known measures for preventing dental caries.

H2. Briefly describe these disorders, stating possible causes of each.

a. Periodontal disease

b. Cirrhosis

■ **H3.** Match the terms with the descriptions.

A. Anorexia nervosa	Dys. Dysphagia
B. Bulimia	F. Flatus
Cho. Cholecystitis	Hem. Hemorrhoids
Colit. Colitis	Hep. Hepatitis
Colos. Colostomy	Htb. Heartburn
Con. Constipation	P. Peptic ulcer
Dia. Diarrhea	

_____ a. Incision of the colon, creating artificial anus

_____ b. Inflammation of the liver

_____ c. Inflammation of the colon

_____ d. Burning sensation in region of esophagus and stomach; probably due to gastric contents in lower esophagus

_____ e. Frequent defecation of liquid feces

_____ f. Inflammation of the gallbladder

_____ g. Infrequent or difficult defecation

_____ h. Craterlike lesion in the GI tract due to acidic gastric juices

_____ i. Excess air (gas) in stomach or intestine, usually expelled through anus

_____ j. Binge-purge syndrome

_____ k. Loss of appetite and self-imposed starvation

_____ l. Difficulty in swallowing

ANSWERS TO SELECTED CHECKPOINTS: CHAPTER 24

A2. Accessory structures.

A3. Ingestion, secretion, movement, digestion, absorption, defecation.

A4. Chemical, mechanical.

A5.

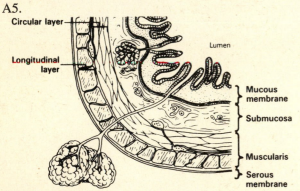

Figure LG 24.2A Gastrointestinal (GI) tract and related gland seen in cross section.

A6. (a) Ser. (b) Muc. (c) Sub. (d) Mus.

A8. (a) F. (b) M. (c) Meso. (d) G. (e) L.

B1.

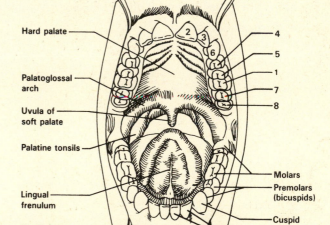

Figure LG 24.3A Mouth (oral cavity).

B4. (a) Parotid. (b) Sublingual. (c) Liters. (d) Dissolving medium for foods, lubrication, waste removal medium, and source of lysozyme, salivary amylase, and salivary lipase. (e) 6.5. (f) Water, Cl^-, HCO_3^-, HPO_4^{2-}, urea, uric acid, lysozyme, salivary amylase. (g) Smell, sight, touch, or memory of food; nausea; swallowing of irritating food; nerve impulses.

B5. (a) C B A. (b) A B C. (c) A B C.

B6. Anchors tooth to bone and acts as a shock absorber.

B7. (a) 20; 32. (c) See Figure LG 24.3A. (d) Third molars or "wisdom teeth."

B8. (a) Mastication; bolus. (b) Chemical; amylase; starch; is not; acidic pH of the stomach. (c) Tongue; triglycerides (or fats).

B9. Swallowing. (a) P. (b) V. (c) E.

B10. (a) 25 (10); stomach; hiatal hernia. (b) Striated, smooth. (c) Achalasia, acidic; heartburn; heart. (d) Vomiting (or emesis); loss of gastric fluids can lead to serious imbalance of electrolytes.

B11. (a) Extrinsic, XII. (b) Parasympathetic. (c) VII, IX; maxillary, V.

C1. (a) Cardia → fundus → body → pylorus. (b) GC; LC; GC. (c) Pylorospasm; vomiting.

C2.

Name of Cell	Type of Secretion	Function of Secretion
a. Chief (zygomatic)	Pepsinogen	Precursor of pepsin
	Gastric lipase	Splits triglycerides
b. Mucous	Mucus	Protects gastric lining from acid and pepsin
c. Parietal (oxyntic)	HCl	Activates pepsinogen to pepsin; inhibits gastrin secretion; stimulates secretion of secretin and CCK; kills microbes in food;
	Intrinsic factor	facilitates absorption of vitamin B_{12}
d. Enteroendocrine (or G) cells	Gastrin	Stimulates secretion of HCl and pepsinogen; contracts lower esophageal sphincter; increases gastric motility and relaxes pyloric and ileocecal sphincters

C3. (a) Arranged in rugae. (b) It has three, rather than two, layers of smooth muscle; the extra one is an oblique layer located inside the circular layer.

C4. (a) Acid. (b) Pepsin is released in the inactive state (pepsinogen); mucus protects the stomach lining from pepsin. (c) Ulcer. (d) Gastric lipase, emulsified fats, as in butter; ineffective; its optimum pH is 5 or 6, and most fats have yet to be emulsified (by bile from liver).

C5. (a) Cephalic, gastric, intestinal; cephalic. (b) Vagus; parasympathetic; sight, smell, or thought of food; presence of food in the stomach; anger, fear, anxiety. (c) Gastric; increase, gastrin; presence of partially digested proteins, alcohol, and caffeine in the stomach; stimulated, relaxed. (d) Stimulate; greater, mast; H_2 blockers decrease HCl production which could otherwise lead to "acid stomach" or peptic ulcer. (e) Enterogastric, inhibits; delay; GIP, secretin, CCK.

C6. Two to four; carbohydrates; fats (triglycerides).

C7. (a) Yes; small intestine. (b) Slowly; fatty foods (such as pizza or nachos). (c) Women usually have a lower level of a gastric enzyme (alcohol dehydrogenase) that converts alcohol to acetaldehyde. So more alcohol remains in chyme that enters (and can be absorbed by) the small intestine.

D2. (a) Stomach. (b) Duodenum, spleen. (c) Exocrine; alkaline. (d) Proteins; enterokinase; chymotrypsin, carboxypeptidase. (e) Amylase digests carbohydrates, including starch; lipase digests lipids. (f) G, H; duodenum. (g) Common bile duct; pancreatic proteases such as trypsin, chymotrypsin, and carboxypeptidase may digest tissue proteins of the pancreas itself. (h) Pancreatic islets (of Langerhans); insulin, glucagon, somatostatin, and pancreatic polypeptide; blood vessels, portal.

D4. (a) Parasympathetic, vagus, small intestine. (b) Contains partially digested fats and proteins; digestive enzymes such as trypsin, amylase, and lipase. (c) Is acidic; HCO_3^-.

D5. (a) 1.4, 3.0; upper right; 4, right. (b) Falciform; umbilical. (c) Hepatic artery, portal vein; sinusoids, hepatic veins.

D6. (a) Superior to inferior. (b) Draw arrows from L and M toward B and then J; superiorly.

D7. (a) Bilirubin, red blood; jaundice; emulsification, absorption. (b) Clotting proteins (prothrombin and fibrinogen), albumin, and globulins. (c) Products of protein digestion, such as ammonia; medications such as sulfa drugs or penicillin; and steroid hormones. (d) Worn-out RBCs, WBCs, and some bacteria. (e) Glycogen, fat. (f) A, D, E, K; B_{12}, D. (g) Does not.

D9. (a) P. (b–c) G. (d) P. (e–g) L.

D10. (a) All ↑. (b) ↓, ↓, ↑, ↓. (c) All ↑ except ↓ gastric emptying. (d) All ↑ except ↓ gastric juice.

E1. (a) A. (b) B. (c) Duodenum → jejunum → ileum.

E2. (a) V. (b) MV. (c) CF. (d) ALF. (e) DG. (f) MALT. (g) BB. (h) L. (i) PC. (j) SKC.

E5. (a) Shorter-chain polysaccharides such as α-dextrins (5–10 glucoses), trisaccharides (such as maltotriose), or disaccharides (such as maltose). (b) Shorter-chain polypeptides → tripeptides, dipeptides, and amino acids. (c) Fatty acids, glycerol.

E6. Lactase, milk and other dairy products; gas (flatus) and bloating because bacteria ferment the undigested lactose present in the GI tract.

E8. (a) Amy, Lip, M. (b) M. (c) Gas, HCl, IF, Lip, M, P. (d) Amy, Glu, Ins, Lip, RD, TCP. (e) CCK, DA, E, GIP, Lys, M, MLSD, NP, S. (f) Alb, B, Chol, Ft, FP, Gly, KB.

E9. (a) CCK, H. (b) S, H. (b) CCK, H. (d) CCK and S. (e) G, H. (f) CCK, GIP, and S. (g) P, TCP. (h) HCl, N. (i) P. (j) E. (k) Alb and FP. (l) Gly. (m) Lip. (n) B, N. (o) Amy. (p) MLSD. (q) NP. (r) MLSD, DA, and NP. (s) DA. (t) Ft. (u) Lys.

E10. (a) Small; epithelial cells lining the intestine, blood or lymph capillaries. (b) Secondary, Na$^+$; facilitated diffusion; facilitated diffusion. (c) Blood, villi; portal, liver. (d) Bile, micelles; thoracic duct; recycled to the liver; enterohepatic. (e) Protein; liver, lipo, HDLs; triglycerides, phospholipids, and cholesterol. (f) 9.3 (9.8); 2, 7; osmosis, small; large. (g) 100–200; diarrhea, electrolytes.

E11. (a) Fat-soluble; A D E K. (b) Examples: A, night blindness and dry skin; D, rickets or osteomalacia due to decreased calcium absorption; K, excessive bleeding.

F1. Diameter; 1.5 (5.0); colon.

F2. C → AsC → HF → TC → SF → DC → SC → R → AnC.

F4. (a–d) LI. (e) Sto. (f) SI. (g–h) LI. (i) SI, LI.

F7. (a) Methane, carbon dioxide, hydrogen; carbohydrates. (b) Indole, skatole; proteins and amino acids. (c) Stercobilin, bilirubin.

F8. 3–10; water, salts, epithelial cells, bacteria and products of their decomposition, as well as undigested foods.

F10. (a) Insoluble; hemorrhoids (or "piles"), constipation; vegetable skins and wheat bran. (b) Lowers blood cholesterol level; apples, broccoli, oats and oat bran.

G1. (a) Decreased taste sensations, gum inflammation (pyorrhea) leading to loss of teeth and loose-fitting dentures, and difficulty swallowing (dysphagia). (b) Decreased muscle tone and neuromuscular feedback may lead to constipation.

G2. (a) E. (b) E. (c) M.

H3. (a) Colos. (b) Hep. (c) Colit. (d) Htb. (e) Dia. (f) Cho. (g) Con. (h) P. (i) F. (j) B. (k) A. (l) Dys.

WRITING ACROSS THE CURRICULUM: CHAPTER 24

1. Contrast the superior and inferior portions of the esophagus, stomach, and small and large intestines with regard to (a) types of epithelium in the mucosa and (b) types of muscle in the muscularis.

2. Place in correct sequence each of these structures in the pathway of food through the GI tract. Then briefly describe each structure: anal canal, ascending colon, cecum, duodenum, esophagus, hepatic flexure, ileocecal sphincter, oropharynx, pyloric sphincter, sigmoid colon.

3. Describe modifications of the small intestinal wall that facilitate digestion and absorption within this organ.

4. Describe nine functions of the liver. Identify which of these are essential to survival.

5. Describe the effects of the following hormones on digestion: GIP, secretin, and CCK.

MASTERY TEST: CHAPTER 24

Questions 1–10: Circle the letter preceding the one best answer to each question.

1. Which of these organs is not part of the GI tract but is an accessory organ?
   - A. Mouth
   - B. Pancreas
   - C. Stomach
   - D. Small intestine
   - E. Esophagus

2. The main function of salivary and pancreatic amylase is to:
   - A. Lubricate foods
   - B. Help absorb fats
   - C. Digest polysaccharides to smaller carbohydrates
   - D. Digest disaccharides to monosaccharides
   - E. Digest polypeptides to amino acids

3. Which enzyme is most effective at pH 1 or 2?
   A. Gastric lipase
   B. Maltase
   C. Pepsin
   D. Salivary amylase
   E. Pancreatic amylase

4. Choose the *true* statement about fats.
   A. They are digested mostly in the stomach.
   B. They are the type of food that stays in the stomach the shortest length of time.
   C. They stimulate release of gastrin.
   D. They are emulsified and absorbed with the help of bile.

5. Which of the following is a hormone, not an enzyme?
   A. Gastric lipase
   B. Gastrin
   C. Pepsin
   D. Trypsin

6. Which of the following is under only nervous (not hormonal) control?
   A. Salivation
   B. Gastric secretion
   C. Intestinal secretion
   D. Pancreatic secretion

7. All of the following are enzymes involved in protein digestion *except:*
   A. Amylase
   B. Trypsin
   C. Carboxypeptidase
   D. Pepsin
   E. Chymotrypsin

8. All of the following chemicals are produced by the walls of the small intestine *except:*
   A. Lactase
   B. Secretin
   C. CCK
   D. Trypsin
   E. Brush border enzymes

9. Which type of movement is used primarily to propel chyme through the intestinal tract, rather than to mix chyme and enzymes?
   A. Peristalsis
   B. Rhythmic segmentation
   C. Haustral churning

10. Choose the *false* statement about layers of the wall of the GI tract.
    A. Most large blood and lymph vessels are located in the submucosa.
    B. The mysenteric plexus is part of the muscularis layer.
    C. Most glandular tissue is located in the layer known as the mucosa.
    D. The mucosa layer forms the peritoneum.

Questions 11–15: Arrange the answers in correct sequence.

_____ _____ _____11. From anterior to posterior:
   A. Palatoglossal arch
   B. Palatopharyngeal arch
   C. Palatine tonsils

_____ _____ _____ _____12. GI tract wall, from deepest to most superficial:
   A. Mucosa
   B. Muscularis
   C. Serosa
   D. Submucosa

_____ _____ _____ _____ _____13. Pathway of chyme:
   A. Ileum
   B. Jejunum
   C. Cecum
   D. Duodenum
   E. Pylorus

_____ _____ _____ _____ _____14. Pathway of bile:
   A. Bile canaliculi
   B. Common bile duct
   C. Common hepatic duct
   D. Right and left hepatic ducts
   E. Hepatopancreatic ampulla and duodenum

_____ _____ _____ _____ _____15. Pathway of wastes:
   A. Ascending colon
   B. Transverse colon
   C. Sigmoid colon
   D. Descending colon
   E. Rectum

Questions 16–20: Circle T (true) or F (false). If the statement is false, change the underlined word or phrase so that the statement is correct.

T F 16. In general, the sympathetic nervous system <u>stimulates</u> salivation and secretions of the gastric and intestinal glands.

T F 17. The principal chemical activity of the stomach is to begin digestion of <u>protein</u>.

T F 18. The esophagus produces <u>no digestive enzymes or mucus</u>.

T F 19. Cirrhosis and hepatitis are diseases of the <u>liver</u>.

T F 20. The greater omentum <u>connects the stomach to the liver</u>.

Questions 21–25: Fill-ins. Complete each sentence with the word or phrase that best fits.

_____ 21. Stomach motility is _____-creased by CCK and GIP, and _____-creased by stomach gastrin and parasympathetic nerves.

_____ 22. Mumps involves inflammation of the _____ salivary glands.

_____ 23. Most absorption takes place in the _____, although some substances, such as _____, are absorbed in the stomach.

_____ 24. Mastication is a term that means _____

_____ 25. Epithelial lining of the GI tract, as well as the liver and pancreas, are derived from _____-derm.

ANSWERS TO MASTERY TEST: ★ CHAPTER 24

Multiple Choice

1. B
2. C
3. C
4. D
5. B
6. A
7. A
8. D
9. A
10. D

Arrange

11. A C B
12. A D B C
13. E D B A C
14. A D C B E
15. A B D C E

True–False

16. F. Inhibits
17. T
18. F. No digestive enzymes, but it does produce mucus
19. T
20. F. Connects the stomach and duodenum to the transverse colon and drapes over the transverse colon and coils of the small intestine.

Fill-ins

21. De, in
22. Parotid
23. Small intestine; alcohol
24. Chewing
25. Endo

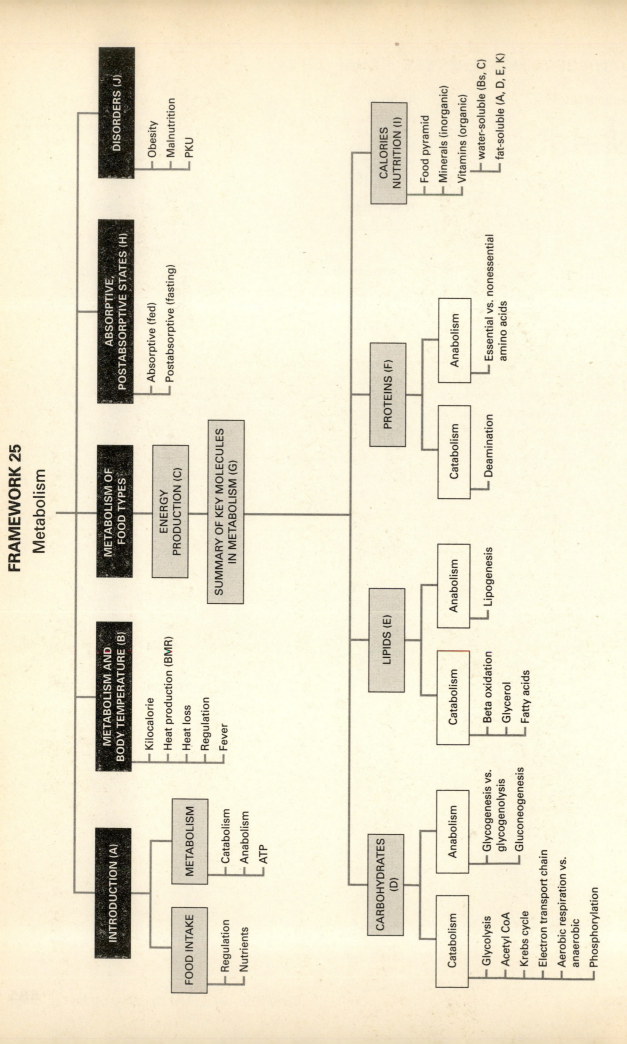

FRAMEWORK 25
Metabolism

INTRODUCTION (A)

FOOD INTAKE
 └ Regulation
 └ Nutrients

METABOLISM
 └ Catabolism
 └ Anabolism
 └ ATP

METABOLISM AND BODY TEMPERATURE (B)
 └ Kilocalorie
 └ Heat production (BMR)
 └ Heat loss
 └ Regulation
 └ Fever

METABOLISM OF FOOD TYPES

ENERGY PRODUCTION (C)

SUMMARY OF KEY MOLECULES IN METABOLISM (G)

ABSORPTIVE, POSTABSORPTIVE STATES (H)
 └ Absorptive (fed)
 └ Postabsorptive (fasting)

DISORDERS (J)
 └ Obesity
 └ Malnutrition
 └ PKU

CARBOHYDRATES (D)

 Catabolism
  └ Glycolysis
  └ Acetyl CoA
  └ Krebs cycle
  └ Electron transport chain
  └ Aerobic respiration vs. anaerobic
  └ Phosphorylation

 Anabolism
  └ Glycogenesis vs. glycogenolysis
  └ Gluconeogenesis

LIPIDS (E)

 Catabolism
  └ Beta oxidation
  └ Glycerol
  └ Fatty acids

 Anabolism
  └ Lipogenesis

PROTEINS (F)

 Catabolism
  └ Deamination

 Anabolism
  └ Essential vs. nonessential amino acids

CALORIES NUTRITION (I)
 └ Food pyramid
 └ Minerals (inorganic)
 └ Vitamins (organic)
   └ water-soluble (Bs, C)
   └ fat-soluble (A, D, E, K)

Metabolism

Ingested foods, once digested, absorbed, and delivered, are used by body cells. This array of biochemical reactions within cells is known as metabolism. Products of digestion may be reassembled, as in formation of human protein from the amino acids of meat or beans, or in synthesis of fats from ingested oils. These reactions are examples of anabolism. The flip side of metabolic currency is catabolism, as in the breakdown of complex carbohydrates stored in liver or muscle to provide simple sugars for quick energy. Catabolism releases energy that maintains body heat and provides ATP to fuel activity. Vitamins and minerals play key roles in metabolism—for example, in enzyme synthesis and function.

The Chapter 25 Framework provides an organizational preview of the metabolism of the major food groups. Study the key terms and concepts it presents. As you begin this chapter, carefully examine the Chapter 25 Topic Outline and check off each objective as you complete it.

TOPIC OUTLINE AND OBJECTIVES

A. Introduction to metabolism

☐ 1. Explain how food intake is regulated.
☐ 2. Define a nutrient and list the functions of the six principal classes of nutrients.
☐ 3. Define metabolism and explain the role of ATP in anabolism and catabolism.

B. Metabolism and body temperature

☐ 4. Define basal metabolic rate (BMR) and explain several factors that affect it.
☐ 5. Explain how normal body temperature is maintained by the hypothalamic thermostat, heat production and conservation, and heat loss.
☐ 6. Describe fever and hypothermia as abnormalities of temperature regulation.

C. Energy production

☐ 7. Describe oxidation-reduction reactions and explain the role of ATP in metabolism.

D. Carbohydrate metabolism

E. Lipid metabolism

F. Protein metabolism

☐ 8. Describe the metabolism of carbohydrates, lipids, and proteins.

G. Summary of key molecules in metabolism

H. Absorptive and postabsorptive states

☐ 9. Compare the metabolic reactions occurring during the absorptive (fed) and postabsorptive (fasting) states and explain their hormonal regulation.

I. Calories and nutrition

☐ 10. Describe how to select foods to include in a healthy diet.
☐ 11. Compare the sources, functions, and importance of minerals and vitamins in metabolism.

J. Disorders: homeostatic imbalances

☐ 12. Define heat cramps, heat exhaustion, heatstroke, obesity, vitamin or mineral overdose, malnutrition, phenylketonuria (PKU), and celiac disease.

WORDBYTES

Now become familiar with the language of this chapter by studying each wordbyte, its meaning, and an example of its use within a term. After you study the entire list, self-check your understanding by writing the meaning of each wordbyte on the line. As you continue through the *Learning Guide,* identify (and fill in) additional terms that contain the same wordbyte.

Wordbyte	Self-check	Meaning	Example(s)
ana-	_____	up	*ana*bolism
calor-	_____	heat	*calor*ie
cata-	_____	down	*cata*bolism
de-	_____	remove, from	*de*amination
-gen	_____	to form	gluconeo*genesis*
-lysis	_____	destruction	hydro*lysis*
neo-	_____	new	gluco*neo*genesis

CHECKPOINTS

A. Introduction to metabolism (pages 807–809)

A1. List three primary functions of nutrients.

■ A2. List the six principal classes of nutrients.

■ A3. Complete this Checkpoint about regulation of food intake.

a. The site within the brain that is the location of feeding and satiety centers is the *(medulla? hypothalamus?).* When the satiety center is active, an individual will feel *(hungry? satiated or full?).*

b. The feeding center is constantly active except when it is inhibited by the

_____ center. Circle all of the answers that will keep the feeding center active and enhance the desire to eat.

↓ Glucose in blood

↓ Amino acids in blood

↓ Fats entering intestine (with related ↓ in CCK release)

↓ Environmental temperature

↓ Stretching of stomach

c. Name the "satiety hormone." _____ Explain the reason for this name.

A4. Explain why metabolism might be thought of as an "energy-balancing act."

■ **A5.** Complete this table comparing catabolism with anabolism.

Process	Definition	Releases or Uses Energy	Examples
a. Catabolism			
b. Anabolism			

A6. ATP stands for a_____ t_____

p_____. Explain how it functions as the "energy currency" (or "money machine cash") of the cell. (For help, refer to Checkpoint D16, page 37 of the *Learning Guide*.)

■ **A7.** About *(1–2%? 10–20%? 40%? 65–70%?)* of the energy released in catabolism is available for cellular activities. The rest of the energy is converted to

_____.

B. Metabolism and body temperature   (pages 809–813)

B1. Define basal metabolic rate (BMR).

■ **B2.** Contrast *calorie* and *kilocalorie.*

If you ingest, digest, absorb, and metabolize a slice of bread, the energy released from the bread equals about 80 to 100 *(cal? kcal?).*

■ **B3.** Answer these questions about catabolism.

Catabolism of foods *(uses? releases?)* energy. Most of the heat produced by the body comes from catabolism by *(oxidation? reduction?)* of nutrients.

■ **B4.** Discuss BMR in this exercise.

a. BMR may be determined by measuring the amount of _____ consumed within a period of time, because oxygen is necessary for the metabolism of foods. Normally, for each liter of oxygen consumed, the body metabolizes enough food to

release about _____ kcal. Oxygen consumption is normally measured on a

_____.

b. Suppose you consume 15 liters of oxygen in an hour. Your body would release

_____ kcal of heat in that hour.

c. A BMR of 20% below the standard value is likely to be due to *(hyper? hypo?)-*thyroidism.

■ **B5.** Do this activity on regulation of body temperature.

a. Which temperature is normally higher? *(Core? Shell?)* Explain how an increase in core temperature, for example, above 112–114°F can be fatal.

b. The body's "thermostat" is located in the preoptic area of the *(eyes? hypothalamus? medulla? skin?).* If the body temperature needs to be raised, nerve impulses from the preoptic area are sent to the heat-*(losing? promoting?)* center, which is primarily *(parasympathetic? sympathetic?).* An example of such a response is vaso-*(constriction? dilation?)* of blood vessels in skin, resulting in heat *(loss or release? conservation?).*

c. Circle the factor in each pair that is likely to lead to increased metabolic rate with increased body temperature.
   1. Age: 12 years old   52 years old
   2. Body temperature: 98.6°F   103.6°F
   3. State of activity: Running 2 miles   Typing at computer
   4. Adrenal medulla or thyroid hormones:
        Increased epinephrine or thyroid hormone
        Decreased epinephrine or thyroid hormone
   5. SDA related to ingestion of: Protein   Carbohydrate

B6. Lil is outside on a snowy day without a warm coat. Her skin is pale and chilled and she begins to shiver. Explain how these responses are attempts of her body to maintain homeostasis of temperature. What other responses might raise her body temperature? Include these words in your answers: *sympathetic, metabolism, adrenal medulla, thermogenesis, thyroid,* and *coat.*

B7. Write the percentage of heat loss by each of the following routes (at room temperature). Then write an example of each of these types of heat loss. One is done for you.

a. Radiation _____ _____

b. Evaporation _____ _____

c. Convection __15__   **Cooling by draft while taking a shower**_____

d. Conduction _____ _____

B8. During an hour of active exercise Bill produces 1 liter of sweat. The amount of heat

loss (cooling) that accompanies this much evaporation is _____ Cal. On a humid day,
(more? less?) evaporation of Bill's sweat occurs, so Bill would be cooled *(more? less?)*
on such a day.

B9. Arrange in order the events believed to occur in the production of fever of 39.4°C

(103°F) and recovery from this state. _____ _____ _____ _____ _____

_____

A. 39.4°C (103°F) temperature is reached and maintained.
B. Prostaglandins (PGs) cause the hypothalamic "thermostat" to be reset from normal
37.5°C (98.6°F) to a higher temperature such as 39.4°C (103°F).
C. A "chill" occurs as skin feels cool (due to vasoconstriction) and shivering occurs—
attempts to raise body temperature to the 39.4°C (103°F) designated by the hypothalamus.
D. Source of pyrogens is removed (for example, bacteria are killed by antibiotics), lowering hypothalamic "thermostat" to 37.5°C (98.6°F).
E. "Crisis" occurs. Obeying hypothalamic orders, the body shifts to heat loss mechanisms
such as sweating and vasodilation, returning body temperature to normal.
F. Infectious organisms cause phagocytes to release interleukin-1, which then causes
hypothalamic release of PGs.

B10. In what ways is a fever beneficial?

B11. Explain why elderly persons are at greater risk for hypothermia.

Write seven or more signs or symptoms of hypothermia.

C. Energy production   (pages 813–814)

C1. Complete this overview of catabolism (energy production).

a. Organic nutrients such as glucose are rich in hydrogen; energy is contained within the
C — H bonds. In catabolism, much of the energy in glucose is ultimately released

and stored in the high-energy molecule named _____. Write the
chemical formulas to show this overall catabolic conversion:

_____ $\rightarrow$ _____ + _____ + energy stored in ATP
   Glucose   *carbon dioxide*   *water*

Note that the inorganic compound carbon dioxide is energy-*(rich? poor?)* because it
lacks C — H bonds.

b. The reaction shown is an oversimplification of the entire catabolic process, which actually entails many steps. Each pair of hydrogen atoms (2H) removed from glucose consists of two components: one hydrogen ion or proton lacking orbiting electrons, expressed as *(H+? H-?)*, and a hydrogen nucleus with two orbiting electrons, known as a *(hydride? hydroxyl?)* ion, and expressed as *(H+? H-?)*.

c. Hydrogens removed from glucose must first be trapped by H-carrying coenzymes, such

as _____, _____, or _____; all of these coenzymes are derivatives of a vita-

min named _____. Ultimately the Hs combine with oxygen to form

_____. In other words, glucose is dehydrogenated (loses H) while oxygen is reduced (gains H).

d. Catabolism involves a complex series of stepwise reactions in which hydrogens are transferred. Because oxygen is the final molecule that serves as an oxidizer, catabolism is often described as biological *(reduction? oxidation?)*.

e. Along the way, energy from the original C — H bonds of glucose (and present in

the electrons of H) is trapped in _____. This process is known as

_____, because it involves the addition of a phosphate to ADP.
A more complete expression of the total catabolism of glucose (shown briefly in (a) above) is included below. Write the correct labels under each chemical: *oxidized, oxidizer, reduced, reducer:*

$$\text{Glucose} \quad + \quad O_2 \quad + \text{ADP} \longrightarrow \quad CO_2 \quad + \quad H_2O \quad + \quad \text{ATP}$$
(_____) (_____)   (_____) (_____)

■ **C2.** Check your understanding of oxidation–reduction reactions and generation of ATP involved in metabolism by circling the correct answer in each case.

a. Which molecule is more reduced and therefore contains more chemical potential energy (can release more energy when catabolized)?
   A. Lactic acid $(C_3H_6O_3)$
   B. Pyruvic acid $(C_3H_4O_3)$

b. Which is more oxidized and therefore now contains less chemical potential energy since it has given up Hs to some coenzyme?
   A. Lactic acid $(C_3H_6O_3)$
   B. Pyruvic acid $(C_3H_4O_3)$

c. Which is more likely to happen when a molecule within the human body is oxidized?
   A. The molecule gives up Hs
   B. The molecule gains Hs

d. Which coenzyme is more reduced, indicating that it has received Hs when some other molecule was oxidized?
   A. NAD+
   B. NADH + H+

e. Which molecule contains more chemical potential energy (can release more energy when catabolized)?
   A. ATP
   B. ADP

f. Which is an example of oxidative phosphorylation?
   A. Creatine-P + ADP → creatine + ATP
   B. Generation of ATP by energy released by the electron transfer chain

g. Which is the substrate in the following example of substrate-level phosphorylation?
   Creatine-P + ADP → creatine + ATP
   A. ATP
   B. Creatine

D. Carbohydrate metabolism   (pages 814–826)

■ **D1.**   Answer these questions about carbohydrate metabolism.

a. The story of carbohydrate metabolism is really the story of _____ metabolism because this is the most common carbohydrate (and in fact the most common energy source) in the human diet.

b. What other carbohydrates besides glucose are commonly ingested? How are these converted to glucose?

c. Just after a meal, the level of glucose in the blood *(increases? decreases?)*. Cells use

some of this glucose; by _____ glucose, they release energy.

d. List three or more mechanisms by which excess glucose is used or discarded.

e. Increased glucose level of blood is known as _____. This can occur after a meal containing concentrated carbohydrate or in the absence of the hormone

_____ because this hormone *(facilitates? inhibits?)* entrance of glucose

into cells. Lack of insulin occurs in the condition known as _____.

f. Name two types of cells that do not require insulin for glucose uptake so that glucose entry is always possible (provided blood glucose level is adequate).

g. As soon as glucose enters cells, glucose combines with _____ to

form glucose-6-phosphate. This process is known as _____ and is

catalyzed by enzymes called _____. What advantage does this process provide?

■ **D2.**   Describe the process of glycolysis in this exercise.

a. Glucose is a _____-carbon molecule (which also has hydrogens and oxygens in it).

During glycolysis each glucose molecule is converted to two _____-carbon mole-

cules named _____. *(A lot? A little?)* energy is released from glucose during glycolysis. This process requires *(one? many?)* step(s) and *(does? does not?)* require oxygen.

b. The initial steps of glycolysis also involve _____-ation of glucose, along with conversion of this carbohydrate to another six-carbon sugar,

_____. These preliminary steps in glycolysis are catalyzed by the

enzyme _____, called the key regulator of glycolysis. This enzyme is most active when the levels of *(ADP? ATP?)* are high and *(ADP? ATP?)* are low (indicating the need for generation of ATP by glycolysis). When ATP levels are already high, then cellular glucose is instead used for *(ana? cata?)*-bolism of *(glycogen? CO_2 + H_2O?)*.

c. The phosphorylated fructose is broken down into two *(2? 3? 6?)*-carbon molecules

named glucose 3-phosphate (_____) and dihydroxyacetone phosphate. The G 3–P molecules are then *(oxidized? reduced?)*, giving up their hydrogens to *(FAD? NAD?)*.

d. The end result of glycolysis is formation of two _____ acids and a net gain of *(2? 3? 4?)* ATP molecules.

e. The fate of pyruvic acid depends on whether sufficient _____ is available. If it is, pyruvic acid undergoes chemical change in phases 2 and 3 of glucose catabolism (see below). Those processes *(do? do not?)* require oxygen; that is, they are *(aerobic? anaerobic?)*.

f. If the respiratory rate cannot keep pace with glycolysis, insufficient oxygen is available to break down pyruvic acid. In the absence of sufficient oxygen, pyruvic acid is tem-

porarily converted to _____. This is likely to occur during active

_____.

g. Several mechanisms prevent the accumulation of an amount of lactic acid that might be

harmful. The liver changes some back to _____. Excess pCO_2 caused by exercise *(stimulates? inhibits?)* respiratory rate, so more oxygen is available for breakdown of the pyruvic acid.

■ **D3.** Review glucose catabolism up to this point and summarize the remaining phases by doing this exercise. Refer to Figure LG 25.1. A six-carbon compound named

(a) _____ is converted to two molecules of (b) _____

via a number of steps, together called (c) _____, occurring in the

(d) _____ of the cell.

In order for pyruvic acid to be further broken down, it must undergo a

(e) _____ step. This involves removal of the one-carbon molecule,

(f) _____. The remaining two-carbon (g) _____

group is attached to a carrier called (h) _____, forming

(i) _____.

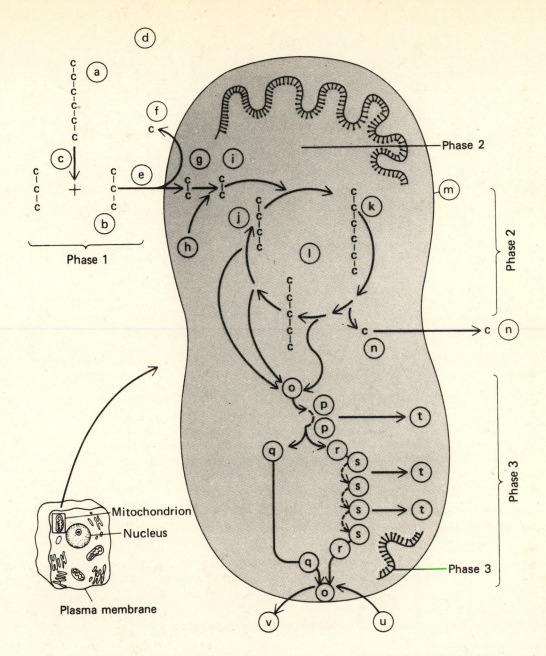

Figure LG 25.1 Summary of glucose catabolism, described in three phases. Phase 1 (glycolysis) occurs in the cytosol. Phase 2 (Krebs cycle) takes place in the matrix of mitochondria. Phase 3 (electron transport) takes place on inner mitochondrial membrane. Letters refer to Checkpoint D3.

This compound hooks on to a four-carbon compound called (j) _____.

The result is a six-carbon molecule of (k) _____. The name of this

compound is given to the cycle of reactions called the (l) _____ cycle,

occurring in the (m) _____.

Two main types of reactions occur in the Krebs cycle. One is the decarboxylation reaction,

in which (n) _____ molecules are removed, and hydrogens removed at these steps combine with coenzymes FAD or NAD. (Notice these locations on Figure 25.6, page 819 of your text.) (What happens to the CO_2 molecules that are removed?) As a result, the six-carbon citric acid is eventually shortened to regenerate oxaloacetic acid, giving the cyclic nature to this process.

The other major type of reaction (oxidation–reduction) involves removal of

(o) _____ atoms during oxidation of compounds such as isocitric and α-ketoglutaric acids. (See Figure 25.6, page 819 in your text.) These hydrogens are carried

off by two coenzymes named (p) _____ and "trapped" for use in phase 3.

In phase 3, hydrogen atoms on coenzymes NAD and FAD (and later coenzyme Q) are

ionized to (q) _____ and (r) _____. Electrons are shuttled along a chain of

(s) _____. (Refer to Checkpoint D4.) During electron transport, energy in these electrons (derived from hydrogen atoms in ingested foods) is "tapped" and

ultimately stored in (t) _____.

Finally, the electrons, depleted of some of their energy, are reunited with hydrogen ions

and with (u) _____ to form (v) _____. Notice why your body requires oxygen—to keep drawing hydrogen atoms off of nutrients (in biological oxidation) so that energy from them can be stored in ATP for use in cellular activities.

D4. Now describe details of the electron transport chain in this activity.

a. The first step in this chain is transfer of high-energy electrons from NADH + H⁺ to the coenzyme *(FMN? Q?)*. As a result this carrier is *(oxidized? reduced?)* to

_____. From which vitamin is this carrier derived? _____

b. As electrons are passed on down the chain, $FMNH_2$ is *(oxidized? reduced?)* back to

_____.

c. Which mineral is the portion of cytochromes on which electrons are ferried?

_____ Cytochromes are positioned toward the *(beginning? end?)* of the transport chain.

d. Circle two other minerals that are known to be involved in the electron transport chain and are hence necessary in the diet.
   Calcium   Copper   Iodine   Sulfur   Zinc

e. The last carrier in the chain is *(coenzyme Q? cytochrome a_3?)*. It passes e^- to

_____, the final oxidizing molecule. This is a key step in the process of *(aerobic? anaerobic?)* cellular respiration.

f. Where are the electron carriers located? Circle the correct answer.
   MM. Mitochondrial matrix
   IMM. Inner mitochondrial membrane (cristae)
   OMM. Outer mitochondrial membrane

g. The carriers are grouped into *(two? three? four?)* complexes. Each complex acts to pump *(protons = H⁺? electrons = e⁻?)* from the *(matrix? space between inner*

and outer mitochondrial membranes?) into the _____.

h. As a result, a gradient of H⁺ is set up across the *(inner? outer?)* mitochondrial membrane, resulting in potential energy known as the _____ force. How do these protons (H⁺) travel back into the mitochondrial matrix?

i. The channels through which H⁺ diffuses contain the enzyme known as ATP

_____. Because the formation of the resulting ATP is driven by an energy source that consists of a gradient and subsequent diffusion of a *chemical*

(specifically _____), this mechanism is known as _____ generation of ATP.

■ **D5.** Summarize aerobic respiration in this activity.

a. Complete the chemical equation showing the overall reaction for aerobic respiration.

$$\underline{\hspace{1cm}} + 6 \underline{\hspace{1cm}} + 36\text{--}38 \text{ ADPs} + 36\text{--}38 \text{ } \textcircled{P} \longrightarrow 6 \underline{\hspace{1cm}} + 6 \underline{\hspace{1cm}} + \underline{\hspace{1cm}} \text{ ATPs}$$

Glucose   Oxygen   Carbon   Water

                                               dioxide

b. Fill in the correct number of high-energy molecules (ATPs or GTPs) yielded from processes in aerobic respiration.

1. Substrate-level phosphorylation via oxidation of glucose to pyruvic acid in

   glycolysis: _____ ATPs.

2. Oxidative phosphorylation in electron transport chain (ETS) of six reduced NADs (NADH + 2H⁺) resulting from glycolysis that takes place under aerobic condi-

   tions: _____ ATPs.

3. Oxidative phosphorylation in electron transport chain (ETS) of two reduced NADs

   (NADH + 2H⁺) yielded from transition step (formation of acetyl CoA): _____ ATPs.

4. Oxidative phosphorylation in electron transport chain (ETS) of six reduced NADs

   (NADH + 2H⁺) yielded from the Krebs cycle: _____ ATPs.

5. Oxidative phosphorylation in electron transport chain (ETS) of two reduced FADs

   yielded from the Krebs cycle: _____ ATPs. (Notice that each FADH₂ yields fewer ATPs than NADH + 2H⁺ because FADH₂ enters the ETS at a lower level.)

6. Substrate-level phosphorylation via oxidation of succinyl CoA to succinic acid in the

   Krebs cycle: _____ GTPs.

7. Grand total of ATPs or GTPs resulting from complete catabolism of glucose by glycolysis, Krebs cycle, electron transport, and resulting phosphorylation of ADP to

   ATP (or GDP to GTP): _____ ATPs and/or GTPs

■ **D6.** Most of the 36 to 38 ATPs generated from the total oxidation of hydrogens derive from (*glycolysis? the Krebs cycle?*). Of those generated by the Krebs cycle, most are formed from the oxidation of hydrogens carried by (*NAD? FAD?*). About what percentage of the energy originally in glucose is stored in ATP after glucose is completely catabolized? (*10%? 40%? 60%? 99%?*)

■ **D7.** Complete the exercise about the reaction shown below.

(1)

Glucose ⇄ Glycogen

(2)

a. You learned earlier that excess glucose may be stored as glycogen. In other words,

glycogen consists of large branching chains of _____. Name the process of glycogen formation by labeling (1) in the chemical reaction.

b. Between meals, when glucose is needed, glycogen can be broken down again to release glucose. Label (2) above with the name of this process. [Note that a number of steps are actually involved in both processes (1) and (2).]

c. Which of these two reactions is anabolic? _____

d. Where is most (75%) of glycogen in the body stored? _____

e. Identify a hormone that stimulates reaction (1). Write its name on the upper arrow in the reaction above. As more glucose is stored in the form of glycogen, the blood level of

glucose _____-creases. Now write below the lower arrow the names of two hormones that stimulate glycogenolysis.

f. In order for glycogenolysis to occur, a _____ group must be added to glucose as it breaks away from glycogen. The enzyme catalyzing this reaction is known

as _____. Does the reverse reaction [shown in (2)] require phosphorylation also? (*Yes? No?*)

D8. Define *gluconeogenesis* and briefly discuss how it is related to other metabolic reactions.

Name three hormones that stimulate gluconeogenesis.

■ **D9.** Circle the processes at left and the hormones at right that lead to increased blood glucose level (hyperglycemia).

Processes	Hormones
Glycogenesis	Insulin
Glycogenolysis	Glucagon
Glycolysis	Epinephrine
Gluconeogenesis	Cortisol
	Thyroid hormone
	Growth hormone

For extra review, refer to Figures LG 18.3 and LG 18.6 pages 373 and 381 of the *Learning Guide.*

E. Lipid metabolism   (pages 826–828)

E1. List six examples of structural or functional roles of fats in the body.

_____   _____   _____

_____   _____   _____

■ **E2.** About 98% of all of the energy reserves stored in the body are in the form of *(fat? glycogen?)*. About *(10%? 25%? 50%?)* of the stored triglycerides (fats) in the body are located in subcutaneous tissues. Is the fat stored in these areas likely to be "the same fat" that was located there two years ago? *(Yes? No?)* Explain.

■ **E3.** Do this exercise about fat metabolism.

a. The initial step in catabolism of triglycerides is their breakdown into

_____ and _____, a process called _____-lysis

and catalyzed by enzymes called _____-ases.

b. Glycerol can then be converted into _____ (G 3-P); this molecule can then enter glycolytic pathways to increase ATP production. If the cell does not need

to generate ATP, then G 3-P can be converted into _____. This

anabolic step is an example of gluco-_____.

c. Recall that fatty acids are long chains of carbons with attached hydrogens and a few oxygens. Two-carbon pieces are "snipped off" of fatty acids by a process called

_____, occurring in the _____.

d. These two-carbon pieces (_____) attach to coenzyme _____.

The resulting molecules, named _____, may enter the Krebs cycle. Complete catabolism of one 16-carbon fatty acid can yield a net of *(36–38? 129?)* ATPs.

e. When large numbers of acetyl CoAs form, they tend to pair chemically:

Acetic acid   +   acetic acid   →   _____
   (2C)   (2C)   (4C)

The presence of excessive amounts of these acids in blood *(raises? lowers?)* blood pH. Because

these acids are called "keto" acids, this condition is known as _____.

f. Slight alterations of acetoacetic acid leads to formation of β-hydroxybutyric acid and

_____. Collectively, these three chemicals are known as

_____ bodies. Formation of ketone bodies is known as

_____; it takes place in the _____.

g. Circle the cells that can use acetoacetic acid for generation of ATP:

Brain cells   Hepatic cells
Cardiac muscle cells   Cells of the renal cortex

h. Explain why the blood level of acetoacetic acid is normally maintained at a low level.

i. Ketogenesis occurs when cells are forced to turn to fat catabolism. State two or more reasons why cells might carry out excessive fat catabolism leading to ketogenesis and ketosis.

j. *A clinical challenge.* Explain why a person whose diabetes is out of control might have sweet breath.

F. Protein metabolism (pages 828–830)

F1. Which of the three major nutrients (carbohydrates, lipids, and proteins) fulfills each function?

a. Most direct source of energy; stored in body least: _____

b. Constitutes almost all of energy reserves, but also used for body-building: _____

c. Usually used least for fuel; used most for body structure and for regulation: _____

F2. Contrast the caloric values of the three major food types.

a. Carbohydrates and proteins each produce _____ Cal/g or (because a pound contains

454 g) _____ Cal/lb.

b. Fats produce _____ Cal/g or _____ Cal/lb.

F3. Throughout your study of systems of the body, you have learned about a variety of roles of proteins. List at least six functions of proteins in the body, using one or two words for each function. Include some structural and some regulatory roles.

_____ _____ _____

_____ _____ _____

F4. Name two hormones that enhance transport of amino acids into cells and also stimulate protein synthesis.

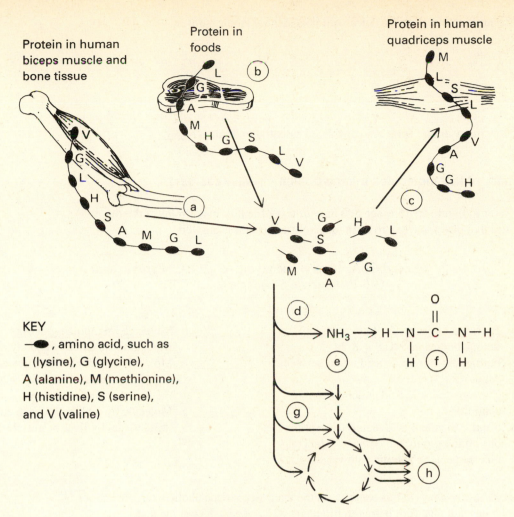

Protein in human
biceps muscle and
bone tissue

Protein in
foods

Protein in human
quadriceps muscle

KEY

—●— , amino acid, such as
L (lysine), G (glycine),
A (alanine), M (methionine),
H (histidine), S (serine),
and V (valine)

$$NH_3 \rightarrow H-N-C-N-H$$

Figure LG 25.2 Metabolism of protein. Lowercase letters refer to Checkpoint F5.

■ **F5.** Refer to Figure LG 25.2 and describe the uses of protein.

Two sources of proteins are shown: (a) _____ and

(b) _____. Amino acids from these sources may undergo a process

known as (c) _____ to form new body proteins. If other energy
sources are used up, amino acids may undergo catabolism. The first step is

(d) _____, in which an amino group (NH_2) is removed and converted

(in the liver) to (e) _____, a component of (f) _____,
which exits in urine. The remaining portion of the amino acid may enter

(g) _____ pathways at a number of points (Figure 25.14 page 829 in

your text). In this way, proteins can lead to formation of (h) _____.

F6. Contrast essential amino acids with nonessential amino acids.

How many amino acids are considered essential to humans? _____

G. Summary of key molecules in metabolism (pages 830–831)

■ **G1.** Refer to Figure 25.15 (page 830 in your text). Identify major roles of the three molecules that play most key roles in metabolism. Use these answers.

> Acetyl CoA. Acetyl coenzyme A   PA. Pyruvic acid
> G 6-P. Glucose 6-phosphate

_____ a. Entrance molecule for fatty acids and ketone bodies into Krebs cycle

_____ b. Formed from pyruvic acid under aerobic conditions

_____ c. Converts to lactic acid under anaerobic conditions

_____ d. Similar to amino acid alanine except for one amino group; molecule by which alanine can enter catabolic pathways

_____ e. Molecule into which glucose is converted when glucose enters any body cell

_____ f. Molecule by which blood glucose is converted to oxaloacetic acid (OAA) to enter the Krebs cycle

_____ g. Molecule by which glycogen can be utilized as fuel by liver or muscle cells

■ **G2.** *For extra review.* Do this exercise on the main steps in metabolic interconversions of carbohydrates and fats. (For help, refer to pages 827 and 830–832 of the text.)

a. Excessive carbohydrate calories may be converted into body fat. Show the pathway by filling in blanks. Use the same answers as for Checkpoint G1.

Glycogen $\rightarrow$ _____ $\rightarrow$ _____ $\rightarrow$

_____ $\rightarrow$ Fatty acid portion of fats

b. Mammals, including humans, *(can? cannot?)* convert acetyl CoA into pyruvic acid. As a result, fatty acids *(can? cannot?)* be used to form glucose. Gluconeogenesis involving fats can occur, however, by the following pathway:

The _____ portion of triglycerides $\rightarrow$ glyceraldehyde

3-phosphate (G 3-P) $\rightarrow$ _____.

H. Absorptive and postabsorptive states (pages 832–835)

■ **H1.** Contrast these two metabolic states by writing *A* if the description refers to the absorptive state and *P* if it refers to the postabsorptive state.

_____ a. Period when the body is fasting and is challenged to maintain an adequate blood glucose level

_____ b. State during which glucose (stored in glycogen in liver and muscle) is released

_____ c. State during which most systems (excluding nervous system) switch over to use of fatty acids as energy sources

_____ d. Time when the body is absorbing nutrients from the GI tract

_____ e. State during which the principal concern is formation of stores of glucose (as glycogen) and fat

_____ f. State when insulin is released under stimulation from gastric inhibitory peptide (GIP); insulin then facilitates transport of glucose and amino acids into cells

_____ g. Period dominated by anti-insulin hormones, such as glucagon

■ **H2.** Do this exercise about maintenance of glucose level between meals.

   a. The normal blood glucose level in the postabsorptive state is about _(30–50? 70–110? 150–170? 600–700?)_ mg glucose/100 ml blood.

   b. Now list four sources of glucose that may be called upon during the postabsorptive state.

   _____

   _____

   _____

   _____

   c. Can fatty acids serve as a source of glucose? _(Yes? No?)_ Explain.

   d. Fats can be used to generate ATP by being broken down into _____, which can then enter the Krebs cycle. Therefore, fatty acids are said to provide a glucose- _(sparing? utilizing?)_ mechanism between meals.

   H3. _For extra review._ Complete the table summarizing how metabolism is hormonally regulated. (Refer to Exhibits 25-3 and 254, pages 833 and 835 of your text.)

Hormone	Source	Action
a.		Stimulates glycogenesis in liver; stimulates glucose uptake in other cells
b.	Alpha cells of islets of Langerhans in pancreas	
c. Human growth hormone (hGH)		
d. Glucocorticoids		
e.	Thyroid gland	
f.	Adrenal medulla	

■ **H4.** Circle the hormones in Checkpoint H3 above that are hyperglycemic, that is, tend to *increase* blood glucose.

I. Calories and nutrition   (pages 835–841)

■ **I1.** Fill in the number of calories needed each day by typical persons in each category.

_____ a. Active women, teen girls, and most men   _____ c. Elderly adults

_____ b. Active men and teen boys   _____ d. Children

■ **I2.** Fill in the percentage of calories needed each day in each category of nutrients.

   a. Carbohydrates: _____%

   b. Fats: no more than _____% with no more than _____% of total calories in the form of saturated fats

   c. Proteins: _____%

■ **I3.** Keep in mind that carbohydrates and proteins each provide about 4 Cal/g and fats provide about 9 Cal/g. Considering the information in Checkpoint I2, determine the number of grams of each of the three major nutrient groups appropriate for a healthy diet of 2000 Cal/day. One is done for you.

   a. Carbohydrates

   b. Fats

   c. Proteins   **2000 Cal × (0.12–0.15) = 240–300 Cal/day. At 4 Cal/g = *60–75 g/day*.**

■ **I4.** Using the Food Pyramid as a guide, identify the number of servings suggested per day for each food group listed.

_____ a. Bread, cereal, rice, pasta   _____ d. Meat, fish, poultry, dry beans, eggs, nuts

_____ b. Vegetables   _____ e. Milk, yogurt, and cheese

_____ c. Fruits   _____ f. Fats, oils, sweets

■ **I5.** Define *minerals*.

   Minerals make up about _____% of body weight and are concentrated in the _____.

■ **I6.** Study Exhibit 25.5 (pages 837–838 in your text). Then check your understanding of minerals by doing this matching exercise.

Ca. Calcium	Fe. Iron	Na. Sodium
Cl. Chlorine	I. Iodine	P. Phosphorus
Co. Cobalt	K. Potassium	S. Sulfur
F. Fluorine	Mg. Magnesium	

_____ a. Main anion in extracellular fluid, part of HCl in stomach; component of table salt

_____ b. Involved in generation of nerve impulse, helps to regulate osmosis, acts in buffer systems

_____ c. Most abundant mineral in the body, found mostly in bones and teeth; necessary for normal muscle contraction and for blood clotting

_____ d. Important component of hemoglobin and cytochromes

_____ e. Main cation inside of cells; used in nerve transmission

_____ f. Essential component of thyroxin

_____ g. Constituent of vitamin B_{12}, so necessary for red blood cell formation

_____ h. Important component of amino acids, vitamins, and hormones

_____ i. Improves tooth structure

_____ j. Found mostly in bones and teeth; important in buffer system and in ATP processes; component of DNA and RNA

■ **I7.** Circle the correct answer. The primary purpose of vitamins is:

A. Synthesis of body structures
B. Regulation of body activities

■ **I8.** Vitamins are *(organic? inorganic?)*. Most vitamins *(can? cannot?)* be synthesized in the body. In general, what are the functions of vitamins?

■ **I9.** Contrast the two principal groups of vitamins, and list the main vitamins in each group.

a. Fat-soluble _____

b. Water-soluble _____

I10. Defend or dispute this statement: "Most persons who eat a balanced diet do need to take vitamin or mineral supplements."

■ **I11.** Select the vitamin that fits each description.

A   B₁   B₂   B₁₂   C   D   E   K

_____ a. This serves as a coenzyme that is essential for blood clotting, so it is called the antihemorrhagic vitamin; synthesized by intestinal bacteria.

_____ b. Its formation depends upon sunlight on skin and also on kidney and liver activation; necessary for calcium absorption.

_____ c. Riboflavin is another name for it; a component of FAD; necessary for normal integrity of skin, mucosa, and eye.

_____ d. This vitamin acts as an important coenzyme in carbohydrate metabolism; deficiency leads to beriberi.

_____ e. Formed from carotene, it is necessary for normal bones and teeth; it prevents night blindness.

_____ f. This substance is also called ascorbic acid; deficiency causes anemia, poor wound healing, and scurvy.

_____ g. This coenzyme, the only B vitamin not found in vegetables, is necessary for normal erythropoiesis; absorption from GI tract depends on intrinsic factor.

_____ h. Also known as tocopherol, it is necessary for normal red blood cell membranes; deficiency is associated with sterility in some animals.

J. Disorders: homeostatic imbalances (pages 842–843)

J1. Contrast these temperature disorders: *heatstroke/heat exhaustion*

J2. Contrast the terms in the following pairs.

a. *Obesity/morbid obesity*

b. *Toxicity of water-soluble vitamins/toxicity of fat-soluble vitamins*

■ **J3.** *A clinical challenge.* Which condition involves deficiency in both protein and calories? *(Kwashiorkor? Marasmus?)* Explain why an individual with kwashiorkor may have a large abdomen even though the person's diet is inadequate.

J4. Individuals with PKU are unable to convert _____ to

_____. What are results of toxic levels of phenylalanine?

J5. Persons with celiac disease require a diet that is lacking in most *(meats? vegetables? grains?)*. The problem stems from a water-insoluble protein named

_____, which causes destruction of the intestinal lining.
Name two grains that are acceptable in the diet of a patient with celiac disease.

_____ , _____

ANSWERS TO SELECTED CHECKPOINTS: CHAPTER 25

A2. Carbohydrates, proteins, lipids, minerals, vitamins, and water.

A3. (a) Hypothalamus; satiated or full. (b) Satiety; all answers. (c) CCK; this hormone, secreted when triglycerides enter the small intestine, causes a feeling of fullness (inhibits eating).

Process	Definition	Releases or Uses Energy	Examples
a. Catabolism	Breakdown of complex organic compounds into smaller ones	Energy released as heat or stored in ATP	Glycolysis, Krebs cycle, glycogenolysis
b. Anabolism	Synthesis of complex organic molecules from smaller ones	Uses energy (ATP)	Synthesis of protein, fats, or glycogen

A7. 40%; heat.

B2. (a) A kilocalorie is the amount of heat required to raise 1000 grams (= 1000 ml = 1 liter) of water 1°C; a kilocalorie (kcal) = 1000 calories (cal); kcal.

B3. Releases; oxidation.

B4. (a) Oxygen; 4.9; spirometer. (b) 73.5 (= 4.9 kcal/liter oxygen × 15 liters/hr). (c) Hypo.

B5. (a) Core; hyperthermia denatures (changes chemical configurations of) proteins. (b) Hypothalamus; promoting, sympathetic; constriction, conservation. (c1) 12 years old; (c2) 103.6°F; (c3) Running 2 miles; (c4) Increased epinephrine or thyroid hormone; (c5) Protein.

B8. 580 (= 0.58 cal/ml water × 1000 ml/liter); less, less.

B9. F B C A D E.

C1. (a) ATP; $C_6H_{12}O_6 \rightarrow CO_2 + H_2O$ + energy stored in ATP; poor. (b) H^+, hydride, H^-. (c) NAD^+, $NADP^+$, FAD, niacin; H_2O. (d) Oxidation. (e) ATP; phosphorylation;

Glucose + O_2 + ADP →
(reducer)   (oxidizer)
   CO_2 + H_2O + ATP
(oxidized)   (reduced)

C2. (a) A. (b) B. (c) A. (d) B. (e) A. (f) B. (g) B.

D1. (a) Glucose. (b) Starch, sucrose, lactose; by enzymes located mostly in liver cells. (c) Increases, oxidizing. (d) Stored as glycogen, converted to fat or protein and stored, excreted in urine. (e) Hyperglycemia; insulin, facilitates; diabetes mellitus. (f) Neurons and liver cells. (g) Phosphate; phosphorylation, kinases; cells can utilize the trapped glucose since phosphorylated glucose cannot pass out of most cells.

D2. (a) 6; 3, pyruvic acid; a little; many, does not. (b) Phosphoryl, fructose; phosphofructokinase; ADP, ATP; ana, glycogen. (c) 3, G 3-P; oxidized, NAD. (d) Pyruvic, 2. (e) Oxygen; do, aerobic. (f) Lactic acid; exercise. (g) Pyruvic acid; stimulates.

D3. (a) Glucose. (b) Pyruvic acid. (c) Glycolysis. (d) Cytosol. (e) Transition. (f) CO_2. (g) Acetyl (h) Coenzyme A. (i) Acetyl coenzyme A. (j) Oxaloacetic acid. (k) Citric acid. (l) Citric acid (or Krebs). (m) Matrix of mitochondria. (n) CO_2, which are exhaled. (o) Hydrogen. (p) NAD and FAD. (q) H^+. (r) Electrons (e^-). (s) Cytochromes or electron transfer system. (t) ATP. (u) Oxygen. (v) Water.

D4. (a) FMN; reduced, $FMNH_2$; B_2 (riboflavin). (b) Oxidized, FMN. (c) Iron (Fe); end. (d) Copper and sulfur. (e) Cytochrome a_3; oxygen; aerobic. (f) IMM. (g) Three; protons = H^+, matrix, space between inner and outer mitochondrial membranes. (h) Inner, proton motive; across special channels in the inner mitochondrial membrane. (i) Synthetase; H^+, chemiosmotic.

D5. (a) $C_6H_{12}O_6 + 6O_2 + 36$ or 38 ADPs

Glucose   Oxygen

$+ 36$ or 38 Ⓟ $\longrightarrow 6CO_2 + 6H_2O +$

Carbon   Water
dioxide

36–38 ATPs. (b1) 2; (b2) 4–6. (b3) 6. (b4) 18. (b5) 4. (b6) 2. (b7) 36–38.

D6. The Krebs cycle; NAD; 40.

D7. (a) Glucose; 1: glycogenesis. (b) 2: Glycogenolysis. (c) Glycogenesis. (d) Muscles. (e) Insulin; de; glucagon and epinephrine. (f) Phosphate; phosphorylase; yes.

D9. Processes: glycogenolysis and gluconeogenesis; hormones: all except insulin.

E2. Fat; 50%; no, it is continually catabolized and resynthesized.

E3. (a) Fatty acids, glycerol, lipo, lip. (b) Glyceraldehyde 3-phosphate; glucose; neogenesis. (c) Beta oxidation, matrix of mitochondria. (d) Acetic acid, A; acetyl CoA (or acetyl coenzyme A); 129. (e) Acetoacetic acid; lowers; ketoacidosis. (f) Acetone; ketone; ketogenesis, liver. (g) Cardiac muscle cells, cells of the renal cortex, and (during starvation) brain cells. (h) Acetoacetic acid is normally catabolized for ATP production as fast as the liver generates it. (i) Examples of reasons: starvation, fasting diet, lack of glucose in cells due to lack of insulin (diabetes mellitus), excess growth hormone (GH) which stimulates fat catabolism. (j) As acetone (formed in ketogenesis) passes through blood in pulmonary vessels, some is exhaled and detected by its sweet aroma.

F1. (a) Carbohydrates. Exception: cardiac muscle and cortex of the kidneys utilize acetoacetic acid from fatty acids preferentially over glucose. (b) Lipids. (c) Proteins.

F2. (a) 4, about 1800. (b) 9, about 4000.

F3. See Exhibit 2.6, page 43 of your text.

F4. Insulin and hGH. (Thyroid hormone also stimulates protein synthesis.)

F5. (a) Worn-out cells as in bone or muscle. (b) Ingested foods. (c) Anabolism. (d) Deamination. (e) Ammonia. (f) Urea. (g) Glycolytic or Krebs cycle. (h) $CO_2 + H_2O + ATP$.

G1. (a–b) Acetyl CoA. (c–d) Pyruvic acid. (e–g) G 6-P.

G2. (a) G 6-P, pyruvic acid, acetyl CoA. (b) Cannot; cannot; glycerol, glucose.

H1. (a–c) P. (d–f) A. (g) P.

H2. (a) 70–110. (b) Liver glycogen (4-hour supply); muscle glycogen and lactic acid during exercise; glycerol from fats; as a last resource, amino acids from tissue proteins. (c) No. See Checkpoint G2b in this chapter. (d) Acetic acids (and then acetyl CoA); sparing.

H4. (b) Glucagon. (c) hGH. (d) Glucocorticoids. (e) Thyroxine. (f) Epinephrine.

I1. (a) 2200. (b) 2800. (c) 1600. (d) 2200.

I2. (a) 50–60. (b) 30, 10. (c) 12–15.

I3. (a) 2000 Cal $\times$ (0.50–0.60) = 1000–1200 Cal/day. At 4 Cal/g = 250–300 g/day. (b) 2000 Cal $\times$ (0.30) = less than 600 cal/day. At 9 Cal/g = 600/9 = 67 g/day of total fat; a maximum of 10% should be saturated fat = 22 g/day.

I4. (a) 6–11. (b) 3–5. (c) 2–4. (d) 2–3. (e) 2–3. (f) Sparingly.

I5. Inorganic substances; 4, skeleton.

I6. (a) Cl. (b) Na. (c) Ca. (d) Fe. (e) K. (f) I. (g) Co. (h) S. (i) F. (j) P.

I7. B.

I8. Organic; cannot; most serve as coenzymes, maintaining growth and metabolism.

I9. (a) A D E K. (b) B complex and C.

I11. (a) K. (b) D. (c) B_2. (d) B_1. (e) A. (f) C. (g) B_{12}. (h) E.

J3. Marasmus; protein deficiency (due to lack of essential amino acids) decreases plasma proteins. Blood has less osmotic pressure, so fluids exit from blood. Movement of these into the abdomen (ascites) increases the size of the abdomen. Fatty infiltration of the liver also adds to the abdominal girth.

J4. Phenylalanine, tyrosine; toxicity to the brain with possible mental retardation.

J5. Grains; gluten; rice, corn.

1. Contrast anabolism and catabolism. Give one example of each in metabolism of carbohydrates, lipids, and proteins.
2. Discuss homeostatic mechanisms of temperature regulation.
3. Explain the significance of B-complex vitamins in normal metabolism. Include roles of niacin (B_1), which forms coenzymes NAD^+ and $NADP^+$, riboflavin (B_2), which forms coenzymes FAD and FMN, and pantothenic acid, which forms coenzyme A.
4. Describe interconversions of carbohydrates, lipids, and proteins, including roles of the three key molecules, acetyl CoA, pyruvic acid, and G 6-P.
5. Describe guidelines for healthy eating and discuss roles foods may play in disorders such as cancer and diabetes. Explain the possible health value of olive oil, canola oil, and peanut oil.

MASTERY TEST: CHAPTER 25

Questions 1–12: Circle the letter preceding the one best answer to each question.

1. Which of these processes is anabolic?
   A. Pyruvic acid $\rightarrow CO_2 + H_2O + ATP$
   B. Glucose $\rightarrow$ pyruvic acid + ATP
   C. Protein synthesis
   D. Digestion of starch to maltose
   E. Glycogenolysis

2. Which of the following can be represented by the equation Glucose $\rightarrow$ pyruvic acids + small amount ATP?
   A. Glycolysis
   B. Transition step between glycolysis and Krebs cycle
   C. Krebs cycle
   D. Electron transport system

3. Which hormone is said to be hypoglycemic because it tends to lower blood sugar?
   A. Glucagon
   B. Glucocorticoids
   C. Growth hormone
   D. Insulin
   E. Epinephrine

4. The complete oxidation of glucose yields all of these products *except:*
   A. ATP
   B. Oxygen
   C. Carbon dioxide
   D. Water

5. All of these processes occur exclusively or primarily in liver cells *except* one, which occurs in virtually all body cells. This one is:
   A. Gluconeogenesis
   B. Beta oxidation of fats
   C. Ketogenesis
   D. Deamination and conversion of ammonia to urea
   E. Krebs cycle

6. The process of forming glucose from amino acids or glycerol is called:
   A. Gluconeogenesis
   B. Glycogenesis
   C. Ketogenesis
   D. Deamination
   E. Glycogenolysis

7. All of the following are parts of fat absorption and fat metabolism *except:*
   A. Lipogenesis
   B. Beta oxidation
   C. Chylomicron formation
   D. Ketogenesis
   E. Glycogenesis

8. Which of the following vitamins is water soluble? A, C, D, E, K?

9. At room temperature most body heat is lost by:
   A. Evaporation
   B. Convection
   C. Conduction
   D. Radiation

10. Choose the *false* statement about vitamins.
    A. They are organic compounds.
    B. They regulate physiological processes.
    C. Most are synthesized by the body.
    D. Many act as parts of enzymes or coenzymes.

11. Choose the *false* statement about temperature regulation.
    A. Some aspects of fever are beneficial.
    B. Fever is believed to be due to a "resetting of the body's thermostat."
    C. Vasoconstriction of blood vessels in skin will tend to conserve heat.
    D. Heat-producing mechanisms which occur when you are in a cold environment are primarily parasympathetic.

12. Which of the following processes involves a cytochrome chain?
    A. Glycolysis
    B. Transition step between glycolysis and Krebs cycle
    C. Krebs cycle
    D. Electron transport system

Questions 13–14: Arrange the answers in correct sequence.

_____ _____ _____ _____ _____ _____ 13. Steps in complete oxidation of glucose:
    A. Glycolysis takes place.
    B. Pyruvic acid is converted to acetyl CoA.
    C. Hydrogens are picked up by NAD and FAD; hydrogens ionize.
    D. Krebs cycle releases CO_2 and hydrogens
    E. Oxygen combines with hydrogen ions and electrons to form water.
    F. Electrons are transported along cytochromes, and energy from electrons is stored in ATP.

_____ _____ _____ 14. Using the Food Pyramid as a guide, arrange these food groups according to number of servings suggested per day, from greatest to least:
    A. Meat, fish, poultry, dry beans, eggs, and nuts
    B. Vegetables
    C. Bread, cereal, rice, pasta

Questions 15–20: Circle T (true) or F (false). If the statement is false, change the underlined word or phrase so that the statement is correct.

T F 15. Complexes of carrier molecules utilized in the electron transport system are located in the <u>mitochondrial matrix.</u>

T F 16. Catabolism of carbohydrates involves <u>oxidation,</u> which is a process of <u>addition of hydrogens.</u>

T F 17. Nonessential amino acids are those that <u>are not used in the synthesis of human protein.</u>

T F 18. The complete oxidation of glucose to CO_2 and H_2O yields <u>4</u> ATPs.

T F 19. Anabolic reactions are <u>synthetic reactions that release energy.</u>

T F 20. <u>Carbohydrates, proteins, and vitamins</u> provide energy and serve as building materials.

Questions 21–25: Fill-ins. Complete each sentence with the word or phrase that best fits.

_____ 21. Catabolism of each gram of carbohydrate or protein results in release of about _____ kcal, whereas each gram of fat leads to about _____ kcal.

_____ 22. Pantothenic acid is used to form the coenzyme named _____.

_____ 23. _____ is the mineral that is most common in extracellular fluid (ECF); it is also important in osmosis, buffer systems, and nerve impulse conduction.

_____ 24. Aspirin and acetaminophen (Tylenol) reduce fever by inhibiting synthesis of _____ so that the body's "thermostat" located in the _____ is reset to a lower temperature.

_____ 25. In uncontrolled diabetes mellitus, the person may go into a state of acidosis due to the production of _____ acids resulting from excessive breakdown of _____.

ANSWERS TO MASTERY TEST: ✩ CHAPTER 25

Multiple Choice

1. C
2. A
3. D
4. B
5. E
6. A
7. E
8. C
9. D
10. C
11. D
12. D

Arrange

13. A B D C F E
14. C B A

True–False

15. F. Mitochondrial inner (cristae) membrane
16. F. Oxidation, removal of hydrogens
17. F. Can be synthesized by the body
18. F. 36 to 38
19. F. Synthetic reactions that require energy
20. F. Carbohydrates and proteins but not vitamins

Fill-ins

21. 4, 9
22. Coenzyme A
23. Sodium (Na^+)
24. Prostaglandins (PG), hypothalamus
25. Acetoacetic (keto-), fats

FRAMEWORK 26
Urinary System

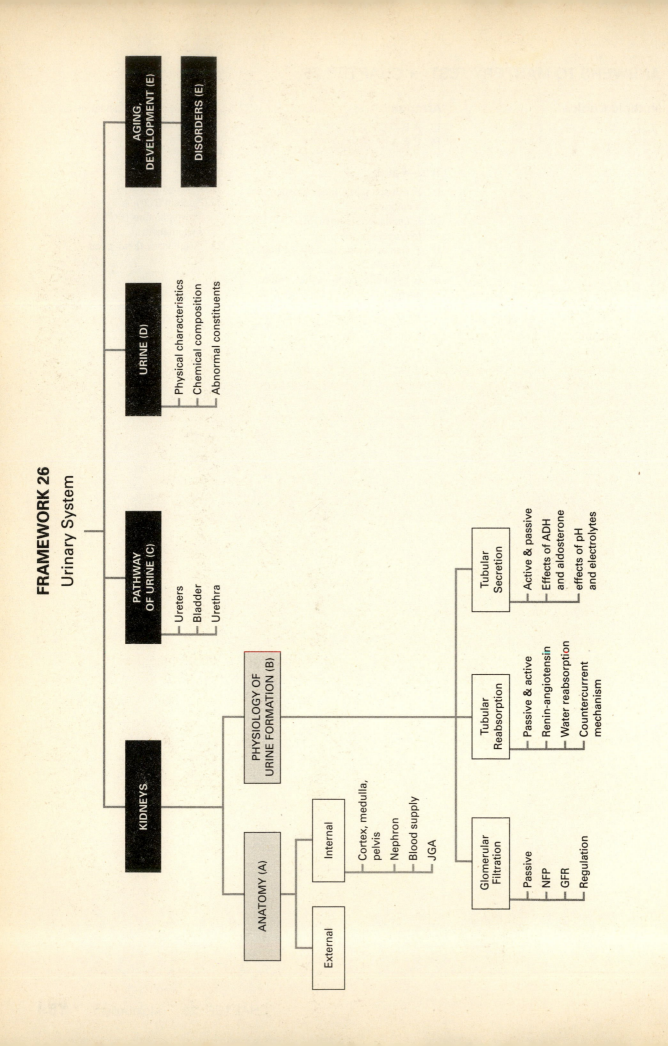

The Urinary System

CHAPTER
26

Wastes like urea or hydrogen ions (H^+) are products of metabolism that must be removed or they will lead to deleterious effects. The urinary system works with lungs, skin, and the GI tract to eliminate wastes. Kidneys are designed to filter blood and selectively determine which chemicals stay in blood or exit in the urine. The process of urine formation in the kidney requires an intricate balance of factors such as blood pressure and hormones (ADH and aldosterone). Examination of urine (urinalysis) can provide information about the status of the blood and possible kidney malfunction. A healthy urinary system requires an exit route for urine through ureters, bladder, and urethra.

Start your study of the urinary system with a look at the Chapter 26 Framework and familiarize yourself with the key concepts and terms. Examine the Chapter 26 Topic Outline and Objectives. Check off each one as you complete it.

TOPIC OUTLINE AND OBJECTIVES

A. Kidneys: anatomy

☐ 1. List the functions of the kidneys.
☐ 2. Identify the external and internal gross anatomical features of the kidneys.
☐ 3. Trace the path of blood flow through the kidneys.
☐ 4. Discuss how the structure of a nephron contributes to regulating the volume, composition, and pressure of blood.

B. Physiology of urine formation

☐ 5. Discuss the process of urine formation through glomerular filtration, tubular reabsorption, and tubular secretion.
☐ 6. Describe how the kidneys produce dilute and concentrated urine.
☐ 7. Explain how hemodialysis is performed and what it accomplishes.

C. Pathway of urine

☐ 8. Discuss the anatomy, histology, and physiology of the ureters, urinary bladder, and urethra.

D. Urine

☐ 9. List and describe the physical characteristics, normal constituents, and abnormal constituents of urine.

E. Aging and the development of the urinary system

☐ 10. Describe the effects of aging on the urinary system.
☐ 11. Describe the development of the urinary system.

F. Disorders, clinical terminology

☐ 12. Discuss the causes of urinary tract infections (UTIs), glomerulonephritis, nephrotic syndrome, renal failure, polycystic kidney disease, and diabetes insipidus.
☐ 13. Define medical terminology associated with the urinary system.

WORDBYTES

Now study the following parts of words that may help you better understand terminology in this chapter.

Wordbyte	Self-check	Meaning	Example
anti-	_____	against	*anti*diuretic
azot-	_____	nitrogen-containing	*azot*emia
calyx	_____	cup	major *calyx*
cyst-	_____	bladder	*cyst*ostomy
glomus	_____	ball	*glom*erular
insipid	_____	without taste	diabetes *insipid*us
juxta-	_____	next to	*juxta*glomerular
-ptosis	_____	drooping	nephro*ptosis*
recta	_____	straight	vasa *recta*
ren-	_____	kidney	*ren*al vein
-ulus	_____	small	glomer*ulus*
urin-	_____	urinary	*urin*alysis

CHECKPOINTS

A. Kidneys: anatomy   (pages 848–861)

A1. Complete this list of body structures and tissues, besides urinary organs that perform excretory functions. Then list the substances they eliminate or detoxify.

Structure or tissue	Substances eliminated or made less toxic
1. _____	Excess H^+ is bound
2. _____	_____
3. _____	_____
4. _____	CO_2, heat, and some water eliminated
5. _____	_____
6. GI tract (via anus)	_____

■ **A2.** Describe functions of the urinary system in this exercise.

    a. List waste products eliminated through urine.

    b. Kidneys (along with liver) contribute to production of "new glucose" made from prod-

       ucts of proteins or fats. This process is called gluco-_____.

    c. Which ions do kidneys excrete when blood pH is too low? _____

    d. Kidneys produce _____, which regulates blood pressure, and

       _____, which is vital to red blood cell formation. Kidneys also help

       to activate vitamin _____.

■ **A3.** Identify the organs that make up the urinary system on Figure LG 26.1 and answer the following questions about them.

    a. The kidneys are located at about *(waist? hip?)* level, between _____ and _____

       vertebrae. Each kidney is about _____ cm (_____ in.) long. Visualize kidney location and size on yourself.

    b. The kidneys are in an extreme *(anterior? posterior?)* position in the abdomen. They are

       described as _____ because they are posterior to the peritoneum.

    c. Identify the parts of the internal structure of the kidney on Figure LG 26.1. Then color all structures as indicated by color code ovals.

■ **A4.** Select terms that match each description of kidneys. Choose from these terms. Not all terms will be used.

Adipose capsule	Renal calyx	Renal fascia
Nephrology	Renal capsule	Renal papilla
Nephroptosis	Renal columns	Renal pyramids
Parenchyma		

Term	Description
a. _____	Floating or drooping kidney
b. _____	Outermost layer of connective tissue that anchors kidneys to abdominal wall
c. _____	Study of kidneys
d. _____	Functional part of organ, such as kidney
e. _____	Makes up most of renal medulla
f. _____	Portion of renal cortex located between renal pyramids

■ **A5.** About what percentage of cardiac output passes through the kidneys each minute?

    _____% This amounts to about _____ ml/min.

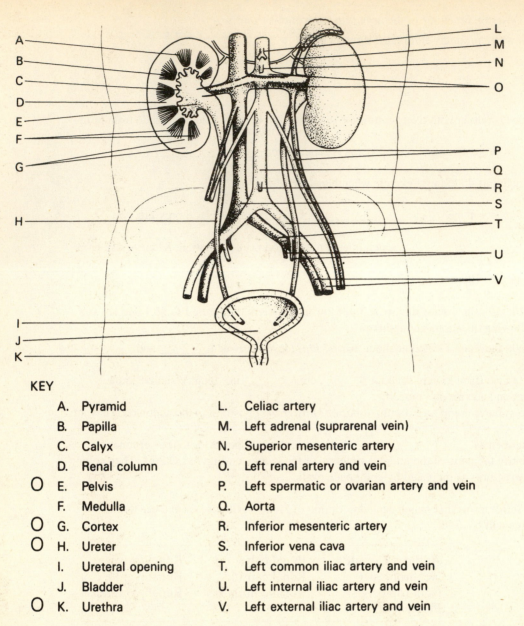

A ─
B ─
C ─
D ─
E ─
F ─
G ─

H ─

I ─
J ─
K ─

L
M
N
O

P
Q
R
S
T

U

V

KEY

	A.	Pyramid	L. Celiac artery
	B.	Papilla	M. Left adrenal (suprarenal vein)
	C.	Calyx	N. Superior mesenteric artery
	D.	Renal column	O. Left renal artery and vein
O	E.	Pelvis	P. Left spermatic or ovarian artery and vein
	F.	Medulla	Q. Aorta
O	G.	Cortex	R. Inferior mesenteric artery
O	H.	Ureter	S. Inferior vena cava
	I.	Ureteral opening	T. Left common iliac artery and vein
	J.	Bladder	U. Left internal iliac artery and vein
O	K.	Urethra	V. Left external iliac artery and vein

Figure LG 26.1 *(Left side of diagram)* Organs of the urinary system. Structures A to G are parts of the kidney. Identify and color as directed in Checkpoint A3. *(Right side of diagram)* Blood vessels of the abdomen. Color as directed in Chapter 21, Checkpoint E8, page 470 of the *Learning Guide*.

■ **A6.** Follow the pathway that blood takes through kidneys by naming vessels 1–8 on Figure LG 26.2. Some arrows are shown; add more to reinforce your understanding of the pathway of blood.

1. _____ 2. _____ 3. _____

4. _____ 5. _____ 6a. _____

6b. _____ 7. _____

Which of these vessels extends most deeply into the renal medulla? _____

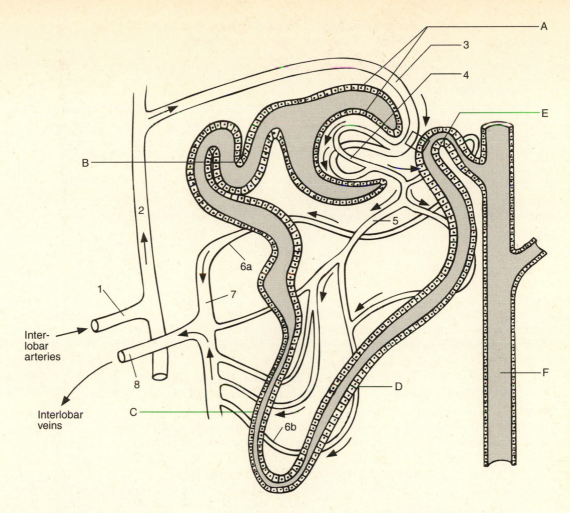

Figure LG 26.2 Diagram of a nephron. Letters refer to Checkpoint A8. Numbers refer to Checkpoints A6 and A8. The boxed area refers to Checkpoint A10.

- **A7.** Explain what is unique about the blood supply of kidneys in this activity.

  a. In virtually all tissues of the body, blood in capillaries flows into vessels named

  _____. However, in kidneys, blood in glomerular capillaries flows

  directly into _____ _____.

  b. Which vessel has a larger diameter? *(Afferent? Efferent?)* arteriole. State one effect of this difference.

- **A8.** Do this exercise about the functional unit of the kidney, a nephron.

  a. On Figure LG 26.2, label parts A–F of the nephron, with letters arranged according to the flow of urine.

  b. Which one of those structures forms part of the renal corpuscle?

  _____

  The other part of the renal corpuscle is a cluster of capillaries known as a

  _____ and numbered *(4? 6b?)* in the figure.

c. Arrange in order these components of the renal corpuscle from innermost to outermost, in other words, along the pathway of "cleaning" (filtration) of blood:

_____ → _____ → _____ → _____ → _____

I. Bloodstream
II. Endothelial cells forming much of the wall of glomerular capillaries, with endothelial pores located between epithelial cells
III. Basement membrane of the glomerulus
IV. Visceral layer of the glomerular (Bowman's) capsule, composed of podocytes with footlike pedicels and tiny filtration slits
V. Capsular (Bowman's) space

d. Structures lettered II, III, and IV above form the _____-capsular membrane. List the functional advantages of the structure of the endothelial-capsular membrane.

e. Water and other substances "cleaned" (filtered) out of glomerular blood pass across this membrane and collect in the _____ space, which lies between

_____ and parietal layers of the capsule. The *(parietal? visceral?)* layer of the glomerular (Bowman's) capsule forms the outermost boundary of the renal capsule.

f. Identify the part of the renal tubule on the figure that consists of cuboidal epithelium with a brush border of many microvilli that increase the absorptive surface area.

_____ Which part consists of simple squamous epithelium?

_____

g. In which regions are *principal cells* located? *(B and C? E and F?)* These cells are

sensitive to two hormones. Name them. _____ and

_____ The role of intercalated cells—also located in E and F—is to

secrete _____.

h. In which part of the renal tubule does most reabsorption of filtered water and solutes

take place? _____

■ **A9.** Contrast two types of nephrons by writing C for cortical nephrons and JM for juxtamedullary nephrons.

_____ a. Also known as short-looped nephrons because these lie only in the renal cortex

_____ b. Constitute about 15–20% of nephrons

_____ c. Have a thin (squamous epithelial) layer in ascending limb of loop of Henle

_____ d. Permit significant variations in concentration of urine

■ **A10.** Describe the juxtaglomerular apparatus (JGA) in this exercise.

a. As shown in the boxed structure in Figure LG 26.2, the JGA consists of *(two? four?)* types of cells. One type is located in the final portion of the *(ascending? descending?)*

limb of the loop of Henle. These cells, known as _____ _____

cells, monitor concentration of _____ in the tubular lumen.

b. The second type of cell is a _____ _____

_____ cell located in the wall of the *(afferent arteriole? glomeru-*

lus?). These cells are known as _____ cells; they secrete the chemical

known as _____, which regulates _____.

B. Physiology of urine formation   (pages 861–878)

■ **B1.** Do this exercise on urine formation.

a. List the three steps of urine formation.

_____ _____ _____

b. What percentage of plasma entering nephrons is filtered out of glomerular blood and passed into the capsular space? About *(1%? 15–20%? 48–60%? 99%?)*. This percent-

age is known as the filtration _____. The actual amount of glomerular filtrate is about *(180? 60? 1–2?)* liters/day.

c. Of this filtrate, about _____ liters/day of fluid are reabsorbed in the

"second step" of urine formation. This is about _____% of the glomerular filtrate.

So the total amount of urine eliminated from kidneys daily is usually about _____ liters (or quarts).

■ **B2.** Refer to Figure LG 26.3. Now describe the first step in urine production in this checkpoint.

a. Glomerular filtration is a process of *(pushing? pulling?)* fluids and solutes out of

_____ and into the fluid known as _____. The direction of glomerular filtration is shown by the thick arrow from W to X on the figure.

b. *(Most substances? Only a few types of substances?)* are forced out of blood in the process of glomerular filtration. Refer to Figure LG 26.3, area W (plasma in glomeruli) and area X (filtrate just beyond the capsule). Compare the compositions of these two fluids (described in the first two columns of Exhibit 26.1, page 866 in your textbook). Note that during filtration all solutes are freely filtered from blood except

_____. Why is so little protein filtered?

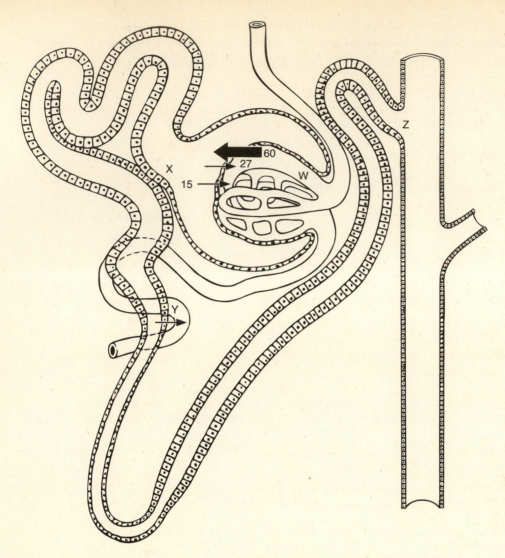

Figure LG 26.3 Diagram of a nephron with renal tubule abbreviated. Numbers refer to Checkpoint B3; letters refer to Checkpoints B2, B10, and B11.

c. Blood pressure (or glomerular blood hydrostatic pressure or GBHP) in glomerular

capillaries is about _____ mm Hg. This value is *(higher? lower?)* than that in other capillaries of the body. This extra pressure is accounted for by the fact that the diameter of the efferent arteriole is *(larger? smaller?)* than that of the afferent arteriole. Picture three garden hoses connected to each other, representing the afferent arteriole, glomerular capillaries, and efferent arteriole. The third one (efferent arteriole) is extremely narrow; it creates such resistance that pressure builds up in the first two. Fluids will be forced out of pores in the middle (glomerular) hose.

d. Describe three structural features of nephrons that enhance their filtration capacity.

■ **B3.** Glomerular blood hydrostatic pressure (GBHP) is not the only force determining the amount of filtration occurring in glomeruli.

a. Name two other forces (shown on Figure. 26.10, page 862 in your textbook).

_____ _____

b. Normal values for these three forces are given on Figure LG 26.3. Write the formula for calculating net filtration pressure (NFP); draw an arrow beneath each term to show the direction of the force in Fig. LG 26.3. Then calculate a normal NFP using the values given.

NFP = (glomerular blood hydrostatic pressure) – (_____ + _____)

Normal NFP = (_____ mm Hg) – (_____ mm Hg + _____ mm Hg)

   = _____ mm Hg

Note that blood (hydrostatic) pressure pushes *(out of? into?)* glomerular blood and is largely counteracted by the other two forces.

■ **B4.** *A clinical challenge.* Show the effects of alterations of these pressures in a pathological situation. Determine NFP of the patient whose pressure values are shown below.

Glomerular blood hydrostatic pressure = 42
Blood colloid osmotic pressure = 27
Capsular filtrate hydrostatic pressure = 15

NFP = _____

Note which value(s) is(are) abnormal and suggest causes.

■ **B5.** Define GFR in this exercise.

a. To what do the letters GFR refer? g _____ f _____

r _____

b. Write a normal value for GFR. _____ ml/min (_____ liters/day)

c. Describe these two causes of decreased GFR. GFR will usually decrease if blood pressure in kidneys _____-creases, for example, during stress when sympathetic nerves

decrease the diameter of the _____-ferent arteriole. Imagine that the first of the three garden hoses in Checkpoint B2c is narrowed, so *(more? less?)* blood flows into the glomerulus (middle hose). Either enlarged prostate gland or kidney

_____ lodged in the ureter back up urine into renal tubules of kidneys, opposing glomerular blood pressure, and this will decrease GFR also.

d. What consequences may be expected to accompany an abnormally low GFR?

■ **B6.** List three mechanisms that regulate GFR:

_____ _____ _____

■ **B7.** *A clinical challenge.* Mrs. K's blood pressure has dropped slightly. Explain how her renal autoregulation mechanism functions to ensure adequate renal blood flow.

a. When Mrs. K's systemic blood pressure decreases, her NFP and GFR are likely to *(increase? decrease?)*, causing filtrate to move more *(rapidly? slowly?)* through the renal tubules. As a result, *(more? less?)* fluid and salts are resorbed in the PCT and loop of Henle.

b. Consequently, the amount of fluid slowly passing the macula densa is *(high? low?)* in volume and concentration of salts. This condition causes JGA cells to produce *(more? fewer?)* of certain vasoconstrictor substances. As afferent arterioles then dilate, *(more? less?)* blood flows into glomerular capillaries, causing NFP and GFR to *(increase? decrease?)*.

c. This restoration of homeostasis by renal autoregulation occurs by a *(positive? negative?)* feedback mechanism. The name autoregulation indicates that neither the autonomic nervous system nor any chemical made outside the kidneys is involved. The control *(is? is not?)* intrinsic to kidneys.

■ **B8.** Now describe chemicals involved in other mechanisms that regulate GFR by completing this checkpoint. Select the chemical that matches each description. Use the following list of answers. (It may help to refer to Figure LG 18.5, page 376 in the *Learning Guide.*)

ACE. Angiotensin converting enzyme	Ang II. Angiotensin II
ADH. Antidiuretic hormone	ANP. Atrial natriuretic peptide
Ald. Aldosterone	Epi. Epinephrine
Ang I. Angiotensin I	R. Renin

_____ a. An enzyme secreted by JGA cells of the kidney

_____ b. Produced by action of renin on a plasma protein made by liver

_____ c. Made by lungs, this enzyme converts angiotensin I to angiotensin II

_____ d. A vasoconstrictor that also stimulates thirst and release of aldosterone and ADH

_____ e. Produced by adrenal cortex, it causes collecting ducts to reabsorb more Na^+ and water

_____ f. Released from the posterior pituitary, this hormone also increases blood volume

_____ g. The "third factor," it is produced by cells of the heart

_____ h. This hormone increases GFR; it opposes actions of renin, aldosterone, and ADH and suppresses secretion of those chemicals

_____ i. May have clinical applications in reducing edema by causing diuresis and natriuresis

_____ j. Released under sympathetic stimulation; leads to arteriolar vasoconstriction and decreased GFR (two answers)

_____ k. Of all chemicals in the above list, the one that tends to decrease blood pressure

B9. *A clinical challenge.* Joyce, an accident victim, was admitted to the hospital 3 hours ago. Her chart indicates that she had been hemorrhaging at the scene of the accident. Joyce's nurse monitors her urinary output and notes that it is currently 12 ml/hr. (Normal output is 30–50 ml/hr.) Explain why a person in such severe stress is likely to have decreased urinary output (oliguria).

■ **B10.** Refer to Figure LG 26.3 and do this exercise about formation of urine.

a. We have already seen that glomerular filtration results in movement of substances from *(blood to filtrate, forming urine? filtrate, forming urine to blood?)*, as shown by the arrow between W and X. If this were the only step in urine formation, all the substances in the filtrate would leave the body in urine. Note from Exhibit 26.1 (page 866 in your

text) that the body would produce _____ liters of urine each day, and it would contain many valuable substances. Obviously some of these "good" substances must be drawn back into blood and saved.

b. Recall that blood in most capillaries of the body flows into vessels named

_____, but blood in glomerular capillaries flows into

_____ and then into _____. This unique arrangement permits blood to recapture some of the substances indiscriminately pushed out during filtration. This occurs during the second step of urine formation, called

_____. This process moves substances in the *(same? opposite?)* direction as glomerular filtration, as shown by the direction of the arrow at area Y of Figure LG 26.3.

■ **B11.** Refer to areas Y and Z on Figure LG 26.3 and to the two columns on the far right of Exhibit 26.1, page 866 of your text. Discuss the effects of tubular reabsorption in this learning activity.

a. Which solutes are 100% reabsorbed into area Y so that virtually none remains in urine (area Z)?

b. About what percentage of water that is filtered out of blood is reabsorbed back into

blood? _____% Identify two ions that are about 99% reabsorbed. _____ _____

c. *(Urea? Uric acid?)* is about 90% reabsorbed, whereas _____ is about 50% reabsorbed.

d. Which ion is 100% reabsorbed but still shows up in urine because some of this ion is

secreted (step 3 in urine formation) from blood to urine? _____

e. Identify the solute that is filtered from blood and not reabsorbed.

_____ State the clinical significance of this fact.

f. Tubular reabsorption is a *(nonselective? discriminating?)* process that enables the body to save valuable nutrients, ions, and water.

■ **B12.** Arrange in correct sequence the steps in reabsorption of Na⁺ in the PCT:

_____ _____ _____

a. Sodium pump actively expels Na⁺ from the base and sides of tubule cells.
b. Na⁺ passively diffuses from negatively charged fluid within the lumen of the tubule into tubule cells.
c. Na⁺ diffuses from spaces between tubule cells into peritubular capillary blood.

B13. *For extra review.* Describe reabsorption mechanisms in this exercise.

a. About _____% of total resting metabolic energy is used for reabsorption of Na⁺ in renal tubules by *(primary? secondary?)* active transport mechanisms. Explain how active transport of Na⁺ also promotes osmosis of water and diffusion of substances such as K^+, Cl^-, HCO_3^-, and urea.

b. Reabsorption of glucose, amino acids, and lactic acid occurs almost completely in the *(PCT? DCT?),* and the process utilizes Na⁺ *(symporters? antiporters?).* This mechanism is an example of *(primary? secondary?)* active transport because it is the sodium pump, not the symporter, that uses ATP. If blood concentration of a substance such as glucose

is abnormally high, the renal _____ is surpassed. In other words, the

blood level of glucose is greater than the T_m (or _____

_____).

■ **B14.** *A clinical challenge.* Eileen and Marlene are both "spilling sugar" in urine; that is,

they have _____-uria. Discuss possible causes of glycosuria in each case, considering their blood sugar levels.

a. Eileen: glucose of 340 mg/100 ml plasma

b. Marlene: glucose of 80 mg/100 ml plasma

■ **B15.** Identify the part of the renal tubule that best fits the description. Use these answers.

A. Ascending limb of the loop of Henle
CD. Collecting duct
D. Descending limb of the loop of Henle

DCT. Distal convoluted tubule
PCT. Proximal convoluted tubule

_____ a. This portion of the renal tubule with a prominent brush border permits tubular reabsorption of most filtered water and almost 100% of filtered nutrients (such as glucose).

_____ b. This is the first site where osmosis of water is not necessarily coupled to reabsorption of other filtrates.

_____ c. Little or no water is reabsorbed here because these cells are impermeable to water.

_____ d. Variable amounts of Na^+ and water may be reabsorbed here under the influence of aldosterone and ADH. (two answers)

_____ e. Absorption of almost all HCO_3^- and at least half of Na^+, K^+, Cl^-, and urea occurs here.

■ **B16.** Do this exercise on hormonal regulation of urine volume and composition.

a. Aldosterone and ADH are hormones that act on _____ cells located in *(PCT and loop of Henle? last part of DCT as well as collecting ducts?)*. Aldosterone is made by the *(adrenal medulla? adrenal cortex? hypothalamus?)*. This hormone stimulates production of a protein that helps pump *(Na^+? K^+?)* and also water from urine to blood. So aldosterone tends to *(in? de?)*-crease blood pressure.

b. ADH causes principal cells to become *(more? less?)* permeable to water; as a result,

reabsorption of water _____-creases, leading to _____-crease of blood pressure.

c. ADH is made by the _____ and released from the *(anterior? posterior?)* pituitary. It is regulated by a *(positive? negative?)* feedback mechanism

because decreased body fluid leads to _____-crease of ADH release and retention of body fluid.

d. Diabetes insipidus involves _____-crease of ADH and _____-crease of urinary output.

■ **B17.** The third step in urine production is _____. It involves movement of substances from *(blood to urine? urine to blood?)*. In other words, tubular secretion is movement of substances in *(the same? opposite?)* direction as movement occurring in filtration. List four or more substances that are secreted by the process of tubular secretion.

■ **B18.** Do this exercise on regulation of blood level of potassium ion (K^+).

a. What percentage of filtered K^+ is normally reabsorbed in PCT, loop of Henle, and DCT?

(*Hint:* Refer to Exhibit 26.1, page 866) _____%

b. Explain why kidneys might need to secrete variable amounts of K^+ from blood to urine.

c. When kidney reabsorption of the cation Na^+ increases, secretion of K^+ (also a cation)

_____-creases. Aldosterone *(enhances? inhibits?)* such processes, causing lowered blood K^+ levels. On the other hand, aldosterone deficiency may lead to *(low? high?)* blood levels of K^+, and even cardiac arrest.

■ **B19.** Explain how tubular secretion helps to control pH. Refer to Figure 26.17 (page 871 in the text) and do this exercise.

   a. Jenny's respiratory rate is slow, and her arterial pH is 7.33. Her blood therefore contains

   high levels of _____. Her blood is therefore somewhat (*acidic? alkaline?*). Explain why.

   b. Jenny's kidneys can assist in restoring normal pH (Figure 26.17a, b). The presence of acid (H^+) stimulates the PCT cells to eliminate H^+. In order for this to happen, another cation present in the tubular lumen must exchange places with H^+. What ion enters

   kidney tubule cells? _____ It joins HCO_3^- to form _____. This process requires (*antiporters? symporters?*).

   c. Figure 26.18c shows another possible mechanism for reducing acidity of blood. Intercalated cells located in (*PCT? collecting ducts?*) pump H^+ out into the lumen, causing urine to be up to (*10? 1000?*) times more acidic than blood. This pump (*does? does not?*) require ATP; the pump is an example of (*primary? secondary?*) active transport.

   The leftover HCO_3^- (from H_2CO_3) may now enter peritubular blood, _____-creasing blood pH even more.

   d. In addition, when blood is acidic, more ammonia (_____) is produced as a byproduct of (*lipid? protein?*) metabolism. H^+ in the collecting duct may then be buffered by

   combining with ammonia to form ammonium (_____), which is excreted into urine.

   Alternatively, H^+ may combine with _____ to form _____, which is also excreted in urine.

■ **B20.** State an example of a situation in which your body needs to produce dilute urine in order to maintain homeostasis.

Now describe how your body can accomplish this by doing the following learning activity.

   a. To form dilute urine, kidney tubules must reabsorb (*more? fewer?*) solutes than usual. Two factors facilitate this. One is permeability of the ascending limb to Na^+, K^+, and Cl^- (*and also to? but not to?*) water. Thus solutes enter interstitium and then capillaries, but water stays in urine.

   b. The second requirement for dilute urine is a (*high? low?*) level of ADH. The function of

   ADH is to _____-crease permeability of distal and collecting tubules to water. Less ADH forces more water to stay in urine, leading to a (*hyper? hypo?*)-osmotic urine.

■ **B21.** State two reasons why it might be advantageous for the body to produce a concentrated urine.

Now refer to Figure LG 26.4 and describe mechanisms for concentrating urine in this exercise.

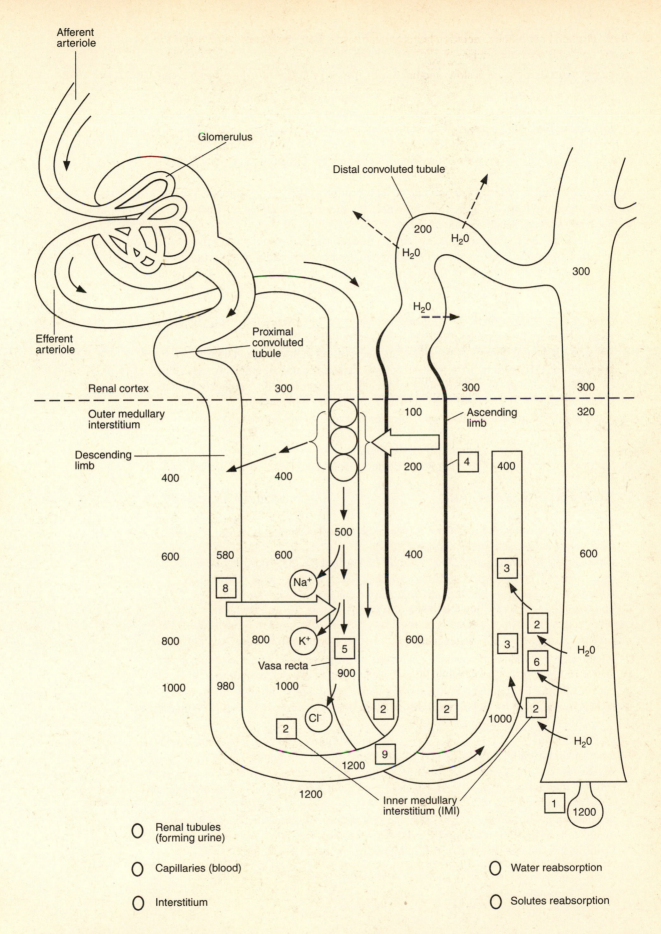

Afferent
arteriole

Glomerulus

Distal convoluted tubule

200

H_2O

H_2O

300

H_2O

Efferent
arteriole

Proximal
convoluted
tubule

Renal cortex   300   300   300

Outer medullary
interstitium   100   Ascending   320
                                            limb

Descending   ⬜4
limb
        400   400   200   400

                                500

        600   580   600   ⬜3
                                400
                        Na⁺
        ⬜8   ⬜2
                                            ⬜2
        800   800   ⬜5   ⬜3
                    K⁺   ⬜6 H_2O

            Vasa recta   900

        1000   980   1000   ⬜2
                    Cl⁻   ⬜2   ⬜2   1000
        ⬜2 H_2O
                    1200 ⬜9
            1200
                                            ⬜1   1200
        1200   Inner medullary
                    interstitium (IMI)

○ Renal tubules
  (forming urine)

○ Capillaries (blood)   ○ Water reabsorption

○ Interstitium   ○ Solutes reabsorption

Figure LG 26.4   The concentration of urine in the nephron. Numbers refer to checkpoint B21. Color as directed in Checkpoints B21 and B22.

a. To help you clearly differentiate microscopic parts of the kidney, first color yellow the renal tubules (containing urine that is forming), color red the capillaries (containing blood), and color green all remaining areas (the interstitium). (Recall that kidneys are composed of over one million nephrons. The *interstitium* is all of the matrix *between* nephrons.) Also draw arrows to show route of "forming urine" within the renal tubule.

b. Now find ⏢1 on the figure. This marks the endpoint of the figure where concentrated urine exits from the nephron. Such urine is known as *(hypo? hyper?)*-tonic (or

_____-osmotic) urine. To produce a small amount of urine that is

highly concentrated in wastes, the rate of reabsorption must _____-crease. Therefore more fluid from the urine forming within the lumen of the collecting duct will pass back into blood. (See ⏢3 on the figure.)

c. This leads to the key question of how to increase tubular reabsorption. The answer lies in establishment of a solute gradient with the highest concentration in the *(outer? inner?)* medullary interstitium (IMI), around the inner portions of the long-loop juxta-medullary tubules. Solute concentration in this region may reach a maximum of

(300? 600? 1200?) mOsm, as compared to only _____ mOsm in the cortical and outer medullary interstitium. (Note these values on the figure.)

d. Because the final segments of the collecting ducts reside in the inner medullary interstitium (IMI) region, urine in these ducts is exposed to this significantly hypertonic environment, shown at ⏢2 on the figure. Much like a desert (or a dry sponge or blotter) surrounding a small stream of fluid, this "dry" (hypertonic) interstitium acts to draw water out of the urine. Ultimately, the fluid in the interstitium will pass into

_____ blood (⏢3 on the figure).

e. So how can this hypertonic interstitium be developed? Two mechanisms are involved.

The first depends on the relative _____ of different portions of the

renal tubule to water and to several solutes. The second is the _____ mechanism (discussed in Checkpoint B22).

f. Consider the first mechanism. On Figure LG 26.4, find the thick portion of the tubule (next to the ⏢4). This is part of the *(ascending? descending?)* tubule. The thick *(squamous? columnar?)* epithelium here causes this region to be *(permeable? impermeable?)* to water in the absence of ADH. However, ions such as Na^+,

_____, and _____ can be moved (by _____-porters) from the tubule into interstitium and then to blood here. Show this by writing names of these chemicals in circles on the figure. (Note that water is forced to stay inside the ascending limb.)

g. As a result, the fluid in the outer medullary and cortical interstitium becomes *(more? less?)* concentrated (or hypertonic). Many of these ions then move into long capillaries

known as vasa _____ (⏢5 on the figure), carrying ions deep into the inner medullary interstitium (IMI) (⏢3) and increasing hypertonicity.

h. As the "desert" (IMI) exerts an osmotic (pulling) force, water is drawn out of urine into IMI and then into capillary blood (⏢2 and ⏢3 on the figure). Of course, this mechanism

depends on the hormone _____, which permits normally impermeable collecting tubule cells to develop water pores and therefore to become more water permeable. By itself this mechanism is effective in increasing urine concentration.

i. But urine concentration can be enhanced even further by the role played by urea. Concentration of urea within these tubules is increased as water moves out of the tubule lumen ([2] on the figure). Because this final portion of the collecting tubule is *(permeable? impermeable?)* to urea, this solute can diffuse out into the IMI, further contributing to the "desert-like" hypertonicity. Show this by writing "urea" near the [6]. The urea factor increases renal capability for water reabsorption and concentration of urine even more.

j. In summary, we have discussed a urine-concentrating mechanism based on differences of specific regions of renal tubule in permeability to water and urea. The resulting concentration gradient consists of a relatively hypotonic renal cortex or outer medullary region (300 mOsm) with an inner medullary interstitium (IMI) highly concentrated in

the waste _____, as well as ions _____, _____, and _____. This IMI "desert" (or blotter) is ready to pull (osmose) water out of urine. But this can

happen only when the hormone _____ is released as the signal that the body needs to

"save fluid." ADH then causes _____ cells in collecting ducts to become more permeable, allowing water to leave them and move into IMI and then into capillary blood. The final outcome (at [1] on the figure) is a *(large? small?)* volume of *(dilute? concentrated?)* urine.

■ **B22.** Refer to Figure LG 26.4 and check your understanding of the countercurrent mechanism in this checkpoint.

a. The term *countercurrent* derives from the fact that the "forming urine" in the descending limbs of renal tubules runs parallel and in the *(same? opposite?)* direction(s) to that in the ascending limb. This is also true of blood flow in the comparable regions of the vasa recta. Draw arrows in the vasa recta to show direction of blood flow there.

b. This mechanism, like that discussed in Checkpoint B21, leads to production of *(dilute? concentrated?)* urine. The countercurrent mechanism *(also? does not?)* depend(s) on the concentration gradient in the interstitium. The *(ascending? descending?)* limb is highly

permeable to water but not to solutes, whereas the _____ limb exhibits just the reverse situation: it is permeable to *(water? solutes?) (and? but not to?)* water. Show this on Figure LG 26.4 by coloring large, hollow arrows to indicate correct locations of water or solute reabsorption.

c. A consequence of movement of water out of the descending limb ([8] in the figure) is

increased concentration of solutes (_____ mOsm) in the renal tubule as that fluid rounds the bend at the "hairpin" ([9] in the figure).

d. Explain why urine becomes more and more dilute, finally falling to *(100? 300? 600?)* mOsm by the top of the ascending tubule.

e. But as urine flows into the DCT and collecting tubules, water will be pulled out (into the IMI "desert") at times when ADH is present. So in the presence of this hormone, it is

possible to produce highly concentrated urine (_____ mOsm) that is *(two? four?*

ten?) times the osmolality of blood plasma and glomerular filtrate (_____ mOsm).

f. The next time you finish a vigorous exercise session, pause for a moment to reflect upon the amazing intricacy of these mechanisms that allow you to limit your fluid loss in

urine because you have just eliminated so much fluid from your _____ glands!

B23. Do this exercise on actions of diuretics.

a. A chemical that mimics the action of ADH is called a(n) *(antidiuretic? diuretic?)*. Such

a chemical _____-creases urinary output.

b. A medication that increases urinary output is called a(n) _____.

Removal of extra body fluid in this way is likely to _____-crease blood pressure.

c. Most diuretics cause loss of *(K⁺? Na⁺?)* from blood plasma; therefore patients must be sure to ingest adequate food or a supplement containing this ion.

B24. *For extra review,* match the names of diuretics with descriptions below.

> Acet. Acetazolamide (Diamox)
> Alc. Alcohol
> Caf. Caffeine (as in coffee, tea, or cola)
> Fur. Furosemide (Lasix)
> Spir. Spironolactone (aldactone)

_____ a. Unlike most diuretics, this diuretic is potassium-sparing because it inhibits action of aldosterone; therefore clients do not normally need to take K⁺ supplements

_____ b. Inhibits secretion of ADH

_____ c. Medication that inhibits Na⁺ (and there-fore water) reabsorption by inhibiting carbonic anhydrase in the brush border of PCT cells

_____ d. Naturally occurring diuretic that inhibits Na⁺ (and water) reabsorption

_____ e. The most potent diuretic drugs; known as loop diuretics

B25. *A clinical challenge.* Mrs. P is taking furosemide (Lasix) to lower her blood pressure. Answer the following questions.

a. Why is this diuretic called a *loop diuretic?*

b. How does the action of this diuretic lower blood pressure?

c. Mrs. P's nurse reminds her to eat a banana or orange each day. Why?

d. Some time later Mrs. P is given a prescription for spironolactone (Aldactone). Speculate as to why Mrs. P's doctor changed her prescription.

B26. *A clinical challenge.* Of the following factors circle the ones that are likely to increase above normal in a patient with kidney failure. State the rationale for your answer.

BUN   Plasma creatinine   GFR   Renal plasma clearance of glucose

■ **B27.** For a chemical to work effectively in renal clearance tests for evaluation of renal function, that chemical *(must? must not?)* be readily filtered and *(must? must not?)* be readily absorbed. Two chemicals that are often used for testing renal clearance are

_____ and _____.

B28. Compare and contrast *hemodialysis* and *continuous ambulatory peritoneal dialysis* (CAPD) as methods of cleansing the blood.

C. Pathway of urine   (pages 878–882)

■ **C1.** Arrange in correct sequence the structures through which urine flows once it is formed in collecting ducts:

MajC. Major calyx	Ureter. Ureter
MinC. Minor calyx	Ureth. Urethra
PapD. Papillary duct	UrinB. Urinary Bladder
RenP. Renal pelvis	

Collecting duct   →   _____   →   _____   →

_____   →   _____   →   _____   →

_____   →   _____   →   outside the body

■ **C2.** Which organs in the list above (Checkpoint C1) are singular in the human body; in other words, only one of that organ is normally present in each person?

_____

■ **C3.** Refer to Figure LG 26.1 and complete the following exercise about the ureters.

a. Ureters connect _____ to _____. Ureters are about

_____ cm (_____ in.) long. What causes urine to flow through these tubes?

b. What prevents urine from moving (retrograde) back into the ureters during contraction of the urinary bladder?

c. Kidneys and ureters are located in the *(anterior? posterior?)* of the abdominopelvic

cavity. They are said to be _____-peritoneal.

d. Kidney stones are known as renal _____. Lithotripsy is a *(surgical? nonsurgical?)* technique for destroying kidney stones by high-frequency

_____ waves. Suggest two or more health practices that may help you to avoid formation of kidney stones.

e. Ureters enter the urinary bladder at two corners of the *(detruser? trigone?)* muscle.

The third corner marks the opening of the _____.

f. Both the ureters and the urinary bladder are lined with _____ epithelium. What advantage does this type of epithelium offer in these organs?

■ **C4.** Refer to Figures LG 28.2 and 28.3 (pages 648 and 651 in your *Guide*), and complete the following exercise about the bladder and the urethra.

a. The urinary bladder is located in the *(abdomen? true pelvis?)*. Two sphincters lie just inferior to it. The *(internal? external?)* sphincter is under voluntary control.

b. Urine leaves the bladder through the _____. In females the length of

this tube is _____ cm (_____ in.); in males it is about _____ cm (_____ in.). Write in correct sequence (according to the pathway of urine) the three portions of the male urethra:

_____ → _____ → _____

In addition to urine, what other fluid passes through the male urethra?

_____

■ **C5.** Ureters, urinary bladder, and urethra are lined with _____ membrane. What is the clinical significance of that fact?

■ **C6.** In the micturition reflex, *(sympathetic? parasympathetic?)* nerves stimulate the

_____ muscle of the urinary bladder and cause relaxation of the internal sphincter. What stimulus initiates the micturition reflex?

C7. Contrast urinary incontinence and retention. List possible causes of each.

D. Urine (pages 882–883)

■ **D1.** Review factors that influence urine volume in this checkpoint.

a. *Blood pressure.* When blood pressure drops, _____ cells of the

kidney release _____, which leads to formation of

_____. This substance *(increases? decreases?)* blood pressure in two
ways: directly, because it serves as a *(vasoconstrictor? vasodilator?)*, and indirectly,

because it stimulates release of _____, which causes retention of both

_____ and _____. Increase in fluid volume will *(increase? decrease?)* blood
pressure.

b. *Blood concentration.* The hypothalamic hormone _____ helps *(conserve? eliminate?)*
fluid. When body fluids are hypertonic, ADH is released, causing *(increased?
decreased?)* tubular reabsorption of water back into blood.

c. *Temperature.* On hot days skin gives up *(large? small?)* volumes of fluid via sweat
glands. As blood becomes more concentrated, the posterior pituitary responds by
releasing *(more? less?)* ADH, so urine volume *(increases? decreases?)*.

d. *Diuretic medications* tend to _____-crease urinary output and _____-crease blood
pressure. For extra review, return to Checkpoint B23.

D2. Describe the following characteristics of urine. Suggest causes of variations from the
normal in each case.

a. Volume: _____ ml/day   _____

b. Color: _____   _____

c. Turbidity: _____   _____

d. Odor: _____   _____

e. pH: _____   _____

f. Specific gravity: _____   _____

D3. The following substances are not normally found in urine. Explain what the presence
of each might indicate.

a. Albumin

b. Bilirubin

c. Casts

d. High levels of hippuric acid

D4. Contrast the following pairs of conditions. Describe what is present in urine in each case and possible causes.

a. *Glucosuria/acetonuria*

b. *Hematuria/pyuria*

c. *Vaginitis due to Candida albicans/vaginitis due to Trichomonas vaginalis*

E. Aging and development of the urinary system (pages 883–885)

■ **E1.** GFR _____-creases with age. Two other common problems associated with aging

are inability to control voiding (_____) and increased frequency of

_____ (UTIs). Explain why UTIs may occur more often in elderly males.

Nocturia may also occur more among the aged population. Define *nocturia*.

■ **E2.** Do this exercise about urinary system development.

a. Kidneys form from _____-derm, beginning at about the

_____ week of gestation.

b. Which develops first? *(Pro? meso? meta?)*-nephros Which extends most superiorly in

location? _____-nephros Which one ultimately forms the kidney? _____-nephros

c. The urinary bladder develops from the original _____, which is

derived from _____-derm.

F. Disorders, clinical terminology   (885–886)

■ **F1.** Match each of the terms with the correct description.

> | | |
> |---|---|
> | A. Azotemia | E. Enuresis |
> | C. Cystitis | GI. Glomerulonephritis |
> | DI. Diabetes insipidus | Po. Polyuria |
> | Diur. Diuresis | Py. Pyelitis |
> | Dys. Dysuria | U. Uremia |

_____ a. Painful urination

_____ b. Inflammation of the urinary bladder

_____ c. Inflammation of the pelvis and calyces of the kidney

_____ d. Bed wetting

_____ e. Excessive urine

_____ f. Inflammation of the kidney involving glomeruli; may follow strep infection

_____ g. Presence of nitrogen-containing chemicals such as creatinine or urea in blood

_____ h. Production of a large volume of urine

_____ i. Production of a large volume of urine due to deficiency of ADH or insensitivity of kidneys to ADH

■ **F2.** List three or more signs or symptoms of urinary tract infections (UTIs).

F3. Discuss and contrast two types of renal failure in this learning activity.

a. *(Acute? chronic?)* kidney failure is an abrupt cessation (or almost) of kidney function. List several causes of acute renal failure (ARF).

b. *(Acute? chronic?)* renal failure is progressive and irreversible. Describe changes that occur during the three stages of CRF.

A2. (a) Nitrogen-containing products of protein catabolism, such as ammonia, urea, uric acid; also certain ions and excessive water. (b) Neogenesis. (c) H^+. (d) Renin, erythropoietin, D.

A3. (a) Waist, T12, L3; 10–12 (4–5). (b) Posterior; retroperitoneal. (c) See KEY to Figure LG 26.1.

A4. (a) Nephroptosis. (b) Renal fascia. (c) Nephrology. (d) Parenchyma. (e) Renal pyramids. (f) Renal columns.

A5. 20–25; 1200.

A6. 1, Arcuate artery; 2, interlobular artery; 3, afferent arteriole; 4, glomerulus; 5, efferent arteriole; 6a, peritubular capillaries; 6b, vasa recta; 7, interlobular vein; 8, arcuate vein. Vasa recta.

A7. (a) Venules; efferent arterioles. (b) Afferent; increases glomerular blood pressure. (See Checkpoint B2c.)

A8. (a) A, glomerular (Bowman's) capsule; B, proximal convoluted tubule (PCT); C, descending limb of the loop of Henle; D, ascending limb of the loop of Henle; E, distal convoluted tubule (DCT); F, collecting duct. (b) A, glomerular (Bowman's) capsule; glomerulus, 4. (c) I → II → III → IV → V. (d) Endothelial; their size and design prevent loss of red blood cells and large- or medium-sized proteins from blood, yet allow smaller wastes to exit into "forming urine" (glomerular filtrate). (e) Capsular (Bowman's) space, visceral; parietal. (f) B (PCT); C (descending limb of loop of Henle) and the first part of D (ascending limb of loop of Henle). (g) E and F; ADH, aldosterone; H^+. (h) B (PCT).

A9. (a) C. (b–d) JM.

A10. (a) Two; ascending; macula densa, NaCl. (b) Modified smooth muscle, afferent arteriole; juxtaglomerular (JG); renin, blood pressure.

B1. (a) Glomerular filtration, tubular reabsorption, tubular secretion. (b) 15–20%; fraction; 180. (c) 178–179; 99; 1–2 (= 180 − 178 or 179).

B2. (a) Pushing, blood, filtrate. (b) Most substances; protein molecules. Those molecules are too large to pass through healthy endothelial-capsular membranes, and the basement membrane restricts passage of proteins. (c) 60; higher; smaller. (Refer to Checkpoint A7b.) (d) Glomerular capillaries are long, thin, and porous and have high blood pressure.

B3. (a) Blood colloid osmotic pressure (BCOP) and capsular hydrostatic pressure (CHP). (b) NFP = (GBHP) − (CHP + BCOP); NFP = (60) − (15 + 27); 18; out of.

B4. (a) NFP = 0 mm Hg = (42) − (27 + 15). Anuria due to low glomerular blood pressure could be related to hemorrhage or stress.

B5. (a) Glomerular filtration rate. (b) 125 (180). (c) De, af; less; stones. (d) Additional water and waste products (such as H^+ and urea) normally eliminated in urine are retained in blood, possibly resulting in hypertension and acidosis.

B6. (a) Renal autoregulation, hormonal regulation, and neural regulation.

B7. (a) Decrease, slowly; more. (b) Low; fewer; more, increase. (c) Negative; is.

B8. (a) R. (b) Ang I. (c) ACE. (d) Ang II. (e) Ald. (f) ADH. (g–i) ANP. (j) R, Epi. (k) ANP.

B10. (a) Blood to filtrate forming urine; 180. (b) Venules; efferent arterioles, capillaries; tubular reabsorption; opposite.

B11. (a) Glucose and bicarbonate (HCO_3^-). (b) 99; Na^+ and Cl^-. (c) Uric acid, urea. (d) K^+. (e) Creatinine; creatinine clearance is often used as a measure of glomerular filtration rate (GFR). One hundred percent of the creatinine (a breakdown product of muscle protein) filtered or "cleared" out of blood will show up in urine because 0% is reabsorbed. (f) Discriminating.

B12. B A C.

B14. Glycos. (a) Eileen's hyperglycemia could be caused by ingestion of excessive amounts of carbohydrates or by hormonal imbalance, such as deficiency in effective insulin (diabetes mellitus) or excess growth hormone. Eileen's glucose level of 340 mg/100 ml plasma exceeds the renal threshold of about 200 mg/100 ml plasma because kidneys have a limited transport maximum (T_m) for glucose. (b) Although Marlene's blood glucose is in the normal (70–110 mg/per 100 ml) range, she may have a rare kidney disorder with a reduced transport maximum (T_m) for glucose.

B15. (a) PCT. (b) D. (c) A. (d) DCT and CD. (e) PCT.

B16. (a) Principal, last part of DCT as well as collecting ducts; adrenal cortex; Na^+; in. (b) More, in, in. (c) Hypothalamus, posterior; negative, in. (d) De, in.

B17. Tubular secretion; blood to urine; the same. K^+, H^+, ammonium (NH_4^+), creatinine, penicillin, and *para*-aminohippuric acid.

B18. (a) 100. (b) Dietary K^+ deficit or loss of K^+ accompanying tissue trauma or use of certain medications (such as most diuretics) leads to need for increased plasma K^+ levels via reabsorption. Dietary K^+ excesses (such as some salt substitutes) require extra elimination (secretion) of K^+. (c) In; enhances; high.

B19. (a) CO_2; acidic; increasing levels of CO_2 tend to cause increase of H^+: $CO_2 + H_2O \rightarrow H_2CO_3 \rightarrow H^+ + HCO_3^-$. (Review content on transport of CO_2 in Chapter 23.) (b) Na^+; $NaHCO_3^-$

(sodium bicarbonate); antiporters. (c) Collecting ducts, 1000; does, primary; in. (d) NH_3, protein; NH_4; HPO_4^{2-}, $H_2PO_4^-$.

B20. During periods of high water intake or decreased sweating. (a) More; but not to. (b) Low; in; hypo.

B21. Water retention by kidneys raises blood pressure as plasma volume is increased and also helps prevent or reverse dehydration, for example, when drinking water is unavailable or when extra water loss occurs via sweating or fever. (a) Refer to Figure LG 26.4A. (b) Hyper, hyper; in. (c) Inner; 1200, 300. (d) Capillary. (e) Permeability; countercurrent. (f) Ascending; columnar, impermeable; K^+, CI^-; sym; see Figure LG 26.4A. (g) More; recta. (h) ADH. (i) Permeable; see Figure LG 26.4A. (j) Urea, Na^+, K^+, CI^-; ADH; principal; small, concentrated.

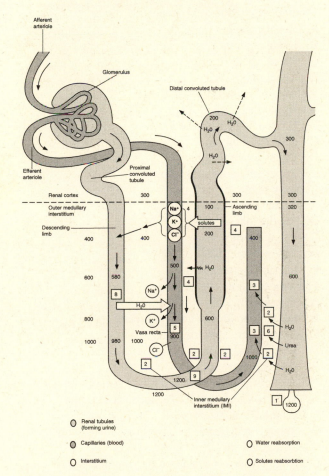

Figure LG 26.4A The concentration of urine in the nephron.

B22. (a) Opposite; see Figure LG 26.4A. (b) Concentrated; also; descending, ascending, solutes, but not to; see Figure LG 26.4A. (c) 1200. (d) 100;

see Checkpoint B21f. (e) 1200, four, 300. (f) Sweat.

B23. (a) Antidiuretic; de. (b) Diuretic (or "water pill"); de. (c) K^+.

B24. (a) Spir. (b) Alc. (c) Acet. (d) Caf. (e) Fur.

B26. BUN and plasma creatinine both increase in blood because less urea and creatinine are filtered from blood. Because GFR decreases, kidneys "clear out" less glucose from blood (renal plasma clearance of glucose decreases), resulting in hyperglycemia.

B27. Must, must not; creatinine, inulin.

C1. PapD → MinC → MajC → RenP → Ureter → UrinB → Ureth.

C2. Urin B, Ureth.

C3. (a) Kidneys, urinary bladder; 25–30 (10–12); gravity, hydrostatic pressure, and peristaltic contractions. (b) No anatomical valve is present, but filling of the urinary bladder compresses urethral openings, thus forcing urine through the open urethral sphincter. (c) Posterior; retro. (d) Calculi; nonsurgical, shock; drink plenty of water (8–10 glasses/day) to keep urine dilute, limit dietary calcium, and discuss with a dietician and/or urologist a diet that minimizes extremes of urine pH. (e) Trigone; urethra. (f) Transitional; the epithelium stretches and becomes thinner as these organs fill with urine.

C4. (a) True pelvis; external. (b) Urethra; 3.8 (1.5), 20 (8); prostatic → embranous → spongy; semen.

C5. Mucous; microbes can spread infection from the exterior of the body (at urethral orifice) along mucosa to kidneys.

C6. Parasympathetic, detrusor; stretching the walls of the bladder by more than 200–400 ml (about 0.8–1.6 cups) of urine.

D1. (a) Juxtaglomerular, renin, angiotensin II; increases, vasoconstrictor, aldosterone, Na^+, H_2O; increase. (b) ADH, conserve; increased. (c) Large; more, decreases. (d) In, de.

E1. De; incontinence, urinary tract infections; increased incidence of prostatic hypertrophy leads to urinary retention; excessive urination at night.

E2. (a) Meso, third. (b) Pro; pro; meta. (c) Cloaca (urogenital sinus portion), endo.

F1. (a) Dys. (b) C. (c) Py. (d) E. (e) Po. (f) Gl. (g) A. (h) Diur. (i) DI.

F2. Burning with urination or painful urination (dysuria); frequency of urination due to irritation of the urinary tract; if the infection spreads up the ureters to kidneys (located at waist level), low back pain may result.

1. Discuss the significance of the structure and arrangement of blood vessels within a nephron.
2. Describe how blood pressure is increased by mechanisms initiated by renin and aided by angiotensin I and II, aldosterone, and ADH.

3. Describe factors that increase risk for urinary tract infections (UTIs). Explain how cystitis is related to UTIs.
4. Discuss signs and symptoms that are likely to be present in a person who has chronic renal failure (CRF).

MASTERY TEST: CHAPTER 26

Questions 1–6: Arrange the answers in correct sequence.

_____ _____ _____ 1. From superior to inferior:
  A. Ureter
  B. Bladder
  C. Urethra

_____ _____ _____ _____ 2. From most superficial to deepest:
  A. Renal capsule
  B. Renal medulla
  C. Renal cortex
  D. Renal pelvis

_____ _____ _____ _____ _____ 3. Pathway of glomerular filtrate:
  A. Ascending limb of loop
  B. Descending limb of loop
  C. Collecting tubule
  D. Distal convoluted tubule
  E. Proximal convoluted tubule

_____ _____ _____ _____ _____ 4. Pathway of blood:
  A. Arcuate arteries
  B. Interlobular arteries
  C. Renal arteries
  D. Afferent arteriole
  E. Interlobar arteries

_____ _____ _____ _____ _____ 5. Pathway of blood:
  A. Afferent arteriole
  B. Peritubular arteries and vasa recta
  C. Glomerular capillaries
  D. Venules and veins
  E. Efferent arteriole

_____ _____ _____ _____ _____ _____ 6. Order of events to restore low blood pressure to normal:
  A. This substance acts as a vasoconstrictor and stimulates aldosterone.
  B. Renin is released and converts angiotensinogen to angiotensin I.
  C. Angiotensin I is converted into angiotensin II.
  D. Decrease in blood pressure or sympathetic nerves stimulate juxtaglomerular cells.
  E. Under influence of this hormone Na^+ and H_2O reabsorption occurs.
  F. Increased blood volume raises blood pressure.

Questions 7–12: Circle the letter preceding the one best answer to each question.

7. Which of these is a normal constituent of urine?
   - A. Albumin
   - B. Urea
   - C. Glucose
   - D. Casts
   - E. Acetone

8. Which is a normal function of the bladder?
   - A. Oliguria
   - B. Nephrosis
   - C. Calculi
   - D. Micturition

9. Which parts of the nephron are composed of simple squamous epithelium?
   - A. Ascending and descending limbs of loop of the nephron
   - B. Glomerular capsule (parietal layer) and descending limb of loop of the nephron
   - C. Proximal and distal convoluted tubules
   - D. Distal convoluted and collecting tubules

10. Choose the one *false* statement.
    - A. An increase in glomerular blood hydrostatic pressure (GBHP) causes a decrease in net filtration pressure (NFP).
    - B. Glomerular capillaries have a higher blood pressure than other capillaries of the body.
    - C. Blood in glomerular capillaries flows into arterioles, not into venules.
    - D. Vasa recta pass blood from the efferent arteriole toward venules and veins.

11. Choose the one *true* statement about renal corpuscle structure and function.
    - A. Afferent arterioles offer more resistance to flow than do efferent arterioles.
    - B. An efferent arteriole normally has a larger diameter than an afferent arteriole.
    - C. The juxtaglomerular apparatus (JGA) consists of cells of the proximal convoluted tubule (PCT) and the afferent arteriole.
    - D. Most water reabsorption normally takes place across the proximal convoluted tubule (PCT).

12. Which one of the hormones listed here causes increased urinary output and naturesis?
    - A. ANP
    - B. ADH
    - C. Aldosterone
    - D. Angiotensin II

Questions 13–20: Circle T (true) or F (false). If the statement is false, change the underlined word or phrase so that the statement is correct.

T F 13. Antidiuretic hormone (ADH) <u>decreases</u> permeability of distal and collecting tubules to water, thus <u>increasing</u> urine volume.

T F 14. A ureter is a tube that carries urine from <u>the bladder to the outside of the body.</u>

T F 15. Diabetes insipidus is a condition caused by a defect in production of <u>insulin</u> or lack of sensitivity to this hormone by principal cells in collecting ducts.

T F 16. A person taking diuretics is likely to urinate <u>less</u> than a person taking no medications.

T F 17. Blood colloid osmotic pressure (BCOP) is a force that tends to <u>push substances from blood into filtrate.</u>

T F 18. The thick portion of the ascending limb of the loop of the nephron is permeable to <u>Na+ and Cl- but not to water.</u>

T F 19. The countercurrent mechanism permits production of <u>small volumes of concentrated urine.</u>

T F 20. In infants 2 years old and under, <u>retention</u> is normal.

Questions 21–25: Fill-ins. Complete each sentence with the word or phrase that best fits.

_____ 21. The glomerular capsule and its enclosed glomerular capillaries constitute a _____.

_____ 22. A _____ consists of a renal corpuscle and a renal tubule.

_____ 23. A toxic level of urea in blood is know as _____.

_____ 24. Sympathetic impulses cause greater constriction of _____ arterioles with resultant _____-crease in GFR and _____-crease in urinary output.

_____ 25. A person with chronic renal failure is likely to be in acidosis because kidneys fail to excrete _____, may be anemic because kidneys do not produce _____, and may have symptoms of hypocalcemia because kidneys do not activate _____.

ANSWERS TO MASTERY TEST: ★ CHAPTER 26

Arrange

1. A B C
2. A C B D
3. E B A D C
4. C E A B D
5. A C E B D
6. D B C A E F

Multiple Choice

7. B   10. A
8. D   11. D
9. B   12. A

True–False

13. F. Increases; decreasing
14. F. A kidney to the bladder
15. F. ADH
16. F. More
17. F. Pull substances from filtrate into blood
18. T
19. T
20. F. Incontinence

Fill-ins

21. Renal corpuscle
22. Nephron
23. Uremia
24. Afferent, de, de
25. H^+, erythropoietic factor, vitamin D

FRAMEWORK 27

Fluid, Electrolyte, Acid–Base Balance

FLUID (A)

- COMPARTMENTS
  - ICF (⅔ of fluids)
  - ECF (⅓ of fluids)
    - interstitial
    - plasma
    - lymph
    - CSF
    - GI

- WATER
  - Percentage body weight
    - 60% of body weight
    - less in elderly, obese
    - more in infants
  - Daily intake: 2500 ml
    - food, drink: 2300 ml
    - metabolic water: 200 ml
  - Daily output: 2500 ml
    - kidneys: 1500 ml
    - skin: 600 ml
    - lungs: 300 ml
    - GI: 100 ml
  - Regulation
    - thirst
    - hormones: ADH, ANP, aldosterone

ELECTROLYTES (B)

- Distribution and roles of ions
  - in ICF:
  - in ECF (mEq/l):
    - Na^+: 136–142
    - Cl^-: 95–103
    - K^+: 3.8–5.0
    - HCO_3^-: 22–26
    - Ca^{2+}: 4.6–5.5
    - HPO_4^{2-}: 1.7–2.6
    - Mg^{2+}: 1.3–2.1

MOVEMENTS OF BODY FLUIDS (C)

- Plasma to IF
  - vesicular transport
  - diffusion
  - bulk flow
- IF to ICF
  - roles of Na^+ and K^+

ACID–BASE BALANCE (D)

- REGULATION
  - Arterial blood pH: 7.35–7.45
  - Mechanisms
    - buffers
    - respirations
    - kidneys

- IMBALANCE
  - Respiratory imbalances (alteration in pCO_2)
    - acidosis, alkalosis
  - Metabolic imbalances (alteration in HCO_3^-)
    - acidosis, alkalosis
  - Compensatory mechanism
    - respirations
    - kidneys
  - Signs/symptoms
    - acidosis: ↓CNS
    - alkalosis: ↑CNS

Fluid, Electrolyte, and Acid–Base Homeostasis

CHAPTER
27

For the maintenance of homeostasis, levels of fluid, ions or electrolytes, and acids and bases must be kept within acceptable limits. Without careful regulation of pH, for example, nerves will become overactive (in alkalosis) or underfunction, leading to coma (in acidosis). Alterations in blood levels of electrolytes can be lethal: excess K^+ can cause cardiac arrest, and decrease in blood Ca^{2+} can result in tetany of the diaphragm and respiratory failure. Systems you studied in Chapter 23 (respiratory) and Chapter 26 (urinary), along with buffers in the blood, normally achieve fluid, electrolyte, and acid–base balance.

   As you begin your study of fluids, electrolytes, and acid–base balance, carefully examine the Chapter 27 Topic Outline and Objectives; check off each one as you complete it. Be sure to refer to the Framework frequently and note relationships among key terms in each section.

TOPIC OUTLINE AND OBJECTIVES

A. Fluid compartments and fluid balance

☐ 1. Compare the locations of intracellular fluid (ICF) and extracellular fluid (ECF) and describe the various fluid compartments of the body.

☐ 2. Describe the sources of water loss and gain and how they are regulated.

B. Electrolytes in body fluids

☐ 3. Contrast the electrolyte concentrations of the three major fluid compartments—plasma, interstitial fluid, and intracellular fluid.

☐ 4. Explain the functions of sodium, chloride, potassium, bicarbonate calcium, phosphate, and magnesium and the regulation of their concentrations.

C. Movement of body fluids

☐ 5. Describe the factors involved in the movement of fluid between plasma and interstitial fluid and between interstitial fluid and intracellular fluid.

D. Acid–base balance and imbalance

☐ 6. Compare the role of buffers, exhalation of carbon dioxide, and kidney excretion of H^+ in maintaining pH of body fluids.

☐ 7. Define acid–base imbalances, their effects on the body, and how they are treated.

WORDBYTES

Now become familiar with the language of this chapter by studying each wordbyte, its meaning, and an example of its use within a term. After you study the entire list, self-check your understanding by writing the meaning of each wordbyte on the line. As you continue through the *Learning Guide,* identify (and fill in) additional terms that contain the same wordbyte.

Wordbyte	Self-check	Meaning	Example(s)
hyper-	_____	above	*hyper*ventilation
hypo-	_____	below	*hypo*tonic
-osis	_____	condition of	alkal*osis*

CHECKPOINTS

A. Fluid compartments and fluid balance   (pages 891–895)

■ **A1.**   Do this exercise about body fluid by answering questions about 40-year-old Alex, who is 193 cm (6′ 4″) tall and weighs 100 kg (220 pounds).

a. Fill in the expected weight in each category:

1. _____ kg of solids (such as fats, protein, and calcium)

2. _____ kg of fluid within cells

3. _____ kg of fluid in interstitial fluid

4. _____ kg of fluid in blood plasma.

b. The total amount of fluid in Alex's body amounts to about _____ kg, whereas the

amount of his extracellular fluid is about _____ kg.

■ **A2.**   About 67% of body fluids are *(inside of? outside of?)* cells. Check your understanding of fluid compartments by circling all fluids that are considered *extracellular fluids (ECF):*

Blood plasma	Lymph
Aqueous humor and vitreous humor of the eyes	Pericardial fluid
Endolymph and perilymph of ears	Synovial fluid
Fluid within liver cells	Cerebrospinal fluid (CSF)
Fluid immediately surrounding liver cells	

For extra review, refer to Figure LG 1.1, page 9 of the *Learning Guide.*

■ **A3.**   *A clinical challenge.* Is a 45-year-old woman who has a water content of 55% likely to be in fluid balance? Explain.

■ **A4.** Circle the person in each pair who is more likely to have a higher proportion of water content.

a. 37-year-old woman/67-year-old woman
b. Female/male
c. Lean person/obese person

■ **A5.** Daily intake and output (I and O) of fluid usually both equal _____ ml (_____ liters). Of the intake, about _____ ml are ingested in food and drink, and _____ ml are a product of catabolism of foods (discussed in Chapter 25).

■ **A6.** Now list four systems of the body that eliminate water. Indicate the amounts lost by each system on an average day. One is done for you.

a. Integument (skin): 600 ml/day

b. _____: _____ ml/day

c. _____: _____ ml/day

d. _____: _____ ml/day
On a day when you exercise vigorously, how would you expect these values to change?

■ **A7.** Describe mechanisms for replacement of fluid that result from reduction of the body's fluid volume.

a. Loss of fluid volume is a condition known as *(de? re? over?)*-hydration. This condition stimulates thirst in three ways. One is _____-creased blood volume (and blood pressure), which stimulates release of _____ from kidneys and leads to production of _____ II. This chemical plays several roles in increasing blood pressure (see Figure LG 18.5, page LG 376), one of which is to *(stimulate? inhibit?)* thirst receptors located in the _____.

b. Identify the other two factors that stimulate thirst receptors during dehydration:

_____-crease in blood osmotic pressure, as well as _____-creased saliva production.

c. List three causes of dehydration.

In these cases, is it advisable to begin fluid replacement before or after onset of thirst? *(Before? After?)*

d. Identify the major inhibitor of thirst:
A. Wetting of oral mucosa
B. Stretching of the stomach or intestines

■ **A8.** Circle the factors that increase fluid output:

A. ADH
B. Aldosterone
C. Diarrhea
D. Increase in blood pressure (BP)
E. Fever
F. Hyperventilation
G. Atrial natriuretic peptide (ANP)

B. Electrolytes in body fluids (pages 895–898)

■ **B1.** Circle all of the answers that are classified as electrolytes:

A. Proteins in solutions
B. Anions
C. Glucose

D. K^+ and Na^+
E. H^+

■ **B2.** Circle the electrolytes that are anions.

Ca^{2+}	HCO_3^- (bicarbonate)	Lactate	Cl^-
HPO_4^{2-} (phosphate)	Na^+	H^+	K^+

B3. Contrast terms in each pair

a. *mEq/liter* and *mOsm/liter*

b. *Osmotic effect of electrolyte/osmotic effect of nonelectrolyte*

c. *Deciliter (dl)/liter*

d. *Osmotic effect of a mole of sodium chloride (NaCl)/osmotic effect of a mole of calcium chloride (CaCl$_2$)*

B4. List four general functions of electrolytes.

_____

_____

_____

_____

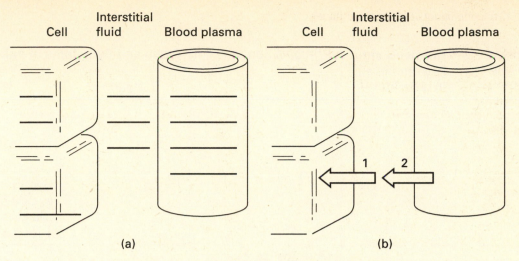

Figure LG 27.1 Diagrams of fluid compartments. (a) Write in electrolytes according to Checkpoint B5 (b) Direction of fluid shift resulting from Na⁺ deficit. Color according to Checkpoint C5

■ **B5.** On Figure LG 27.1a, write the chemical symbols for electrolytes that are found in greatest concentrations in each of the three compartments. Write one symbol on each line provided. Use the following list of symbols. (*Hint:* Refer to Figure 27.4, page 895 of the text.)

Cl^-. Chloride	Mg^{2+}. Magnesium
HCO_3^-. Bicarbonate	Na^+. Sodium
HPO_4^{2-}. Phosphate	Protein⁻. Protein
K^+. Potassium	

■ **B6.** Analyze the information in Figure LG 27.1 by answering these questions.

a. Circle the two compartments that are most similar in electrolyte concentration.
    cell   blood plasma   interstitial fluid (IF)
b. List two major differences in electrolyte content of these two similar compartments.

c. Most body protein is located in which compartment? _____
d. Protein anions in intracellular fluid are more likely to be bound to the cation *(Na⁺? K⁺?)*.

B7. Check your understanding of the functions and disorders related to major electrolytes by completing Table LG 27.1 (next page).

■ **B8.** *A clinical challenge.* Norine is having her serum electrolytes analyzed. Complete this exercise about her results.

a. Her Na⁺ level is 128. Norine's value indicates that she is likely to be in a state of *(hyper? hypo?)*-natremia. Write two or more signs or symptoms of this electrolyte imbalance.

Table LG 27.1 Summary of Major Electrolytes

Electrolyte (Name and Symbol)	Normal Range (Serum Level), in mEq/liter	Principal Functions	Signs and Symptoms of Imbalances:	
			Hyper–	Hypo–
a.	136–142			
b.				Hypochloremia: muscle spasms, alkalosis, ↓ respirations, coma
c.		Most abundant cation in ICF; helps maintain fluid volume of cells; functions in nerve transmission		
d.			Hypercalcemia: weakness, lethargy, confusion, stupor, and coma	
e.	1.7–2.6	Helps form bones, teeth; important components of DNA, RNA, ATP, and buffer system		
f. Magnesium (Mg^{2+})				

b. A diagnosis of hypochloremia would indicate that Norine's blood level of _____

ion is lower than normal. A normal range for serum chloride (Cl$^-$) level is _____

mEq/liter. One cause of low chloride is excessive _____.

c. A normal range for K$^+$ in serum is 3.8–5.0 mEq/liter. This range is considerably *(higher? lower?)* than that for Na$^+$. This is reasonable since K$^+$ is the main cation in *(ECF? ICF?)*, yet lab analysis of serum electrolytes is examining K$^+$ in *(ECF? ICF?)*.

d. Norine's potassium (K$^+$) level is 5.6 mEq/liter, indicating that she is in a state of hyper-

_____. If Norine's serum K$^+$ continues to increase, for example to 8.0 mEq/l, her most serious risk, possibly leading to death, is likely to be

_____.

e. A normal range for Mg^{2+} level of blood is 1.3–2.1 mEq/liter. Norine's electrolyte report indicates a value of 1.1 mEq/liter for Mg^{2+}. This electrolyte imbalance is known as

_____. List three or more factors related to nutritional state that could lead to this condition.

f. Norine's Ca^{2+} level is 3.9 mEq/liter. A normal range is 4.6–5.5 mEq/liter. This elec-

trolyte imbalance is known as _____. This condition is often associ-ated with (decreased? elevated?) phosphate levels, because blood levels of calcium and phosphate are (directly? inversely?) related. Low blood levels of both Ca^{2+} and Mg^{2+} cause overstimulation of the central nervous system and muscles. List two symptoms of these electrolyte imbalances.

■ **B9.** *The Big Picture: Looking Back* and *A clinical challenge.* Refer to figures in earlier chapters of the *Learning Guide* or the text and complete this Checkpoint about regulation and imbalances of electrolytes.

a. On Figure LG 3.2, page LG 46, you identified the major anions and cations inside of cells and in fluids surrounding cells. When significant tissue trauma occurs, injured cells are likely to release their contents, leading to increased blood levels of the major ICF cation *(K⁺? Na⁺?)*. This condition is know as hyper-*(calc? kal? natr?)*-emia.

b. In Figure LG 10.1, page LG 178, you identified sarcoplasmic reticulum (SR) in muscle tissue. Which electrolyte is sequestered inside of SR and released into sarcoplasm sur-rounding thick and thin myofilaments only when this ion is needed to trigger muscle contraction? *(Ca²⁺? Cl⁻? K⁺ Na⁺?)*

c. On Figure LG 12.2, page LG 233, you demonstrated the major anions and cations in and around nerve cells. Which major ECF cation is permitted to pass through voltage-gated channels in neuron plasma membranes to initiate an action potential? *(K⁺? Na⁺?)*

d. On Figure LG 12.3, page LG 235, you identified that the voltage of a resting membrane

potential is typically _____ mV. The major ICF cation in neurons is *(K⁺? Na⁺?)*. In hypokalemia, nerve cells (as well as body fluids) would be deficient in K⁺ and so would be more likely to have a resting membrane potential of *(−65? −80?)*. This means a greater stimulus would be necessary to raise the membrane potential to threshold of

about _____ mV. In hypokalemia, nerves and muscles are therefore likely to be *(hyper? hypo?)*-active. Write two or more signs or symptoms of hypokalemia related to nerves or muscles.

e. Figure LG 18.4, page LG 377, shows that *(calcitonin? PTH?)* enhances movement of calcium into bones; this occurs by stimulation of osteo-*(blasts? clasts?)* in bone. PTH

_____-creases calcium in blood by drawing calcium from several sources. Name

three. _____ _____ _____
In fact, excessive PTH can lead to *(hyper? hypo?)*-calcemia.

f. Figure LG 18.5, page LG 378, emphasizes the effect of aldosterone on blood levels of several electrolytes. Aldosterone tends to increase blood levels of $(H^+? \ K^+? \ Na^+?)$ and decrease blood levels of $(H^+? \ K^+? \ Na^+?)$. An overproduction of aldosterone (as in primary aldosteronism) is likely to lead to *(hypo? hyper?)*-natremia and *(hypo? hyper?)*-kalemia.

g. The same figure points out that ADH _____-creases reabsorption of water from urine into blood. The more water reabsorbed, the *(more? less?)* concentrated Na^+ will be in blood. What effect would overproduction of ADH (as in the syndrome of inappropriate ADH production [SIADH]) have upon blood level of Na^+? Leads to *(hypo? hyper?)*-natremia. What effect would deficient production of ADH (as in diabetes insipidus)

have upon blood level of Na^+? Urine output would greatly _____-crease, leading to dehydration and *(hypo? hyper?)*-natremia.

h. Refer to text Figure 23.23 (page737) and complete the chemical formulas showing formation of bicarbonate.

$$CO_2 \ + \ H_2O \ \rightarrow \ \underline{\hspace{4cm}} \ \rightarrow \ \underline{\hspace{3cm}} \ + H^+$$
$$\text{(carbonic acid)} \qquad\qquad \text{(bicarbonate)}$$

In fact, this equation indicates how most (_____%) of CO_2 in blood is transported as bicarbonate. Bicarbonate also serves as a buffer for *(acids? bases?)*. An arterial blood level of bicarbonate of 32 mEq/ml is *(high? normal? low?)*. Excessive blood levels of

bicarbonate can be reduced as bicarbonate is excreted by _____.

C. Movement of body fluids   (pages 898–899)

■ **C1.** Match mechanisms for exchange of substances between plasma and interstitial fluid with descriptions below. Use answers in the box.

> D. Diffusion
> FR. Filtration and reabsorption
> VT. Vescular transport

_____ a. Most transport across capillaries occurs by this process

_____ b. Only lipid-soluble substances can cross the blood–brain barrier by this process

_____ c. Only a fraction of transport across capillary walls occurs by this process involving endocytosis and exocytosis

_____ d. Known as bulk flow

_____ e. This process leaves some fluid and proteins in interstitial spaces and this is ultimately picked up by lymphatic capillaries

■ **C2.** Edema is an abnormal increase of *(intracellular fluid? interstitial fluid? plasma?)*. Describe three or more factors that may lead to edema.

■ **C3.** Electrolyte levels of the body *(do? do not?)* affect fluid levels. Explain.

■ **C4.** Which hormone is the major regulator of K+? _____

■ **C5.** Show on Figure LG 27.1b the shift of fluid resulting from loss of Na+, for example by excessive loss of sweat. Color arrows (1 and 2) according to color code ovals.

D. Acid–base balance and imbalance   (pages 900–904)

■ **D1.** Complete this Checkpoint about acid–base regulation.

a. The pH of blood and other ECF should be maintained between _____ and _____.
b. Name the three major mechanisms that work together to maintain acid–base balance.

_____ _____ _____

c. List four important buffer systems in the body.

_____ _____

_____ _____

■ **D2.** Refer to Figure LG 27.2 and complete this exercise about buffers.

a. In Figure LG 27.2, you can see why buffers are sometimes called "chemical sponges." They consist of an anion (in this case bicarbonate, or HCO_3^-) that is combined with a cation, which in this buffer system may be either H^+ or Na^+. When H^+ is drawn to the

HCO_3^- "sponge," the weak acid _____ is formed. When Na^+ attaches

to HCO_3^-, the weak base _____ is present. As you color H^+ and Na^+, imagine these cations continually trading places as they are "drawn in or squeezed out of the sponge."

b. Buffers are chemicals that help the body to cope with *(strong? weak?)* acids or bases that are easily ionized and cause harm to the body. When a strong acid, such as hydrochloric acid (HCl), is added to body fluids, buffering occurs, as shown in Figure LG 27.2b. The easily ionized hydrogen ion (H^+) from HCl is "absorbed by the sponge" as Na^+ is released from the buffer/sponge to combine with Cl^-. The two products,

_____ and _____, are weak (less easily ionized).

c. Now complete Figure LG 27.2c by writing the names of products resulting from buffering of a strong base, H^+ (and its arrow) and Na^+ (and its arrow) to show this exchange reaction.

d. When the body continues to take in or produce excess strong acid, the concentration of

the _____ member of this buffer pair will decrease as it is used up in the attempt to maintain homeostasis of pH.

e. Though buffer systems provide rapid response to acid–base imbalance, they are limited since one member of the buffer pair can be used up, and they can convert only strong acids or bases to weak ones; they cannot eliminate them. Two systems of the body that

can actually eliminate acid or basic substances are _____ and

_____.

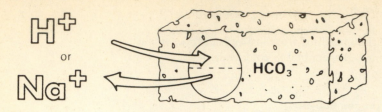

a. H⁺ and Na⁺: only one of you can join me at any one time.

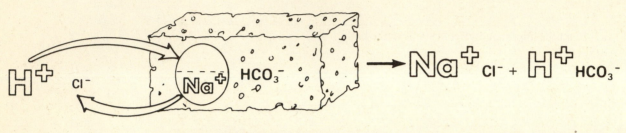

b. Na⁺ is squeezed out as H⁺ is drawn in.

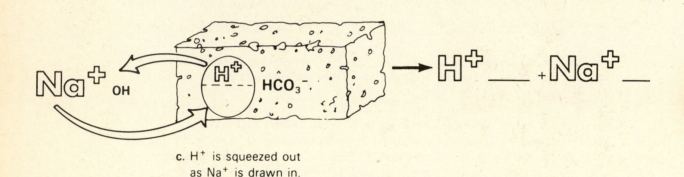

c. H⁺ is squeezed out as Na⁺ is drawn in.

○ H⁺ and arrows showing movement of H⁺

○ Na⁺ and arrows showing movement of Na⁺

Figure LG 27.2 Buffering action of carbonic acid–sodium bicarbonate system. Complete as directed in Checkpoint D2.

■ **D3.** *For extra review.* Check your understanding of the carbonic acid–bicarbonate buffer in this Checkpoint. When a person actively exercises, *(much? little?)* CO_2 forms, leading to H_2CO_3 production. Consequently, *(much? little?)* H⁺ will be added to body tissues. In order to buffer this excess H⁺ (to avoid drop in pH and damage to tissues), the *(H_2CO_3? NaHCO₃?)* component of the buffer system is called upon. The excess H⁺ then replaces the

_____ ion of $NaHCO_3$, in a sense "tying up" the potentially harmful H⁺.

■ **D4.** Phosphates are more concentrated in *(ECF? ICF?)*. (For help, see Figure 27.4, page 895 in the text.) Name one type of cell in which the phosphate buffer system is most

important. _____

D5. Write the reaction showing how phosphate buffers strong acid.

■ **D6.** Which is the most abundant buffer system in the body? _____
Circle the part of the amino acid shown below that buffers acid. Draw a square around the part of the amino acid that buffers base.

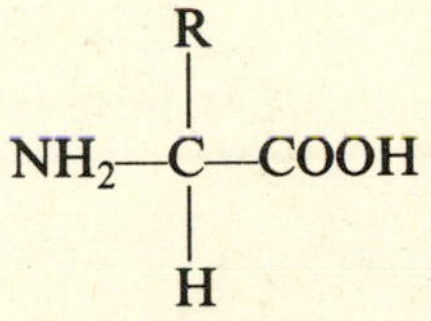

$$NH_2—C—COOH$$

with R above C and H below C.

D7. Explain how the hemoglobin buffer system buffers carbonic acid, so that an acid even weaker than carbonic acid (that is, hemoglobin) is formed.

For extra review. See Chapter 23, Checkpoint E4, pages 523–524 in the *Learning Guide* and the diagram on page 737 in the text.

■ **D8.** Complete this Checkpoint about respiration and kidneys related to pH.

a. Hyperventilation will tend to *(raise? lower?)* blood pH, because as the person exhales

CO_2, less CO_2 is available for formation of _____ acid and free hydrogen ion.

b. A slight decrease in blood pH will tend to *(stimulate? inhibit?)* the respiratory center and so will *(increase? decrease?)* respirations.

c. The kidneys regulate acid–base balance by altering their tubular secretion of _____

or elimination of _____ ion.

■ **D9.** An acid–base imbalance caused by abnormal alteration of the respiratory system is classified as *(metabolic? respiratory?)* acidosis or alkalosis. Any other cause of acid–base imbalance, such as urinary or digestive tract disorder, is identified as *(metabolic? respiratory?)* acidosis or alkalosis.

■ **D10.** *A clinical challenge.* Indicate which of the four categories of acid–base imbalances listed in the box is most likely to occur as a result of each condition described below.

MAcid. Metabolic acidosis	RAcid. Respiratory acidosis
MAlk. Metabolic alkalosis	RAlk. Respiratory alkalosis

_____ a. Decreased blood level of CO_2 as a result of hyperventilation

_____ b. Decreased respiratory rate in a patient taking overdose of morphine

_____ c. Excessive intake of antacids

_____ d. Prolonged vomiting of stomach contents

_____ e. Ketosis in uncontrolled diabetes mellitus

_____ f. Decreased respiratory minute volume in a patient with emphysema or fractured rib

_____ g. Excessive loss of bicarbonate from the body, as in prolonged diarrhea or renal dysfunction

■ **D11.** Refer to Table LG 27.2 and do this exercise on diagnosis of acid–base imbalances.

Patient	Arterial pH	pCO$_2$ (mm Hg)	HCO$_3^-$ (m Eq/liter)
Normal	acidosis 7.35–7.45 alkalosis	alkalosis 35–45 acidosis	acidosis 22–26 alkalosis
Patient W	← 7.29	52* →	30** →
Patient X	7.50 →	← 28*	23**
Patient Y	← 7.31	← 32**	← 19*
Patient Z	7.44	53** →	35* →

Arrows indicate which values have increased above normal (→) or decreased below normal (←). Absence of an arrow means the value is normal. * indicates *primary value* associated with the initial acid–base problem. ** indicates *secondary value* which may be due to compensatory mechanisms for acid–base imbalances.

a. First refer to normal values for arterial blood pH, pCO_2, and HCO_3^-. Acidosis is indicated by arterial blood pH less than *(7.35? 7.45?)*. Keep in mind that acidity and pH are *(directly? inversely?)* related; that is, as acidity of blood increases, its pH

_____-creases. Which patients have blood pH that indicate acidosis? Patients *(W? X? Y? Z?)*

b. Look at the blood chemistry values for patient W, a 61-year-old man with diagnoses of emphysema and pneumonia. The table indicates that a pCO_2 value greater than *(35? 45?)* may be an indicator of acidosis. Patients with chronic respiratory problems retain CO_2, as indicated by this patient's pCO_2 of 52, which is *(higher? lower?)* than normal. The arrow drawn to the right of the 52 indicates that such high CO_2 levels have led this patient into acidosis.

c. Patient X is a 27-year-old woman who has just arrived in the emergency room. She is hyperventilating and in a state of high anxiety after an accident that has critically injured her child. Hyperventilation tends to *(increase? decrease?)* blood pCO_2, as

shown by her value of _____ mm Hg. The arrow next to the 28 points left to show the effect (alkalosis) of this lowered blood pCO_2.

d. Both patients W and X have acid–base imbalances caused by changes in respiration. Therefore patient W's imbalance is classified as *(respiratory? metabolic?)* acidosis, and

patient X's imbalance is known as _____ _____.

e. In *respiratory* acid–base imbalances, the value that is altered is always *(pCO_2? HCO_3^-?)*. For clarity, call this the *primary* value associated with respiratory acid–base imbalances; this value is marked by a * in Table LG 27.2. The alteration is very logical;

for example, increased pCO_2 is associated with _____-osis because CO_2 leads to production of carbonic acid, and decreased pCO_2 indicates

_____-osis because less carbonic acid is made.

f. In order to maintain homeostasis, the body will try to compensate for acid–base imbalances by altering the *secondary* (or "other") value in the *same direction* as the primary value, so that the two values are more matched or in balance. The secondary value is marked with ** in the table. Where compensation is occurring, you will see the arrow for the secondary value (**) going in the *same direction* as the primary value (*). For

respiratory acidosis, in which pCO_2 (the primary value) _____-creases, we can

expect compensatory mechanisms to _____-crease the secondary value (HCO_3^-), possibly to a value greater than normal, in an attempt to make pH more alkaline. It is even possible that a person in respiratory acidosis could move into metabolic alkalosis due to overcompensation, for example, by reduced kidney excretion of sodium bicarbonate ($NaHCO_3$).

g. Does it appear that patient W has done some compensating for his respiratory acidosis? *(Yes? No?)* This is indicated by his HCO_3^- value, which is *(within? outside?)* the normal range. This patient has *(fully? partially?)* compensated respiratory acidosis because his pH is still abnormal.

h. If patient W had his HCO_3^- value within the normal range, this would indicate *(full? partial? no?)* compensation. If patient W's values were pH of 7.39, pCO_2 of 52 mm Hg, and HCO_3^- of 30, this return to the normal pH range would indicate *(fully? partially? un-?)* compensated respiratory acidosis.

i. Patient X's HCO_3^- value (23 mEq) is *(high? normal? low?)*, indicating that she has *(fully? partially? un-?)* compensated respiratory alkalosis. Although she may eventually

excrete HCO_3^- from her _____ to match the decreased pCO_2 level, this has not yet occurred because her anxiety and hyperventilation have been relatively short-term.

j. Patient Y is in diabetic ketoacidosis, is lethargic, and is hyperventilating. The lowered HCO_3^- level indicates that the patient is using bicarbonate to buffer ketoacids resulting from uncontrolled diabetes. The primary value (*) affected is *(pCO_2? HCO_3^-?)*, indicating *(respiratory? metabolic?)* acid–base imbalance. What factor indicates partial compensation?

k. Patients with ketoacidosis or other metabolic acidosis (such as from excessive aspirin

intake) may even move into _____ alkalosis by compensatory hyperventilation. In that case, the pCO_2 would continue to drop, and pH would return to normal and go right on to a value greater than *(7.35? 7.45?)*.

l. Patient Z has a history of peptic ulcer and had been vomiting for days preadmission. The patient states that she had taken a "bunch of antacids" for her ulcer. Does her pH indicate a pH imbalance? *(Yes? No?)* Are her other two values normal? *(Yes? No?)* What type of acid–base imbalances have normal pH but abnormal pCO_2 and HCO_3^-? *(Fully? Partially? Un-?)* compensated imbalances.

m. Patient Z's ulcer, vomiting (loss of gastric HCl), and ingestion of antacids, as well as the level of her pH (7.44) so close to the alkaline end of the normal range, all suggest meta-

bolic _____-osis. Her compensatory mechanisms apparently include *(hypo? hyper?)*-ventilation, with retention of CO_2 which has helped to *(raise? lower?)* her pH back into the normal range.

n. Complete this Checkpoint by writing complete diagnoses (including degree of compensation) for each patient in Table LG 27.2.

■ **D12.** Acidosis has the effect of *(stimulating? depressing?)* the central nervous system (CNS). Write one or more signs of acidosis.

Alkalosis is more likely to cause *(spasms? lethargy?)* because this acid–base imbalance

_____ the CNS.

ANSWERS TO SELECTED CHECKPOINTS: CHAPTER 27

A1. (a1) 40; (a2) 40; (a3) 16; (a4) 4. (b) 60, 20.
A2. Inside of; all answers except fluid within liver cells.
A3. The percentage is within the normal range; however, more information is needed because the fluid must be distributed correctly among the three major compartments: cells (ICF) and ECF in plasma and interstitial fluid. For example, excessive distribution of fluid to interstitial space occurs in the fluid imbalance known as edema.
A4. (a) 37-year-old woman. (b) Male. (c) Lean person.
A5. 2500 (2.5); 2300, 200.
A6. (Any order) (b) Kidney, 1500. (c) Lungs, 300. (d) GI, 100. Output of fluid increases greatly via sweat glands and slightly via respirations. To compensate, urine output decreases and intake of fluids increases.
A7. (a) De; de, renin, angiotensin; stimulate, hypothalamus. (b) In, de. (c) Sweating, vomiting, diarrhea; before. (d) B.

A8. C–G.
B1. All except C (glucose).
B2. Cl^-, HCO_3^- (bicarbonate), HPO_4^{2-} (phosphate), lactate.
B5.

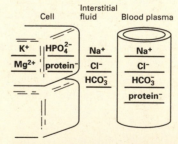

Figure LG 27.1A   Diagram of fluid compartments.

B6. (a) Blood plasma and interstitial fluid.
     (b) Plasma contains more plasma protein, and slightly more sodium and less chloride than IF. (c) Intracellular (ICF). (d) K^+.

B8. (a) Normal serum Na^+ ranges from 136 to 142; hypo; examples: headache, low blood pressure, tachycardia, and weakness, possibly leading to confusion, stupor, and coma if her serum Na^+ continues to drop. (b) Cl^-; 95–103; emesis, dehydration or use of certain diuretics. (c) Lower; ICF, ECF. (d) Kalemia; ventricular fibrillation resulting in cardiac arrest. (e) Hypomagnesemia; malnutrition, malabsorption, diarrhea, alcoholism, or excessive lactation. (f) Hypocalcemia; elevated, inversely; tetany, convulsions.

B9. (a) K^+; kal. (b) Ca^{2+}. (c) Na^+. (d) –70; K^+; –80; –55; hypo; flaccid paralysis, fatigue, changes in cardiac muscle and GI muscle function. (e) Calcitonin, blasts; in; bones, intestine, kidneys; hyper. (f) Increases Na^+ and decreases H^+ and K^+; hyper, hypo. (g) In; less; hypo; in, hyper. (h) H_2CO_3, HCO_3^-; 70; acids; high; kidneys.

C1. (a–b) D. (c) VT. (d–e) FR.

C2. Interstitial fluid; hypertension (which increases filtration at arteriolar ends of capillaries), excessive NaCl intake (which causes fluid retention), congestive heart failure (which backs up blood into venules and capillary beds), or blockage of lymph flow (for example, by cancer metastasized to lymph nodes).

C3. Do, especially Na^+ (the major cation in ECF), which "holds" (retains) water.

C4. Aldosterone.

C5. 1, Shift of fluid leading to overhydration or water intoxication of brain cells, possibly leading to convulsions and coma; 2, shift of fluid (from plasma to replace IF that entered cells), possibly leading to hypovolemic shock.

D1. (a) 7.35–7.45. (b) Buffers, respiration, and kidney excretion. (c) Carbonic acid–bicarbonate, phosphate, hemoglobin–oxyhemoglobin, protein.

D2. (a) Carbonic acid, or H_2CO_3; sodium bicarbonate, or $NaHCO_3$. (b) Strong; the salt sodium chloride (NaCl), the weak acid H_2CO_3 ($H \cdot HCO_3$). (c) Water (H_2O or $H \cdot OH$) and sodium bicarbonate ($NaHCO_3$). (d) $NaHCO_3$ or weak base. (e) Respiratory (lungs), urinary (kidneys).

D3. Much; much; $NaHCO_3^-$; Na^+.

D4. ICF; kidney cells.

D6. Protein;

$$\text{Buffers acid} - \overset{R}{\underset{H}{\overset{|}{\underset{|}{(NH_2) - C - \boxed{COOH}}}}} - \text{Buffers base}$$

D8. (a) Raise, carbonic. (b) Stimulate, increase. (c) H^+, HCO_3^-.

D9. Respiratory; metabolic.

D10. (a) RAlk. (b) RAcid. (c) MAlk. (d) MAlk. (e) MAcid. (f) RAcid. (g) MAcid.

D11. (a) 7.35; inversely; de; W and Y. (b) 45 (except when pCO_2 is high due to hypoventilation as compensation for metabolic alkalosis: see Patient Z); higher. (c) Decrease, 28. (d) Respiratory, respiratory alkalosis. (e) pCO_2; acid, alkal. (f) In, in. (g) Yes; outside; partially. (h) No; fully. (i) Normal, un; kidneys. (j) HCO_3^-, metabolic; pCO_2 is decreased also related to hyperventilation. Respiratory; 7.45. (l) No; no; fully. (m) Alkal; hypo, lower. (n) Patient W: respiratory acidosis, partially compensated; X: respiratory alkalosis uncompensated; Y: metabolic acidosis, partially compensated; Z: metabolic alkalosis, fully compensated.

D12. Depressing; lethargy, stupor, coma; spasms, stimulates.

WRITING ACROSS THE CURRICULUM: CHAPTER 27

1. Contrast the chemical composition of intracellular fluid with that of plasma and interstitial fluid.

2. Explain the importance of maintaining fluid balance among the three major fluid compartments of the body.

3. Explain how electrolytes and nonelectrolytes differ in their effects on osmotic pressure.

4. Explain why changes in fluids result in changes in concentration of electrolytes in interstitial and intracellular fluids. Be sure to state which two electrolytes normally exert the greatest effect on this balance.

5. Describe effects of ADH, ANP, and aldosterone on fluid and electrolyte balance.

6. Describe how you might be able to distinguish two persons with acid–base imbalances: one in acidosis and one in alkalosis.

7. Contrast causes of respiratory acidosis and metabolic acidosis. State several examples of causes.

Questions 1–13: Circle T (true) or F (false). If the statement is false, change the underlined word or phrase so that the statement is correct.

T F   1. Hyperventilation tends to <u>raise</u> pH.

T F   2. Chloride is the major <u>intracellular anion.</u>

T F   3. Under normal circumstances fluid intake each day <u>is greater than</u> fluid output.

T F   4. During edema there is increased movement of fluid <u>out of plasma and into interstitial fluid.</u>

T F   5. Parathyroid hormone causes an increase in the blood level of <u>calcium.</u>

T F   6. When aldosterone is in <u>high</u> concentrations, sodium is conserved (in blood) and potassium is excreted (in urine).

T F   7. In general, electrolytes cause <u>greater</u> osmotic effects than nonelectrolytes.

T F   8. The body of an infant contains a <u>higher</u> percentage of water than the body of an adult.

T F   9. <u>Starch, glucose, HCO_3^-, Na^+, and K^+</u> are all electrolytes.

T F 10. Most transport across capillaries occurs by <u>endocytosis and exocytosis.</u>

T F 11. <u>Protein</u> is the most abundant buffer system in the body.

T F 12. Sodium plays a much <u>greater</u> role than magnesium in osmotic balance of ECF because <u>more</u> mEq/liter of sodium than magnesium are in ECF.

T F 13. The pH of arterial blood is <u>higher</u> than that of venous blood and should have a pH of <u>7.40 to 7.45.</u>

Questions 14–20: Circle the letter preceding the one best answer to each question.

14. Three days ago Ms. Weldon fractured several ribs in an auto accident, and her breathing has been shallow since then. Her arterial blood pH is 7.29, pCO_2 is 50 mm Hg, and HCO_3^- is 30 mEq/liter. Her acid–base imbalance is most likely to be:
    A. Respiratory acidosis
    B. Respiratory alkalosis
    C. Metabolic acidosis
    D. Metabolic alkalosis

15. Ms. Weldon's acid–base imbalance (question 14) appears to be:
    A. Fully compensated
    B. Partially compensated
    C. Uncompensated

16. Mr. Star is in cardiac care post-myocardial infarction. His arterial blood pH is 7.31, pCO_2 is 39 mm Hg, and HCO_3^- is 20 mEq/liter. His acid–base imbalance is most likely to be:
    A. Metabolic acidosis, fully compensated
    B. Metabolic acidosis, partially compensated
    C. Metabolic acidosis, uncompensated

17. On an average day the greatest volume of fluid output is via:
    A. Feces
    B. Sweat
    C. Expiration
    D. Urine

18. Which term refers to a lower than normal blood level of sodium?
    A. Hyperkalemia
    B. Hypokalemia
    C. Hypernatremia
    D. Hyponatremia
    E. Hypercalcemia

19. Which two electrolytes are in greatest concentration in ICF?
    A. Na^+ and Cl^-
    B. Na^+ and K^+
    C. K^+ and Cl^-
    D. K^+ and HPO_4^{2-}
    E. Na^+ and HPO_4^{2-}
    F. Ca^{2+} and Mg^{2+}

20. All of the following factors will tend to increase fluid output *except:*
    A. Fever
    B. Vomiting and diarrhea
    C. Hyperventilation
    D. Increased glomerular filtration rate
    E. Decreased blood pressure

_____ 21. List four factors that increase production of aldosterone.

_____ 22. About _____% of body fluids form part of _____-cellular fluid, and _____% are in ECF.

_____ 23. _____ are chemicals that have at least one ionic bond and dissociate into positive and negative ions.

_____ 24. To compensate for acidosis, kidneys will increase secretion of _____, and in alkalosis kidneys will eliminate more _____.

_____ 25. Normally blood serum contains about _____ mEq/liter of Na^+ and _____ mEq/liter of K^+.

ANSWERS TO MASTERY TEST: ✯ CHAPTER 27

True–False

1. T
2. F. Extracellular anion
3. F. Equals
4. T
5. T
6. T
7. T
8. T
9. F. HCO_3^-, Na^+, K^+ (Starch and glucose are nonelectrolytes.)
10. F. Diffusion
11. T
12. T
13. T

Multiple Choice

14. A
15. B
16. C
17. D
18. D
19. D
20. E

Fill-ins

21. Decreased blood volume or cardiac output, decrease of Na^+ or increase of K^+ in ECF, or physical stress.
22. 67, intra-, 33
23. Electrolytes
24. H^+, HCO_3^-
25. 136 to 142, 3.8 to 5.0.

UNIT V

Continuity

FRAMEWORK 28

The Reproductive Systems

- **MALE REPRODUCTION (A)**
  - **TESTES**
    - In scrotum
    - Seminiferous tubules
    - Interstitial endocrinocytes
    - Sustenacular cells
  - **DUCTS**
    - Epididymis
    - Ductus (vas) deferens
    - Ejaculatory duct
    - Urethra
  - **ACCESSORY SEX GLANDS**
    - Seminal vesicles
    - Prostate
    - Bulbourethral glands
  - **PENIS**
    - Corpus cavernosum & spongiosum
    - Glans
    - Prepuce

- **FEMALE REPRODUCTION**
  - **REPRODUCTIVE ORGANS (B)**
    - Ovaries
    - Uterine tubes
    - Uterus
      - endometrium
      - myometrium
      - Pap smear
    - Vagina
    - Vulva
    - Perineum
    - Mammary glands
  - **REPRODUCTIVE CYCLE (C)**
    - **Hormonal Control**
      - Hypothalamus
      - Anterior pituitary
      - Ovary
    - **Menstrual Cycle**
      - Menstrual phase (days 1–5)
      - Preovulatory phase (days 6–13)
      - Ovulation (day 14)
      - Postovulatory phase (days 15–28)

- **PHYSIOLOGY OF SEXUAL INTERCOURSE, BIRTH COTROL (D)**
  - Sexual activity
    - erection
    - lubrication
    - orgasm
  - Birth control

- **AGING, DEVELOPMENT (E)**
  - MENARACHE, MENOPAUSE
  - **DISORDERS (F)**

The Reproductive Systems

CHAPTER

28

Unit V introduces the systems in females and males that provide for continuity of the human species, not simply homeostasis of individuals. The reproductive systems produce sperm and ova that may unite and develop to form offspring. In order for sperm and ova to reach a site for potential union, they must each pass through duct systems and be nourished and protected by secretions. External genitalia include organs responsive to stimuli and capable of sexual activity. Hormones play significant roles in the development and maintenance of the reproductive systems. The variations of hormones in ovarian and uterine (menstrual) cycles are considered in detail, along with methods of birth control. As we have done for all other systems, we discuss development of the reproductive system and normal aging changes.

As you begin your study of reproduction, carefully examine the Chapter 28 Topic Outline and Objectives. Check off each objective as you complete it. The Chapter 28 Framework provides an organizational preview of the reproductive systems. Be sure to refer to the Framework frequently and note relationships among key terms in each section.

TOPIC OUTLINE AND OBJECTIVES

A. Male reproduction

☐ 1. Define reproduction and classify the organs of reproduction by function.

☐ 2. Explain the structure, histology, and functions of the testes, ducts, accessory sex glands, and penis.

B. Female reproductive organs

☐ 3. Describe the location, histology, and functions of the ovaries, uterine (Fallopian) tubes, uterus, vagina, vulva, and mammary glands.

C. Female reproductive cycle

☐ 4. Compare the principal events of the ovarian and uterine cycles.

D. Physiology of sexual intercourse; birth control

☐ 5. Describe the similarities and differences in the male and female sexual response.

☐ 6. Contrast the various kinds of birth control (BC) and their effectiveness.

E. Aging and development of the reproductive systems

☐ 7. Describe the effects of aging on the reproductive systems.

☐ 8. Describe the development of the reproductive systems.

F. Disorders; medical terminology

☐ 9. Explain the symptoms and causes of sexually transmitted diseases (STDs) such as chlamydia, gonorrhea, syphilis, genital herpes, trichomoniasis, and genital warts.

☐ 10. Describe the symptoms and causes of male disorders (testicular cancer, prostate disorders, impotence, and infertility) and female disorders (amenorrhea, dysmenorrhea, premenstrual syndrome [PMS], toxic shock syndrome [TSS], ovarian cysts, endometriosis, infertility, breast cysts and benign tumors, cervical cancer, pelvic inflammatory disease [PID], and vulvovaginal candidiasis).

☐ 11. Define medical terminology associated with the reproductive systems.

WORDBYTES

Now become familiar with the language of this chapter by studying each wordbyte, its meaning, and an example of its use within a term. After you study the entire list, self-check your understanding by writing the meaning of each wordbyte on the line. As you continue through the *Learning Guide,* identify (and fill in) additional terms that contain the same wordbyte.

Wordbyte	Self-check	Meaning	Example(s)
andro-	_____	man	*andro*gen
-arche	_____	beginning	men*arche*
cervix	_____	neck	*cervix* of uterus
corp-	_____	body	*corp*us cavernosum
crypt-	_____	hidden	*crypt*orchidism
-genesis	_____	formation	spermato*genesis*
gyn-	_____	woman	*gyn*ecologist
hyster(o)-	_____	uterus	*hyster*ectomy
labia	_____	lips	*labia* majora
mamm-	_____	breast	*mamm*ogram
mast-	_____	breast	*mast*ectomy
meio-	_____	smaller	*meio*sis
men-	_____	month	*men*opause, *men*ses
-metrium	_____	uterus	endo*metrium*
mito	_____	thread	*mito*sis
oo-, ov-	_____	egg	*oo*genesis, *ov*um
orchi-	_____	testis	*orchi*dectomy
rete	_____	network	*rete* testis
salping-	_____	tube, trumpet	*salping*itis
troph-	_____	nutrition	*troph*oblast

CHECKPOINTS

A. Male reproduction (pages 909–923)

A1. Define each of these groups of reproductive organs.

a. Gonad

b. Ducts

c. Accessory glands

d. Supporting structures

■ **A2.** As you go through the chapter, note which structures of the male reproductive system are included in each of the above categories.

■ **A3.** Describe the scrotum and testes in this exercise.

a. Arrange coverings around testes from most superficial to deepest:

_____ _____ _____
D. Dartos: smooth muscle that wrinkles skin
TA. Tunica albuginea: dense fibrous tissue
TV. Tunica vaginalis: derived from peritoneum

b. The temperature in the scrotum is about 3°C *(higher? lower?)* than the temperature in the abdominal cavity. Of what significance is this?

c. During fetal life, testes develop in the *(scrotum? posterior of the abdomen?)*.

Undescended testes is a condition known as _____. This condition

places the male at higher risk for _____.

d. Skeletal muscles that elevate testes closer to the warmth of the pelvic cavity are known as *(cremaster? dartos?)* muscles.

■ **A4.** Discuss spermatogenesis in this exercise.

a. Spermatozoa are one type of gamete; name the other type: _____

Gametes are *(haploid? diploid?)*. Human gametes contain _____ chromosomes.

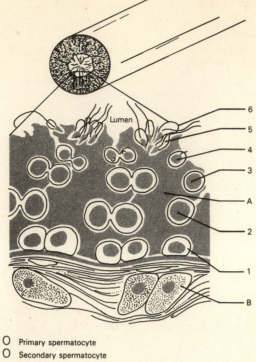

O Primary spermatocyte
O Secondary spermatocyte

Figure LG 28.1 Diagram of a portion of a cross section of a seminiferous tubule showing the stages of spermatogenesis. Color and label as directed in Checkpoint A5.

b. Fusion of gametes produces a cell called a _____. This cell, and all cells of the organism derived from it, contain the *(haploid? diploid?)* number of chromosomes, written as *(n? 2n?)*. These chromosomes, received from both sperm and

ovum, are said to exist in _____ pairs.

c. Gametogenesis assures production of haploid gametes from otherwise diploid humans. In males,

this process, called _____-genesis, occurs in the _____.

■ **A5.** Refer to Figure LG 28.1 and do the following Checkpoint. Use the following list of cells for answers.

Early spermatid	Secondary spermatocyte
Interstitial endocrinocyte (cell of Leydig)	Spermatogonium
Late spermatid	Spermatozoon (sperm)
Primary spermatocyte	Sustentacular (Sertoli) cell

a. Label cells A and B and 1–6.
b. Mark cells 1–6 according to whether they are diploid (2n) or haploid (n).

c. Meiosis is occurring in cells numbered _____ and _____. Color these meiotic cells according to color ovals on the figure. Which cell undergoes reduction division

(meiosis I)? _____ Which undergoes equatorial division (meiosis II)? _____
For extra review of the process of meiosis, review Chapter 3, Checkpoint F7, pages 58–59 in the *Learning Guide.*

d. From this figure, you can conclude that as cells develop and mature, they move *(toward? away from?)* the lumen of the tubule.

e. Which cells are in the process of spermiogenesis? _____ and _____

f. Which cells secrete testosterone? _____
g. Which cells produce nutrients for developing sperm, phagocytose excess spermatid

cytoplasm, form the blood-testis barrier, and produce the hormone inhibin? _____
h. Which cells lie dormant near the basement membrane until they divide mitotically beginning at puberty? They also serve as a reservoir of stem cells available for

spermatogenesis throughout life. _____

A6. Explain how cytoplasmic bridges between cells formed during spermatogenesis may facilitate survival of Y-bearing sperm and therefore permit generation of male offspring.

A7. Describe spermatozoa (sperm) in this Checkpoint.
a. Number produced in a normal male: _____/day

b. Life expectancy of sperm once inside female reproductive tract: _____ hours

c. Portion of sperm containing nucleus: _____

d. Part of sperm containing mitochondria: _____
e. Function of acrosome:

f. Function of tail: _____

A8. Match the hormones listed in the box with descriptions below. Answers may be used more than once.

DHT	Inhibin
FSH	LH
GnRH	Testosterone

a. Released by the hypothalamus; stimulates release of two anterior pituitary hormones:

_____

b. Inhibits release of GnRF: _____
c. Co-stimulate sustantacular cells to secrete ABP, which keeps testosterone near cells undergoing spermatogenesis (two hormones):

_____

d. Stimulates release of testosterone:

_____

e. Principal male hormone and precursor of dihydro-

testosterone (DHT): _____

f. Testicular hormone that inhibits FSH production:

_____

g. Before birth, stimulates development of external

genitals: _____
h. At puberty, facilitate development of male sexual organs and secondary sex characteristics; contribute to libido (two hormones):

_____

i. Anabolic hormones, because they stimulate protein synthesis (two hormones):

_____

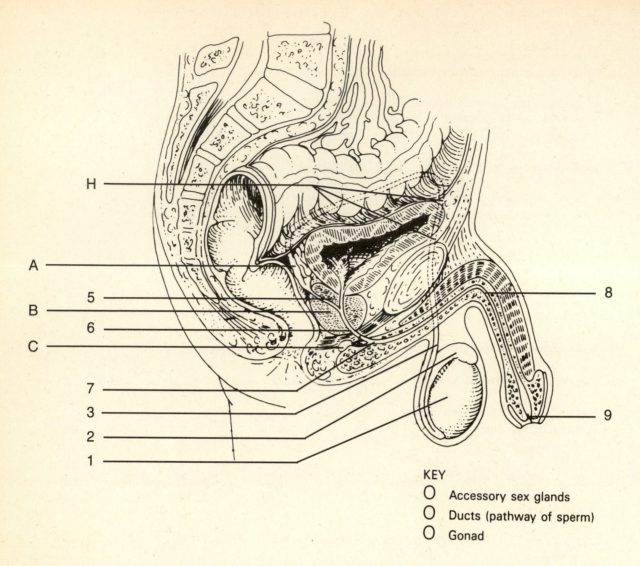

KEY

O Accessory sex glands
O Ducts (pathway of sperm)
O Gonad

Figure LG 28.2 Male organs of reproduction seen in sagittal section. Structures are numbered in order along the pathway taken by sperm. Color and label as directed in Checkpoints A10, A11, A15, and A20.

A9. List five functions of androgens.

■ **A10.** Refer to Figure LG 28.2. Structures in the male reproductive system are numbered in order along the pathway that sperm take from their site of origin to the point where they exit from the body. Match the structures in the figure (using numbers 1 to 9) with descriptions below.

_____ a. Ejaculatory duct

_____ b. Epididymis

_____ c. Prostatic urethra

_____ d. Spongy (cavernous) urethra

_____ e. Membranous urethra

_____ f. Ductus (vas) deferens (portion within the abdomen)

_____ g. Ductus (vas) deferens (portion within the scrotum)

_____ h. Urethral orifice

_____ i. Testis

■ **A11.** Which structures numbered 1–9 in Figure LG 28.2 are paired? Circle their numbers:

      1   2   3   4   5   6   7   8   9

    Which structures are single? Write their numbers: _____

■ **A12.** Answer these questions about the ductus (vas) deferens.
a. The ductus deferens is about _____ cm (_____ inches) long.
b. It is located *(entirely? partially?)* in the abdomen. It enters the abdomen via the

    _____. Protrusion of abdominal contents through this weakened area

    is known as _____.
c. What structures besides the vas compose the seminal cord?

d. A vasectomy is performed at a point along the section of the vas that is within the

    _____. Must the peritoneal cavity be entered during this proce-
dure? *(Yes? No?)* Will the procedure affect testosterone level and sexual desire of the
male? *(Yes? Not normally?)* Why?

■ **A13.** Which portion of the male urethra is longest? *(Prostatic? Membranous? Spongy or
cavernous?)*

■ **A14.** Match the accessory sex glands with their descriptions.

> | B. Bulbourethral glands | P. Prostate gland | S. Seminal vesicles |

_____ a. Paired pouches posterior to the urinary
            bladder
_____ b. Structures that empty secretions (along
            with contents of vas deferens) into
            ejaculatory duct
_____ c. Single doughnut-shaped gland
            located inferior to the bladder;
            surrounds and empties secretions into
            urethra

_____ d. Contribute about 60% of seminal fluid

_____ e. Pair of pea-sized glands located in the
            urogenital diaphragm
_____ f. Produces clotting enzymes and fibrino-
            lysin
_____ g. Forms 25% of semen

_____ h. Produces PSA

■ **A15.** Color parts of the male reproductive system on Figure LG 28.2 according to color
code ovals. Label the *seminal vesicle, prostate gland,* and *bulbourethral gland* on the
figure.

■ **A16.** *A clinical challenge.* Explain the clinical significance of the shape and location of the
prostate gland.

■ **A17.** Complete the following exercise about semen.

a. The average amount per ejaculation is _____ ml.

b. The average range of number of sperm is _____/ml. When the count

falls below _____/ml, the male is likely to be sterile. Only one sperm
fertilizes the ovum. Explain why such a high sperm count is required for fertility.

c. The pH of semen is slightly *(acid? alkaline?)*. State the advantage of this fact.

d. What is the function of *seminalplasmin?*

e. What is the function of *fibrinolysin* in semen?

A18. List six or more characteristics of semen studied in a semen analysis.

■ **A19.** Refer to Figure 28.10 (page 922 in the text) and do this exercise.

a. The _____, or foreskin, is a covering over the

_____ of the penis. Surgical removal of the foreskin is known

as _____.

b. The names *corpus* _____ and *corpus* _____ indi-
cate that these bodies of tissue in the penis contain spaces that distend in the presence

of excess _____. As a result, blood is prohibited from leaving the
penis because *(arteries? veins?)* are compressed. This temporary state is known as the
(erect? flaccid?) state.

c. The urethra passes through the corpus _____. The urethra functions

in the transport of _____ and _____. During
ejaculation, what prevents sperm from entering the urinary bladder and urine from
entering the urethra?

A20. Label these parts of the penis on Figure LG 28.2: *prepuce, glans, corpus
cavernosum, corpus spongiosum,* and *bulb of the penis.*

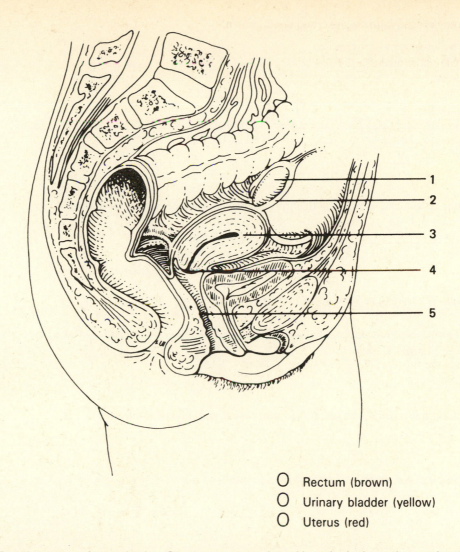

O Rectum (brown)
O Urinary bladder (yellow)
O Uterus (red)

Figure LG 28.3 Female organs of reproduction. Structures are numbered in order in the pathway taken by ova. Color and label as directed in Checkpoints B1, B2, B7, B9, and B11.

B. Female reproductive organs (pages 923–936)

■ **B1.** Refer to Figure LG 28.3 and match the structures numbered 1–5 with their descriptions. Note that structures are numbered in order along the pathway taken by ova from the site of formation to the point of exit from the body.

_____ a. Uterus (body)   _____ d. Uterine (Fallopian) tube

_____ b. Uterus (cervix)   _____ e. Vagina

_____ c. Ovary

■ **B2.** Indicate which of the five structures in Figure LG 28.3 are paired. Circle the numbers.

      1   2   3   4   5

Which structures are singular? Write their numbers: _____

■ **B3.** Describe the ovaries in this Checkpoint.

a. The two ovaries most resemble:
  A. Almonds (unshelled)
  B. Peas
  C. Pears
  D. Doughnuts

b. Position of the ovaries is maintained by *(ligaments? muscles?)*. Name two or more of these structures.

c. Arrange in sequence according to chronological appearance, from first to develop to

  last. _____ _____ _____   CL. Corpus luteum
  OF. Ovarian follicle
  VOF. Vesicular ovarian (Graafian) follicle

■ **B4.** Color the four cells on Figure LG 28.4 according to the color code ovals.

■ **B5.** Describe exactly what is expelled during ovulation. Place an asterisk ([*]) next to that cell on Figure LG 28.4.

What determines whether the second meiotic (equatorial) division will occur?

■ **B6.** Complete this exercise about the uterine tubes.

a. Uterine tubes are also known as _____ tubes. They are about

  _____ cm (_____ inches) long.

b. Arrange these parts of the uterine tubes in order from most medial to most lateral:

  _____ _____ _____   A. Ampulla
  B. Infundibulum and fimbriae
  C. Isthmus

c. What structural features of the uterine tube help to draw the ovum into the tube and propel it toward the uterus?

d. List two functions of uterine tubes.

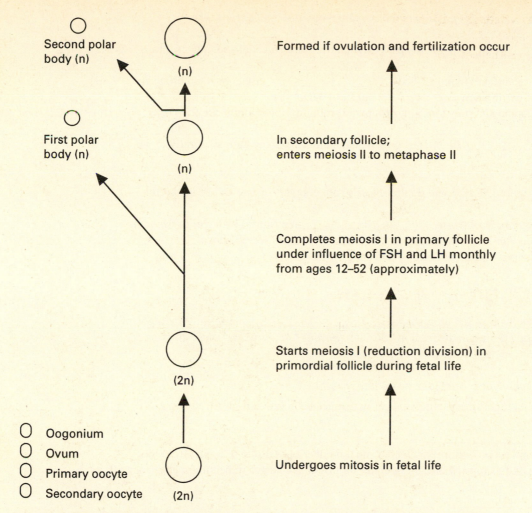

Second polar
body (n)

First polar
body (n)

(n)

(n)

(2n)

○ Oogonium
○ Ovum
○ Primary oocyte
○ Secondary oocyte

(2n)

Formed if ovulation and fertilization occur

In secondary follicle;
enters meiosis II to metaphase II

Completes meiosis I in primary follicle
under influence of FSH and LH monthly
from ages 12–52 (approximately)

Starts meiosis I (reduction division) in
primordial follicle during fetal life

Undergoes mitosis in fetal life

Figure LG 28.4   Oogenesis. Color and complete as directed in Checkpoints B4 and B5.

e. Define *ectopic pregnancy* and list possible causes of this condition.

B7.   Color the pelvic organs on Figure LG 28.3. Use the color code ovals on the figure.

■ **B8.**   Answer these questions about uterine position.

a. The organ that lies anterior and inferior to the uterus is the _____.

The _____ lies posterior to it.

b. The rectouterine pouch lies *(anterior? posterior?)* to the uterus, whereas the

_____-uterine pouch lies between the urinary bladder and the uterus.

c. The fundus of the uterus is normally tipped *(anteriorly? posteriorly?)*. If it is
malpositioned posteriorly, it would be *(anteflexed? retroflexed?)*.

d. The _____ ligaments attach the uterus to the sacrum. The

_____ ligaments pass through the inguinal canal and anchor into

external genitalia (labia majora). The _____ ligaments are broad,
thin sheets of peritoneum extending laterally from the uterus.
e. Important ligaments in prevention of drooping (prolapse) of the uterus are the

_____ ligaments, which attach the base of the uterus laterally to the
pelvic wall.

■ **B9.** Refer to Figure LG 28.3 and do this exercise about layers of the wall of the uterus.

a. Most of the uterus consists of _____-metrium. This layer is *(smooth muscle? epithelium?)*. It should appear red in the figure (as directed above).
b. Now color the peritoneal covering over the uterus green. This is a *(mucous? serous?)*

membrane known as the _____.

c. The innermost layer of the uterus is the _____-metrium. Which portion of it is shed during menstruation? Stratum *(basalis? functionalis?)*. Which arteries supply the stratum functionalis? *(Spiral? Straight?)*

■ **B10.** *A clinical challenge.* Name the procedures described.
a. Direct examination of the vaginal and cervical mucosa by low-power microscope:

_____.

b. Relatively painless procedure in which a small number of cells are removed from the cervical area and examined microscopically for possible changes indicating malignancy:

_____

c. Surgical removal of the uterus: _____.

■ **B11.** Refer to Figure LG 28.3. In the normal position, at what angle does the uterus join

the vagina? _____

B12. Describe these vaginal structures.
a. Fornix

b. Rugae

c. Hymen

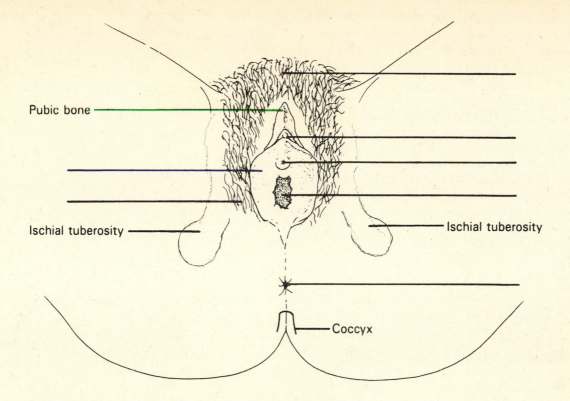

Pubic bone

Ischial tuberosity

Ischial tuberosity

Coccyx

Figure LG 28.5 Perineum. Label as directed in Checkpoint B14.

■ **B13.** The pH of vaginal mucosa is *(acid? alkaline?)*. What is the clinical significance of this fact?

■ **B14.** Refer to Figure LG 28.5 and do the following exercise.
   a. Label these perineal structures: *anus, clitoris, labia major, labia minor, mons pubis, urethra,* and *vagina.*
   b. Draw the borders of the perineum, using landmarks labeled on the diagram. Draw a line between the right and left ischial tuberosities. Label the triangles that are formed. In

   which triangle are all of the female external genitalia located? _____
   c. Draw a dotted line where an *episiotomy* incision would be performed for childbirth.

■ **B15.** Describe mammary glands in this Checkpoint.

Each breast consists of 15–20 _____ composed of lobules. Milk-secret-

ing glands, known as _____, empty into a duct system that terminates at the *(areola? nipple?).*

■ **B16.** *A clinical challenge.* Mrs. Rodriguez, a 55-year-old woman, tells the nurse that she is concerned about having breast cancer. Do this exercise about the nurse's conversation with Mrs. Rodriguez.

    a. The nurse asks some questions to evaluate Mrs. Rodriguez's risk of breast cancer. Which of the following responses indicate an increased risk of breast cancer?

        A. "My older sister had cancer when she was 40 and she had the breast removed."

        B. "I've never had any kind of cancer myself."

        C. "I have two daughters; they were born when I was 36 and 38."

        D. "I don't smoke and I never did."

        E. "I do need to cut back on fatty foods; I know I eat too many of them."

    b. The nurse tells the client that the American Cancer Society recommends three methods of early detection of breast cancer. These are listed below. On the line next to each of these, write the frequency that the nurse is likely to suggest to Mrs. Rodriguez.

        1. Self breast exam (SBE) _____

        2. Mammogram _____

        3. Breast exam by physician _____

C. Female reproductive cycle   (pages 937–941)

    C1. Contrast *menstrual cycle* and *ovarian cycle*.

■ **C2.** Select hormones produced by each of the endocrine glands listed below.

E. Estrogen	LH. Luteinizing hormone
FSH. Follicle-stimulating hormone	P. Progesterone
GnRH. Gonadotropin-releasing hormone	R. Relaxin
I. Inhibin	

Hypothalamus: _____   Anterior pituitary: _____

Ovary: _____

■ **C3.** Now select the hormones listed in C2 that fit descriptions below.

_____ a. Stimulates release of FSH   _____ f. Prepares endometrium for implantation by fertilized ovum

_____ b. Stimulates release of LH   _____ g. Dilates cervix for labor and delivery

_____ c. Inhibits release of FSH   _____ h. β-Estradiol, estrone, and estriol are forms of this hormone

_____ d. Inhibits release of LH   _____ i. Helps control fluid and electrolyte balance and lowers blood cholesterol level

_____ e. Stimulates development and maintenance of uterus, breasts, and secondary sex characteristics; enhances protein anabolism

■ **C4.** Refer to Figure LG 28.6, and check your understanding of events in female cycles in this checkpoint.

a. The average duration of the menstrual cycle is _____ days. The menstrual phase

usually lasts about _____ days. During this period about _____ ml of blood, cells, mucus, and other fluid exits from the uterus in menstrual flow. On Figure LG 28.6d, color red the endometrium in the menstrual phase, and note the changes in its thickness. Also write *menses* on the line in that phase on Figure LG 28.6d.

b. Following the menstrual phase is the _____ phase. During both of

these phases, _____ develop in the ovary. Usually only _____ matures each month, becoming the dominant follicle, and later the

_____ (Graafian) follicle. Normally about _____ days are

required (_____ days of the previous cycle plus _____ days of the current cycle) for a secondary follicle to develop into a mature follicle. The mature follicle contains a developing ovum that has reached metaphase of meiosis *(I? II?)* of oogenesis.

c. Follicle development occurs under the influence of the hypothalamic hormone

_____, which regulates the anterior pituitary hormones _____ and later _____.

d. What are the two main functions of follicles?

e. How do estrogens affect the endometrium?

For this reason, the preovulatory phase is also known as the _____ phase. Because follicles reach their peaks during this phase, it is also known as the

_____ phase. Write those labels on Figures LG 28.6d and 28.6b.

f. A moderate increase in estrogen production by the growing follicles initiates a *(positive? negative?)* feedback mechanism that *(increases? decreases?)* FSH level. It is this change in FSH level that causes atresia of the remaining partially developed follicles. *Atresia* means *(degeneration? regeneration?)*.

g. The principle of this negative feedback effect is utilized in oral contraceptives. By starting to take "pills" (consisting of target hormones estrogen and progesterone) early in the cycle (day 5), and continuing until day 25, a woman will maintain a very low level of

the tropic hormones _____. Without these hormones, follicles and

ova will not develop, and so the woman will not _____.

h. The second anterior pituitary hormone released during the cycle is _____. The surge of this hormone results from the *(high? low?)* level of estrogens present just at the end of the preovulatory phase. This is an example of a *(positive? negative?)* feedback mechanism.

i. The LH surge causes the release of the ovum (actually secondary oocyte), the event

known as _____. The LH surge is the basis for a home test for *(ovulation? pregnancy?)*. Color all stages of follicle development through ovulation on Figure LG 28.6b according to the color code ovals.

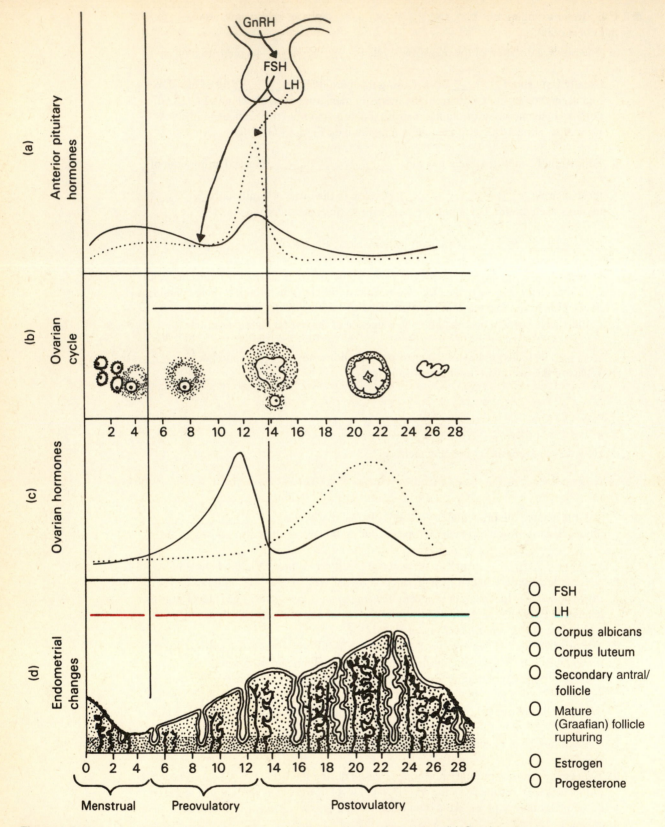

O	FSH
O	LH
O	Corpus albicans
O	Corpus luteum
O	Secondary antral/ follicle
O	Mature (Graafian) follicle rupturing
O	Estrogen
O	Progesterone

Menstrual   Preovulatory   Postovulatory

Figure LG 28.6 Correlations of female cycles. (a) Anterior pituitary hormones. (b) Ovarian changes: follicles and corpus luteum. (c) Ovarian hormones. (d) Uterine (endometrial) changes. Color and label as directed in Checkpoint C4.

j. Following ovulation is the _____ phase. It lasts until day _____ of a 28-day cycle. Under the influence of tropic hormone _____, follicle cells are changed into the corpus _____. These cells secrete two hormones: _____ and _____. Both prepare the endometrium for _____. What preparatory changes occur?

k. What effect do rising levels of target hormones estrogens and progesterone have upon the GnRH and LH?

l. One function of LH is to form and maintain the corpus luteum. So as LH _____-creases, the corpus luteum disintegrates, forming a white scar on the ovary known as the corpus _____. Color these two final stages of the original follicle on Figure LG 28.6b. Also write alternative names for the postovulatory phase on lines in Figure LG 28.6b and d.

m. The corpus luteum had been secreting _____ and _____. With the demise of the corpus luteum, the levels of these hormones rapidly *(increase? decrease?)*. Because these hormones were maintaining the endometrium, endometrial tissue now deteriorates and will be shed during the next _____ phase.

n. If fertilization should occur, estrogens and progesterone are needed to maintain the endometrial lining. The _____ around the developing embryo secretes a hormone named _____.

It functions much like LH in that it maintains the corpus _____, even though LH has been inhibited (step k above). The corpus luteum continues to secrete estrogens and progesterone for several months until the placenta itself can secrete sufficient amounts. Incidentally, hCG is present in the urine of pregnant women only and so is routinely used to detect _____.

o. Review hormonal changes by coloring blood levels of hormones in Figure LG 28.6a and c, according to the color code ovals. Then note uterine changes by coloring red the remainder of the endometrium in Figure LG 28.6d.

■ **C5.** To check your understanding of the events of these cycles, list letters of the major events in chronological order beginning on day 1 of the cycle. To do this exercise it may help to list on a separate piece of paper days 1 to 28 and write each event next to the

approximate day on which it occurs. _____ _____ _____ _____ _____

_____ _____ _____ _____ _____ _____

A. Ovulation occurs
B. Estrogen and progesterone levels drop
C. FSH begins to stimulate follicles
D. LH surge is stimulated by high estrogen levels
E. High levels of estrogens and progesterone inhibit LH
F. Estrogens are secreted for the first time during the cycle
G. Corpus luteum degenerates and becomes corpus albicans
H. Endometrium begins to deteriorate, leading to next month's menses
I. Follicle is converted into corpus hemorrhagicum and then into corpus luteum
J. Rising level of estrogen inhibits FSH secretion
K. Corpus luteum secretes estrogens and progesterone

■ **C6.** Describe signs of ovulation in this Checkpoint.

a. Ovulation is likely to occur during the 24-hour period after the basal temperature

_____-creases by about 0.4–0.6°F. This temperature change is related to increase in *(estrogens? progesterone?)*.

b. Under the influence of *(estrogens? progesterone?)* cervical mucus increases in quantity and becomes *(more stretchy? thicker?)* at midcycle. The cervix also becomes *(firmer? softer?)*

c. Some women experience a pain around ovulation known as _____.

D. Physiology of sexual intercourse; birth control   (pages 942–945)

D1. Define *sexual intercourse* or *coitus*.

■ **D2.** Answer these questions about the process.

a. *(Sympathetic? Parasympathetic?)* impulses from the *(brainstem? sacral cord?)* cause dilation of arteries of the penis or the vagina, leading to the *(flaccid? erect?)* state.

b. *(Sympathetic? Parasympathetic?)* impulses cause sperm to be propelled into the urethra of the male. Expulsion from the urethra to the exterior of the body is called

_____.

D3. Compare and contrast the phases of sexual activity in females and males by completing this table.

Phase	Female	Male
a. Erection		
b. Lubrication		
c. Orgasm		

■ **D4.** Match the methods of birth control listed in the box with descriptions below.

Ch.	Chemical methods	OC.	Oral contraceptive
CI.	Coitus interruptus	R.	Rhythm
Con.	Condom	RU 486.	RU 486
D.	Diaphragm	T.	Tubal ligation
IUD.	Intrauterine device	V.	Vasectomy
N.	Norplant		

_____ a. Spermicidal foams, jellies, and creams

_____ b. "The pill," a combination of progesterone and estrogens, causing decrease in GnRH, FSH, and LH so ovulation does not occur

_____ c. Removal of a portion of the ductus (vas) deferens

_____ d. Tying off of the uterine tubes

_____ e. A natural method of birth control involving abstinence during the period when ovulation is most likely

_____ f. A mechanical method of birth control in which a dome-shaped structure is placed over the cervix; accompanied by use of a jelly or cream

_____ g. Small object, an intrauterine device, placed in the uterus by physician

_____ h. A rubber sheath over the penis

_____ i. Withdrawal

_____ j. Progestin-containing capsules placed under the skin for up to 5 years

_____ k. Competitive inhibitor of progesterone; a chemical that induces miscarriage

D5. Contrast benefits and risks associated with oral contraceptives.

E. Aging and development of the reproductive system (pages 946–949)

E1. Contrast changes occurring at puberty in females and males by completing this table.

	Female	Male
a. Age of onset		
b. Hormonal changes		
c. Changes in reproductive organs		
d. Secondary sex characteristics developed		

E2. Contrast these terms: *menarche/menopause*

E3. Describe normal aging changes that involve the reproductive system in:

a. Females

b. Males

■ **E4.** Which type of cancer is more common in women over 65? *(Cervical? Uterine?)*

■ **E5.** Name the reproductive structures (if any) that develop from the following embryonic ducts.

a. Mesonephric duct in females: _____

b. Mesonephric duct in males: _____

c. Paramesonephric (Müller's) duct in females: _____

d. Paramesonephric (Müller's) duct in males: _____

■ **E6.** Write E (endoderm) or M (mesoderm) next to each structure below to indicate its embryonic origin.

_____ a. Ovaries and testes

_____ b. Prostrate and bulbourethral glands

_____ c. Greater (Bartholin's) and lesser vestibular glands

■ **E7.** Identify male homologues for each of the following.

a. Labia majora: _____

b. Clitoris: _____

c. Lesser vestibular (Skene's) glands: _____

d. Greater vestibular (Bartholin's) glands: _____

F. Disorders, medical terminology   (pages 949–953)

F1. Define *sexually transmitted disease* and *venereal disease*.

F2. Describe the main characteristics of these sexually transmitted diseases by completing the table.

Disease	Causative Agent	Main Symptoms
a. Gonorrhea		
b.	*Treponema pallidum*	
c.	Type II herpes simplex virus	
d.	*Trichomonas vaginalis*	
e. Chlamydia		
f.	Human papillomavirus (HPV)	

F3. Discuss effects of syphilis on children born to mothers with untreated syphilis.

■ **F4.** Check your understanding of the disorders listed in the box by matching them with the correct descriptions.

```
A.    Amenorrhea        HPV.  Human papillomavirus
C.    Chlamydia           S.  Syphilis
D.    Dysmenorrhea     Trich.  Trichomoniasis
E.    Endometriosis      TSS.  Toxic shock syndrome
G.    Gonorrhea
```

_____ a. Painful menstruation, partly due to
            contractions of uterine muscle
_____ b. Possible cause of infection and possible
            blindness in newborns if bacteria transmit-
            ted to eyes during birth (2 answers)
_____ c. Spreading of uterine lining into
            abdominopelvic cavity via uterine tubes
_____ d. Caused by bacterium *Treponema*
            pallidum; may involve many systems in
            tertiary stage

_____ e. Absence of menstrual periods

_____ f. Most associated with *Staphylococcus*
            aureus toxin and tampons
_____ g. Caused by flagellated protozoan

_____ h. Genital warts

■ **F5.** Fill in the term (*impotence* or *infertility*) that best fits the description. Then list several causes of each condition.

a. Inability to fertilize an ovum: _____.

b. Inability to attain and maintain an erection: _____.

F6. List six symptoms of *premenstrual syndrome (PMS)*.

_____ _____

_____ _____

_____ _____

F7. Name the microorganism associated with *toxic shock syndrome (TSS)*.

_____ Relate this condition to use of tampons.

F8. What is PID?

It is most commonly caused by the sexually transmitted diseases (STDs) known as

_____.

■ **F9.** Oophorectomy is removal of a(n) _____, whereas

_____ refers to removal of a uterine (Fallopian) tube.

ANSWERS TO SELECTED CHECKPOINTS: CHAPTER 28

A2. (a) Testes. (b) Seminiferous tubules, epididymis, vas deferens, ejaculatory duct, urethra. (c) Prostate, seminal vesicles, bulbourethral glands. (d) Scrotum, penis.

A3. (a) D TV TA. (b) Lower; lower temperature is required for normal sperm production. (c) Posterior of the abdomen; cryptorchidism; sterility and testicular cancer. (d) Cremaster.

A4. (a) Ova; haploid; 23. (b) Zygote; diploid, 2n; homologous. (c) Spermato, seminiferous tubules of testes.

A5. (a–b) See Figure LG 28.1A. (c) 2, 3; 2; 3. (d) Toward. (e) 4, 5. (f) B. (g) A. (h) 1.

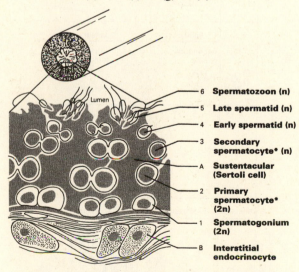

Figure LG 28.1A Diagram of a portion of a cross section of a seminiferous tubule showing the stages of spermatogenesis.

A7. (a) 300 million. (b) 48. (c) Head. (d) Midpiece. (e) A specialized lysosome that releases enzymes that help the sperm to penetrate the secondary oocyte. (f) Propels sperm.

A8. (a) GnRF. (b) Testosterone. (c) FSH and testosterone. (d) LH. (e) Testosterone. (f) Inhibin. (g) DHT. (h–i) Testosterone and DHT.

A10. (a) 5. (b) 2. (c) 6. (d) 8. (e) 7. (f) 4. (g) 3. (h) 9. (i) 1.

A11. Paired: 1–5; singular: 6–9.

A12. (a) 45, (18). (b) Partially; inguinal canal and rings; inguinal hernia. (c) Testicular artery, veins, lymphatics, nerves, and cremaster muscle. (d) Scrotum; no; not normally; hormones exit through testicular veins, which are not cut during vasectomy.

A13. Spongy or cavernous.

A14. (a) S. (b) S. (c) P. (d) S. (e) B. (f) P. (g) P. (h) P.

A15. Accessory sex glands: A (seminal vesicles), B (prostate), C (bulbourethral glands); ducts, 2–9; gonad, 1.

A16. Because it surrounds the urethra (like a doughnut), an enlarged prostate can cause painful and difficult urination (dysuria) and bladder infection (cystitis) due to incomplete voiding.

A17. 2.5–5 ml. (b) 50–150 million; 20 million; a high sperm count increases the likelihood of fertilization, because only a small percentage of sperm ever reaches the secondary oocyte; in addition, a large number of sperm are required to release sufficient enzyme to digest the barrier around the secondary oocyte. (c) Alkaline; protects sperm against acidity of the male urethra and the female vagina. (d) Serves as an antibiotic that destroys bacteria. (e) It liquefies the clotted semen within 20 minutes of the entrance of semen to the vagina, thereby permitting movement of sperm into the uterus.

A19. (a) Prepuce, glans; circumcision. (b) Cavernosum, spongiosum, blood; veins; erect. (c) Spongiosum; urine, sperm; sphincter action at base of bladder.

B1. (a) 3. (b) 4. (c) 1. (d) 2. (e) 5.

B2. 1, 2: paired; 3–5, singular.

B3. (a) A. (b) Ligaments; mesovarian, ovarian, and suspensory ligament. (c) OF VOF CL.

B4.

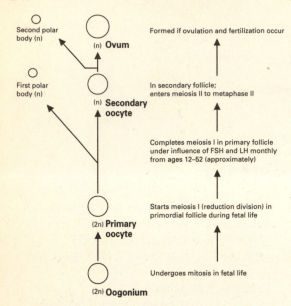

Figure LG 28.4A Oogenesis.

B5. The secondary oocyte (see Figure LG 28.4A) along with the first polar body and some surrounding supporting cells are expelled. Meiosis II is completed only if the secondary oocyte is fertilized by a spermatazoon.

B6. (a) Fallopian; 10 (4). (b) C A B. (c) Currents produced by movements of fimbriae, ciliary action of columnar epithelial cells lining tubes, and peristalsis of the muscularis layer of the uterine tube wall. (d) Passageway for secondary oocyte, and sperm toward each other and site of fertilization. (e) Pregnancy following implantation at any location other than in the body of the uterus, for example, tubal pregnancy, or pregnancy on ovary, intestine, or in cervix of the uterus.

B8. (a) Bladder; rectum. (b) Posterior, vesico. (c) Anteriorly; retroflexed. (d) Uterosacral; round; broad. (e) Cardinal.

B9. (a) Myo; smooth muscle. (b) Serous perimetrium. (c) Endo; functionalis; spiral.

B10. (a) Colposcopy. (b) Pap (Papanicolaou) smear. (c) Hysterectomy.

B11. Close to 90° angle.

B13. Acid; retards microbial growth but may harm sperm cells.

B14. (a and c) See Figure LG 28.5A. (b) Urogenital triangle.

B15. Lobes; alveoli; nipple.

B16. (a) A, C, and E. (b) 1: monthly; 2 and 3: annually or more often based on the decision made by physician and client.

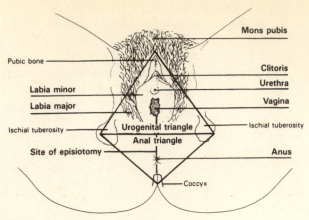

Figure LG 28.5A Perineum.

C2. Hypothalamus: GnRH; anterior pituitary: FSH, LH; Ovary: E, I, P, R.

C3. (a–b) GnRH. (c–d) I (E and P indirectly because inhibit GnRH). (e) E. (f) P and E. (g) R. (h) E. (i) E.

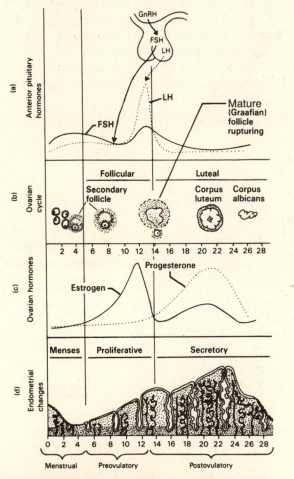

Figure LG 28.6A Correlations of female cycles.

(a) Anterior pituitary hormones. (b) Ovarian changes: follicles and corpus luteum. (c) Ovarian hormones. (d) Uterine (endometrial) changes. Refer to Checkpoint C4.

C4. (a) 28; 5; 50–150; see Figure LG 28.6A. (b) Pre-ovulatory; secondary follicles; 1; mature; 20 (6 + 14); II. (c) GnRH, FSH, LH. (d) Secretion of estrogens and development of the ovum. (e) Estrogens stimulates replacement of the sloughed off stratum functionalis by growing (proliferating) new cells; proliferative; follicular. See Figure LG 28.6A. (f) Negative, decreases; degeneration. (g) FSH and LH; ovulate. (h) LH; high; positive. (i) Ovulation; ovulation. See Figure LG 28.6A. (j) Postovulatory; 28; LH, luteum; progesterone, estrogens; implantation; thickening by growth and secretion of glands and increased retention of fluid. (k) Inhibition of these hormones. (l) De; albicans. See Figure LG 28.6A. (m) Progesterone, estrogens; decrease; menstrual. (n) Placenta, human chorionic gonadotropin (hCG); luteum; pregnancy. (o) See Figure LG 28.6A.

C5. C F J D A I K E G B H.

C6. (a) In; progesterone. (b) Estrogen, more stretchy; softer. (c) Mittelschmerz.

D2. (a) Parasympathetic, sacral cord, erect. (b) Sympathetic; ejaculation.

D4. (a) Ch. (b) OC. (c) V. (d) T. (e) R. (f) D. (g) IUD. (h) Con. (i) Cl. (j) N. (k) RU 486.

E4. Uterine.

E5. (a) None. (b) Ducts of testes and epididymides, ejaculatory ducts and seminal vesicles. (c) Uterus and vagina. (d) None.

E6. (a) M. (b) E. (c) E.

E7. (a) Scrotum. (b) Penis. (c) Prostate. (d) Bulbourethral glands.

F4. (a) D. (b) G, C. (c) E. (d) S. (e) A. (f) TSS. (g) Trich. (h) HPV.

F5. (a) Infertility (or sterility). (b) Impotence.

F9. Ovary, salpingectomy.

WRITING ACROSS THE CURRICULUM: CHAPTER 28

1. Explain how gonads function as both endocrine glands and exocrine glands.

2. Discuss the roles of hormones in male puberty.

3. Contrast the processes of oogenesis and spermatogenesis.

4. Define menopause and list several physiological effects noted by many women during this change. Discuss whether menopause can be considered a normal age-related change.

5. Discuss incidence of and risk factors for (a) ovarian cancer and (b) uterine prolapse.

MASTERY TEST: CHAPTER 28

Questions 1–10: Arrange the answers in correct sequence.

_____ _____ _____ 1. From most external to deepest:
   A. Endometrium
   B. Serosa
   C. Myometrium

_____ _____ _____ 2. Portions of the urethra, from proximal (closest to bladder) to distal (closest to outside of the body):
   A. Prostatic
   B. Membranous
   C. Spongy

_____ _____ _____ 3. Thickness of the endometrium, from thinnest to thickest:
   A. On day 5 of cycle
   B. On day 13 of cycle
   C. On day 23 of cycle

_____ _____ _____ 4. From anterior to posterior:
                                         A. Uterus and vagina
                                         B. Bladder and urethra
                                         C. Rectum and anus

_____ _____ _____ _____ 5. Order in which hormones begin to increase level, from day 1 of menstrual cycle:
    A. FSH
    B. LH
    C. Progesterone
    D. Estrogen

_____ _____ _____ _____ 6. Order of events in the female monthly cycle, beginning with day 1:
    A. Ovulation
    B. Formation of follicle
    C. Menstruation
    D. Formation of corpus luteum

_____ _____ _____ _____ 7. From anterior to posterior:
    A. Anus
    B. Vaginal orifice
    C. Clitoris
    D. Urethral orifice

_____ _____ _____ _____ 8. Pathway of milk in breasts:
    A. Alveoli
    B. Secondary tubules
    C. Mammary ducts
    D. Lactiferous ducts

_____ _____ _____ _____ _____ 9. Pathway of sperm:
    A. Ejaculatory duct
    B. Testis
    C. Urethra
    D. Ductus (vas) deferens
    E. Epididymis

_____ _____ _____ _____ _____ _____ 10. Pathway of sperm entering female reproductive system:
     A. Uterine cavity
     B. Cervical canal
     C. External os
     D. Internal os
     E. Uterine tube
     F. Vagina

Questions 11–12: Circle the letter preceding the one best answer to each question.

11. Eighty to 90% of seminal fluid (semen) is secreted by the combined secretions of:
    A. Prostate and epididymis
    B. Seminal vesicles and prostate
    C. Seminal vesicles and seminiferous tubules
    D. Seminiferous tubules and epididymis
    E. Bulbourethral glands and prostate

12. In the normal male, there are two of each of the following structures *except:*
    A. Testis   D. Ductus deferens
    B. Seminal vesicle   E. Epididymis
    C. Prostate   F. Ejaculatory duct

Questions 13–15: Circle letters preceding all correct answers to each question.

13. Which of the following are produced by the testes?
   A. Spermatozoa
   B. Testosterone
   C. Inhibin
   D. GnRH
   E. FSH
   F. LH

14. Which of the following are produced by the ovaries and then *leave* the ovaries?
   A. Follicle
   B. Secondary oocyte
   C. Corpus luteum
   D. Corpus albicans
   E. Estrogen
   F. Progesterone

15. Which of the following are functions of LH?
   A. Begin the development of the follicle.
   B. Stimulate change of follicle cells into corpus luteum cells.
   C. Stimulate release of secondary oocyte (ovulation).
   D. Stimulate corpus luteum cells to secrete estrogens and progesterone.
   E. Stimulate release of GnRH.

Questions 16–20: Circle T (true) or F (false). If the statement is false, change the underlined word or phrase so that the statement is correct.

T F 16. Spermatogonia and oogonia both continue to divide throughout a person's lifetime to produce new primary spermatocytes or primary oocytes.

T F 17. The menstrual cycle refers to a series of changes in the uterus, whereas the ovarian cycle is a series of changes in the ovary.

T F 18. Meiosis is a part of both oogenesis and spermatogenesis.

T F 19. Each oogonium beginning oogenesis produces four mature ova.

T F 20. An increased level of a target hormone (such as estrogen) will inhibit release of its related releasing factor and target hormone (GnRH and FSH).

Questions 21–25: Fill-ins. Complete each sentence with the word or phrase that best fits.

_____ 21. _____ is a term that means undescended testes.

_____ 22. Oral contraceptives consist of the hormones _____ and _____.

_____ 23. _____ cells in testes phagocytose degenerating spermatogenic cells, secrete inhibin, and also form the blood-testis barrier.

_____ 24. Menstrual flow occurs as a result of a sudden _____-crease in the hormones _____ and _____.

_____ 25. Three main functions of estrogens are promotion of growth and maintenance of _____, _____ balance, and stimulation of protein _____-bolism.

ANSWERS TO MASTERY TEST: ☆ CHAPTER 28

Arrange
1. B C A
2. A B C
3. A B C
4. B A C
5. A D B C
6. C B A D
7. C D B A
8. A B C D
9. B E D A C
10. F C B D A E

Multiple Choice
11. B
12. C

Multiple Answers
13. A, B, C
14. B, E, F
15. B, C, D

True–False
16. F. Spermatogonia; primary spermatocytes (Oogonia do not divide after birth of female baby.)
17. T
18. T
19. F. One mature ovum (or secondary oocyte) and three polar bodies
20. T

Fill-ins
21. Cryptorchidism
22. Progesterone and estrogens
23. Sustentacular
24. De, progesterone, and estrogens
25. Reproductive organs, fluid and electrolyte, ana

FRAMEWORK 29
Development & Inheritance

DEVELOPMENT DURING PREGNANCY (A)

HORMONES OF PREGNANCY, GESTATION, DIAGNOSTIC TECHNIQUES (B)

- Hormones
  - estrogens
  - progesterone
  - hCG and hCS
  - relaxin
- Gestation
- Diagnostic texts
  - ultrasonography
  - amniocentesis
  - CVS

LABOR, ADJUSTMENTS AT BIRTH, LACTATION (C)

- Stages of labor
  - dilation
  - expulsion
  - placental
- Adjustments of infant at birth
  - respiratory
  - cardiovascular
- Lactation
  - PRH, PRL, OT
  - colostrum
  - advantages

INHERITANCE (D)

- Genome
- Genotype, phenotype
- Alleles
- Homozygous, heterozygous
- Punnett squares
- Dominant, recessive
- Multiple alleles
- Polygenic inheritance
- Autosomes, sex chromosomes
- Sex-linked
- Environmental effects
- Disorders

FERTILIZATION, CLEAVAGE

- Capacitation
- Syngamy
- Morula
- Blastocyst
- Alternative methods

CLEAVAGE, IMPLANTATION

- Germ layers
  - endoderm
  - mesoderm
  - ectoderm

EMBRYONIC DEVELOPMENT

- Months 1–2
- Embryonic membranes
  - yolk sac
  - amnion
  - chorion
  - allantois
- Placenta and umbilical cord
  - decidua
  - chorionic villi
  - afterbirth

FETAL GROWTH & DEVELOPMENT

- Months 3–9
- Fetal circulation
  - umbilical arteries (2)
  - placenta
  - umbilical vein (1)
  - ductus venosus
  - foramen ovale
  - ductus arteriosus

Development and Inheritance

Twenty-eight chapters ago you began a tour of the human body, starting with the structural framework and varieties of buildings (chemicals and cells) and moving on to the global communication systems (nervous and endocrine), transportation routes (cardiovascular), and refuse removal systems (respiratory, digestive, and urinary). This last chapter synthesizes all the parts of the tour: A new individual, with all those body systems, is conceived and develops during a nine-month pregnancy. Adjustments are made at birth for survival in an independent (outside of mother) world. But parental ties are still evident in inherited features, signs of intrauterine care (mother's avoidance of alcohol or cigarettes), and provision of nourishment, as in lactation.

As you begin your study of development and inheritance, carefully examine the Chapter 29 Topic Outline and Objectives. Check off each objective as you complete it. The Chapter 29 Framework provides an organizational preview of development and inheritance. Be sure to refer to the Framework frequently and note relationships among key terms in each section. And congratulations on having completed your first tour through the wonders of the human body.

TOPIC OUTLINE AND OBJECTIVES

A. Development and growth during pregnancy

- ☐ **1.** Explain the processes associated with fertilization, formation of a morula, development of a blastocyst, and implantation.
- ☐ **2.** Describe several alternative methods of fertilization.
- ☐ **3.** Discuss the principal maternal body changes associated with embryonic and fetal growth.

B. Hormones of pregnancy; gestation; diagnostic techniques

- ☐ **4.** Compare the sources and functions of the hormones secreted during pregnancy.
- ☐ **5.** Describe the anatomical and physiological changes associated with gestation.

C. Parturition, labor, adjustments at birth, physiology of lactation

- ☐ **6.** Describe the stages of labor and the changes that occur during the puerperium.

- ☐ **7.** Explain the respiratory and cardiovascular adjustments that occur in an infant at birth.
- ☐ **8.** Discuss the physiology and control of lactation.

D. Inheritance; disorders

- ☐ **9.** Define inheritance and describe the inheritance of several traits.
- ☐ **10.** Discuss potential hazards to an embryo and fetus associated with chemicals and drugs, irradiation, alcohol, and cigarette smoking.
- ☐ **11.** Define and describe Down syndrome, Turner's syndrome, metafemale syndrome, Klinefelter's syndrome, and fragile X syndrome.

WORDBYTES

Now become familiar with the language of this chapter by studying each wordbyte, its meaning, and an example of its use within a term. After you study the entire list, self-check your understanding by writing the meaning of each wordbyte on the line. As you continue through the *Learning Guide,* identify (and fill in) additional terms that contain the same wordbyte.

Wordbyte	Self-check	Meaning	Example(s)
-blast	_____	germ, to form	*blast*ula
-cele	_____	hollow	blasto*cele*
-centesis	_____	puncture	amino*centesis*
ecto-	_____	outside of	*ecto*derm
endo-	_____	inside of	*endo*derm
lact-	_____	milk	*lact*ation
meso-	_____	middle	*meso*derm
syn-	_____	together	*syn*gamy
-troph	_____	nourish	syncytio*troph*oblast

CHECKPOINTS

A. Development and growth during pregnancy (pages 958–969)

■ **A1.** Do this exercise about events associated with fertilization.

a. How many sperm are ordinarily introduced into the vagina during sexual intercourse?

_____ About how many reach the area where the developing ovum

is located? _____

b. The name given to the developing ovum at the time of fertilization is the

_____. List four or more mechanisms in the female and/or male that favor the meeting of sperm with oocyte so that fertilization can occur.

Female   **Male**

c. How long must sperm be in the female reproductive tract before they can fertilize the female gamete? Several *(minutes? hours? days?)* Describe the changes sperm undergo during this time?

These changes in the sperm are known as _____.

d. Name the usual site of fertilization. _____.
    What do the corona radiata and zona pellucida have to do with fertilization?

e. An _____ pregnancy is one that takes place in a site outside of the

    uterine cavity. Where do most of those pregnancies occur? _____
    List several other possible sites of such pregnancies.

f. How many sperm normally penetrate the developing egg? _____ What is the name of

    this event? _____ What prevents further sperm from entering the
    female gamete?

■ **A2.** Describe roles of the following structures in fertilization.

a. A male pronucleus present in a *(secondary spermatocyte? spermatozoon?)* combines with a female pronucleus from a *(secondary oocyte? ovum?)*. Each pronucleus contains the *(n? 2n?)* number of chromosomes.

b. The product of fusion of the two pronuclei is known as a _____

    nucleus, containing _____ chromosomes. Completion of meiosis II by the secondary oocyte results in the final product of fertilization, consisting of segmentation nucleus, cytoplasm, and zona pellucida. This product of fertilization is known as the

    _____.

■ **A3.** Contrast types of multiple births by choosing the answer that best fits each description.

DTwin. Dizygotic twins	MTwin. Monozygotic twins

_____ a. Also known as fraternal twins

_____ b. Two infants derived from a single fertilized ovum that divides once

_____ c. Two infants derived from two secondary oocytes each fertilized by a different sperm

A4. Summarize the main events during the week following fertilization. Describe each of the following stages. Be sure to note how many days after fertilization each stage occurs. (For help, refer to Figures 29.2 to 29.4, pages 959–962 of the text.)

a. Cleavage (days _____)

b. Morula (days _____)

c. Blastocyst (days _____)

A5. Draw a diagram of a blastocyst. Label: *blastocele, inner cell mass,* and *trophoblast.* Color red the cells that will become the embryo. Color blue the cells that will become part of the placenta. (Refer to Figure 29.3b, page 961 of the text, for help.)

■ **A6.** Describe the process of implantation in this exercise.

   a. Implantation is likely to occur about _____ days after ovulation, or about day

      _____ of a 28-day cycle. Note that the developing baby is implanted in the uterus even before the mother "misses" the first day of her next menstrual cycle.

   b. During implantation, cells of the *(inner cell mass? trophoblast?)* secrete enzymes that enable the blastocyst to burrow into the uterine lining.

   c. The trophoblast is composed of two layers known as the _____-

      trophoblast and the _____-trophoblast. The trophoblast secretes the

      hormone _____, which maintains the corpus *(albicans? luteum?)*. Ultimately the trophoblast becomes part of the *(amnion? chorion?)* portion of the placenta.

   d. Because the trophoblast is in contact with the mother's endometrial lining, it is reasonable that the developing embryo might be rejected as a "foreign graft" by the mother's immune system. State one mechanism that may explain why it is not rejected.

A7. Describe the following aspects of in vitro fertilization (IVF).

   a. In what year was the first live birth as a result of this procedure? *(1962? 1970? 1978? 1986?)*

   b. Briefly describe this procedure.

   c. Contrast these two forms of IVF: *embryo transfer* and *GIFT*.

A8. Describe how the conditions of being underweight or overweight may affect reproductive hormones and fertility.

■ **A9.** The word *embryo* refers to the developing individual during the *(first two? last seven?)* months of development in utero. Define *fetus*.

■ **A10.** Complete this exercise about the embryonic period. Refer to Figure 29.5 (pages 963–964 in your text).

a. Identify the structures that ultimately form from these two parts of the implanted blastocyst.

Trophoblast   **Inner Cell Mass**

Between these two portions of the blastocyst is the _____ cavity.

b. The inner cell mass first forms the embryonic _____ consisting of two layers. The upper cells (closer to the implantation site toward the right on

Figure 29.5b, page 964 of the text) are called _____-derm cells. The

lower cells (toward the left in the figure) are called _____-derm

cells. Eventually, a layer of _____-derm will form between these.

Collectively the three layers are called _____ layers.

c. The process of formation of the primary germ layers is known as

_____, and it occurs by the end of the _____ week following fertilization.

d. _____-derm cells line the amniotic cavity. The delicate fetal membrane formed from these cells that surrounds the fetus and is broken at birth is called

the _____. Another name for this membrane is the

"_____." (Ectoderm cells also form a number of fetal structures: see Checkpoint A12.)

e. The lower cells of the embryonic disc, called _____ cells, line the

primitive _____ and for a while extend out to form the

_____ sac. This sac *(is? is not?)* as significant in humans as it is in birds.

f. Some mesoderm cells move around to line the original trophoblast cells (Figure 29.5,

page 964 of the text). Together these cells form the _____ layer of the placenta. Which membrane is more superficial? *(Amnion? chorion?)* The space between

these layers is called the _____.

A11. Each of the primary germ layers is involved with formation of both fetal organs and fetal membranes. Name the major structures derived from each primary germ layer: endoderm, mesoderm, and ectoderm. Consult Exhibit 29.1 (page 965 in your text).

Endoderm	Mesoderm	Ectoderm

■ **A12.** *For extra review.* Do this matching exercise on primary germ layer derivatives.

Ecto. Ectoderm Endo. Endoderm Meso. Mesoderm

_____ a. Epithelial lining of all of digestive, respiratory, and genitourinary tracts except near openings to the exterior of the body

_____ b. Epidermis of skin, epithelial lining of entrances to the body (such as mouth, nose, and anus), hair, nails

_____ c. All of the skeletal system (bone, cartilage, joint cavities)

_____ d. All skeletal and cardiac muscle; most smooth muscle

_____ e. Blood and all blood and lymphatic vessels

_____ f. Entire nervous system, including posterior pituitary

_____ g. Thyroid, parathyroid, thymus, and pancreas

■ **A13.** Write the name of each fetal membrane next to its description.

a. Originally formed from ectoderm, this membrane encloses fluid that acts as a shock

   absorber for the developing baby: _____

b. Derived from mesoderm and trophoblast, this membrane becomes the principal part of

   placenta: _____

c. Endoderm-lined membrane serving as exclusive nutrient supply for embryos of some

   species: _____

d. A small membrane that forms umbilical blood vessels: _____

■ **A14.** Further describe the placenta in this exercise.

a. The placenta has the shape of a _____ embedded in the wall of the

uterus. Its fetal portion is the _____ fetal membrane. Fingerlike

projections, known as _____, grow into the uterine lining. Consequently maternal and fetal blood are *(allowed to mix? brought into close proximity but
do not mix?)*. Fetal blood in umbilical vessels is bathed by maternal blood in

_____ spaces.

b. The maternal aspect of the placenta is derived from the *(chorion? decidua basalis?)*.

This is a portion of the _____-metrium of the uterus.

c. What is the ultimate fate of the placenta?

■ **A15.** *A clinical challenge.* Following a birth, vessels of the umbilical cord are carefully

examined. A total of _____ vessels should be found, surrounded by a mucous jellylike
connective tissue. The vessels should include *(1? 2?)* umbilical artery(-ies) and *(1? 2?)*
umbilical vein(s). Lack of a vessel may indicate a congenital malformation.

A16. List three uses of human placentas following birth.

A17. *A clinical challenge.* Describe the following conditions in this exercise. Define the
condition, state the frequency of its occurrence, suggest possible causes, and list potential
consequences for fetus and/or mother.

a. Placenta previa

b. Umbilical cord prolapse and entrapment

■ **A18.** Study Exhibit 29.2 (page 969 in your text). Write the number of the month when each of the following events occurs.

_____ a. Heart starts to beat.

_____ b. Heartbeat can be detected.

_____ c. Backbone and vertebral canal form.

_____ d. Eyes are almost fully developed; eyelids are still fused.

_____ e. Eyelids separate; eyelashes form.

_____ f. Limb buds develop.

_____ g. Ossification begins.

_____ h. Limb buds become distinct as arms and legs; digits are well formed.

_____ i. Nails develop.

_____ j. Fine (lanugo) hair covers body.

_____ k. Lanugo hair is shed.

_____ l. Fetus is capable of survival.

_____ m. Subcutaneous fat is deposited.

_____ n. Testes descend into scrotum.

B. Hormones of pregnancy; gestation; diagnostic techniques (pages 970–974)

■ **B1.** Complete this exercise about the hormones of pregnancy.

a. During pregnancy, the level of progesterone and estrogens must remain *(high? low?)* in order to support the endometrial lining. During the first three or four months or so, these

hormones are produced principally by the _____ located in the

_____, under the influence of the tropic hormone _____. (It might be helpful to review the section on endocrine relations in Chapter 28.)

b. Human chorionic gonadotropin (hCG) reaches its peak between the _____ and

_____ month of pregnancy. Because it is present in blood, it will be filtered into

_____, where it can readily be detected as an indication of

pregnancy as early as _____ days after fertilization. Sometimes the urine of a woman who is not pregnant might erroneously indicate that she is pregnant. List several factors which could account for such a "false pregnancy test."

c. In addition, the chorion produces the hormone hCS; these initials stand for

_____ chorionic _____. List two functions of this hormone.

Peak levels of hCS occur at about week *(9? 20? 32?)* of gestation.

d. _____ is a hormone that relaxes pelvic joints and helps dilate the

cervix near the time of birth. This hormone is produced by the _____

and _____.

■ **B2.** The normal human gestation period is about _____ weeks, counting from *(fertilization? the beginning of the last menstrual period?)* until birth.

■ **B3.** Complete this exercise about maternal adaptations during pregnancy. Describe normal changes that occur.

a. Pulse _____-creases.

b. Blood volume _____-creases by about _____%.

c. Tidal volume _____-creases about _____%.

d. Gastrointestinal (GI) motility _____-creases, which may cause *(diarrhea? constipation?)*.

e. Pressure upon the inferior vena cava _____-creases, which may lead to

_____.

f. Increase in weight of the uterus from _____ grams in nonpregnant state to about

_____ grams at term.

■ **B4.** About 10–15% of all pregnant women in the United States experience PIH, which

stands for p _____-i_____

h _____. This condition apparently results from impaired function of

_____. Write three or more signs or symptoms of PIH.

B5. Defend or dispute this statement: "All women who are pregnant or lactating should avoid exercise because of its deleterious effects."

■ **B6.** Complete this exercise about prenatal diagnostic tests.

a. Amniocentesis involves withdrawal of _____ fluid. This fluid is constantly recycled through the fetus (by drinking and excretion); it *(does? does not?)* contain fetal cells and products of fetal metabolism. For what purposes is amniotic fluid collected and examined?

For _____

(if performed at 14–16 weeks). For _____
(if performed at about 35 weeks).

b. *(Amniocentesis? Ultrasound?)* is by far the most common test for determining true
fetal age if the date of conception is uncertain. The technique *(is also? is not?)* used for
diagnostic purposes. Fetal ultrasonography involves *(use of a transducer on? a puncture
into?)* the abdomen. For this procedure the mother's urinary bladder should be
(full? empty?).

c. CVS (meaning c_____ v_____

s_____) is a diagnostic test that can be performed as early as *(4? 8?
16? 22?)* weeks of pregnancy. It *(does? does not?)* involve penetration of the uterine

cavity. Cells of the _____ layer of the placenta embedded in the
uterus are studied. These *(are? are not?)* derived from the same sperm and egg that
united to form the fetus. Therefore this procedure, like amniocentesis, can be used to

diagnose _____.

C. Parturition, labor, adjustments at birth, physiology of lactation (pages 974–977)

C1. Contrast these pairs of terms.

a. *Labor/parturition*

b. *False labor/true labor*

■ **C2.** Do the following hormones stimulate (S) or inhibit (I) uterine contractions during the
birth process?

_____ a. Progesterone   _____ c. Oxytocin

_____ b. Estrogens

C3. What is the "show" produced at the time of birth?

■ **C4.** Identify descriptions of the three phases of labor (first, second, or third).

_____ a. Stage of expulsion: from complete cervi-   _____ c. Time from onset of labor to complete
cal dilation through delivery of the baby   dilation of the cervix; the stage of dilation

_____ b. Time after the delivery of the baby until
the placenta ("afterbirth") is expelled; the
placental stage

C5. Define these terms related to birth and the postnatal period.

a. *Dystocia*

b. *Involution*

c. *Lochia*

d. *Puerperium*

C6. What adjustments must the newborn make at birth in attempting to cope with the new environment? Describe adjustments of the following.

a. Respiratory system (What serves as a stimulus for the respiratory center of the medulla?)

The respiratory rate of a newborn is usually about *(one-third? three times?)* that of an adult.

b. Heart and blood vessels (*For extra review* go over changes in fetal circulation, Chapter 21, pages LG 468–469.)

c. Blood cell production

■ **C7.** A premature infant ("preemie") is considered to be one who weighs less than

_____ grams (_____ pounds _____ ounces) at birth. These infants are at high risk

for RDS (or _____). *For extra review* of RDS, refer to Chapter 23, Checkpoint H3, page LG 526.

C8. Explain why as many as 50% of all normal newborns may experience a temporary state of jaundice soon after birth.

■ **C9.** Complete this exercise about lactation.

a. The major hormone promoting lactation is _____, which is secreted

by the _____. During pregnancy the level of this hormone increases somewhat, but its effectiveness is limited while estrogens and progesterone are *(high? low?)*.

b. Following birth and the loss of the placenta, a major source of estrogens and progesterone is gone, so effectiveness of prolactin *(increases? decreases?)* dramatically.

c. The sucking action of the newborn facilitates lactation in two ways. What are they?

d. What hormone stimulates milk ejection (let-down)? _____

C10. Contrast *colostrum* with *maternal milk* with respect to time of appearance, nutritional content, and presence of maternal antibodies.

C11. Defend or dispute this statement: "Breast-feeding is an effective, reliable form of birth control."

C12. List five or more advantages offered by breast-feeding.

D. Inheritance; disorders; medical terminology (pages 977–984)

D1. Define these terms.

a. *Inheritance*

b. *Genetic counseling*

c. *Homologues (homologous chromosomes)*

D2. Compare and contrast.

a. *Genotype/phenotype*

b. *Dominant allele/recessive allele*

c. *Homozygous/heterozygous*

d. *Angelman syndrome/Prader-Willi syndrome*

■ **D3.** Answer these questions about genes controlling PKU. (See Figure 29.13, page 978 in the text.)

a. The dominant gene for PKU is represented by *(P? p?)*. The dominant gene is for the *(normal? PKU?)* condition.

b. The letters PP are an example of a genetic makeup or *(genotype? phenotype?)*. A person with such a genotype is said to be *(homozygous dominant? homozygous recessive? heterozygous?)*. The phenotype of that individual would *(be normal? be normal, but serve as a "carrier" of the PKU gene? have PKU?)*.

c. The genotype for a heterozygous individual is _____. The phenotype is

_____.

d. Use a Punnett square to determine possible genotypes of offspring when both parents are Pp.

■ **D4.** Do this exercise about genes and inheritance.

a. Define *genome.*

A human cell is estimated to have about _____ genes. *(Most? A small number?)* of these have been identified on the various human chromosomes.

b. In most cases, the gene for *(having a severe disorder? being normal?)* is dominant.

State one or more exceptions to this rule. _____

c. What is the meaning of the term *aneuploid?*

One example is *trisomy 21,* in which the individual has *(2? 3? 4?)* of chromosome 21. This alteration in the genome occurs in the condition known as

_____.

d. Sickle cell anemia (SCA) is a condition that is controlled by *(complete? incomplete?)* dominance. Match the correct genotype below with the related description. Choose from these genotypes:

> A. Homozygous recessive
> B. Homozygous dominant
> C. Heterozygous

1. Genotype present in persons who have SCA _____

2. Genotype present in persons who have the sickle cell trait _____

e. List the three possible alleles for inheritance of the ABO group. _____

_____ _____ Which of these is/are dominant? _____

List the possible genotypes of persons who produce A antigens on red blood cells.

_____ Describe the phenotype of persons with the OO genotype.

f. ABO blood group inheritance is known as *(multiple allele? polygenic?)* inheritance. List two or more examples of characteristics governed by polygenic inheritance.

D5. Contrast *autosome* and *sex chromosome*.

■ **D6.** Answer these questions about sex inheritance.
  a. All *(sperm? secondary oocytes?)* contain the X chromosome. About half of the sperm

  produced by a male contain the _____ chromosome and half contain the _____.
  b. Every cell (except those forming ova) in a normal female contains *(XX? XY?)* genes on sex chromosomes. All male cells (except those forming sperm) are said to be *(XX? XY?)*.
  c. Who determines the sex of the child? *(Mother? Father?)* Show this by using a Punnett square.

  d. Name the gene on the Y chromosome that determines that a developing individual will be male.

■ **D7.** Complete this exercise about X-linked inheritance.
  a. Sex chromosomes contain other genes besides those determining sex of an individual.

  Such traits are called _____ traits. Y chromosomes are shorter than X chromosomes and lack some genes. One of these is the gene controlling ability to

  _____ red from green colors. Thus this ability is controlled entirely

  by the _____ gene.
  b. Write the genotype for females of each type: color-blind, _____; carrier,

  _____; normal, _____.
  c. Now write possible genotypes for males: color-blind, _____; normal,

  _____.

  d. Determine the results of a cross between a color-blind male and a normal female.

e. Now determine the results of a cross between a normal male and a carrier female.

f. Color blindness and other X-linked, recessive traits are much more common in *(males? females?)*.

D8. Name several other examples of X-linked traits.

Discuss *fragile X syndrome.*

■ **D9.** Complete this exercise about potential hazards to the embryo and fetus. Match the terms in the box with the related descriptions.

C. Cigarette smoking	FAS. Fetal alcohol syndrome	T. Teratogen

_____ a. An agent or influence that causes defects in the developing embryo

_____ b. May cause defects such as small head, facial irregularities, defective heart, retardation

_____ c. Linked to a variety of abnormalities such as lower infant birth weight, higher mortality rate, GI disturbances, and SIDS

A1. (a) Approximately 300 to 500 million; less than 1%. (b) Secondary oocyte; mechanisms in the female: peristaltic contractions and ciliary action of uterine tubes, uterine contractions promoted by prostaglandins in semen, production of a sperm-attracting chemical by the secondary oocyte; mechanisms in the male: presence of flagella on sperm and enhancement of motility by the acrosomal enzyme acrosin. (c) Hours; increased fragility of the acrosome, which permits release of enzymes that can penetrate the coverings around the oocyte; capacitation. (d) Distal portion of the uterine tube; they are the coverings over the oocyte which sperm must penetrate. (e) Ectopic; uterine tube; ovaries, abdomen, cervix portion of the uterus, and broad ligament. (f) One; syngamy; sperm penetration of the secondary oocyte, which triggers release of calcium and then granules from the oocyte. These block entrance of other sperm through the zona pellucida.

A2. (a) Spermatozoon, secondary oocyte; n. (b) Segmentation, 46; zygote.

A3. (a) DTwin. (b) MTwin. (c) DTwin.

A6. (a) 6, 20. (b) Trophoblast. (c) Syncytio, cyto; hCG, luteum; chorion. (d) Mother may produce antibodies that mask the antigens derived from paternal components of the developing baby.

A9. First two; developing individual during the last seven months of human gestation.

A10. (a) Trophoblast: part of chorion of placenta; inner cell mass: embryo and parts of placenta; amniotic. (b) Disc; ecto; endo; meso; primary germ. (c) Gastrulation, third. (d) Ecto; amnion; bag of waters. (e) Endoderm, gut, yolk; is not. (f) Chorion; chorion; extraembryonic coelom.

A12. (a) Endo. (b) Ecto. (c) Meso. (d) Meso. (e) Meso. (f) Ecto. (g) Endo.

A13. (a) Amnion. (b) Chorion. (c) Yolk sac. (d) Allantois.

A14. (a) Flattened cake (or thick pancake); chorion; chorionic villi; brought into close proximity but do not mix; intervillous. (b) Decidua basalis; endo. (c) It is discarded from mother at birth (as the "afterbirth"); separation of the decidua from the remainder of endometrium accounts for bleeding during and following birth.

A15. 3; 2; 1.

A18. (a) 1. (b) 3. (c) 1. (d) 3. (e) 6. (f) 1. (g) 2. (h) 2. (i) 3. (j) 5. (k) 9. (l) 7. (m) 8. (n) 7.

B1. (a) High; corpus luteum, ovary, hCG. (b) First, third; urine, 14; excess protein or blood in her urine, ectopic production of hCG by uterine cancer cells, or presence of certain diuretics or hormones in her urine. (c) Human (chorionic) somatomammotropin; stimulates development of breast tissue and alters metabolism during pregnancy (see page 970 in the text); 32. (d) Relaxin; placenta, ovaries.

B2. 38, fertilization.

B3. (a) In. (b) In, 30–50. (c) In, 30–40. (d) De, constipation. (e) In, varicose veins. (f) 60–80, 900–1200.

B4. Pregnancy-induced hypertension; kidneys; hypertension, fluid retention leading to edema with excessive weight gain, headaches and blurred vision, and proteinuria.

B6. (a) Amniotic; does; detecting suspected genetic abnormalities (14–16 weeks); assessing fetal maturity (35 weeks). (b) Ultrasound; is also; use of a transducer on; full. (c) Chorionic villus sampling, 8; does not; chorion; are; genetic and chromosomal defects.

C2. (a) I. (b) S. (c) S.

C4. (a) Second. (b) Third. (c) First.

C7. 2500 grams (5 pounds, 8 ounces); respiratory distress syndrome.

C9. (a) Prolactin (PRL), anterior pituitary; high. (b) Increases. (c) Maintains prolactin level by inhibiting PIH and stimulating release of PRH; stimulates release of oxytocin. (d) Oxytocin.

D3. (a) P; normal. (b) Genotype; homozygous dominant; be normal. (c) Pp; normal but a carrier of PKU gene. (d) 25% PP, 50% Pp, and 25% pp.

D4. (a) The complete genetic makeup of a particular organism; 70,000–100,000; a small number. (b) Being normal; Huntington's disease (HD) or other disorders on the left side of Exhibit 29.3, page 979 in the text. (c) Having a chromosome number other than normal (which is 46 for humans); 3; Down syndrome. (d) Incomplete; 1, A; 2, C. (e) I^A, I^B, and i; I^A and I^B; $I^A I^A$ or $I^A i$; produces neither A nor B antigens on red blood cells. (f) Multiple allele; skin color, eye color, height, or body build.

D6. (a) Secondary oocytes; X, Y. (b) XX; XY. (c) Father, as shown in the Punnett square (Figure 29.17, page 981 in the text), because only the male can contribute the Y chromosome. (d) SRY (sex-determining region of the Y chromosome).

D7. (a) X-linked; differentiate; X. (b) $X^c X^c$, $X^c X^c$, $X^c X^c$. (c) $X^c Y$, $X^c Y$. (d) All females are carriers and all males are normal. (e) Of females, 50% are normal and 50% are carriers; of males, 50% are normal and 50% are color-blind. (f) Males.

D9. (a) T. (b) FAS. (c) C.

1. Summarize the three major events in the embryonic period following implantation.
2. Describe the layers that form the placenta and outline its protective functions.
3. Describe effects of the following maternal or fetal hormones upon labor and parturition: progesterone, estrogens, oxytocin, and relaxin (maternal); cortisol and epinephrine (fetal).
4. Discuss effects of embryonic or fetal environment upon polygenic traits. Include effects of cigarettes, alcohol and other drugs, and radiation.

MASTERY TEST: CHAPTER 29

Questions 1–2: Arrange the answers in correct sequence.

_____ _____ _____ _____   1. From most superficial to deepest (closest to embryo):
   A. Amnion
   B. Amniotic cavity
   C. Chorion
   D. Decidua

_____ _____ _____ _____ _____   2. Stages in development:
   A. Morula
   B. Blastocyst
   C. Zygote
   D. Fetus
   E. Embryo

Questions 3–11: Circle the letter preceding the one best answer to each question.

3. During early months of pregnancy, hCG is produced by the _____ and it mimics the action of _____.
   A. Follicle; progesterone
   B. Corpus luteum; FSH
   C. Chorion of placenta; LH
   D. Amnion of placenta: estrogens

4. These fetal characteristics—weight of 8 ounces to 1 pound, length of 10 to 12 inches, lanugo covers skin, and brown fat is forming—are associated with which month of fetal life?
   A. Third month   C. Fifth month
   B. Fourth month   D. Sixth month

5. All of the following structures are developed from mesoderm *except:*
   A. Aorta   D. Humerus
   B. Biceps muscle   E. Sciatic nerve
   C. Heart   F. Neutrophil

6. Which part of the structures surrounding the fetus is developed from maternal cells (not from fetal)?
   A. Allantois   D. Decidua
   B. Yolk sac   E. Amnion
   C. Chorion

7. All of the following events occur during the first month of embryonic development *except:*
   A. Endoderm, mesoderm, and ectoderm are formed.
   B. Amnion and chorion are formed.
   C. Limb buds develop.
   D. Ossification begins.
   E. The heart begins to beat.

8. Implantation of a developing individual (blastocyst stage) usually occurs about _____ after fertilization.
   A. Three weeks   D. Seven hours
   B. One week   E. Seven minutes
   C. One day

9. The high level of estrogens present during pregnancy is responsible for all of the following *except:*
   A. Stimulates release of prolactin
   B. Prevents ovulation
   C. Maintains endometrium
   D. Prevents menstruation
   E. Stimulates uterine contractions

10. Which of the following is a term that refers to discharge from the birth canal during the days following birth?

A. Autosome   D. Lochia
B. Cautery   E. PKU
C. Karyotype

11. Formation of the three primary germ layers is known as:

A. Implantation   C. Gastrulation
B. Blastocyst   D. Morula
    formation   formation

Questions 12–20: Circle T (true) or F (false). If the statement is false, change the underlined word or phrase so that the statement is correct.

T F 12. Oxytocin and progesterone both stimulate uterine contractions.

T F 13. Of the primary germ layers, only the ectoderm forms part of a fetal membrane.

T F 14. Maternal blood mixes with fetal blood in the placenta.

T F 15. Amniocentesis is a procedure in which amniotic fluid is withdrawn for examination at about 8–10 weeks during the gestation period.

T F 16. Stage 2 of labor refers to the period during which the "after-birth" is expelled.

T F 17. Maternal pulse rate, blood volume, and tidal volume all normally increase during pregnancy.

T F 18. The decidua is part of the endoderm of the fetus.

T F 19. Inner cell mass cells of the blastocyst form much of the chorion of the placenta.

T F 20. A phenotype is a chemical or other agent that causes physical defects in a developing embryo.

Questions 21–25: Fill-ins. Complete each sentence with the word or phrase that best fits.

_____ 21. The newborn infant generally has a respiratory rate of _____, pulse of _____, and a WBC count that may reach _____, all values that are _____ than those of an adult.

_____ 22. As a prerequisite for fertilization, sperm must remain in the female reproductive tract for at least several hours so that _____ can occur.

_____ 23. hCG is a hormone made by the _____; its level peaks at about the end of the _____ month of pregnancy, and up to this time it can be used to detect _____.

_____ 24. Determine the probable genotypes of children of a couple in which the man has hemophilia and the woman is normal. _____

_____ 25. This is the _____ mastery test question in this book.

ANSWERS TO MASTERY TEST: ★ CHAPTER 29

Arrange
1. D C A B
2. C A B E D

Multiple Choice
3. C
4. C
5. E
6. D
7. D
8. B
9. A
10. D
11. C

True–False
12. F. Oxytocin but not progesterone stimulates.
13. F. All three layers form parts of fetal membranes: ectoderm (trophoblast) forms part of chorion; mesoderm forms parts of the chorion and the allantois; endoderm lines the yolk sac.
14. F. Does not mix but comes close to fetal blood
15. F. Amniocentesis, 14–16
16. F. Fetus
17. T
18. F. Endometrium of the uterus
19. F. Trophoblast
20. F. Teratogen

Fill-ins
21. 45, 120–160, up to 45,000, higher
22. Capacitation (dissolving of the covering over the ovum by secretion of acrosomal enzymes of sperm)
23. Chorion of the placenta, second, pregnancy
24. All male children normal; all female children carriers
25. Last or final. Congratulations! Hope you learned a great deal.